T0140094

Advances in Intelligent Systems and Computing

Volume 1059

The series "Advances in Intelligent Systems and Computing" contains publications on theory, applications, and design methods of Intelligent Systems and Intelligent Computing. Virtually all disciplines such as engineering, natural sciences, computer and information science, ICT, economics, business, e-commerce, environment, healthcare, life science are covered. The list of topics spans all the areas of modern intelligent systems and computing such as: computational intelligence, soft computing including neural networks, fuzzy systems, evolutionary computing and the fusion of these paradigms, social intelligence, ambient intelligence, computational neuroscience, artificial life, virtual worlds and society, cognitive science and systems, Perception and Vision, DNA and immune based systems, self-organizing and adaptive systems, e-Learning and teaching, human-centered and human-centric computing, recommender systems, intelligent control, robotics and mechatronics including human-machine teaming, knowledge-based paradigms, learning paradigms, machine ethics, intelligent data analysis, knowledge management, intelligent agents, intelligent decision making and support, intelligent network security, trust management, interactive entertainment, Web intelligence and multimedia.

The publications within "Advances in Intelligent Systems and Computing" are primarily proceedings of important conferences, symposia and congresses. They cover significant recent developments in the field, both of a foundational and applicable character. An important characteristic feature of the series is the short publication time and world-wide distribution. This permits a rapid and broad dissemination of research results.

** Indexing: The books of this series are submitted to ISI Proceedings, EI-Compendex, DBLP, SCOPUS, Google Scholar and Springerlink **

More information about this series at http://www.springer.com/series/11156

Ashish Khanna · Deepak Gupta ·
Siddhartha Bhattacharyya ·
Vaclav Snasel · Jan Platos ·
Aboul Ella Hassanien
Editors

International Conference on Innovative Computing and Communications

Proceedings of ICICC 2019, Volume 2

 Springer

Editors
Ashish Khanna
Computer Science and Engineering
Maharaja Agrasen Institute of Technology
Delhi, India

Deepak Gupta
Computer Science and Engineering
Maharaja Agrasen Institute of Technology
Delhi, India

Siddhartha Bhattacharyya
Department of Information Technology
RCC Institute of Information Technology
Kolkata, West Bengal, India

Vaclav Snasel
Department of Computer Science
VŠB - Technical University of Ostrava
Ostrava-Poruba, Czech Republic

Jan Platos
Department of Computer Science
VŠB - Technical University of Ostrava
Ostrava-Poruba, Czech Republic

Aboul Ella Hassanien
Faculty of Computers and Information
Cairo University
Giza, Egypt

ISSN 2194-5357   ISSN 2194-5365   (electronic)
Advances in Intelligent Systems and Computing
ISBN 978-981-15-0323-8   ISBN 978-981-15-0324-5   (eBook)
https://doi.org/10.1007/978-981-15-0324-5

This Springer imprint is published by the registered company Springer Nature Singapore Pte Ltd.
The registered company address is: 152 Beach Road, #21-01/04 Gateway East, Singapore 189721, Singapore

Dr. Ashish Khanna would like to dedicate this book to his mentors Dr. A. K. Singh and Dr. Abhishek Swaroop for their constant encouragement and guidance and his family members including his mother, wife, and kids. He would also like to dedicate this work to his (late) father Sh. R. C. Khanna with folded hands for his constant blessings.

Dr. Deepak Gupta would like to dedicate this book to his father Sh. R. K. Gupta, his mother Smt. Geeta Gupta, his mentors Dr. Anil Kumar Ahlawat and Dr. Arun Sharma for their constant encouragement, his family members including his wife, brothers, sisters, kids, and his students close to his heart.

Professor (Dr.) Siddhartha Bhattacharyya would like to dedicate this book to his late father Ajit Kumar Bhattacharyya, his late mother Hashi Bhattacharyya, his beloved wife Rashni, and his colleagues Anirban, Hrishikesh, Indrajit, Abhijit, Biswanath, and Hiranmoy, who has been beside him through thick and thin.

Professor (Dr.) Jan Platos would like to dedicate this book to his wife Daniela and his daughters Emma and Margaret.

Professor (Dr.) Aboul Ella Hassanien would like to dedicate this book to his beloved wife Azza Hassan El-Saman.

ICICC-2019 Steering Committee Members

General Chairs

Prof. Dr. Vaclav Snasel, VSB - Technical University of Ostrava, Czech Republic.
Prof. Dr. Siddhartha Bhattacharyya, Principal, RCC Institute of Information Technology, Kolkata.

Honorary Chair

Prof. Dr. Janusz Kacprzyk, FIEEE, Polish Academy of Sciences, Poland.

Conference/Symposium Chair

Prof. Dr. Maninder Kaur, Director, Guru Nanak Institute of Management, Delhi, India.

Technical Program Chairs

Dr. Pavel Kromer, VSB - Technical University of Ostrava, Czech Republic.
Dr. Jan Platos, VSB - Technical University of Ostrava, Czech Republic.
Prof. Dr. Joel J. P. C. Rodrigues, National Institute of Telecommunications—Inatel, Brazil.
Prof. Dr. Aboul Ella Hassanien, Cairo University, Egypt.
Dr. Victor Hugo C. de Albuquerque, Universidade de Fortaleza, Brazil.

Conveners

Dr. Ashish Khanna, Maharaja Agrasen Institute of Technology, India.
Dr. Deepak Gupta, Maharaja Agrasen Institute of Technology, India.

Publicity Chairs

Dr. Hussain Mahdi, National University of Malaysia.
Prof. Dr. Med Salim Bouhlel, University of Sfax, Tunisia.
Dr. Mohamed Elhoseny, Mansoura University, Egypt.
Dr. Anand Nayyar, Duy Tan University, Vietnam.
Dr. Andino Maseleno, STMIK Pringsewu, Lampung, Indonesia.

Publication Chairs

Dr. D. Jude Hemanth, Associate Professor, Karunya University, Coimbatore.
Dr. Nilanjan Dey, Techno India College of Technology, Kolkata, India.
Gulshan Shrivastava, National Institute of Technology, Patna, India.

Co-conveners

Dr. Avinash Sharma, Maharishi Markandeshwar (Deemed to be University), India.
P. S. Bedi, Guru Tegh Bahadur Institute of Technology, Delhi, India.
Mr. Moolchand Sharma, Maharaja Agrasen Institute of Technology, India.

ICICC-2019 Advisory Committee

Prof. Dr. Vincenzo Piuri, University of Milan, Italy.
Prof. Dr. Valentina Emilia Balas, Aurel Vlaicu University of Arad, Romania.
Prof. Dr. Marius Balas, Aurel Vlaicu University of Arad, Romania.
Prof. Dr. Mohamed Salim Bouhlel, University of Sfax, Tunisia.
Prof. Dr. Aboul Ella Hassanien, Cairo University, Egypt.
Prof. Dr. Cenap Ozel, King Abdulaziz University, Saudi Arabia.
Prof. Dr. Ashiq Anjum, University of Derby, Bristol, UK.
Prof. Dr. Mischa Dohler, King's College London, UK.
Prof. Dr. Sanjeevikumar Padmanaban, University of Johannesburg, South Africa.
Prof. Dr. Siddhartha Bhattacharyya, Principal, RCC Institute of Information Technology, Kolkata, India.
Prof. Dr. David Camacho, Associate Professor, Universidad Autonoma de Madrid, Spain.
Prof. Dr. Parmanand, Dean, Galgotias University, UP, India.
Dr. Abu Yousuf, Assistant Professor, University Malaysia Pahang, Gambang, Malaysia.
Prof. Dr. Salah-ddine Krit, University Ibn Zohr – Agadir, Morocco.
Dr. Sanjay Kumar Biswash, Research Scientist, INFOCOMM Lab, Russia.
Prof. Dr. Maryna Yena S, Senior Lecturer, Medical University of Kiev, Ukraine.
Prof. Dr. Giorgos Karagiannidis, Aristotle University of Thessaloniki, Greece.
Prof. Dr. Tanuja Srivastava, Department of Mathematics, IIT Roorkee.
Dr. D. Jude Hemanth, Associate Professor, Karunya University, Coimbatore.
Prof. Dr. Tiziana Catarci, Sapienza University of Rome, Italy.
Prof. Dr. Salvatore Gaglio, Universita Degli Studi di Palermo, Italy.
Prof. Dr. Bozidar Klicek, University of Zagreb, Croatia.
Dr. Marcin Paprzycki, Associate Professor, Polish Academy of Sciences, Poland.
Prof. Dr. A. K. Singh, NIT Kurukshetra, India.
Prof. Dr. Anil Kumar Ahlawat, KIET Group of Institutes, India.
Prof. Dr. Chang-Shing Lee, National University of Tainan, Taiwan.
Dr. Paolo Bellavista, Associate Professor, Alma Mater Studiorum - Università di Bologna.

Prof. Dr. Frede Blaabjerg, President, IEEE Power Electronics Society, University of Aalborg, Denmark.

Prof. Dr. Jens Bo Holm Nielson, University of Aalborg, Denmark.

Prof. Dr. Venkatadri Marriboyina, Amity University, Gwalior, India.

Dr. Pradeep Malik, Vignana Bharathi Institute of Technology (VBIT), Hyderabad, India.

Dr. Ahmed A. Elngar, Assistant Professor, Faculty of Computers and Information, Beni Suef University, Beni Suef, Egypt.

Prof. Dr. Dijana Oreski, Faculty of Organization and Informatics, University of Zagreb, Varazdin, Croatia.

Prof. Dr. Dhananjay Kalbande, Professor and Head, Sardar Patel Institute of Technology, Mumbai, India.

Prof. Dr. Avinash Sharma, Maharishi Markandeshwar Engineering College, MMDU Campus, India.

Dr. Sahil Verma, Lovely Professional University, Phagwara, India.

Dr. Kavita, Lovely Professional University, Phagwara, India.

Prof. Prasad K. Bhaskaran, Professor and Head of Department, Ocean Engineering & Naval Architecture, IIT Kharagpur.

Preface

We hereby are delighted to announce that VSB - Technical University of Ostrava, Czech Republic, Europe, has hosted the eagerly awaited and much-coveted International Conference on Innovative Computing and Communication (ICICC-2019). The second version of the conference was able to attract a diverse range of engineering practitioners, academicians, scholars, and industry delegates, with the reception of abstracts including more than 2200 authors from different parts of the world. The committee of professionals dedicated toward the conference is striving to achieve a high-quality technical program with tracks on innovative computing, innovative communication network and security, and Internet of things. All the tracks chosen in the conference are interrelated and are very famous among the present-day research community. Therefore, a lot of research is happening in the above-mentioned tracks and their related sub-areas. As the name of the conference starts with the word "innovation," it has targeted out-of-box ideas, methodologies, applications, expositions, surveys, and presentations helping to upgrade the current status of research. More than 550 full-length papers have been received, among which the contributions are focused on theoretical, computer simulation-based research, and laboratory-scale experiments. Among these manuscripts, 129 papers have been included in the Springer proceedings after a thorough two-stage review and editing process. All the manuscripts submitted to the ICICC-2019 were peer-reviewed by at least two independent reviewers, who were provided with a detailed review pro forma. The comments from the reviewers were communicated to the authors, who incorporated the suggestions in their revised manuscripts. The recommendations from two reviewers were taken into consideration while selecting a manuscript for inclusion in the proceedings. The exhaustiveness of the review process is evident, given the large number of articles received addressing a wide range of research areas. The stringent review process ensured that each published manuscript met the rigorous academic and scientific standards. It is an exalting experience to finally see these elite contributions materialize into two book volumes as ICICC-2019 proceedings by Springer entitled *International Conference on Innovative Computing and Communications*. The articles are organized into two volumes in some broad categories covering subject matters on machine learning, data mining, big data, networks, soft computing, and cloud computing, although given the diverse areas of research reported it might not have been always possible.

ICICC-2019 invited five keynote speakers, who are eminent researchers in the field of computer science and engineering, from different parts of the world. In addition to the plenary sessions on each day of the conference, five concurrent technical sessions are held every day to assure the oral presentation of around 129 accepted papers. Keynote speakers and session chair(s) for each of the concurrent sessions have been leading researchers from the thematic area of the session. A technical exhibition is held during all the 2 days of the conference, which has put on display the latest technologies, expositions, ideas, and presentations. The delegates were provided with a sovereign to quickly go through the contents, participants, and intended audience. The research part of the conference was organized in a total of 35 special sessions. These special sessions provided the opportunity for researchers conducting research in specific areas to present their results in a more focused environment.

An international conference of such magnitude and release of the ICICC-2019 proceedings by Springer has been the remarkable outcome of the untiring efforts of the entire organizing team. The success of an event undoubtedly involves the painstaking efforts of several contributors at different stages, dictated by their devotion and sincerity. Fortunately, since the beginning of its journey, ICICC-2019 has received support and contributions from every corner. We thank them all who have wished the best for ICICC-2019 and contributed by any means toward its success. The edited proceedings volumes by Springer would not have been possible without the perseverance of all the steering, advisory, and technical program committee members.

All the contributing authors owe thanks from the organizers of ICICC-2019 for their interest and exceptional articles. We would also like to thank the authors of the papers for adhering to the time schedule and for incorporating the review comments. We wish to extend our heartfelt acknowledgment to the authors, peer reviewers, committee members, and production staff whose diligent work put shape to the ICICC-2019 proceedings. We especially want to thank our dedicated team of peer reviewers who volunteered for the arduous and tedious step of quality checking and critique on the submitted manuscripts. We wish to thank our faculty colleagues Mr. Moolchand Sharma and Ms. Prerna Sharma for extending their enormous assistance during the conference. The time spent by them and the midnight oil burnt are greatly appreciated, for which we will ever remain indebted. The management, faculties, and administrative and support staff of the college have always been extending their services whenever needed, for which we remain thankful to them.

Lastly, we would like to thank Springer for accepting our proposal for publishing the ICICC-2019 conference proceedings. Help received from Mr. Aninda Bose, Acquisition Senior Editor, in the process has been very useful.

Delhi, India   Ashish Khanna
                                                                                       Deepak Gupta
                                                                          Organizers, ICICC-2019

About This Book (Volume 2)

International Conference on Innovative Computing and Communication (ICICC-2019) was held on 21–22 March at VSB - Technical University of Ostrava, Czech Republic, Europe. This conference was able to attract a diverse range of engineering practitioners, academicians, scholars, and industry delegates, with the reception of papers including more than 2200 authors from different parts of the world. Only 129 papers have been accepted and registered with an acceptance ratio of 23% to be published in two volumes of prestigious Springer Advances in Intelligent Systems and Computing (AISC) series. Volume 2 includes the accepted papers of networks and cloud computing tracks. This volume includes a total of 50 papers from these two tracks.

Contents

About the Editors

Dr. Ashish Khanna has expertise in Teaching, Entrepreneurship, and Research & Development He received his Ph.D. degree from National Institute of Technology, Kurukshetra in March 2017. He has completed his M. Tech. in 2009 and B. Tech. from GGSIPU, Delhi in 2004. He has completed his PDF from Brazil. He has around 75 accepted and published research papers and book chapters. He has authored and edited 15 books. His research interest includes Distributed Systems and its variants (MANET, FANET, VANET, IoT), Machine learning, Evolutionary computing and many more. His Cumulative Impact Factor (CIF) is around 90.

Deepak Gupta received his Ph.D. in Computer Science and Engineering from Dr. A.P.J. Abdul Kalam Technical University (AKTU), Master of Engineering (CTA), from Delhi University, and B.Tech. (IT) from GGSIP University in 2017, 2010, and 2005, respectively. He has completed his postdoc from Inatel, Brazil. He is currently working as an Assistant Professor in the Department of Computer Science and Engineering, Maharaja Agrasen Institute of Technology, GGSIP University, Delhi, India. He has 79 publications in reputed international journals and conferences. In addition, he has authored/edited 35 books with international publishers.

Siddhartha Bhattacharyya did his bachelor's in Physics and Optics and Optoelectronics and master's in Optics and Optoelectronics from University of Calcutta, India, in 1995, 1998, and 2000, respectively. He completed his Ph.D. in Computer Science and Engineering from Jadavpur University, India, in 2008. He is currently the Principal of RCC Institute of Information Technology, Kolkata, India. In addition, he is also serving as the Professor of Information Technology and Dean (Research and Development) of the institute. He is a co-author of 5 books and the co-editor of 30 books and has more than 220 research publications in international journals and conference proceedings to his credit.

Vaclav Snasel is the Dean at the Faculty of Electrical Engineering and Computer Science of the Mining University, Technical University of Ostrava. He has experience of almost 30 years in academia and research with industrial cooperation. He works in a multidisciplinary environment involving social network, formal concept analysis, information retrieval, semantic web, knowledge management, data compression, machine intelligence, neural network, web intelligence, bio-inspired computing, data mining, and applied to various real-world problems.

Jan Platos did his bachelor's in Computer Science in 2005; master's in 2006 in Computer Science; and Ph.D. in Computer Science (2010). He is currently a Head of the Department of Computer Science at Faculty of Electrical Engineering and Computer Science, VSB - Technical University of Ostrava. He is the author of more than 190 papers in international journals and conferences. He has also organized 12 international conferences. He also served as a reviewer for more than 30 journals. His research interests include data mining, optimization techniques, soft computing methods, text processing, and data compression.

Aboul Ella Hassanien received his B.Sc. with honours in 1986 and M.Sc. degree in 1993, both from Ain Shams University, Faculty of Science, Pure Mathematics and Computer Science Department, Cairo, Egypt. On September 1998, he received his doctoral degree from the Department of Computer Science, Graduate School of Science & Engineering, Tokyo Institute of Technology, Japan. He is a Full Professor at Cairo University, Faculty of Computer and Artificial Intelligence, IT Department. Professor Aboul Ella is the founder and chair of the Scientific Research Group in Egypt. He has authored over 1000 research publications in peer-reviewed reputed journals, book chapters, and conference proceedings.

Implementation of Square-Odd Scanning Technique in WBAN for Energy Conservation

Rani Kumari, Parma Nand, Rani Astya and Suneet Chaudhary

Abstract The increasing population needs large medical staff for the excellent healthcare services. By the introduction of wireless sensor networks in the field of medical world, we hereby solve the problem of shortage of medical staff across the world. The WBAN gives an excellent opportunity to improve the quality of medical healthcare system. Establishing a wireless network in the field of medical is a very difficult issue as the protocol used for the adhoc network doesn't perform efficiently in the mobile WBAN. This needs a scanning policy for the WBAN to be added in routing to improve the network lifetime and to reduce the errors of the existing protocols for WBAN. The nodes of the sensor network remains active at all times whereas the utilization period of the sensor nodes is only 20% of the total time. This results in high energy consumption. This results in need of an efficient scanning technique for WBAN with dynamic active period. Wireless sensing network uses very light sensors which have very low power backup. So power saving is very important in such type of network. Square-Odd scanning is used to save significant power in wireless sensors. It periodically switches the sensors between sleeping and awake mode. Square-Odd scanning is an improved method for scan the object with increase network lifetime. It focuses on reduction in energy consumption and it improves the life time of sensor. The performance of the Square-Odd approach is better than all other previous scanning algorithms in terms of network lifetime. In this paper we describe existing scanning techniques and proposed scanning algorithm for power saving in WBAN.

R. Kumari (✉)
Computer Science & Engineering, SCSE, Galgotias University, Gr. Noida, India
e-mail: ranichoudhary04@gmail.com

P. Nand · R. Astya
Computer Science & Engineering, Sharda University, Gr. Noida, India
e-mail: parmaastya@gmail.com

R. Astya
e-mail: astyarani@gmail.com

S. Chaudhary
Computer Science & Engineering, MMU, Ambala, India
e-mail: suneetcit81@gmail.com

© Springer Nature Singapore Pte Ltd. 2020
A. Khanna et al. (eds.), *International Conference on Innovative Computing and Communications*, Advances in Intelligent Systems and Computing 1059,
https://doi.org/10.1007/978-981-15-0324-5_1

Keywords WSN · WBAN · Network lifetime · SO

1 Introduction

WBANs include variety of heterogeneous biological sensors. These sensors are unit placed in numerous components of the body and may be wearable or established below the user skin. Every of them has specific needs and is employed for various missions.

In the wireless body area network all nodes send the information to the base station (as shown in Fig. 1) and then base station send this information to the server through different interfaces. Many type of sensors can be used for measuring different parameters. The wireless body area network having large number of sensor nodes as compare to wireless sensor network.

WBANs for social insurance applications are primarily utilized as a part of patient observing assignments. In this kind of system, the sensors are circulated on the human body estimating diverse physiological parameters, which speak to the most generally utilized arrangement inside this space. Sensors hubs around the body with remote abilities are of unique enthusiasm to this sort of WBAN, since they give an agreeable and easy to use approach to screen a patient's wellbeing status over expanded timeframes, keeping away from the utilization of links wired around the patient.

The primary contrast is the region secured by WBAN is less when contrasted with the zone secured by the WBAN. It isn't important to convey the hubs inside the assemblage of patient; the hubs can be embedded over or around the body.

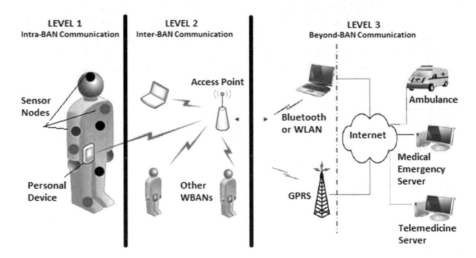

Fig. 1 General architecture for wireless body area networks

The increasing population needs large medical staff for the excellent healthcare services. However, creating a WBAN platform is a very difficult issue as the protocol used for the adhoc network does not perform efficiently in the mobile WBAN. This needs an optimal scanning policy for the WBAN to be added in routing to increase the performance of existing scanning algorithms for WBAN.

Moreover, the sensor nodes remains active at all times whereas the utilization period of the sensor nodes is only 20% of the total time. This results in high energy consumption. This results in need of an efficient scanning algorithm for WBAN with dynamic active period.

Therefore, enhancements within the current technologies and higher solutions to those challenges are needed. These two phenomenon time dependent and energy effective are necessary for every WBAN.

Among all the previous algorithms, the scanning WBAN is non-energy effective and time dependent. Screening the human body an important for everyone so that every human try to require disease free body. Wireless sensor networks (WSNs) became one in all the foremost attention-grabbing areas of analysis in the last few years. A WSN consists of variety of wireless sensor nodes that kind a sensor field and a sink. These giant numbers of nodes, having the skills to sense their surroundings, perform restricted computation and communicate wirelessly kind the WSNs. Recent advances in wireless and electronic technologies have enabled a large vary of applications of WSNs in healthcare monitoring. There are several new challenges that have create problems for the designers of WSNs, so as to fulfill the wants of assorted applications like perceived quantities, size of nodes, and node's autonomy. So, enhancements within the present approaches and higher solutions to those challenges are needed. These two phenomenon time dependent and energy effective are necessary for every WBAN.

2　Square-Odd Scanning Algorithm

In this paper we propose an approach for scanning sensor network for a healthcare monitoring. To increase the network lifetime, they can mainly observe the sensing schedule of every sensor.

Wireless sensing network uses very light sensors which have very low power backup. An algorithm "Square Odd(SO) scanning" is used to detect any object efficiently, effectively and it also saves significant power in wireless sensors. It periodically switches the sensors between sleeping and awake mode. It also saves energy consumption and it improves the life time of sensor.

WBAN systems have main focuses on sensors for the moving object where we apply the scanning technique for target detection which is the main feature of the network. In this, for w seconds nodes goes into wake up state simultaneously and n sensors are linearly placed at the body segment.

Algorithm

There are several sensor nodes which shown in Fig. 2, it may be describe the process of the examine graph. User can draw a Scanning process as a sub graph. In this user describe the scanning process:

Step 1: Here user may choose the one entrance point and one protection point from the different points of entrance and protection points in the graph.

$v1 \rightarrow$ Source point or node (protection point)

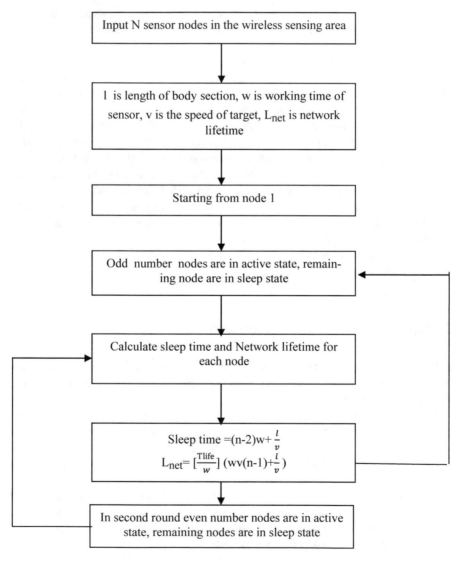

Fig. 2 Square—odd scanning algorithm for energy conservation

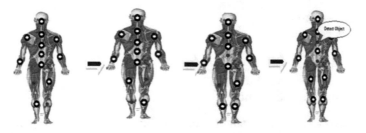

Fig. 3 Square-odd (SO) scanning

$v13 \rightarrow$ Destination point or node (Entrance point)

Step 2: Fig. 3 represent the inquire process of Square-Odd.

Step 3: The scanning mechanism as follows, it may commencement from the right side of protection point and the sensor active first for $v1$ and $v3$ at a time since all other are in sleeping mode.

Step 4: The scanning time for each sensor is w (fixed time) and for scanning complete nodes, it just move on the active for $v2$ and $v4$ sensor node then after $v3$ and $v5$ sensor active.

Two sensor remains in wake up position at the same time so the object is detect definitely, and it will increase the performance and energy level. As well as it will increase the performance of Square-Odd scan method.

So, for this we have create a sensor network with 13 nodes, in which only two nodes are active at a time and remaining nodes are in sleep state. The complete sensor network for scanning is shown in Fig. 4:

Mathematical Formulation

The terms used in this approach as:-

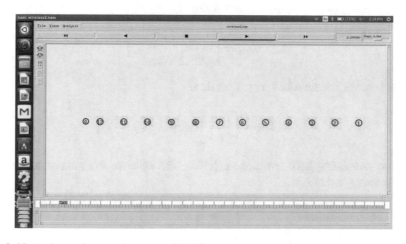

Fig. 4 Network area for scanning sensor detection

- n_i denotes sensor nodes
- D denotes destination
- S denotes Source
- w denote the constant time to sensor active time
- T_{net} denote the network lifetime
- T_{life} denote the sensor lifetime

In Always—Awake Scanning method the life period of network L_{net} is similar to the L_{life} because nodes of sensor network constantly work without stopping.

In Duty Cycling (DC) Screening technique the life period of network L_{net} is the number of span $[L_{life}/w]$ combined with the duration of the period L_{period} (the summation of the sleep time l/v and the waking time w).

$$L_{net} = \left[\frac{L_{life}}{w}\right]\left(\frac{l}{v} + w\right) \tag{1}$$

For Virtual Screening approach the life span of network L_{net} is the number of spans $[L_{life}/w]$ concatenate with the session length L_{period}. L_{period} is addition of the scan period **nw** and sleep period l/v.

$$L_{net} = L_{net} = \left[\frac{L_{life}}{w}\right](L_{silent} + L_{scan})$$

$$L_{net} = \left[\frac{L_{life}}{w}\right]\left(\frac{l}{v} + nw\right) \tag{2}$$

So, with the help of Eqs. (1) and (2) and from the above figure we can compute network lifetime for square-odd algorithm. In this technique we describe the life period of L_{net} is the number of spans $[L_{life}/w]$ combined with the span length L_{period}. L_{period} is the addition of the wake up period wv $(n-1)$ and sleep period l/v.

So network life time during the silent time is

$$L_{net} = \left[\frac{L_{life}}{w}\right]\left(\frac{l}{v}\right) \tag{3}$$

The network life time in during scan time

$$L_{net} = \left[\frac{L_{life}}{w}\right](wv(n-1)) \tag{4}$$

By concatenating both terms, we calculate the network life time of square odd (SO) scanning technique.

$$L_{net} = \left[\frac{L_{life}}{w}\right](L_{silent} + L_{scan})$$

$$L_{\text{net}} = \left[\frac{L_{\text{life}}}{w}\right]\left(\frac{l}{v} + wv(n-1)\right) \tag{5}$$

So from above Eq. (5) we describe the sensor network lifetime for square-odd scanning technique.

Increase Network Lifetime:

There are many strategies and technique use to boost the network time period. All the strategies have completely different sleeping time.

The average network lifespan for Duty Cycling is $\left[\frac{L_{\text{life}}}{w}\right]\left(w + \frac{L_{\text{life}}}{w}\right)$ and for virtual screening is $\left[\frac{L_{\text{life}}}{w}\right]\left(nw + \frac{l}{v}\right)$.

In square-odd (SO) scanning method, only two sensor nodes are active at a time. Once the n sensor nodes are entered in the body segment whose length is l and the speed of target is v. The average network life time for each sensor can be achieved by the quantity $\left[\frac{L_{\text{life}}}{w}\right]\left(wv(n-1) + \left(\frac{l}{v}\right)\right)$. The sleeping time for AA technique is zero. Duty-Cycling scanning has l/v sleeping time. In our implementations, we have shown that the SO scanning technique has the largest sleeping time for nodes, so the network lifetime will be long. SO Scanning has the better network time period as compared to other scanning techniques.

For example, we assume that there are 20 sensor nodes with lifetime of 1 h and the target speed is 2 m/second and the length of body segment is 5 m.

So, $n = 20$ sensors, $L_{\text{life}} = 1$ h $= 3600$ s, $v = 2$ m/s, $l = 5$ m.

Now we calculate the network life time L_{net} for different values of w (working time of sensor node) (Table 1).

In this Square-Odd (SO) scanning method to improve the energy efficiency compares always-awake (AA), duty cycling (DC) and virtual scanning (VS) method. This graph is showing network lifetime or energy efficiency for all scanning methods.

For implementation, we assume that $w = 0.1$ s, square-odd has the network lifespan of 63 h, virtual scanning VS taking the lifespan of 45 h, Duty Cycling (DC) consider 26 h, and Always-awake(AA) has 1 h.

For $w = 0.2$ s, square-odd has the network lifetime of 50.5 h, virtual scanning(VS) has the lifespan of network of 32.5 h, Duty Cycling(DC) has 13.5 h, and Always-awake taking 1 h.

Similarly, for $w = 0.3$ s and $w = 0.4$ s, we get the highest network lifetime for Square-Odd scanning technique as compare to other three scanning algorithms.

Table 1 Network lifetime Comparison of four approaches

W	Always-awake	Duty-cycling	Virtual scanning	Square-odd
0.1	3600	93,600	162,000	226,800
0.2	3600	48,600	117,000	181,800
0.3	3600	33,600	102,000	166,800
0.4	3600	26,100	94,500	159,300

Fig. 5 Network lifetime for different scanning techniques

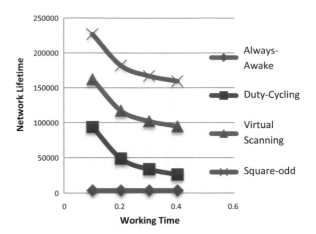

So, according to this performance Square-Odd scanning method is best in terms of network lifetime. This result is shown in Fig. 5. The square-odd scanning has the longest network lifetime as compare to other three scanning techniques. This will be typically as a results of the assorted energy savings throughout the screening technique.

3 Conclusion

According to this performance Square-Odd scanning method is best in terms of network lifetime. There are several methods for scanning like Square-Odd screening, Duty Cycling (DC), Virtual Screening (VS) and Always-Awake (AA) may provide guarantee of the detection of object, but square-odd scanning technique gives the better performance result in the field of network lifespan as compare to other all scanning technique. The implementation of our proposed algorithm is depicting with the help of graphs and tables.

Experimental and simulation implementation represent that the our proposed approach provide better in the fields of increase network lifetime as compare to other existing techniques. Performance results of Square-odd scanning technique are shown in graphs and tables for different values of sensor's working time. So with Square-odd scanning technique, we can easily scan the sensing area with increase network lifetime because in SO technique some nodes are in sleep condition and some nodes are in active condition. Therefore, the power consumption is low so, the total network lifetime will be increase.

4 Future Scope

It can be extended to detect more than one disease or multiple diseases at a time for early recovery of patient. This algorithm can be used to detect gas leakage in gas plants where sensors are applicable to detect the leakage position of gas at early stage. To detect disease in moving body is more complex as compare to static body. This work can be further used in detection of any disease in moving object or in moving body.

References

1. Abrams Z, Goel A, Plotkin S (2004) Set k-cover algorithms for energy efficient monitoring in wireless sensor networks. In IPSN. ACM/IEEE
2. Cardei M, Thai MT, Li Y, Wu W (2005) Energy-efficient target coverage in wireless sensor networks. In: IEEE. INFOCOM
3. Kumar S, Lai TH, Balogh J (2011) On k-coverage in a mostly sleeping sensor network. In: MOBICOM. ACM
4. Tian D, Georganas N (2011, May) A node scheduling scheme for energy conservation in large wireless sensor networks. Wireless Commun Mobile Comput J
5. Jeong J, Gu Y, He T, Du D (2009, April) VISA: virtual scanning algorithm for dynamic protection of body networks. In: Proceedings of 28th IEEE conference on computer communications (INFOCOM 09), Rio de Janeiro, Brazil
6. Lu G, Sadagopan N, Krishnamachari B, Goel A (2006) Delay efficient sleep scheduling in wireless sensor networks. In: INFOCOM. IEEE
7. Cao Q, Abdelzaher T, He T, Stankovic J (2005) Towards optimal sleep scheduling in sensor networks for rare event detection. In IPSN. ACM/IEEE
8. Kumar S, Lai T, Arora A (2005, August) Barrier coverage with wireless sensors. In: MOBICOM. ACM, Cologne, Germany
9. Abolhasan M, Wysocki T, Dutkiewicz E (2004) a review of routing protocols for mobile ad hoc networks. Ad Hoc Netw 2(2):1–22
10. Cheng X, Huang X, Du DZ (2006) Ad hoc wireless networking. Kluwer Academic Publishers, Boston, pp 319–364
11. Kumari R, Nand P (2016) Performance comparison of various routing protocols in WSN and WBAN. In IEEE conference on computing, communication and automation held on 29th–30th April 2016 at Galgotias University, Greater Noida (ICCCA 2016). https://doi.org/10.1109/CCAA.2016.7813814
12. Kumari R, Nand P (2017) An optimized routing algorithm for BAN by considering Hop-count, residual energy and link quality for route discovery. IEEE conference on computing, communication and automation held on 5th–6th May 2017 at Galgotias University, Greater Noida (ICCCA 2017). https://doi.org/10.1109/CCAA.2017.8229884
13. Kumari R, Nand P (2017, September) To improve the performance of routing protocol in mobile WBAN by optimizing the scheduling mechanism. Int J Emerg Res Manag Technol (IJERMT) 6(9). ISSN–2278-9359. (UGC approved Journal)
14. Kumari R, Nand P (2017, October) Performance analysis of existing routing protocols. Int J Sci Res Comput Sci Eng (IJSRCSE) 5(5). ISSN-2320-7639. (UGC approved and Thomson Reuters indexed Journal) https://doi.org/10.26438/ijsrcse/v5i5.4750
15. Kumai R, Nand P (2017) Performance Analysis of existing MAC and routing protocols for WBAN. IEEE sixth international Conference on system modeling and Advancement in system trends held on 29th–30th December at Teerthanker Mahaveer University, Moradabad (SMART-2017)

16. Kumari R, Nand P (2017) Secure Communication using PFS in a distributed environment. On-line international conferences on ancient mathematics and science for computing held on 24th–25th November 2017
17. Kumari R, Nand P (2018) Recent research on wireless body area networks: a survey. Int J Sci Res Comput Sci Eng Info Technol (IJSRCSEIT 3(1).ISSN- 2456-3307. (UGC approved and Thomson Reuters indexed Journal)
18. Kumari R, Nand P (2018) Performance analysis for MANETs using certain realistic mobility models: NS-2. Int J Sci Res Comput Sci Eng (IJSRCSE) 6(1).ISSN- 2320-7639. (UGC approved and Thomson Reuters indexed Journal). https://doi.org/10.26438/ijsrcse/v6i1.7077
19. Kumari A, Tanwar S, Tyagi S, Kumar N (2018) Fog computing for healthcare 4.0 environment: opportunities and challenges. Comput Electrical Eng 72:1–13
20. Tyagi S, Tanwar S, Gupta SK, Kumar N, Rodrigues JJPC (2014) Selective cluster based energy efficient routing protocol for homogeneous wireless sensor network. Wireless BANs for pervasive healthcare & smart environments. ZTE Commun 12(3):26–33
21. Vora J, Tanwar S, Tyagi S, Kumar N, Rodrigues JPC (2017) Home-based exercise system for patients using IoT enabled smart speaker. In: IEEE 19th international conference on e-health networking, applications and services (Healthcom-2017), Dalian University, Dalian, China, 12–15 October 2017, pp 1–6
22. Vora J, Tanwar S, Tyagi S, Kumar N, Rodrigues JPC (2017) FAAL: fog computing-based patient monitoring system for ambient assisted living. In: IEEE 19th international conference on e-health networking, applications and services (Healthcom-2017), Dalian University, Dalian, China, 12–15 October 2017, p 106
23. Khanna A, Singh AK, Swaroop A (2016) A token based solution to group local mutual exclusion problem in mobile ad hoc networks. Arab J for Sci Eng (Springer) (SCI) 41(12):5181–5194. IF 1.092. ISSN 2193-567X
24. Khanna A, Singh AK, Swaroop A (2014) A leader-based k-local mutual exclusion algorithm using token for MANETs. J Info Sci Eng 30(5):1303–1319. (SCI) (IF 0.54). ISSN: 1016-2364
25. Khanna A, Singh AK, Swaroop A (2016) Group local mutual exclusion algorithm in MANETs. CSI Trans ICT (Springer) 227–234. ISSN 2277-9078
26. Varshney S, Kumar C, Swaroop A, Khanna A, Gupta D, Rodrigues J, Pinheiro P, de Albuquerque V (2018) Energy efficient management of pipelines in buildings using linear wireless sensor networks Sensors 18(8):2618. https://doi.org/10.3390/s18082618. SCIE (IF 2.4)
27. Gochhayat SP, Kaliyar P, Conti M, Tiwari P, Prasath VBS, Gupta D, Khanna A (2019) LISA: lightweight context-aware IoT service architecture. J Cleaner Prod SCIE (IF 5.6). [In-Press]
28. Doss S, Nayyar A, Suseendran G, Tanwar S, Khanna A, Thong PH (2018) APD-JFAD: accurate prevention and detection of jelly fish attack in MANET. IEEE Access, in press (SCIE, 2017 IF = 3.557)

Experimental Analysis of OpenStack Effect on Host Resources Utilization

Pericherla S. Suryateja

Abstract Cloud computing is one of the frontier technologies, which over the last decade has gained a widespread commercial and educational user base. OpenStack is one of the popular open-source cloud management platforms for establishing a private or public Infrastructure-as-a-Service (IAAS) cloud. Although OpenStack started with very few core modules, it now houses nearly 38 modules and is quite complex. Such a complex software bundle is bound to have an impact on the underlying hardware utilization of the host system. This paper analyzes the effect of OpenStack on the host machine's hardware. For this purpose, an extensive empirical evaluation has been done on different types of hardware, different virtualization levels and with different flavors of operating systems comparing the CPU utilization, and memory consumption. OpenStack was deployed using Devstack on a single node. From the results it is evident that standalone machine with Ubuntu server operating system is the least affected by OpenStack and thereby has more available resources for computation of user workloads.

Keywords Cloud computing · Openstack · Resource utilization · Devstack

1 Introduction

Over the past decade, cloud computing [1] has become the de facto standard for dynamic provisioning of resources. Its other features like on-demand access, ubiquitous nature, elasticity, and pay-per-use model also lead to its success. Cloud computing provides services like Software-as-a-Service (SaaS), Platform-as-a-Service (PaaS), Infrastructure-as-a-Service (IaaS), DataBase-as-a-Service (DBaaS) and more. Cloud computing can also be deployed in various ways like public cloud, private cloud, community cloud, and hybrid cloud. Nowadays companies are using services from clouds [2] created by different vendors, making it a multi-cloud.

P. S. Suryateja (✉)
Department of CSE, Vishnu Institute of Technology, Kovvada, Andhra Pradesh, India
e-mail: suryateja.pericherla@gmail.com

© Springer Nature Singapore Pte Ltd. 2020
A. Khanna et al. (eds.), *International Conference on Innovative Computing and Communications*, Advances in Intelligent Systems and Computing 1059,
https://doi.org/10.1007/978-981-15-0324-5_2

OpenStack is a cloud operating system for managing resources in a data center or an organization and for establishing a private or public cloud. It is the most popular choice nowadays for providing Infrastructure-as-a-Service (IaaS). It was initially developed by NASA and Rackspace with few components. Now, many companies and developers are supporting the development of OpenStack. Although OpenStack contains approximately 38 components or modules, there are only few core components like nova, neutron, glance, keystone, swift and cinder. OpenStack also provides a dashboard through horizon module which allows administrator to easily monitor and provision resources to the clients.

There are five standard ways for deploying OpenStack: (1) using Juju Charms, (2) using Ansible in Docker containers, (3) using Ansible, (4) using TripleO and (5) using Devstack or Packstack. Both devstack and packstack are scripts which automatically deploys OpenStack on a single node. They are best suited for working with OpenStack for development purpose. Devstack works with Debian versions of Linux like Ubuntu, and packstack works with Fedora and RedHat versions and their derivatives. Devstack was used for experimentation as Ubuntu operating system was selected.

The rest of the paper is organized as follows. Section 2 presents the related work involving OpenStack. Section 3 describes the experimental design, the hardware configuration, and operating system choice. Section 4 presents the results obtained from observation. Section 5 concludes the paper.

2 Related Work

Sefraoui et al. [3] provide a comparative study of cloud platforms like Eucalyptus, OpenNebula, and OpenStack. They aim to signify the importance of open source solutions over commercial solutions. They stress on the role of OpenStack in establishing private clouds. Nasim and Kassler [4] demonstrated the effect of underlying infrastructure on the performance of OpenStack by conducting experiments on two separate testbeds. They deployed OpenStack on virtual environment and also on dedicated hardware. From the experimental results, they conclude that performance of OpenStack on dedicated hardware is far greater than performance on virtual infrastructure. They measured the performance of CPU, data transfer rate and bandwidth in the OpenStack installations in the two testbeds. Gebreyohannes [5] studied the performance of OpenStack Neutron, the network component in OpenStack. Using the IPERF benchmarking tool, the internal network performance of OpenStack was analyzed based on the parameters like packet loss, packet delay, and throughput. From the experiments conducted, author concluded that OpenStack Neutron was scalable and it offers performance with no network bandwidth bottleneck.

Xu and Yuan [6] measured the performance of Quantum (later renamed to Neutron) which was the network component in OpenStack. They measured the performance by deploying Openstack using single-host plan and on multi-host plan and by designing three test cases. All the tests were conducted on the installations after the deployment of Hadoop. Authors concluded that the performance of multi-host plan has significantly improved when compared with single-host plan. Mohammed and Kiran [7] present an experimental report in which they provide guidelines for setting up a cloud computing environment. They used the commodity infrastructure in their university's lab to setup a cloud computing environment. OpenStack was used as the cloud management platform. They used three nodes to setup the environment.

Grzonka [8] focuses on the problem of utilizing the physical and virtual resource in OpenStack cloud platform effectively. Many tests were conducted to measure the utilization of CPU and memory. Matrix multiplication was used for testing the performance of virtual resource. One important discovery in the experimentation was, increase in the number of virtual resource beyond a certain threshold, decreased the performance of the system. Raja [9] establish Iaas cloud using Openstack. Although the details of environment setup are not clear, they measured the utilization of CPU, memory, and network. Tornyai et al. [10] conducted performance evaluation on a private cloud established using OpenStack. The aim of their study was to determine the internal behavior of OpenStack with respect to virtual machine's computing and networking capabilities. Authors used OpenStack component named Rally for automatic performance evaluation of OpenStack internals. Using different test cases they measured concurrency, stress and disk performance in the cloud.

Sha et al. [11] explain the effect of cloud deployment architecture on the cloud's performance by providing a high-level formalization method called Performance Evaluation Process Algebra (PEPA), for modeling, analyzing and evaluating the cloud environment. The performance analysis of the cloud is carried out with respect to response time and the level of resource utilization. Callegati et al. [12] conducted several experiments on a cloud, managed using OpenStack, for evaluating the performance of Network Function Virtualization (NFV) and highlight its potentials and limitations. In the experiments they took care of both single-tenant and multi-tenant scenarios. They also developed a visual tool which plots the different functional blocks created by OpenStack Neutron component. Through the experiments conducted, it was concluded that the Linux Bridge is a bottle-neck in the architecture, while Open vSwitch showed optimal behavior.

Shetty et al. [13] provided an empirical evaluation for measuring the performance of workloads over hypervisor-based virtualization, container-based virtualization and on bare metal machine. For hypervisor-based virtualization OpenStack was selected, for container-based virtualization Docker was used. Phoronix test suite was used for conducting different experiments for measuring the performance of CPU, RAM, and disk in each of the aforementioned scenarios. The final conclusion was, bare machine provided maximum performance, followed by the container virtualization, followed by hypervisor-based virtualization. Ahmad and Qazi [14] investigated the performance of CPU in OpenStack compute node and on a standalone system providing kernel-level virtualization. The experimentation was conducted with no VMs, one, two, three, and four VMs. Later more experiments were conducted by varying the load and changing the operating systems and number of VMs. The operating systems that were tested are Ubuntu and CentOS. The experimental configurations were not provided which makes it unclear to understand the results of the experimentation done.

Sajjad et al. [15] conducted various experiments on the two most popular cloud platforms namely OpenStack and Windows Azure. The features that were measured are CPU, memory, disk, and network. For conducting experiments and evaluating the performance, several benchmark suites were used. Geekbench and LINPACK for CPU, RAMSpeed and STREAM for memory performance evaluation, IOzone for disk performance and finally for network performance, Iperf was used. The conclusion was, both OpenStack and Azure presented similar performance. But, OpenStack was considered as a better choice as it was free. Husain et al. [16] provided performance evaluation of OpenStack and Eucalyptus cloud platforms. The features measured were processor, memory, disk, and network. For benchmarking, Linpack for processor, Stream for memory, Bonnie++ for disk I/O and IPerf for network were used. The results show that OpenStack gives better performance for disk I/O and Eucalyptus provides better performance for processor, memory and network operations. There is no research available in the existing literature that had conducted detailed experiments for finding the effect of OpenStack on underlying hardware.

3 Experimental Design

OpenStack can be installed in several ways as discussed previously. All the experiments were conducted on OpenStack distribution named Queens which was installed using Devstack on a single node. So, all the OpenStack modules reside on a single machine (node). The modules installed are nova, neutron, glance, cinder, keystone, horizon and ceilometer. The experiments include two main scenarios.

3.1 Scenario 1 (S1)

Resource utilization is measured across a VM and a standalone machine with the different underlying hardware. Resources monitored are CPU and memory. CPU utilization is monitored using top command (Linux) and memory usage is retrieved using free command (Linux). Hardware configuration of the machine containing the VM is different from the standalone machine. The host operating system of machine with VM is Windows 8.1 and the guest operating system is Ubuntu Server 16.04. VMW are Workstation was used for creating the VM. The operating system of standalone machine is Ubuntu Server 16.04. Hardware configuration details of both VM and the stand-alone machine are given in Table 1.

Table 1 Hardware configuration

	VM	Standalone
Architecture	x86_64	x86_64
# CPU(s)	1	8
Thread(s) per core	1	2
CPU Speed	1.70 GHz	3.40 GHz
Motherboard	Intel Corporation	HP
L1 Cache	16 KB	256 KB
L2 Cache	–	1 MB
L3 Cache	–	8 MB
RAM	6 GB	8 GB
Disk	40 GB	931 GB
Ethernet	1 Gb/s	100 Mb/s

3.2 Scenario 2 (S2)

Resource utilization is measured across a VM and a standalone machine with the same underlying hardware. Resources monitored are CPU and memory. CPU utilization is monitored using dstat command (Linux) and memory usage is retrieved using free command (Linux). The operating systems used are Ubuntu Server 16.04 and Ubuntu Desktop 16.04. The VM is also created on the standalone machine with the host operating system as Ubuntu Desktop 16.04 and guest operating system as Ubuntu Server 16.04. The configuration of VM created in this scenario is similar to that of the VM created in scenario 1. VM was created using VirtualBox. Hardware configuration details are same as mentioned for standalone machine in scenario 1.

4 Results and Discussion

The results are described separately for the two scenarios mentioned in the previous section.

4.1 Scenario 1 Results

For scenario 1 (as mentioned in Sect. 3), top command is used to monitor the CPU utilization. 2500 frames or instances are collected and plotted. Memory utilization is calculated using the free command and is given in Table 2.

A combined plot of CPU utilization in VM and the standalone machine before installing OpenStack is given in Fig. 1 and a combined plot of CPU utilization in VM and the standalone machine after installing OpenStack is given in Fig. 2. On x-axis, we can see number of frames and y-axis represents percentage of CPU utilized.

Table 2 Memory utilization in scenario 1

Memory	VM		Standalone	
	Before	After	Before	After
Total (MB)	5949	5949	7892	7892
Free (MB)	5568	888	7543	2874
Utilization %	1.24	79.14	1.15	58.55

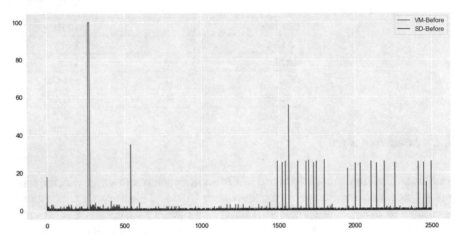

Fig. 1 CPU utilization in VM and standalone machine before installing OpenStack

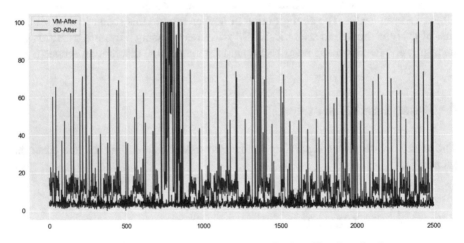

Fig. 2 CPU utilization in VM and standalone machine after installing OpenStack

Table 3 Memory utilization in scenario 2

Memory	VM		Standalone	
	Before	After	Before	After
Total (MB)	5967	5967	7892	7892
Free (MB)	5596	1685	7543	2860
Utilization %	1.2	65.79	1.12	57.24

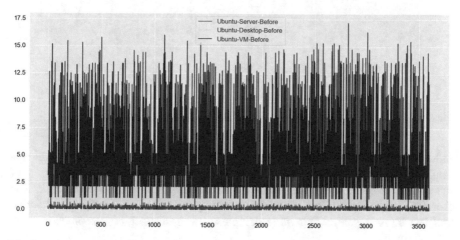

Fig. 3 Ubuntu Server, Desktop, and VM CPU util. before OpenStack installation

4.2 Scenario 2 Results

For scenario 2 (as mentioned in Sect. 3), dstat command is used to monitor the CPU utilization, disk I/O, network, and I/O requests. This command was executed approximately for one hour and 3600 instances or frames were collected. The underlying hardware for this entire scenario is the same. Memory utilization is calculated using the free command and is given in Table 3.

The grand comparison of CPU utilization in standalone machine with Ubuntu Server OS, Ubuntu Desktop OS and in Ubuntu VM before installing OpenStack is presented in Fig. 3 and the comparison of CPU utilization in standalone machine with Ubuntu Server OS, Ubuntu Desktop OS and in Ubuntu VM after installing OpenStack is presented in Fig. 4.

5 Conclusions

The proliferation of cloud computing in the industrial sector for resource provisioning has lead to a profound increase in research and development activities in both

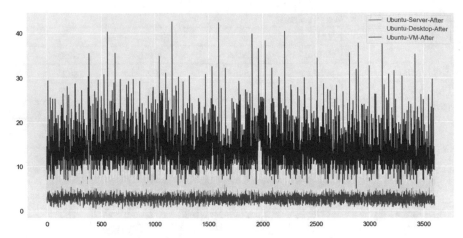

Fig. 4 Ubuntu Server, Desktop, and VM CPU util. after OpenStack installation

industry and academia. Generally, many people in academia can't afford industry-grade servers and other resources for testing cloud computing environment. One of the popular ways for deploying cloud on commodity resources is OpenStack. The Devstack deployment allows a user to install OpenStack cloud on a single machine and test it. As OpenStack is quite complex, the host machine resources may be affected by it.

In this paper, the effect of OpenStack on underlying resources of the host machine is studied by conducting empirical analysis in various scenarios like different hardware configurations, different flavors of OSs and different virtualization levels. From the results, it can be concluded that the best choice for deploying and testing an Open-Stack cloud environment is to use a standalone machine with Ubuntu Server operating system. Unless you are compelled to do so, never use a VM for testing OpenStack cloud environment, as much of the resources are consumed by the OpenStack services which may lead to degraded performance later.

References

1. Garg P, Sharma M, Agrawal S, Kumar Y (2019) Security on cloud computing using split algorithm along with cryptography and steganography. Int Conf Innov Comput Commun 55:71–79
2. Saxena A, Chaurasia A, Kaushik N, Kaushik N (2019) Handling big data using map reduce over hybrid cloud. Int Conf Innov Comput Commun 55:135–144
3. Sefraoui O, Aissaoui M, Eleuldj M (2012) OpenStack: toward an open-source solution for cloud computing. Int J Comput Appl 55:38–42
4. Nasim R, Kassler AJ (2014) Deploying OpenStack: virtual infrastructure or dedicated hardware. In: Proceedings—IEEE 38th annual international computer software and applications conference on work. COMPSACW 2014, pp 84–89
5. Gebreyohannes MB (2014) Network performance study on OpenStack cloud computing. 124

6. Xu Q, Yuan J (2014) A Study on service performance evaluation of openstack. In: Proceedings—2014 9th international conference in broadband wireless computer communications and applications on BWCCA 2014, pp 590–593
7. Mohammed B, Kiran M (2014) Experimental report on setting up a cloud computing environment at the University of Bradford, pp 1–19
8. Grzonka D (2015) The analysis of OpenStack cloud computing platform: features and performance. J. Telecommun Inf Technol 2015:52–57
9. Raja JB (2016) Iaas for private and public cloud using Openstack. Int J Eng Res Technol 5:99–103
10. Tornyai R, Pflanzner T, Schmidt A, Gibizer B, Kertesz A (2016) Performance analysis of an OpenStack private cloud. In: Proceedings 6th international conference on cloud computer service science, pp 282–289
11. Sha L, Ding J, Chen X, Zhang X, Zhang Y, Zhao Y (2015) Performance modeling of Openstack cloud computing platform using performance evaluation process algebra. Proceedings—2015 international conference on cloud computer big data, CCBD 2015, pp 49–56
12. Callegati F, Cerroni W, Contoli C (2016) Virtual networking performance in OpenStack platform for network function virtualization. J Electr Comput Eng
13. Shetty J, Upadhaya S, Rajarajeshwari HS, Shobha G, Chandra J (2017) An empirical performance evaluation of docker container, OpenStack virtual machine and bare metal server. Indones J Electr Eng Comput Sci 7:205–213
14. Ahmad W, Qazi AS (2018) Analysis of interactive utilization of CPU between host and guests in a cloud setup. Comput Sci Eng 8:7–15
15. Sajjad M, Ali A, Khan AS (2018) Performance evaluation of cloud computing resources. Int J Adv Comput Sci Appl 9:187–199
16. Husain A, Zaki MH, Islam S (2018) Performance evaluation of private clouds: OpenStack versus eucalyptus. Int J Distrib Cloud Comput

Effect of Dropout and Batch Normalization in Siamese Network for Face Recognition

Nilagnik Chakraborty, Anirban Dan, Amit Chakraborty and Sarmistha Neogy

Abstract The paper focuses on maximizing feature extraction and classification using one-shot learning (meta-learning). The present work discusses how to maximize the performance of the Siamese Neural Network using various regularization and normalization techniques for very low epochs. In this paper we perform multi-class Face Recognition. A unique pairing of face images helps us to understand the generalization capacity of our network which is scrutinized on AT&T-ORL face databases. We performed experiments to see how learning can be made to converge within a few epochs, and the approach has also made a telling performance on unseen test data which is about 96.01%. Besides, we discuss the ways to speed up learning particularly for a Siamese network and achieve convergence within 5 epochs. We found one of the better regularization techniques for fast reduction of the loss function. It is apparent from our findings that only normalization is the effective approach while working within less epochs. Also, Dropout After Batch Normalization configuration results in smooth loss reduction.

Keywords Regularization · Siamese network · Deep learning · Face recognitionMeta-learning

N. Chakraborty (✉) · A. Dan · A. Chakraborty
Department of Computer Science & Engineering, Government College of Engineering & Ceramic Technology, Kolkata, India
e-mail: chakrabortynilagnik@gmail.com

A. Dan
e-mail: dan.anirban20@gmail.com

A. Chakraborty
e-mail: amitc8250@gmail.com

S. Neogy
Department of Computer Science & Engineering, Jadavpur University, Kolkata, India
e-mail: sarmisthaneogy@gmail.com

© Springer Nature Singapore Pte Ltd. 2020
A. Khanna et al. (eds.), *International Conference on Innovative Computing and Communications*, Advances in Intelligent Systems and Computing 1059,
https://doi.org/10.1007/978-981-15-0324-5_3

Abbreviations

BD   Dropout After Batch Normalization
DB   Batch Normalization After Dropout
BN   Batch normalization,
FV   Face Verification

1   Introduction

Increase in data and computation power has caused a huge bump in Computer vision. This huge progress is the reason for a paradigm shift in the traditional image processing and machine learning methods. Earlier methods for Face recognition were driven like the work in [1], where local binary feature learning method was proposed. In [2] authors create surveys of sparse coding and dictionary learning algorithms used for face recognition systems. Machine learning showed some interesting results and paved way for improved algorithms. Previously the tasks of FV and face identification were considered separate problems but with more and more ensemble-type leaning algorithms in place, the gap has narrowed.

A generic machine learning tactic to this exhaustive type of problem is application of distance-related methods [3] in which a correlational metric is calculated based on the distance between a subject pattern and a target pattern. Traditional methods sheltered discriminative architectures, used classifiers such as NN, SVM [3] to calculate the semblance of two input images. However, when subjected to non-linear categorization problem like FV these techniques have certain rigidity like (a) categories (subjects) are not flexible post-training, (b) properly labeled data should be generatable for all individuals under consideration, and (c) can handle only limited categories (<100) [4].

To save processing speed and resource utilization authors in [5] focused on creating systems for mobile devices, where the technique was required to handle the constraints of a typical low resource and low power architecture. But the results could not still reach standards set by deep learning. Throughout the paper, we attempt to tackle the complexity of FV and understand the limitations and scope of improvement of the current methodologies as discussed before. We propound a progressive regularization schema for Siamese Convolutional Neural Network (Siamese CNN) [6] architecture to maximize feature extraction and regulate the bias-variance tradeoff at very low epochs in the domain of FV task. Our goal was to design an optimized and efficient system for successful implementation in low resource and low processing speed systems. A Computation graph trained using backward propagation [7] usually is a very computationally expensive and time taking process. Recently practitioners are using large number of epoch to increase chances of hitting the global minima of the distribution and thereby convergence, requiring intense computation. However, setting large epochs is an expensive operation which researchers want to avoid. This paper illustrates a way to redefine our learning process by taking a step towards better optimization while considering a fixed resource limit. Here, we detect the face and prioritize the area of interest with appropriate preprocessing. Then deep learning-based methods are applied to compute unique features of the input image. In conclusion, we compare the detected face with the ones in the database by finding Euclidean distance of the distributions. The experiments demonstrate very good performances in a smaller number of epochs.

2 Related Work

2.1 Face Recognition-Via Deep Learning

Latest developments constitute significant advancement in computer vision mainly by virtue of Convolutional Neural Network (CNN). Popular implementations include special architectures like GoogleNet and VGG [8, 9] revolutionized the domain. Sadly, all these architectures are several layers deep and are difficult to optimize. A pre-trained set of weights being applied still takes significant resources to train further. Our architecture was also inspired from DeepFace [10]. It was designed by Facebook researchers. It comprises of a nine-layer each deep siamese net exceeding 119 million weights. Over 4 million input images constituted the training operation with acquired data from Facebook 97% accuracy was claimed to be achieved.

2.2 Siamese Convolutional Network

Two or more identical computation graphs with similar parameters constitute a brand of neural networks known as "Siamese" [6]. Mirroring takes place across the branches of the network when the optimizer updates the parameters. Importance of this kind of network is significant where the objective involves finding correlation between input and the target. We shall discuss them in subsequent sections.

2.3 Computationally Effective Methods for Face Recognition

Authors in [5] focus on creating systems for mobile devices, where they need to solve the problem with limited resources.

2.4 Dropout

Dropout [11] can be visualized as a way of regularizing a network by introducing noise to its hidden neurons. Bernoulli distributed random variables undergoes multiplications with activations which are generally hidden. It is obvious that training and testing philosophies are different. During the training phase, a subnetwork of a layer is generated and activations from previous layer are fed into it. However, dynamic subnetwork generation does not take place during testing.

2.5 Batch Normalization

Ioffe and Szegedy [12] proposes a deterministic information flow by normalizing each neuron into zero mean and unit variance. The normalization of activations that depends on the mini-batch allows efficient training but is neither necessary nor desirable during inference. Therefore, BN accumulates the moving averages of neural means and variances during learning to track the accuracy of a model as it trains. This restricts the effects of drastic changes in previous layers to affect weights in the next layer. This feature is particularly important to handle noise in data. For a layer with d-dimensional input $x = (x\hat{}(1) \dots x\hat{}(d))$, we will normalize each dimension where the expectation and variance are computed over the training data set. The mathematical expression of BN is given below:

$$\hat{x}^{(k)} = \frac{x^{(k)} - \mathrm{E}\left[x^{(k)}\right]}{\sqrt{\mathrm{Var}\left[x^{(k)}\right]}} \tag{1}$$

We were also highly inspired by application of NN in *Hand Gesture recognition* [13] and *Leaf Identification* [14]. Both showed us the possibility that neural network can be applied in wide variety of field with certain tuning or variations in architecture.

3   Outlines of Present Work

Figure 1 shows the different phases of the workplan. First, we will collect the data and process it. Then we will design the architecture and tune in the parameters to get the optimum output. Figure 1 demonstrates all the steps, respectively.

Fig. 1 Phases of the
workplan

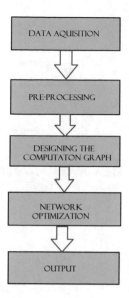

3.1 Data Acquisition

Our dataset is made up of 368 training faces and 161 test faces. The training and test data are further divided into subject's face data and others (data from 2 of our teammate and AT & T's ORL dataset [15]). The black and white images have a dimension of 224 * 224. The dataset is heavily focused on face being covering 75% area in an image. All possible facial expressions and lightening conditions have been considered with various augmentation (e.g. adding noise) to make the dataset more robust. We have merged AT &T's ORL dataset our own dataset. [MENTION ABOVE] We had an objective of building a user face authentication system. Thus, we divided our training set and validation set into user and faces of others (non-users). The dimension of the images was set to 224 * 224 after a lot of experimentation.

3.2 Data Preprocessing

The performance of Neural Network models depends on distinct factors of the dataset. We can't afford to work with the infinite amount of data, which shall increase {chances of divergence for our optimizer before reaching the absolute minima. We undergo (i) Collection of images with face detected (ii) Face Alignment, and (iii) Resize data & image augmentation (as we don't have an infinite amount of data).

Image Collection and Face Detection: Standardization of face prior to analyzing it through a neural network is one of the most important parts of the process. Without proper normalization of the face, it becomes really challenging for the network to

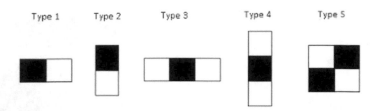

Fig. 2 Rectangle features shown relative to the enclosing detection window

Fig. 3 Finding the sum of
the shaded rectangular area

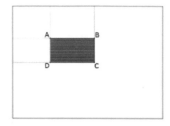

comprehend subtle changes in orientation. The network may recognize it as random
noise instead. Firstly, a KNN based Face Detection algorithm [16] was trained. We
used the KNN with a scale factor of 3.27 and mean size of (30, 30) and the number
of mean neighbors 3 which gave us about 95% accuracy. For detecting the AOI in
the face image, Haar features were extracted for classification. Haar features (Fig. 2)
are applied repeatedly on the input image to detect the face area. The following
mathematical equation is used to compute Haar features.

SUMMATION $=$ Image(C) $+$ Image(A) $-$ Image(B) $-$ Image(D) where points A,
B, C, D refer to integral image in Fig. 3 [17] given below.

Face Alignment (face normalization): After detection of our AOI, face normalization comes to play. Face normalization is done to discard changes in orientation
of input face data and develop a uniform correlation between various pixels in space.
We obtain key 2D landmark locations from the input image. Delaunay's Triangulation is applied to this 2D landmark. Delaunay's Triangulation minimizes the maximum angle of the edges generated from the key landmark points and results in apt
alignment.

Resize data & image augmentation: After generating normalized face data the
quality of our dataset has already enhanced. Now we resize these images into a fixed
dimension to meet the network requirements. We add a bit of noise to a few samples
to remove class bias, if any. All the pre-processing done till far are summarized in
Fig. 4.

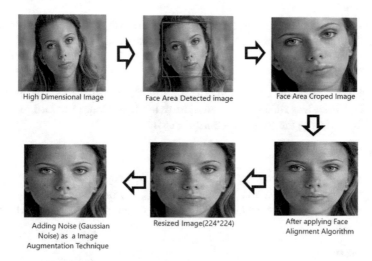

High Dimensional Image   Face Area Detected image   Face Area Croped Image

Adding Noise (Gaussian   Resized Image(224*224)   After applying Face
Noise) as a Image   Alignment Algorithm
Augmentation Technique

Fig. 4 Pre-processing stages to make data ready for subsequent stages

3.3 Designing the Computation Graph

Conventionally simple CNNs are used for understanding the image features and breaking them down for machine comprehension. But the problem with that approach is that more layers of computation graph are needed for finding out basic facial features and then slowly deriving the more complex and unique features in somebody's face for classification. With more nodes, the space complexity increases tremendously and it's obvious that the training time shall increase significantly.

The answer to this problem is a sophisticated approach where we facilitate parallel computation so that the nodes in our graph work more as comparators than as actual feature extractors in an overall sense. The network works on the principles of "true" and "false" pairings. And subsequently updates the weights in order to increase the spatial distance between the two true and false distributions (if images are considered as a distribution of three dimensions). Spatial distances are calculated by finding the Euclidean distance. The resulting networks are of shorter depth with fewer nodes capable of inferencing the covariance of the two pairs in a broader sense by backpropagation.

(a) *Nothing Without CNN*

The fundamental concepts behind CNNs are used in a parallel computation graph as well. Each of the branches of the network has the same trivial Convolution-MaxPool setup. The network learns by reducing the loss function which in turn, tries to find the Euclidean distance. Higher value of Euclidean distance signifies the fact that both the images have higher spatial covariance and are liable to be images from different people. During backpropagation it is ensured that, for pairs in which Euclidean distance is greater than our threshold, value of the weights is adjusted to further increase their spatial distance for next forward iteration of the algorithm.

(b) *Triplet Loss*

We are indebted to Chopra-Lecun [4] for their work on triplet loss. We have incorporated the triplet loss function for calculating the loss during forward propagation and thereby updating the weights according to it.

(c) *Datasets*

ORL [15] is yet another popular dataset of faces of people (mainly grayscale). We have augmented the dataset and added our own data alongside. The images (224 * 224) that were added to the existing dataset were taken using a simple 8MP camera in RGB.

(d) *Rmsprop*

Rmsprop [18] was introduced first by Geoffrey Hinton. It perfectly adapts to the size of each step for each weight along with the nature of the gradient.

The building blocks of our architecture as depicted in Fig. 5 are Convolutional layers, Activation layers, Batch normalization layers, and Max-pooling layers. This architecture has resulted in highest accuracy compared to others we had implemented and experimented with. We have given a detailed analytical study of our observations of various architectures in the next sections. Clearly, there are two parallel computation graphs and we discuss this architecture keeping a single stem in mind. As said before the input dimension of our network is 224 * 224. The initial operation done on every raw input is convolution. Firstly, a 2 * 2 convolution is done using 128 filters and no padding was used. There were a total of 57,472 trainable parameters in

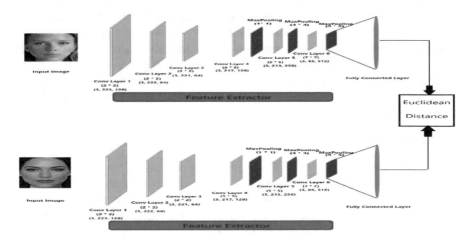

Fig. 5 Is a pictorial overview of our neural network architecture which constitutes two parallel but identical networks

this layer. The number of parameters was kept large to maximize feature extraction from the image. This was followed by Batch-normalization layer with 512 trainable parameters. Rectified linear unit is used as activation function for this layer after batch normalization.

Again, a set of 2 * 2 convolution with 64 filters and Relu activation was implemented but this time without the batch normalization layer. The set was repeated twice in our architecture. There was a total of 24,704 (= 16,448 + 8256) trainable parameters. Here, we focus to enrich our network with more auto-generated features and hence we didn't opt for dimensionality reduction.

Following the above, we implemented a 5 * 5 convolution layer with 128 filters. The total number of trainable parameters was 41,088. This layer was followed up by a batch normalization layer and then by Relu activations. Auto-generated features are not scaled down here too. After that, we focus more on convergence and finding out more unique vectors. Hence, we incorporate max-pooling in our next layers. First, we implement a 5 * 5 convolution with 256 filters. It consisted of 164,096 trainable parameters and was followed by BN, and then activation layer. The 3 * 3 max-pooling was done on the output for dimensionality reduction and comprehending key facial characteristics. The final sets of layers include a 7 * 7 convolution with 512 filters followed by BN, and then activation. A 5 * 5 max-pooling was computed on the output for significant dimensionality reduction. A total of 920,064 (= 918,016 + 2048) parameters were trained in this set. The computed output vector is then flattened into a 1-d array and is fed into a set of fully connected neurons. 128 nodes are used in this case to train massive 852,096 parameters in this final neural computation layer. The learning algorithm or optimizer used is Rmsprop [18] with learning rate = 0.001, decay = $e\hat{\ } (-10)$ and batch size being 8.

SoftMax activation was used here and no batch normalization was used. The above computation is done for both the images in comparison. We can easily see that distribution is obtained as an output from the fully connected layer of both the input images. We compute the Euclidean of both this distribution. The distance is a metric of similarity/dissimilarity of the images. If the distance is more than a threshold value, then we can conclude that the images are from different subjects or if they are within threshold then they are from the same subject. The threshold value is hyperparameter which must be chosen with extreme caution, we found 0.5 to be an acceptable value.

3.4 Network Optimization

Grid search was implemented to discover the optimal hyperparameters. He-initialization, [19] instead of random initialization of weights is found to be effective for convergence in very low epoch. We used Nvidia instance via Colab cloud from Google [20] for training several models at a given time. This immensely helped our optimization strategies.

4   Experimental Results and Analysis

The paper is focused on finding the best regularization, normalization, and feature extraction techniques for a Siamese network working in a very low epoch scenario. Thus, we conducted extensive experiments on our network.

TensorFlow (python 3.6 API) was used to fabricate the model and training was conducted with cloud GPUs.

4.1   Loss Variation with Changing Dropout Values

We had 4 Dropout layers in our network which was first implemented in place of the batch normalization layers of our proposed architecture. We experimented by changing the Dropout rate and finding respective change in losses. We varied the 4 dropout layers in 72 different ways and plotted them using bar graph (Fig. 6). On plotting the losses (blue bars) with the initial and final dropout layers (orange bars) we obtained the graph (Fig. 8). It showcases certain insights like the change in loss is unpredictable with the change in the dropout values after a convolution operation. It shows that regularization by dropout is quite difficult as selecting the right choice of dropout values is really an exhaustive work. Neither very high nor very low values can reduce losses faster, and, moreover, it doesn't guarantee maximum feature extraction as we see below.

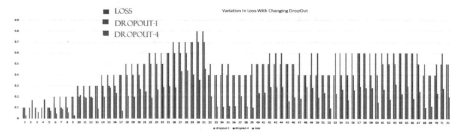

Fig. 6 Loss variation with changing dropout values

4.2 Dropout Versus Batch Normalization

Next, we tried to compare the performances of our regularization technique. We did a comparative study of the loss minimized after each epoch for both the techniques (Figs. 7 and 8). We plotted a scaled graph of both observations in Fig. 9. Ultimately,

Fig. 7 Loss minimization in Dropout after 5 epochs of training

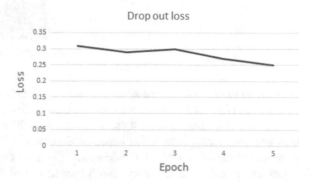

Fig. 8 Loss minimization of BN after 5 epochs

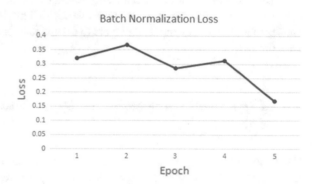

Fig. 9 Loss variation between Dropout and BN

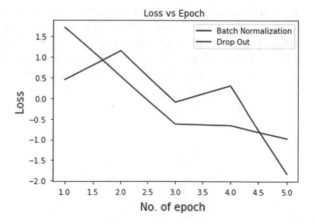

Fig. 10 Accuracy of
Normalization and Dropout
after 5 epochs

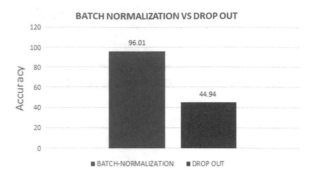

we tried to measure the accuracy of our network on unseen test data after 5 epochs of training. Figure 10 shows the accuracy of both the techniques.

Figure 9 is a scaled curve of curves 7 and 8 with zero mean and unit variance. It is very interesting to note that losses erode faster in BN and network gains more generalization properties by identifying key distinguishing facial features. This fact is proven by the overall accuracy after 5 epochs of training in Fig. 10.

From this observation, we can easily understand that if dropout layers are used then feature extraction from image data takes a real hit. Random switching of nodes causes reduction of key inputs for layers to follow and learning from less parameter's backfires.

4.3   Experiment Using Serialized Dropout and Normalization Layers

We extended our experimentation and tried to implement both the techniques in tandem. We did two experiments, first where we put batch normalization layers after dropout and second where we did the reverse. The results are depicted in Fig. 11. Then we combined and graphed both the results under the same scale. Their respective accuracy after 5 epochs are listed in Fig. 12. We observed that it was beneficial to use dropout after the normalization layer. It gave better performance not only in loss reduction but also in key feature extraction and thereby performance on test data. It is safe to say that if dropout is used after batch normalization, we achieve a balanced regularization and feature extraction compared to the opposite (DB), the higher accuracy levels in Fig. 12 further justify the statement.

Fig. 11 Loss minimization
by using Dropout and BN in
varied order

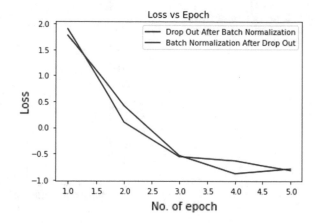

Fig. 12 Accuracy of BN
And Dropout after 5 epochs

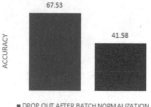

4.4 Comparing Normalization with Dropout After Batch Normalization

After the above experiments, the last comparison is to understand the better way to
rule out overfitting and give the highest stability to the network. Figure 13 gives us a
clear intuition between BN and BD from a unit-scaled graph. Figure 14 depicts the
accuracy graph between both. We conclude that refraining from the use of dropout
for low epoch learning is wise. However, at higher epochs using BD can result in
a much smoother loss reduction, sudden abnormal increase in loss can be avoided
totally by using BD.

Fig. 13 Comparing loss
minimization of Dropout
after BN with

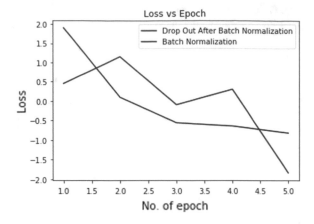

Fig. 14 Accuracy of
Dropout after BN with BN

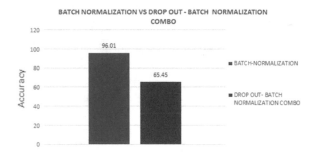

4.5 Deciding the Order of Batch Normalization and Activation Layer

The final piece of the jigsaw is the positioning of the activation layers in our architecture. We put the activation before and after batch normalization and tried to understand the unit-scaled loss curve. The loss curves are shown in Fig. 15 and corresponding accuracy in Fig. 16.

Fig. 15 Loss variation curve
via changing positions
around

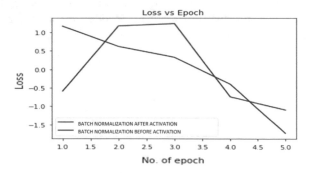

Fig. 16 Accuracy of BN
with activation placed BN
before and after it

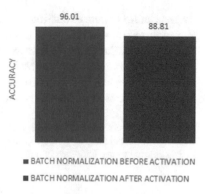

The results were interesting. Batch normalization when applied before activation seems to work more efficiently than the reverse. The loss reduction is smoother with less abrupt behavior. The accuracy of our system was more when we applied the activation later.

5 Key Outcomes

5.1 Using only Batch Normalization gives better results under low epoch restriction scenario.
5.2 Feature extraction is enhanced if only Batch Normalization is used.
5.3 Feature extraction is reduced if only Dropout is used.
5.4 Use of Batch Normalization +Dropout can have some drawbacks in low epoch study.
5.5 Dropout after Batch Normalization results in smoother loss reduction. This is an optimal strategy for rigorous training with large datasets when the number of epochs is not a constraint.
5.6 Generalizing capability is enhanced if initial convolution layers are not followed by max-pooling.
5.7 Batch normalization should be put before Activation.
5.8 Maximizing the number of trainable parameters in the computation graph is very essential while designing the network.

6 Conclusion

We achieved 96.01% accuracy on our own augmented dataset with only 5 epochs of training. This was possible due to our architecture and use of proper regularization and normalization techniques. We tried to maximize the number of trainable parameters for maximum feature extraction and found that use of Batch normalization was vital. We analyzed how to efficiently reduce the triplet loss in our Siamese network. In the future, we plan to employ statistical methods to establish a mathematical harmony with our observations and use Mahalanobis distance which is known to outperform other metrics when it comes to accuracy in those algorithms that use distances. Our main objective is to make supervised learning as efficient as possible.

Acknowledgements We acknowledge AT&T Laboratories Cambridge for images used in this work. We also acknowledge Amit Chakraborty and Anirban Dan, coauthors of this paper, for giving consent to use their data in this work.

References

1. Duan T, Lu J, Feng J, Zhou J (2017) Context-aware local binary feature learning for face recognition. IEEE Trans Pattern Analysis Mach Intell PP(99):1–1
2. Xu Y, Li Z, Yang J, Zhang D A survey of dictionary learning algorithms for face recognition. IEEE Access 5:8
3. Guo G, Li SZ, Chan K (2000) Face recognition by support vector machines. In: 4th IEEE international conference on automatic face and gesture recognition, pp 196
4. Chopra S, Hadsell R, LeCun Y (2005) Learning a similarity metric discriminatively, with application to FV. In: IEEE computer society conference on computer vision & pattern recognition, CVPR 2005, pp 539–546
5. Hassan G, Elgazzar K (2016) The case of face-recognition on mobile devices. In: 2016 IEEE wireless communications and networking conference, pp 1–6
6. Koch G (2015) Siamese neural networks for one-shot image recognition. Accessed August 2018
7. David et al., "Learning Representations by back propagating errors", Nature 1986 pp 533
8. Szegedy et al Going deeper with convolutions CVPR' 15
9. Simonyan K, Zisserman A (2015) Very deep convolutional networks for large-scale image recognition published at ICLR conference
10. Taigman Y et al (2014) DeepFace-Closing the gap to human-level performance in FV. In: Conference on Computer Vision and Pattern Recognition (CVPR)
11. Srivastava N (2015) Dropout: a simple way to prevent neural networks from overfitting
12. Ioffe S, Szegedy C (2015) Batch normalization: accelerating deep network training by reducing internal covariate shift. ICML 37:448
13. Ahlawat S et al (2018) Hand gesture recognition using convolutional neural network. In: International conference on innovative computing and communications, pp 179–186
14. Sharma P et al (2018) Leaf Identification Using HOG, KNN, and neural networks. In: International conference on innovative computing and communications pp 83–91
15. The Database of Face available at: https://www.cl.cam.ac.uk/research/dtg/attarchive/facedatabase.html
16. Viola P et al (2001) Rapid object detection using a boosted cascade of simple features. CVPR 2001
17. Haar like features, Available at wiki: https://en.m.wikipedia.org/wiki/Haar-like_feature. Accessed Dec 2018
18. Rmsprop available at coursera: https://www.coursera.org/learn/neural-networks/lecture/YQHki/rmsprop-divide-the- gradient-by-a-running-average-of-its-recent-magnitude

19. Xavier G, Bengio Y (2010) Understanding the difficulty of training deep feed forward neural networks. Aistats 9
20. Google Colab available at: https://colab.research.google.com/. Accessed Aug–Dec 2018

A Flag-Shaped Microstrip Patch Antenna for Multiband Operation

Purnima K. Sharma, Dinesh Shrama and Ch. Jyotsna Rani

Abstract A high-gain multiband flag-shaped microstrip patch antenna is proposed and developed in this paper. The shape of the antenna is very simple to design with dimensions 2.3 × 1.9 cm on FR4 substrate leading to high-gain and good bandwidth. The designed antenna works at multiple frequencies lying in C-band, X-band and Ku-band. The simple structure of this antenna design allows for an easy fabrication process, covers many applications such as radar, satellite, and wireless communication. The experiment results indicate that the proposed antenna design, having 7.3, 8.7 and wideband at 13.9 GHz used in different areas of communication. The Suggested antenna is modelled and successfully simulated using HFSS. The obtained results are compared and presented to demonstrate the performance of the designed antenna.

Keywords HFSS · FR4 substrate · Return loss · C · X and Ku-band · VSWR

1 Introduction and Literature Survey

Nowadays wireless communication plays a vital role almost in all our daily needs. So antenna is important component for all wireless applications. An antenna is a passive device used to convert an RF signal, propagating on a conductor material, into an electromagnetic signal in free space. Reciprocity is the fundamental property of antenna. Due to this property, antenna characteristics such as antenna gain, radiation pattern, frequency of operation and polarization remains same whether the particular antenna is transmitting or receiving [1]. For long-distance communication satellites are used in which most of the antennas are microstrip patch antennae. Microstrip patch antennas are used due to their compact size and high fidelity. A Microstrip

P. K. Sharma (✉) · Ch. J. Rani
Department of ECE, Sri Vasavi Engineering College, Tadepalligudem, A.P., India
e-mail: purnima.kadali@gmail.com

D. Shrama
Department of ECE, CCET, Punjab, Chandigarh, India

© Springer Nature Singapore Pte Ltd. 2020
A. Khanna et al. (eds.), *International Conference on Innovative Computing and Communications*, Advances in Intelligent Systems and Computing 1059,
https://doi.org/10.1007/978-981-15-0324-5_4

antenna consists of copper patch mounted on a dielectric material and ground plane is connected to the feed line on the bottom side of the dielectric material.

Mr. S. Maci and Mr. Bifji Gentili have designed the antenna at dual-frequency patch antennas that have been carried out, with special emphasis on configurations that are particularly attractive for their simplicity and design flexibility[2]. Mrs. Midasala designed an antenna and array of antenna at Ku-band with gain of 6.8 and 8.5 dB [3]. Mr. Antara Ghosal designed a patch antenna which rejection bandwidth increased from 13.5 to 18.2% [4]. Mr. Settapong Malisuwan designed an E-Shaped patch antenna working at Ku-Band which can be used for different applications [5]. Hasan Sahariar, Henry Soewardiman, and Jesse S. Jur, inclusion of holes (or gaps) in the conductive patch and ground areas progress the mouldability and breathability in the structure [6]. Mr. Fan Yang designed and fabricated an E-shaped microstrip patch antenna and compared with the conventional wide-band patch antennas and which is simple and small in size [7]. Praful Ranjan designed a capacitive coupled patch antenna to increase the bandwidth which is important for satellite communication [8]. K. Kumar Naik developed an antenna which operates at 13.65 GHz, 15.19 GHz with return loss of −44.88 dB, −55.83 dB, and gain 5.89 dBi, 7.22 dBi, respectively [9]. The main advantage of patch antenna array with the series-fed network is compact size a very small antenna spacing which results in a higher unambiguous angle range in a radar system. B. Datta et al. in his paper compared his proposed microstrip patch antenna results with a standard microstrip antenna. It focuses on simulating monopole single feed slotted patch antenna with operating frequency in the range of X-band and Ku-band [10]. The proposed flag-shaped antenna is a multiband antenna used for different applications in X-band and Ku-Band of radar communication, satellite communication, and some wireless computer networks.

2 Proposed Antenna Design

In present paper flag-shaped patch antenna is designed at different bands and simulated on ANSYS HFSS Electronics Desktop R 17.2 version. The patch is the dominant figure of a microstrip patch antenna and remaining components are on the sides of the antenna. The proposed antenna is designed with the substrate material of relative permittivity of 4.4 and dielectric loss tangent of 0.02. The edge feed technique is used in the proposed antenna which is shown in Fig. 1.

The specifications of substrate, patch, and feeding are given in Tables 1, 2 and 3 respectively. Figure 1 specifies the design of single flag-shaped patch antenna. The design is good since it works at different frequencies. It gives the data like Reflection loss, VSWR, Gain. For an antenna to be good the Reflection loss must be below −10 dB and the VSWR value must lie between 1 and 2. Figure 2 shows the simulation results (return loss and VSWR) of proposed flag-shaped antenna.

Fig. 1 Design of single patch antenna

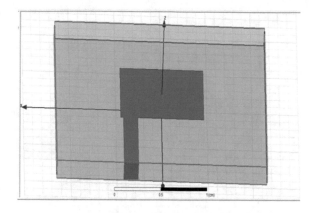

Table 1 Specification of substrate

Position	Sub $X/2$	Sub $Y/2$ (cm)	0 cm
	1.15 cm	0.95	0 cm
X-size	Sub X	2.3	–
Y-size	Sub Y	1.9	–
Z-size	Sub H	62	–

Table 2 Specifications of patch

Position	Sub $X/2$	Sub $Y/2$ (cm)	Sub H
	0.455 cm	0.335	62 mm
Axis	Z	–	–
X-size	Patch X	0.91	–
Y-size	Patch Y	0.67	–

Table 3 Specification of feeding

Position	Sub $X/2$	Sub $Y/2$ (cm)	Sub H
	0.425 cm	0.335	0.15478 cm
Axis	Z	–	–
X-size	Feed X	−0.1555	–
Y-size	Feed Y	0.615	–

3 Experimental Results and Discussions

Figure 2 shows the return loss and VSWR of flag-shaped microstrip patch antenna which has −22.5 dB at 7.3 GHz, −14 dB at 8.7 GHz and −33 dB at 13.9 GHz frequencies wide bandwidth at 13.9 GHz. The design parameters are satisfied by the designed antenna. The design is good so that it works at different frequencies.

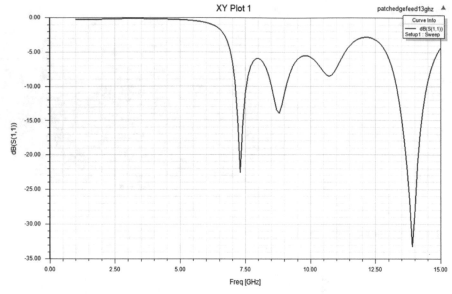

(a) Simulated Returns Loss

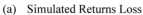

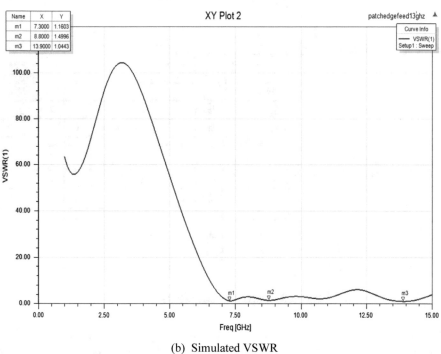

(b) Simulated VSWR

Fig. 2 **a** Simulated return loss and **b** VSWR of proposed antenna

Formulated on the ideal dimensions in Fig. 1, a sampled of the X-band and Ku-band flag-shaped antenna is designed, simulated, fabricated, and experimentally investigated. Figure 3 show top view and bottom views of the proposed antenna correspondingly. S-parameters are measured by the Agilent vector network analyzer (VNA) for the proposed antenna. Figure 4 depicts the comparison between simulated and measured S-parameters of the prototyped antenna. It is clear from Fig. 4 that the simulated and the practical measured returns loss are almost same. The small deviations are mainly caused by the mismatching between the connector and the antenna feed line. Figure 5 shows the gain and radiation pattern of the flag-shaped microstrip patch antenna. These results indicate that this resulted antenna is worth for C-band, X-band, and Ku-band applications.

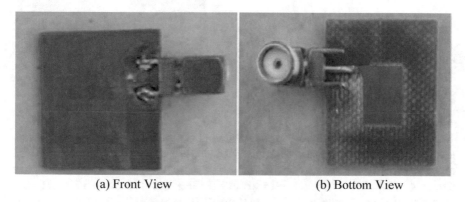

(a) Front View   (b) Bottom View

Fig. 3 Physical design of single patch antenna on FR4 epoxy substrate

Fig. 4 Variation of simulated and measured S-parameter

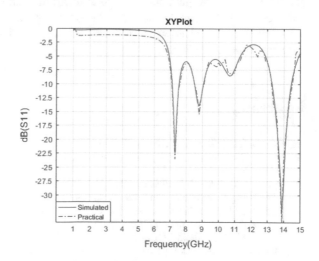

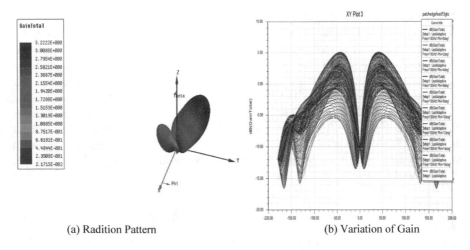

(a) Radition Pattern   (b) Variation of Gain

Fig. 5 Radition pattern and gain of the proposed flag-shaped Antenna

4 Conclusion

A compact flag-shaped microstrip patch antenna is developed. The antenna, characterized by a simple structure and compact size of 2.3 × 1.9 cm, satisfies the 10-dB reflection coefficient requirement for X-band and Ku-band applications. The designed antenna −22.5 dB at 7.3 GHz, −14 dB at 8.7 GHz and −33 dB at 13.9 GHz frequencies wide bandwidth at 13.9 GHz. The VSWR values of 7.3 GHz, 8.7 GHz, and 13.9 GHz are 1.2, 1.3 and 1.2, respectively. To use the antenna for different applications with a better gain and lower side lobes this antenna is designed. From the obtained results we can conclude that the proposed antenna system is very suitable for satellite communication, radar communication, and also some applications for wireless computer networks.

References

1. Balanis A (1992) Antenna theory. IEEE 80(1)
2. Maci S, BifJiGentili G (1997) Slotted rectangular patch dual-frequency antenna. IEEE Antennas and Propag Magazine 39(6)
3. Midasala V, Siddaiah P, Bhavanam SN (2015) Rectangular patch antenna design at 13 GHz using HFSS. Lecture Notes in Bioengineering. Springer, India
4. Ghosal A, Mujindar A, Das K, Das A (2016) Design of ground plane of a slotted microstrip patch antenna for EMI rejection. In: IEEE international conference on electromagnetic interference & compatibility (INCEMIC), 8–9 December 2016
5. Malisuwan S, Sivaraks J, Madan N, Suriyakrai N (2014) Design of microstrip patch antenna for Ku-band satellite communication applications. Int J Comput Commun Eng 3(6)
6. Sahariar H, Soewardiman H, Jur JS (2017) Fabrication and packaging of flexible and breathable patch antennas on textiles. In: Southeast Con 2017 (IEEE Conference), 30 March–2 April 2017

7. Yang F, Zhang XX, Ye X, Rahmat-Samii Y (2001) Wide-band E-shaped patch antennas for wireless communication. IEEE Trans Antennas Propag 49(7)
8. Ranjan P, Tomar GS, Gowri R (2017) Capacitive coupled rectangular microstrip patch antenna for Ku Band. In: 2017 4th IEEE international conference on electrical, computer and electronics (UPCON) GLA University, Mathura, Oct 26–28 2017
9. Naik KK, Sri PAV, Yasasvini N, Anjum M, Dattatreya G (2017) Compact dual-band hex-adecagon circular patch antenna with DGS for Ku band applications. In: 2017 progress in electromagnetics research symposium—fall (PIERS—FALL), 19–22 November 2017
10. Datta B, Sarkar GS, Das A (2015) Compact monopole patch antenna for X & Ku band microwave communication. In: 2015 international conference on communication networks (ICCN), 19–21 Novemeber 2015

Design of Low-Power and High-Frequency Operational Transconductance Amplifier for Filter Applications

Amita Nandal, Arvind Dhaka, Nayan Kumar and Elena Hadzieva

Abstract In this work, a two-stage operational transcoductance amplifier (OTA) has been designed and is used to design various active filters as an application. The main contribution of this work is in the direction of achieving high gain, high bandwidth, high PSRR, and low noise for the proposed OTA. Over the years, different methodologies have been proposed by researchers to enhance the performance of OTA. In this work, the proposed results have been analytically verified with theory and compared with the related work. In this work, 90 nm technology is used for simulations which are carried out using Tanner EDA 16.0 tool and these results are compared with related work performed using 180 nm technology. With the help of this work, a two-stage OTA can be designed having high gain, high bandwidth, high PSRR, etc., and various active filters can also be designed with the help of this OTA, for filtering purposes.

Keywords Active filters · Channel length · Bandwidth · Operational amplifier · Scaling MOSFET · Slew rate and transconductance

A. Nandal (✉) · A. Dhaka
Department of Computer and Communication Engineering, Manipal University Jaipur, Jaipur, India
e-mail: amita_nandal@yahoo.com

A. Dhaka
e-mail: arvind.neomatrix@gmail.com

N. Kumar
Department of Electronics and Communication Engineering, National Institute of Technology, Hamirpur, India
e-mail: 619nayan@gmail.com

E. Hadzieva
University of Science and Technology, Ohrid, Republic of Macedonia
e-mail: elena.hadzieva@uist.edu.mk

© Springer Nature Singapore Pte Ltd. 2020
A. Khanna et al. (eds.), *International Conference on Innovative Computing and Communications*, Advances in Intelligent Systems and Computing 1059, https://doi.org/10.1007/978-981-15-0324-5_5

1 Introduction

OTA is an important building block of any analog processing system. The optimization of gain, power consumption, and bandwidth product for OTA is still developing. The main challenges of these devices are high CMRR, low noise, high input impedance, low power, and high gain [1]. During the last decade, due to metal-oxide-semiconductor field-effect transistor (MOSFET) scaling, the techniques to improve performance of integrated circuits are in high demand for health monitoring systems [2, 3]. The high-frequency limit of op-amps makes their design difficult [4]. Therefore, at high frequencies, operational transconductance amplifiers (OTAs) can be used instead of op-amps. The most demanding application of operational amplifiers is filters [5, 6]. The performance criteria usually include computation of several factors such as gain, bandwidth, slew rate, voltage swing, etc. Op-amps have a wide range of applications like filters [5, 6]. CMOS technology is a prominent candidate for voltage-controlled current devices like OTA [7]. Several designs for OTAs are reported in literature [8–13]. In literature, various researchers have proposed OTA designs using 0.13-μm, 90-nm, 65-nm, and even 45-nm technologies [14, 15]. Presently, 0.18-μm CMOS technology with (f_{max}) up to 40 GHz is used [16]. The major issue in CMOS OTA is optimization of capacitive parasitic [4, 6]. In recent years, considerable attention is given to linearization. A reduction in main signal is observed due to nonlinear terms because of the distortion of the transconductance that results in harmful intermodulation products at the output of the OTA [15, 17, and 18]. In literature, several techniques have been proposed to remove this problem of nonlinearity [7, 19]. A class-AB OTA has an adaptive bias circuit to reduce current flow which makes it different a common OTA [16, 20, 21, and 22]. Nowadays, analog IC researchers are dedicated to improve the performance of continuously scaled commercial ICs. This paper presents an OTA that can work up to several hundred MHz and can be used for various filter applications. The work done in this paper has been compared with [23–26]. In [23], Gm-assisted OTA-RC technique enhances linearity and speed of operational amplifier. However, when it is designed in a 0.18 μm CMOS process, it dissipates more power with some added noise. The high-pass and low-pass filters are cascaded to design a new UWB band-pass filter [24]. In [25], an active-RC channel selection filter is designed with tunable cut-off frequency from 6 to 20 MHz for IEEE802.11a (20 MHz) including the effect of process, voltage, and temperature variations. For higher bandwidth of band-reject filters, a significant work has been carried out in [26]. In this work, a folded cascode OTA has been presented having high transconductance, less noise, and gain bandwidth product (GBW). The need of high gain can be achieved by using other topologies such as cascode, regulated cascode, or folded cascode OTA which can be used to design various analog circuits like voltage-controlled filters, variable frequency oscillators, continuous time active filters, and variable gain amplifiers.

The rest of the paper is organized as follows; Sect. 2 describes design methodology. Computation of parameters is given in Sect. 3. Simulation results for various OTA applications have been explained in Sect. 4. Finally, conclusion is drawn in Sect. 5.

2 Design Methodology

The following notations have been used throughout this work:

(1) $M_1, M_2, M_3 \dots M_n$ and so on represents the transistor notations.
(2) $I_1, I_2, I_3 \dots I_n$ and so on represents the currents for specified transistor $M_1, M_2, M_3 \dots$ and so on, respectively. Here, I_5 denotes bias current source.
(3) C_L and C_C represent load capacitor and compensating capacitor.
(4) V_{DD} and V_{SS} represent power supply.

In the proposed design shown in Fig. 1, three PMOS and five NMOS transistors are used. We have modeled the values of parameters for these transistors based upon filtering requirements and our main focus is to improve gain and bandwidth. One current source I_5 is used to bias the circuit. C_L is load capacitor and C_C is compensation capacitor. Here, for proper current mirroring, the aspect ratio of transistors forming current mirror should be the same; hence, $\left(\frac{W}{L}\right)_3 = \left(\frac{W}{L}\right)_4$ and $\left(\frac{W}{L}\right)_5 = \left(\frac{W}{L}\right)_8$. For proper inputs applied, the necessary condition is $\left(\frac{W}{L}\right)_1 = \left(\frac{W}{L}\right)_2$. We have considered Fig. 1 for two-stage OTA to calculate values of various parameters like DC voltage gain (A_v), gain bandwidth (GB), input common-mode range (ICMR), load capacitance (C_L), slew rate (SR), output voltage swing, and power dissipation (P_{diss}).

The design procedure includes the computation of device channel length. The channel length is dependent on the value of channel length modulation parameter λ which is an important parameter for computation of amplifier gain. After choosing suitable channel length, the next step is to calculate the minimum value of the compensation capacitance (C_C). For phase margin of $60°$, $C_C \geq 0.22 C_L$. On the basis of

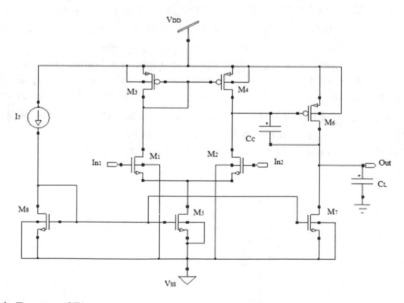

Fig. 1 Two-stage OTA

slew rate, we determine the minimum value of the tail current (I_5) which is computed as $I_5 = \text{SR}(C_C)$. Now, the aspect ratio of transistor M_3 can be determined from the positive maximum input common-mode range requirements. The equation for $\left(\frac{W}{L}\right)_3$ can be given as,

$$\left(\frac{W}{L}\right)_3 = \frac{I_5}{K_p\left[V_{DD} - V_{in(max)} - \left|V_{T3(max)}\right| + V_{T1(min)}\right]^2}. \tag{1}$$

If value of $\left(\frac{W}{L}\right)_3$ is chosen in such a way that minimizes the product of W and L, this results in minimization of gate region area, gate capacitance, and phase margin. The transconductance of the input transistors can be determined from the values of C_C and GB. The transconductance g_{m1} can be calculated from $g_{m1} = \text{GB}(C_C)$. The aspect ratio $\left(\frac{W}{L}\right)_1$ can be calculated for known value of g_{m1} as, $\left(\frac{W}{L}\right)_1 = \frac{g_{m1}^2}{(K_n)(I_5)}$.

From above considerations, we can calculate the saturation voltage of transistor M_5. Using negative ICMR equation, calculate V_{DS5} as

$$V_{DS5} = V_{in(min)} - V_{SS} - \left(\frac{I_5}{K_n}\right)^{\frac{1}{2}} - V_{T1(max)}. \tag{2}$$

If the value of V_{DS5} is less than about 100 mV, then it may result in higher values of aspect ratio which is not acceptable practically. If the value of V_{DS5} is less than zero, then this is because of the strict values of ICMR. These effects can be neglected either by reducing I_5 or by increasing $\left(\frac{W}{L}\right)_1$. Now, V_{DS5} is known, and we can determine $\left(\frac{W}{L}\right)_5$ by using $\left(\frac{W}{L}\right)_5 = \frac{2I_5}{K_n(V_{DS5})^2}$. The design of first stage of a two-stage OTA completes here. Now, the design of output stage is considered. For phase margin of $60°$, transconductance g_{m6} is written as, $g_{m6} = 2.2(g_{m2})\left(\frac{C_L}{C_C}\right)$. For phase margin close to $60°$, the value of g_{m6} is approximately ten times the input stage transconductance g_{m1}. One possible approach to complete the design of M_6 is to achieve proper matching of first-stage current mirror loads M_3 and M_4, which requires the condition that $V_{SG4} = V_{SG6}$. Using this condition, we can write, $\left(\frac{W}{L}\right)_6 = \left(\frac{W}{L}\right)_4\left(\frac{g_{m6}}{g_{m4}}\right)$ (Table 1). Now, dc current I_6 can be written as $I_6 = \frac{g_{m6}^2}{2K_p\left(\frac{W}{L}\right)_6}$.

Another approach for designing the output stage is by using the value of g_{m6} and required value of $V_{DS6(sat)}$ of M_6 according to output range specification. Hence, we

Table 1 Aspect ratios of two-stage OTA	

Aspect ratios of transistors	Value
M_1, M_2	3
M_3, M_4	8
M_5, M_8	7
M_6	70
M_7	27

Parameters	Value
V_{DD}	+0.9 V
V_{SS}	−0.9 V
GB	30 MHz
Slew rate	10 V/μS
+ICMR = $V_{in(max)}$	+0.7 V
−ICMR = $V_{in(min)}$	−0.2 V
A_v	>30 dB
C_L	10 F

Table 2 Design specifications

can determine the design of M_6 as, $\left(\frac{W}{L}\right)_6 = \frac{g_{m6}}{K_p V_{DS(sat)}}$. Now, the design of M_7 can be calculated from equation, $\left(\frac{W}{L}\right)_7 = \left(\frac{W}{L}\right)_5 \left(\frac{I_6}{I_5}\right)$. The values of all the parameters required to design a two-stage OTA are known and hence the required circuitry can be designed. The design specifications as given in Table 2 are considered for the calculation of design parameters of a two-stage OTA.

3 Computation of Parameters

1. For phase margin of 60°, the minimum value of compensation capacitor C_C can be calculated by using $C_C \geq 0.22 C_L$. Putting $C_L = 10\,pF$ from the design specifications, we get $C_C \geq 0.22 \times 10\,pF$. Therefore, $C_C \geq 2.2\,pF$.

2. We can now calculate tail current I_5 from known values of C_C and slew rate (SR) specifications by $I_5 = SR(C_C)$. Now, $I_5 = 10 \times 10^6 (3 \times 10^{-12})$ and we get $I_5 = 30\,\mu A$.

3. From maximum input voltage specifications, we can calculate $\left(\frac{W}{L}\right)_3$ as follows,

$$\left(\frac{W}{L}\right)_3 = \frac{I_5}{K_p \left[V_{DD} - V_{in(max)} - \left|V_{T3(max)}\right| + V_{T1(min)}\right]^2}. \tag{3}$$

4. $\left(\frac{W}{L}\right)_3 = \frac{30 \times 10^{-6}}{50 \times 10^{-6}[0.9-0.7-0.33+0.4]^2}$. After solving, we get $\left(\frac{W}{L}\right)_3 = 8.23$ and $\left(\frac{W}{L}\right)_4 = \left(\frac{W}{L}\right)_3 = 8.23$.

5. The next step is to calculate g_{m1} by using $g_{m1} = GB(C_C)$. On solving, $g_{m1} = 30 \times 10^6 (3 \times 10^{-12})$. Therefore, $g_{m1} = 90\,\mu s.\left(\frac{W}{L}\right)_1$ can be calculated as $\left(\frac{W}{L}\right)_1 = \frac{g_{m1}^2}{K_n I_5}$. By substituting values we get, $\left(\frac{W}{L}\right)_1 = \frac{(90 \times 10^{-6})^2}{(110 \times 10^{-6})(30 \times 10^{-6})}$. After solving, we get $\left(\frac{W}{L}\right)_1 = 2.727$ and $\left(\frac{W}{L}\right)_2 = \left(\frac{W}{L}\right)_1 = 2.727$.

6. V_{DS5} can be calculated as $V_{DS5} = V_{in(min)} - V_{SS} - \left(\frac{I_5}{K_n}\right)^{\frac{1}{2}} - V_{T1(max)}$. By substituting values we get, $V_{DS5} = -0.2 + 0.9 - \left(\frac{30 \times 10^{-6}}{110 \times 10^{-6}}\right)^{1/2} - 0.45$. After solving, we get $V_{DS5} = -0.272$.

7. This value of V_{DS5} can be used to calculate $\left(\frac{W}{L}\right)_5$ as $\left(\frac{W}{L}\right)_5 = \frac{2I_5}{K_n(V_{DS5})^2}$. By substituting values we get, $\left(\frac{W}{L}\right)_5 = \frac{2(30 \times 10^{-6})}{(110 \times 10^{-6})(0.272)^2}$. After solving, we get $\left(\frac{W}{L}\right)_5 = 7.3$ and $\left(\frac{W}{L}\right)_8 = \left(\frac{W}{L}\right)_5 = 7.3\ 73$.

8. Using $g_{m1} = g_{m2}$, we can calculate $g_{m6} = 2.2(g_{m2})\left(\frac{C_L}{C_C}\right)$ as, $g_{m6} = 2.2(90 \times 10^{-6})\left(\frac{10 \times 10^{-12}}{2.2 \times 10^{-12}}\right)$. After solving, we get $g_{m6} = 900\mu s$). Now, we can calculate $\left(\frac{W}{L}\right)_6 = \left(\frac{W}{L}\right)_4\left(\frac{g_{m6}}{g_{m4}}\right)$. By substituting values we get, $\left(\frac{W}{L}\right)_6 = 8.23 \times \left(\frac{900 \times 10^{-6}}{110 \times 10^{-6}}\right)$. On solving we get, $\left(\frac{W}{L}\right)_6 = 67.336$.

9. Now, the value of dc current I_6 can be calculated using $I_6 = \frac{g_{m6}^2}{2K_p\left(\frac{W}{L}\right)_6}$. By substituting values we get, $I_6 = \frac{(900 \times 10^{-6})^2}{2(50 \times 10^{-6})(67.33)}$. After solving, we get $I_6 = 120.292\,\mu A$.

10. Finally, we can calculate $\left(\frac{W}{L}\right)_7$ by using $\left(\frac{W}{L}\right)_7 = \left(\frac{W}{L}\right)_5\left(\frac{I_6}{I_5}\right)$. By substituting values we get, $\left(\frac{W}{L}\right)_7 = 7.3 \times \left(\frac{115 \times 10^{-6}}{30 \times 10^{-6}}\right)$. On solving $\left(\frac{W}{L}\right)_7 = 27.983$, now the aspect ratio of all the transistors used in the design of a two-stage OTA is known and these specifications can be used to design desired filters.

4   OTA Applications

OTA-C filters are constructed using OTAs and capacitors. They are also called $G_m - C$ filters. Such filters are one of the most popular applications of OTAs at low frequencies [1, 4, 18]. OTA$-C$ filters can be used to design all filter functions, i.e., low-pass, band-pass, high-pass, and band-stop. The frequency tuning of OTA-C filters is flexible which makes them superior to passive filters. This property of OTA-C filters makes them suitable for high frequencies operation.

Figure 2 shows the design of low-pass filter (LPF) where input is applied at non-inverting terminal. A feedback resistance R is connected from the output terminal to inverting terminal of OTA. One terminal of capacitor C is connected to inverting terminal of OTA. The other terminal of the capacitor is grounded. The 3 dB cut-off frequency of LPF can be given by expression, $f_{3\,db} = \frac{g_m}{2\pi C}$. Low-pass filter passes the signals with a frequency lower than certain cut-off frequency and attenuates the frequency higher than that cut-off frequency. Cut-off frequency for the designed low-pass filter comes out to be 31 MHz) after simulation for $R = 10\,K\Omega$ and $C = 0.1$ pF.

Figure 3 shows the design of high-pass filter (HPF) where input is applied at the

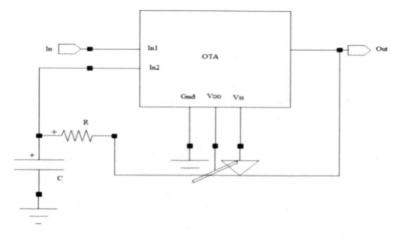

Fig. 2 Low-pass filter

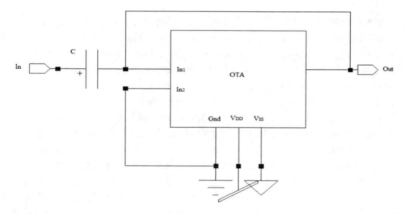

Fig. 3 High-pass filter

inverting terminal. A capacitor C is placed whose on terminal is connected to the input and other terminal is connected to the feedback from the output terminal of HPF. Response of HPF is opposite to the response of LPF. High-pass filter passes the signals with a frequency higher than certain cut-off frequency and attenuates the frequency lower than that cut-off frequency. Cut-off frequency for the designed high-pass filter comes out to be 1 GHz after simulation for $C = 1\,\mathrm{pF}$.

Figure 4 shows the design of band-pass filter (BPF) where non-inverting terminal is connected to ground. A resistor R is connected as feedback from the output terminal to one terminal of capacitor C_1. Input is applied at the inverting terminal of OTA through capacitor C_1. Another capacitor C_2 is connected between output terminal of OTA and ground. The different between frequencies at 3 dB point determines the bandwidth of filter. The signals whose frequency falls outside the bandwidth range

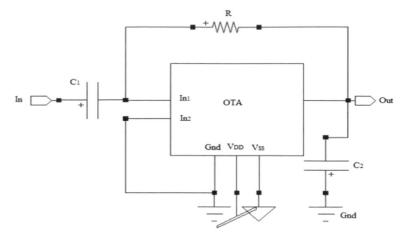

Fig. 4 Band-pass filter

of filter are attenuated. The cut-off frequency of low-pass filter (LPF) is greater than the cut-off frequency of the high-pass filter (HPF). The 3 dB frequencies of BPF are given as, $f_{3\,dB1} = \frac{1}{2\pi RC_1}$ and $f_{3\,dB2} = \frac{1}{2\pi RC_2}$. The upper cut-off frequency for band-pass filter is calculated as 10.2 GHz and lower cut-off frequency is 517 KHz after simulation for $R = 1\,K\Omega$, $C_1 = 0.01\,pF$ and $C_2 = 10\,pF$.

Figure 5 shows the design of band-reject filter (BRF) which is a series connection of LPF and HPF designed above. The required band of frequencies to be rejected by this band-reject filter can be selected by adjusting the cut-off frequencies of low-pass filter and high-pass filter which are cascaded to get band-reject operation. Band-reject filter is the opposite of band-pass filter. It passes almost all the frequencies but attenuates some frequencies in the specified range. The band-reject operation is performed in the range 60 KHz to 70 KHz. Simulation is done for $R = 1\,K\Omega$, $C_1 = 1\,pF$ and $C_2 = 50\,pF$. All the parameters of two-stage OTA are known and

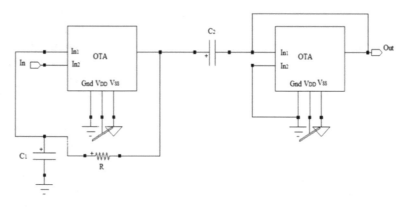

Fig. 5 Band-reject filter

Table 3 Performance comparison of two-stage OTA

S. No.	Parameters	Zhao et al. [22]	This work
1.	Technology	180 nm	90 nm
2.	Supply voltage	1 V	0.9 V
3.	Gain	43 dB	40 dB
4.	Phase margin	83°	54°
5.	Bandwidth	2.9 MHz	40 MHz
6.	Slew rate	3.95 V/μS	3.1 V/μS
7.	Bias current	20 μA	30 μA
8.	Noise	–	47 nV/$\sqrt{Hz}$
9.	Power dissipation	–	283 μW

Table 4 Filter applications

S. No.	Filters	Frequency calculated from this work	Applications from literature
1.	Low pass	31 MHz (cut off)	5th order Chebyshev filter [23]
2.	High pass	1 GHz (cut off)	UWB Band-pass filter [24]
3.	Band pass	0.5 GHz (center)	Direct conversion receiver [25]
4.	Band reject	50 − 60 KHz (center)	Noise reduction in MSF signals [26]

their results comparison is presented in Table 3. Table 4 shows the comparison for various filters designed in literature which are closely related to the work done in this work.

5 Conclusion

The goal of this work is to design a high gain, high bandwidth OTA. A two-stage OTA with 40 dB gain and 40 MHz unity gain frequency has been analyzed and experimentally demonstrated with 90 nm technology. In this paper, a high-gain OTA is designed which can be used as active filters and is also simulated successfully. By simulation results, we have observed that proposed OTA has 40 dB gain, 1.4 V output swing, 47 $\mu V/\sqrt{Hz}$ noise, 3.1V/μS slew rate values, which are better than conventional OTA, where filter application is considered. Filters designed using proposed OTA have cut-off frequencies of 31 MHz for low-pass filter, and 1 GHz high-pass filter and center frequencies of 0.5 GHz for band-pass filter and 50 − 60 KHz for band-reject filter. Using the present sub-micron technologies, very high bandwidth could be realized. Some work can be done to improve this aspect of the design. Also, the slew rate could be increased by the use of slew rate boosting circuits. It should be possible to scale down this design to the present day technologies without making large-scale modifications. Scaling might need some minor recalculations. Some output stages could be applied to achieve higher gain which may also provide higher output swing. Using various output stages, it is possible to obtain a high swing amplifier with high gain and high bandwidth at the same time. However, this will require the use of proper compensation techniques which will increase the complexity of the circuit.

References

1. Ferreira LHC, Sonkusale SR (2014) A 60 dB gain OTA operating at 0.25 V power supply in 130 nm digital CMOS process. IEEE Trans Circuits Syst I 61(6):1609–1617
2. Nayyar A, Puri V, Nhu NG, BioSenHealth 1.0: a novel internet of medical things (IoMT)-based patient health monitoring system. In: International conference on innovative computing and communications, Proceedings of ICICC 2018, vol 1. pp 155–164. https://doi.org/10.1007/978-981-13-2324-9_16
3. Rajbhandari S, Singh A, Mittal M (2018) Big data in healthcare. In: International Conference on Innovative Computing and Communications, Proceedings of ICICC 2018, vol 2. pp 261–269. https://doi.org/10.1007/978-981-13-2354-6_28
4. Lo T, Hung C (2007) A 1 V 50 MHz pseudo differential OTA with compensation of the mobility reduction. IEEE Trans Circuits Syst II 54(12):1047–1051
5. Ferreira LHC, Pimenta TC, Moreno RL (2007) An ultra low voltage ultra low power CMOS Miller OTA with rail-to-rail input/output swing. IEEE Trans Circuits and Syst II 54(10):843–847
6. Zhang X, El-Masry EI (2007) A novel CMOS OTA based on body driven MOSFETs and its applications in OTA-C filters. IEEE Trans Circuits and Syst I 54(6):1204–1212
7. Sinencio ES, Martinez JS (2000) CMOS transconductance amplifiers, architectures and active filters, A tutorial. IEEE Proc Circuits, Devices Syst 147(1):3–12
8. Lin T, Wu C, Tsai M (2007) A 0.8 V 0.25 mW current mirror OTA with 160 MHz GBW in 0.18 μm CMOS. IEEE Trans Circuits Syst II 54(2):131–135
9. Grasso D, Palumbo G, Pennisi S (2006) Three stage CMOS OTA for large capacitive loads with efficient frequency compensation scheme. IEEE Trans Circuits Syst II 53(10):1044–1048
10. Baruqui FAP, Petraglia A (2006) Linear tunable CMOS OTA with constant dynamic range using source degenerated current mirrors. IEEE Trans Circuits and Syst I 53(9):797–801
11. Huang W, Sinencio ES (2006) Robust highly linear high frequency CMOS OTA with IM3 below 70 dB at 26 MHz. IEEE Trans Circuits Syst I Fundam Theory Appl 53(7):1433–1447
12. Lewinski A, Martinez JS (2004) OTA linearity enhancement technique for high frequency applications with IM3 below 65 dB. IEEE Trans Circuits Syst II 51(10):542–548
13. Kuhn W, Stephenson F, Riad AE (1996) A 200 MHz CMOS Q enhanced LC bandpass filter. IEEE J Solid State Circ 31(8):1112–1122
14. Taiwan Semiconductor Manufacturing Company (TSMC) Limited (2008) TSMC platform technology portfolio. Online: http://www.tsmc.com/english/b_technology/b01_platform/b01_platform.htm
15. International Business Machines Corp. (IBM) (2008) IBM Semiconductor solutions. Online: http://www-03.ibm.com/technology/ges/semiconductor
16. Burghartz JN, Hargrove M, Webster CS, Groves RA, Keene M, Jenkins KA et al (2000) RF potential of a 0.18 μm CMOS logic device technology. IEEE Trans Electron Devices 47(4):864–870
17. Martin JL, Baswa S, Angulo JR, Carvajal RG (2005) Low voltage super class-AB CMOS OTA cells with very high slew rate and power efficiency. IEEE J Solid State Circ 40(5):1068–1077

18. Lee H (1998) The design of CMOS radio-frequency integrated circuits. Cambridge University Press, Cambridge, UK
19. Sinencio ES, Andreou AG (1999) Low voltage/low power integrated circuit and systems. IEEE Press, New York
20. Baswa S, Martin AJL, Angulo JR, Carvajal RG (2004) Low voltage micropower super class AB CMOS OTA. Electron Lett 40(4):216–217
21. Baswa S, Martin AJL, Carvajal RG, Angulo JR (2004) Low voltage power efficient adaptive biasing for CMOS amplifiers and buffers. Electron Lett 40(4):217–219
22. Zhao X, Zhang Q, Deng M (2015) Super class-AB bulk-driven OTAs with improved slew rate. Electron Lett 51(19):1488–1489
23. Thyagarajan SV, Pavan S, Sankar P (2010) Low distortion active filters using the Gm-assisted OTA-RC technique. In: 36th European solid state circuits conference, Seville, 14–16 September 2010, pp 162–165
24. Hao ZC, Hong JS (2010) UWB bandpass filter using cascaded miniature high-pass and low-pass filters with multilayer liquid crystal polymer technology. IEEE Trans Microwave Theory Tech 58(4)
25. Lee MY, Lee YH (2014) A 1.8-V operation analog CMOS baseband for direct conversion receiver of IEEE 802.11a. Int J Control Automation 7(9):155–164
26. Fathelbab WM, Jaradat HM, Reynolds D (2010) Two novel classes of band-reject filters realizing broad upper pass bandwidth. In: IEEE MTT-S international microwave symposium digest, Anaheim, CA, 23–28 May 2010, pp 217–220

Design of Low Power Operational Amplifier for ECG Recording

Arvind Dhaka, Amita Nandal, Ajay Gangwar
and Dijana Capeska Bogatinoska

Abstract In this work, a folded cascode topology has been used to design an op-amp in order to optimize biopotential amplifier. This paper aims to design a fully differential amplifier in the direction of achieving high gain, high CMRR, high PSRR and low noise. In this work, 90 nm technology is used for simulations which are carried out using Tanner EDA 16.0 tool and these results are compared with related work performed using 180 and 350 nm technology. With the help of this work, an instrumentation amplifier can be designed which has low noise, low common mode gain, higher stability with supply variation, etc. and biopotential amplifier can also be designed for ECG machine, neural recording, etc. In this work the proposed results have been analytically verified with theory and compared with the related work.

Keywords Biopotential signals · Channel length · Bandwidth · Operational amplifier · Scaling MOSFET · Slew rate and transconductance

A. Dhaka (✉) · A. Nandal
Department of Computer and Communication Engineering, Manipal University Jaipur, Jaipur, India
e-mail: arvind.neomatrix@gmail.com

A. Nandal
e-mail: amita_nandal@yahoo.com

A. Gangwar
Department of Electronics and Communication Engineering, National Institute of Technology Hamirpur, Hamirpur, India
e-mail: gangwar.ajay27@gmail.com

D. C. Bogatinoska
University of Science and Technology, Ohrid, Republic of Macedonia
e-mail: dijana.c.bogatinoska@uist.edu.mk

© Springer Nature Singapore Pte Ltd. 2020
A. Khanna et al. (eds.), *International Conference on Innovative Computing and Communications*, Advances in Intelligent Systems and Computing 1059,
https://doi.org/10.1007/978-981-15-0324-5_6

1 Introduction

Nowadays, a lot of researchers are paying attention in biomedical instruments due to their main challenges such as low-frequency range, high CMRR, low noise, high input impedance, low power and high gain present in its design. Over the years, different methodologies have been proposed by researchers to enhance the performance of a biopotential amplifier. To observe biopotential signal is one of the most vital innovations in clinical practice these days for high quality, low noise and ultra-low power biopotential acquisition devices [1–8]. For biomedical applications, generally available research has been performed on 180 nm, 350 nm technology. During the last decade due to metal-oxide-semiconductor field-effect transistor (MOSFET) scaling the techniques to improve performance of integrated circuits are in high demand. In general biopotential signals have amplitude from a range few tens of microvolts to a few millivolts. Based on the type of biopotential signal to be monitored, the different frequency band can be selected which ranges from sub hertz to few hundred hertz. Electromyogram (EMG), Electromyogram (ECoG), Local field potential (LPF) and Electroencephalogram (EEG) are the other medicinal norms that depend on biopotential signals [9–13]. Figure 1 shows an ECG monitoring system. The main components of this system are electrostatic discharge (ESD) and defibrillator protection, multiplexer (MUX), instrumentation amplifier (IA), analog-to-digital converter (ADC) and at last a wireless transceiver [9–11].

The ECG signal that originates from the body through the electrode is made out of three principal segments; the real differential ECG signal, the differential electrode offset (DEO) and the common-mode signals [14–19]. There are so many issues with ECG signal such as signal bandwidth, low amplitude and need for rejection of common-mode signal. In this work a novel architecture of analog front end amplifier for ECG recording is proposed. The amplifier used has low power, high CMRR, high gain, low noise which is suitable to amplify ECG signal. To accomplish the above problem, the following objectives have been worked upon,

1. To design a two-stage operational amplifier suitable for ECG signal.
2. To design of fully differential operational amplifier for low noise in ECG device.

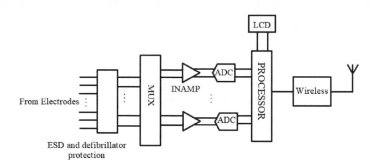

Fig. 1 ECG monitoring system

3. To design an amplifier with High gain and phase margin, High CMRR, High PSRR, High Slew rate, Low noise and High range of Input Common-Mode Range (ICMR).

Due to drastic changes and developments in very large scale integration (VLSI) and internet of things (IoT), the medical technology also needs considerable advancements [20, 21]. This paper provides a solution to design ECG recording system as proposed front end amplifier for future health monitoring system. The rest of the paper is organized as follows: Sect. 2 describes design methodology and Sect. 3 presents design procedure for folded cascode and fully differential op-amp. Simulation results are presented in Sect. 4 and finally, conclusion is drawn in Sect. 5.

2 Design Methodology

In this paper, we propose an unbuffered operational amplifier which can drive capacitive load. We have used the nulling transistor approach to design our model to achieve good stability. After designing the op-amp we have designed fully differential op-amp. The main advantages of fully differential op-amp are low noise, high CMRR, low offset, etc. The values for these parameters are calculated and validated using simulations as provided in Sect. 3. The notations used throughout this paper are as follows:

- $I_1, I_2, I_3 \ldots I_n$ represent specified current for $M_1, M_2, M_3 \ldots M_n$ transistors, respectively.
- g_{mI} and g_{mII} are transconductance of stage 1st and 2nd, respectively.
- $g_{m1}, g_{m2}, g_{m3} \cdots g_{mn}$ are transconductance for $M_1, M_2, M_3 \ldots M_n$ transistors, respectively.
- R_I and R_{II} are resistance of 1st and 2nd stage, respectively.
- $r_{ds1}, r_{ds2}, r_{ds3}, \ldots r_{dsn}$ are small signal drain source resistance for $M_1, M_2, M_3 \ldots M_n$ transistors.
- C_C and C_L are compensation capacitance and load capacitance, respectively.
- V_{dd} and V_{ss} represents power supply.

2.1 Transistor Level Design: Folded Cascode Operational Amplifier

In Fig. 2, the design of two-stage folded cascode operational amplifier is shown. In this design, M_3, M_4 and M_5 should be designed in such a way that the dc current in cascode mirror never goes to zero. Due to parasitic capacitances the switching of mirror is delayed therefore, cascode mirror takes some time so that the current becomes zero. If current through M_4 and M_5 are not greater than M_3 then current

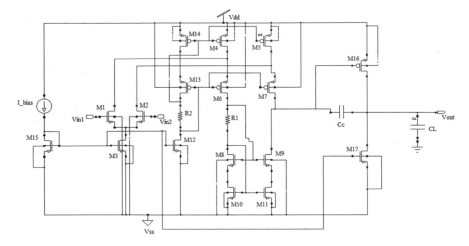

Fig. 2 Architecture of two-stage folded cascode op-amp

Table 1 Aspect ratio (W/L) for two-stage folded cascode op-amp

Name of transistor	Value of W/L
M_1 and M_2	5
M_3 and M_{15}	23
M_4, M_5, M_{13} and M_{14}	48
M_6 and M_7	32
M_8, M_9, M_{10} and M_{11}	8
M_{12}	36
M_{16}	77
M_{17}	55
R_1	3.3 KΩ
R_2	7 KΩ
I_{bias}	50 μA

through M_6 and M_7 will be zero. To avoid this condition current through M_4 and M_5 are normally greater than current through M_3 and typical value of this current is one to two times the current in M_3. Table 1 shows the aspect ratios for various transistors in the folded cascode amplifier.

2.2 Modified Architecture Using Nulling Resistor Approach

Amplifier shown in Fig. 2 has right half pole (RHP) which cannot be ignored because of stability reason. A nulling resistor is placed in series with compensation capacitor

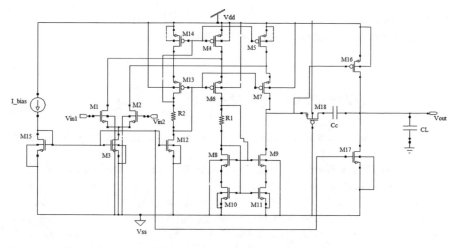

Fig. 3 Modified architecture of folded cascode op-amp for good phase margin

C_C in order to control the RHP of amplifier which is designed with transistor M_{18} as shown in Fig. 3. By using the nulling resistor in compensation scheme, the poles (p_1, p_2) and zero (z) of the amplifier are as follows:

$$p_1 \cong \frac{-g_{mI}}{A_v C_C} \tag{1}$$

where $A_v \approx g_{mI} g_{mII} R_I R_{II}$

$$p_2 \cong \frac{-g_{mII}}{C_L} \tag{2}$$

$$z \cong \frac{-1}{R_z C_C - C_C/g_{mII}} \tag{3}$$

where $R_z \cong \frac{1}{g_{mII}} \left(\frac{C_L + C_C}{C_C} \right)$.

3 Design Procedure for Folded Cascode and Fully Differential Op-Amp

The specifications for various parameters given below are considered to design the model given in Fig. 3.

1. Unity Gain Bandwidth (GB)
2. Input common-mode range (ICMR)[$V_{in(min)}$ and $V_{in(max)}$]
3. Load Capacitance (C_L)

4. Slew rate (SR)
5. Output Voltage Swing [$V_{out(min)}$ and $V_{out(max)}$]

3.1 Design of Folded Cascode Op-Amp

The specifications for various parameters for folded cascade Op-Amp are given in Table 2.

For 75° phase margin $C_C = 0.62C_L$ and for 60° phase margin $C_C = 0.22C_L$. The value for tail current I_3 is calculated as $I_3 = SR * C_L$. The value of bias current in cascode stage to avoid zero current in the cascade is $I_4 = I_5 = 1.2I_3$ to $1.5I_3$. The $\left(\frac{W}{L}\right)_n$ ratio for nth transistor is written as S_n. The S_5 and S_7 from the maximum output voltage can be calculated as $S_4 = S_{14} = S_5 = S_{13}$ and $S_6 = S_7$ such that $S_5 = \left(\frac{W}{L}\right)_5 = \frac{2I_5}{K_P*(V_{sd5})^2}$ and $S_7 = \left(\frac{W}{L}\right)_7 = \frac{2I_7}{K_P*(V_{sd7})^2}$, where $V_{sd5(sat)} = V_{sd7(sat)} = \frac{V_{dd}-V_{out(max)}}{2}$. S_9 and S_{11} from the minimum output voltage can be equated as $S_{11} = S_{10}$ and $S_8 = S_9$ such that $S_9 = \left(\frac{W}{L}\right)_9 = \frac{2I_9}{K_n*(V_{sd9})^2}$ and $S_{11} = \left(\frac{W}{L}\right)_{11} = \frac{2I_{11}}{K_n*(V_{sd11})^2}$, where $V_{ds9(sat)} = V_{ds11(sat)} = \frac{V_{out(min)}-|V_{ss}|}{2}$. The resistances R_1 and R_2 are calculated as $R_1 = \frac{V_{sd13(sat)}}{I_{12}}$ and $R_2 = \frac{V_{sd8(sat)}}{I_6}$. The values of S_1 and S_2 in terms of unity gain bandwidth (GB) and load capacitance (C_L) is calculated as $S_1 = S_2 = \frac{GB^2 C_L^2}{K_n I_3}$. S_3 from the minimum input common mode voltage is given as $S_3 = \left(\frac{W}{L}\right)_3 = \frac{2I_3}{K_n\left[V_{in(min)}^2 - V_{ss} - \sqrt{\frac{I_3}{K_n.S_1}} - V_{t1}\right]}$. S_4 and S_5 for maximum input common mode voltage is given as $S_4 = S_5 = \frac{2I_4}{K_p[V_{dd}-V_{in(max)}-V_{t1}]^2}$, such that the values of S_4 and S_5 satisfy the maximum input common mode voltage. The value of S_{12} from tail current and bias current $S_{12} = \frac{I_4}{I_3} * S_3$. The value of S_{16} is calculated as $S_{16} = \frac{g_{m16}}{g_{m4}} * S_4$ where $g_{m16} \geq 6.1\frac{C_L}{C_C} * g_{m1}$ and $g_{m4} = \sqrt{2K_p S_4 I_D}$. The value of S_{17} is calculated as $S_{17} = \frac{I_{17}}{I_5} * S_3$. These calculated values of various parameters are provided in Table 3.

Table 2 Design specifications of various parameters for folded cascade Op-Amp	Parameter	Value
	Slew rate	10 V/μs
	Load capacitance	5 pF
	The maximum and minimum output voltages	−0.2 V
	Unity gain bandwidth (UGB)	5 MHz
	ICMR (input common mode range)	−0.2 to 0.4 V
	Differential voltage gain (A_d)	> 60 dB
	Power dissipation	< 5 mW

Table 3 The calculated value of various parameters for folded cascode Op-Amp

Parameter	Value
I_3	$50\,\mu\text{A}$
$I_4 = I_5$	$75\,\mu\text{A}$
$S_4 = S_{14} = S_5$	48
$S_6 = S_7 = S_{13}$	32
$S_{11} = S_{10} = S_8 = S_9$	7.42
$S_1 = S_2$	4.48
R_1	$3.3\,\text{k}$
R_2	$7\,\text{k}$
$S_3 = S_4 = S_5$	18.75
S_{12}	34.5
S_{16}	76.8
S_{17}	55.2

3.2 Design of Fully Differential Amplifier

Two-stage fully differential folded cascode amplifier is shown in Fig. 4. The output stage is miller compensated class-A amplifier. To design a fully differential amplifier, current mirror load of single-ended amplifier is replaced by current source load. Fully differential amplifier has to be symmetric about tail current source to avoid the disturbance in common-mode level. The symmetric design of single-ended amplifier as shown in Fig. 4. Every second stage has a compensation capacitance C_C along with nulling resistor. The advantage of fully differential amplifier is that it provides high swing, high CMRR, low noise and rejection of common-mode signal but gain of the amplifier is approximately equal to the gain of single-ended amplifier. The

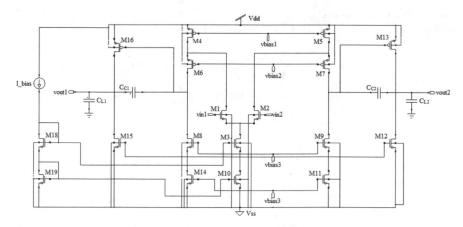

Fig. 4 Architecture of fully differential amplifier

main disadvantage is that it needs common-mode stabilization to avoid disturbance in common-mode level otherwise clipping occurs at the output of amplifier.

4 Simulation Results

We have simulated the proposed folded cascode op-amp and fully differential amplifier and calculated various performance metrics which are tabulated in Tables 4 and 5 respectively. Gain is defined as the ability of an amplifier to increase amplitude of a signal from input to output. The large value of open loop gain of an amplifier enhances the accuracy of the feedback system as well as suppresses the non-linearity. In folded cascode op-amp the gain is 82 dB and phase margin 85°. The ratio of differential gain to common mode gain is called common mode rejection ratio (CMRR) of an amplifier such that $CMMR = \frac{A_{DM}}{A_{CM}}$, where A_{DM} is the differential mode gain and A_{CM} is the common mode gain. Ideally A_{DM} should be infinite and A_{CM} should be zero. The CMRR of folded cascode op-amp is 63.39°. Due to change in supply voltage of an amplifier there is change in its output voltage. The PSRR is defined as $PSRR(s) = 20 \log\left(\frac{A_V(s)}{A_P(s)}\right) dB$. If there is dual power supply in an op-amp such as positive power supply V_{DD} and negative power supply V_{SS}, then each power node has separate power supply gain. In this case, $V_{DD}(A_P, V_{SS})$ is called the transfer function from the $V_{DD}(V_{SS})$ node to the output node where by the $V_{SS}(V_{DD})$ is ac ground. The PSRR of each power supply can be defined as $PSRR_{V_{DD}} = \left(\frac{A_V}{A_P, V_{DD}}\right)$ and $PSRR_{V_{SS}} = \left(\frac{A_V}{A_P, V_{SS}}\right)$.

The positive and negative PSRR of folded cascode op-amp are equal and its value is 50 dB. The input referred noise of folded cascode op-amp is calculated as $6 \mu V/\sqrt{Hz}$

Table 4 Results of proposed folded cascode op-amp

Parameter	Simulated values
Gain	82 dB
Phase margin	85°
3 dB bandwidth	990 Hz
Unity gain bandwidth (GB)	8.5 MHz
CMRR	63.39 dB
PSRR(+)	50 dB
PSRR(−)	50 dB
Noise	$6 \mu V/\sqrt{Hz}$
Slew rate	4.5 V/μs
Swing	1.7 V (Peak to Peak)
ICMR	−0.3 to 0.8 V
Power dissipation	0.37 mW

Table 5 Results of proposed fully differential amplifier

Parameter	Simulated values
Gain	70 dB
Phase margin	85°
3 dB bandwidth	4.5 kHz
Unity gain bandwidth(GB)	17 MHz
CMRR	83 dB
PSRR(+)	63 dB
PSRR(−)	63 dB
Noise	$0.6 \mu V/\sqrt{Hz}$
Slew rate	2.82 V/μs
Swing	1.6 V(Peak to Peak)
ICMR	−0.1 to 0.9 V
Power dissipation	0.54 mW

Table 6 Performance comparison of fully differential amplifier

S. No.	Parameter	Zhang et al. [22]	Hasan and Lee [23]	This work
1.	Technology	130 nm	350 nm	90 nm
2.	Supply voltage (V)	1	3.3	0.9
3.	Gain (dB)	40.5	46.3	70
4.	Phase margin	–	–	85°
5.	Bandwidth (kHz)	8.5	10	4.5
6.	CMRR (dB)	60	85	83
7.	PSRR (dB)	60	83.2	63
8.	Noise $\left(\mu V/\sqrt{Hz}\right)$	3.5	5.16	0.6
9.	Power Dissipation	–	18.5 μW	0.54 mW

at 3 dB frequency. The input common mode range is defined as the maximum voltage range within which all transistors remain in saturation region. The measured range of ICMR is −0.1 to 0.8 V. Slew rate (SR) defines the fastest possible rate of change of the—output voltage. It depends on charging and discharging of capacitor C and it is written as SR $= \left|\frac{dV_{out}}{dt}\right|_{max} = \left|\frac{I}{C}\right|_{max}$, where I is supply current to capacitor. The slew rate in the proposed folded cascode op-amp is 4.5 V/μs.

It is observed that in fully differential amplifier the gain and phase margin of amplifier are obtained as 70 dB and 85° respectively. The CMRR 83 dB of fully differential amplifier is coming up to the frequency range of 1 MHz. The positive and negative PSRR of fully differential amplifier are equal and its value is 63 dB. The output referred noise of fully differential amplifier is calculated as $0.6 \mu V/\sqrt{Hz}$ at 1 kHz. As long as M_3 is in saturation, ICMR can be considered as the range where output varies linearly with respect to input. The measured range of ICMR is -0.1 to 0.9 V. Performance comparison of designed model of fully differential amplifier with Zhang et al. [22] and Hasan and Lee [23] is shown in Table 6.

5 Conclusion

In this work, a two-stage folded cascode amplifier and fully differential amplifier for ECG recording have been designed. It implements a fully differential amplifier to achieve low power, low noise, high CMRR, and high gain. All the blocks of instrumentation amplifier have been designed on 90 nm CMOS process technology. The folded cascode amplifier has mid-band gain of 82 dB with bandwidth of 990 Hz while consuming 0.37 mW of power. The integrated noise is $6 \mu V/\sqrt{Hz}$ at 1 kHz, CMRR is 63.39 dB and PSRR is 50 dB. The fully differential amplifier has a gain of 70 dB with bandwidth of 4.5 kHz while consuming 0.54 mW of power. The integrated noise $0.6 \mu V/\sqrt{Hz}$ at 1 kHz and CMRR is 83 dB and PSRR is 63 dB. By simulation results we have observed that values of various parameters comprehend it over other amplifier topologies, where biopotential signal monitoring is considered as an application. The proposed design of amplifier has high gain, low noise and high CMRR. However, some work can be done in the future to optimize power dissipation, slew rate and rail to rail input range. Another extension of this work could be to design other parts of ECG recording system as a front end amplifier.

References

1. Yazicioglu RF, Hoof CV, Puers R (2009) Biopotential readout circuits for portable acquisition systems. Springer Science, Boston
2. Gyselinckx B, Hoof CV, Ryckaert J, Yazicioglu RF, Fiorini P, Leonov V (2005) Human++: autonomous wireless sensors for body area networks. In: Proceedings of IEEE custom integrated circuits conference, pp 13–19
3. Gyselinckx B, Hoof CV, Ryckaert J, Yazicioglu RF, Fiorini P, Leonov V (2006) Human++: emerging technology for body area networks. In: IFIP international conference very large scale integration, nice, pp 175–180
4. Mundt CW, Montgomery KN et al (2005) A multi parameter wearable physiologic monitoring system for space and terrestrial applications. IEEE Trans Inf Technol Biomed 9:382–391
5. Paradiso R, Loriga G, Taccini N (2005) A wearable health care system based on knitted integrated sensors. IEEE Trans Inf Technol Biomed 9:337–344
6. Sungmee P, Jayaraman S (2003) Enhancing the quality of life through wearable technology. IEEE Eng Med Biol Mag 22:41–48
7. Waterhouse E (2003) New horizons in ambulatory electroencephalography. IEEE Eng Med Biol Mag 22:74–80
8. Yazicioglu RF (2011) "Readout circuits" Bio-Medical CMOS ICs, 1st edn. Springer, New York, pp 125–157
9. Zywietz C (1888) A brief history of electrocardiography progress through technology. Biosigna Institute for Biosignal Processing and Systems Research, Hannover

10. Braunwald E (1997) Heart disease: a textbook of cardiovascular medicine. W.B. Saunders Co., Philadelphia
11. Mieghem CV, Sabbe M, Knockaert D (2004) The clinical value of the ECG in noncardiac conditions. Chest, Northbrook
12. American heart association guidelines for cardiopulmonary resuscitation and emergency cardiovascular care—Part 8: stabilization of the patient with acute coronary syndromes, Circulation, pp 89–110, 2005. Available: http://circ.ahajournals.org/content/112/24_suppl/IV-89.full
13. Electrocardiography [online]. Available: http://en.wikipedia.org/wiki/Electrocardiography
14. Joshi SRA, Miller A (2011) EKG-based heart-rate monitor implementation on the Launch Pad Value Line Development Kit using the MSP430G2452 MCU. Application Report, Texas Instruments, Dallas, Mar 2011
15. Melnyk MD, Silbermann JM (2004) Wireless electrocardiogram system. A Design Project Report, Cornell University, Ithaca, NY, May 2004
16. Bobbie PO, Arif CZ (2004) Electrocardiogram (EKG) data acquisition and wireless transmission. In: ICOSSE Proceedings of WSEAS, CD-Volume-ISBN
17. Soundarapandian MBK (2009) Analog front-end design for ECG systems using delta-sigma ADCs. Application Notes, Texas Instruments, Dallas, Mar 2009
18. Harrison RR, Charles C (2003) A low-power low-noise CMOS amplifier for neural recording applications. IEEE J Solid-State Circuits 38:958–965
19. Winter BB, Webster JG (1983) Driven-right-leg circuit design. IEEE Trans Biomed Eng (BME) 30:62–66
20. Nayyar A, Puri V, Nhu NG (2018) BioSenHealth 1.0: a novel internet of medical things (IoMT)-based patient health monitoring system. In: Proceedings of international conference on innovative computing and communications, ICICC 2018, vol 1, pp 155–164. https://doi.org/10.1007/978-981-13-2324-9_16
21. Rajbhandari S, Singh A, Mittal M (2018) Big data in healthcare. In: Proceedings of international conference on innovative computing and communications, ICICC 2018, vol 2, pp 261–269. https://doi.org/10.1007/978-981-13-2354-6_28
22. Zhang F, Holleman J, Otis BP (2012) Design of ultra-low power biopotential amplifiers for biosignal acquisition applications. IEEE Trans Biomed Circuits Syst 6(4):344–355
23. Hasan MN, Lee KS (2015) A wide linear output range biopotential amplifier for physiological measurement frontend. IEEE Trans Inst Meas 64(1)

Evasion Attack for Fingerprint Biometric System and Countermeasure

Sripada Manasa Lakshmi, Manvjeet Kaur, Awadhesh Kumar Shukla and Nahita Pathania

Abstract Currently, biometrics is being widely used for authentication and identification of an individual. The biometric systems itself needs to be more secured and reliable so it they can provide secure authentication in various applications. To optimize the security, it is vital that biometric authentication frameworks are intended to withstand various sources of attack. In security sensitive applications, there is a shrewd adversary component which intends to deceive the detection system. In a well-motivated attack scenario, in which there exists an attacker who may try to evade a well-established system at test time by cautiously altering attack samples, i.e., Evasion Attack. The aim of this work is to demonstrate that machine learning can be utilized to enhance system security, if one utilizes an adversary-aware approach that proactively intercept the attacker. Also, we present a basic but credible gradient based approach of evasion attack that can be exploited to methodically acquire the security of a Fingerprint Biometric Database.

Keywords Adversarial machine learning · Evasion attacks · Convolution neural networks · Biometrics · Biometrics security · Fingerprint · Classifier

1 Introduction

Biometrics is "any quantifiable characteristic of a human's physiology that can be authentically captured and used as a unique identifier for an individual within a specified population" [1].

Biometrics defines a user by analyzing a person's physiological or behavioral characteristics. These features are distinctive to individuals, so they can be employed to verify or identify a person. With the extensive establishment of biometric frameworks

S. M. Lakshmi (✉) · A. K. Shukla · N. Pathania
Lovely Professional University, Phagwara, India
e-mail: srimanlak94@gmail.com

M. Kaur
Punjab Engineering College (Deemed to be University), Chandigarh, India
e-mail: manvjeet@pec.ac.in

© Springer Nature Singapore Pte Ltd. 2020
A. Khanna et al. (eds.), *International Conference on Innovative Computing and Communications*, Advances in Intelligent Systems and Computing 1059,
https://doi.org/10.1007/978-981-15-0324-5_7

in various applications, there are many concerns about the security and secrecy of biometric technology. Public affirmation of biometrics technology relies on the capability of system designers to show that these frameworks are sturdy, have low error rates, and are tamper-proof.

Due to the swift magnification in sensing and computing technologies, biometric systems have become affordable and are facilely implanted in a diverse customer contrivances (e.g., mobile phones, key fobs, etc.), making this technology vulnerably susceptible to the malevolent designs of terrorists and malefactors. For the authentication of one's individuality, PIN codes or passwords were utilized interiorly. There are several issues related to them, one is that they can be distributed and easily exploited.

There are many types of biometrics used in everyday life such as retina/iris, fingerprints, facial recognition, hand veins pattern and voice, etc. Biometric systems are extensively installed in sundry applications, so there is a plethora of concern with reference to their privacy and security technology. In the more voyaged and organized world we require more robust and secure model of our frameworks. Consequently to enhance the security of the biometrics confirmation frameworks they ought to be planned to such an extent that they withstand diverse wellsprings of attack. Till now the attacks were handled on ordinary biometric frameworks. Making secure systems is daring task and it is essential to evaluate the execution and reliability of any biometric framework with a specific end goal to distinguish and ensure against dangers, attacks, and exploitable vulnerabilities. Adversary exploits the adaptive feature of a machine learning framework to generate it to fail. The adversary has the capability to craft training data that causes the learning framework to process rules that misclassify the inputs. Failure consists of generating the learning system to exhibit errors, misidentifies antagonistic input as genuine, antagonistic input is allowed through the security hurdles and misidentifies genuine input as antagonistic, required input is rejected. If users detect these kind of failures, they could be unable trust in the system and neglect it. If users do not counter the failure, then the danger can be even significant.

Machine learning allows a system to respond to the developing real-world material, both antagonistic and benevolent, and learn to decline unacceptable behavior [2].

There are seven potential attacks against learning systems as describes in Fig. 1 [3].

i.   *Causative Attacks*: In which they alter the training process, i.e., causative attacks affect learning with mastery over training data.
ii. *Exploratory Attacks*: Exploit existing weaknesses, i.e., exploit misclassifications but do not influence the training data as causative attacks. For example, *Evasion Attack*.
iii. *Integrity Pointed at False Negatives*: Allows antagonistic input into a system, i.e., Compromise assets via false negatives. Integrity violations result in misclassifying malicious samples as legitimate.

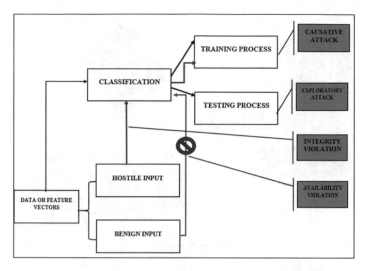

Fig. 1 Attacks in machine learning-oriented biometric system

iv. ***Availability Pointed at False Positives***: Prevents genuine input from penetrating into a system, i.e., denial of service. Availability breached can also cause genuine samples to be misclassified as hostile.

v. ***Targeted Attacks***: Aimed at a specific input or any instance.

vi. ***Indiscriminate Attacks***: Encompasses at wide class of instances.

vii. ***Privacy Violation***: In this it permits the attacker to acquire secret data from the classifier.

The further work is focused on evasion attack, i.e., an exploratory attack.

Evasion is bypassing data security by exploiting and attacking the targeted model or framework, without getting detected. Here, the adversary's aim is to alter hostile data to evade detection. It belongs to the class of integrity, indiscriminate violation, and exploratory Attacks. In these attacks, attacker aims to manipulate the test samples of the data set. In this, attacker's aim is to alter at least one sample which can be misclassified in the testing set. Also, as this attack takes place the false-negative rates increases as in Fig. 2. Furthermore, it alters the dispersion of hostile samples of test data to look like as genuine samples.

Evasion attacks should be considered as a potential scenario to explore vulnerabilities of classifiers learnt on reduced feature sets and to properly design more secure, adversary-aware feature selection algorithms.

For this work, An Adversary needs to be modeled on the basis of attacker's goal, knowledge, capability, and taxonomy [4–6]. Taxonomy of Attacker deals with the influence of Attack over the data, Attack specificity ranging from targeted to indiscriminate and Security violation lead by the attack. Goal of Attacker deals with the objective of the attacker. Here in this work our attacker's goal is to minimize the accuracy and maximize the loss. Knowledge of Attacker includes Perfect and

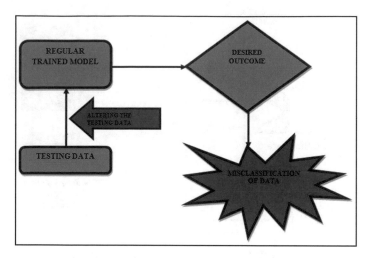

Fig. 2 Evasion attack basic model

Limited knowledge in which Perfect Knowledge basically comprises of Decision Function, Classification Function, Training parameters and the available feedback of the model and in Limited Knowledge, it is assumed that attacker doesn't have knowledge of any of the parameters in the system or model. At last, Capability of Attacker mentions how the adversary can affect training and testing data. For the Evaluation of Security, A security designer needs to predict the adversary by simulating a process called Arms Race [7, 8]. There are two types of Arms Race.

First, *Reactive Arms Race*, *in* which the system designer and attacker try to attain their objectives by responding to the altering etiquette of the opponent. Figure 3 shows the framework for a reactive arms race.

Second, in **Proactive Arms Race**, the System designer attempts to predict the attacker's master plan. Figure 4 shows the framework for proactive arms race.

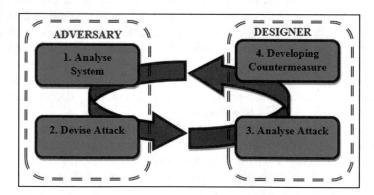

Fig. 3 Reactive arms race

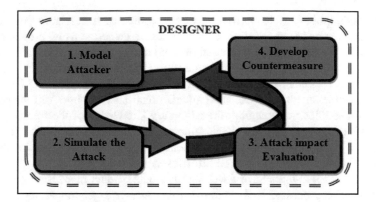

Fig. 4 Proactive arms race

In this work, we have opted for Proactive Arms Race as the system does not wait for the action of attack and loss of the system as the security improvability is in the hands of system designer [9]. When attacker encounters this type of system he may have to exert greater effort to find the novel loopholes. However, this system needs rigorous and structured emendation.

First, a sturdy learning model is proposed to handle performance deterioration due to adversary attack.

Second, we discuss an attacker's model simulating evasion attack on CNN and also, propose a countermeasure towards the attack. Thirdly, experimentally it is proven that the proposed model convincingly conserves its robustness under evasion attack.

The rest of the paper is organized as follows. Section 2 reviews the Summary of literature. Section 3 presents Proposed Methodology, In Sect. 4, Results and Inferences are discussed. Finally, Sect. 5 describes Conclusion and Future Work of the paper.

2 Summary of Literature

Biometric systems are extensively installed in diverse, so there is a lot of concern adjudge to their privacy and security technology. Biometric technology will be entirely globally be endorsed when these systems have low error rates, deface proof and are sturdy.

Subsequently, to refine the security of the biometrics frameworks they ought to be composed such an extent so that they stand up firm against distinctive sources of assaults. Till now the assaults were handled on ordinary biometric frameworks. So, planning secure Machine learning frameworks is difficult and it is vital to evaluate the execution and security of any biometric framework in order to identify and protect in resistance threats, assaults, and exploitable vulnerabilities.

In Khorshidpour et al. [2] have proposed a secure and firm learning model over evasion assaults on the application of PDF malware detection using proactive arms race. The results show that the proposed method notably enhances the robustness of the learning system against altering data and evasion performance with SVM with RBF and linear kernels. The paper of Biggio et al. [5] have proposed a framework for security assessment, they executed attack situations that show various risk levels for the classifier by expanding the adversary knowledge of the framework and adversary's capacity to alter the attack samples. They also evaluated their technique on significant security assignment of malware detection in PDF data and demonstrated that such frameworks can be effortlessly evaded and sidestepped. In paper Fumera et al. [10] authors have developed a model of data dispensation adversarial classification processes, and exploited it to conceive a general strategy (i.e., not tied to a specific application or attack) for designing sturdy classifiers, which are robust against adversarial alteration of input samples at testing phase. The experimental outcomes demonstrated that their procedure can enhance the sturdiness of the classifier to exploratory integrity assaults. Zhang et al. [11] focused on a proficient, wrapper based implementation technique, and validated experimentally its effectiveness on different utilization fields such as spam and malware detection by proposing a novel attacker-aware feature selection framework that can enhance classifier security against evasion attacks. According to the paper of Demontis et al. [12] a basic and adaptable secure learning pattern was proposed that mitigates the effect of evasion attacks, while just marginally intensifying the detection rate without the attack. The main contribution of the authors in this work is that they defined a new learning algorithm to train the linear classifiers with more systematically distributed feature weights. Arslan et al. [3] examined the machine learning methods utilized in biometric applications that were researched for the security point of view along with their advantages and disadvantages. The examinations in the literature between 2010 and 2016 years, utilized algorithms, innovations, measurements, utilization zones, the machine learning methods used for various biometric modalities were researched and evaluated. In the study, machine learning techniques utilized by biometric authentication and identification procedure for the security point of view were inspected in detail.

The paper of Biggio et al. [4] proposed a model that permits the classification of familiar and new threats of biometric recognition frameworks, along with different attacks, their preventions and defense processes. Also, a new prevention technique based on template sanitization was proposed. According to Roli et al. [7] the goal of their work was to provide functional instructions for enhancing the security of pattern recognition in adversarial settings. Fumera et al. [13] gave an exploratory assessment on a spam filtering of the class boundary randomization for making a classifier harder to evade. In the paper, Sadeghi et al. [14] formalized a hypothetical strategy for evaluating the invulnerability of a machine learning-based cyber forensic framework against proving the manipulation attack. They examined on an MLCF framework that utilizes EEG signals which will be a major advance towards bringing computer crime scene investigation into practice. Roberts et al. [15] presented a more extensive and sharper perspective of biometric framework assault vectors. The approach in the research helps and assists operational demands in an organization to control the risk and increase the security of the systems. Ratha et al. [16] pinpointed some loopholes and vulnerabilities to biometrics systems and described problems related to the trade-off between security and comfort.

3 Proposed Methodology

A biometric system is nowadays more prone to different types of zero-day attacks. As the technology is increasing the need for security is in demand. So, a designer needs to get updated by various security attacks that can be done to compromise a system. Mostly a system can be compromised by either the attacker who is outside the organization or by an insider. When an insider attacks a system, he/she has perfect knowledge of the system. Till now, Evasion attack has been simulated on MNIST digits and also against systems consisting of SVM's, KNN's, etc. Hence, the aim is to proactively discover flaws in the system, simulate evasion attack against CNN's onto the system, and determine the deterioration on the system or database, countermeasure the attack using optimization technique. The security model and assumptions for this experiment are mentioned in Fig. 5. The flow diagram of the proposed methodology has been described in Fig. 6. In this proposed framework, the data is first segregated into Testing and Training database in specified ratio. Then, The Convolution neural network is then fed with both Training and Testing database and executed. As discussed above that the proactive arms race is used in which the designer behaves like an attacker and also as the designer itself with perfect knowledge of the testing set, database, Machine learning Classifier.

Eventually, when the system starts its learning and testing then the Evasion attack is simulated. In this the data set of testing is compromised and deteriorated so that the

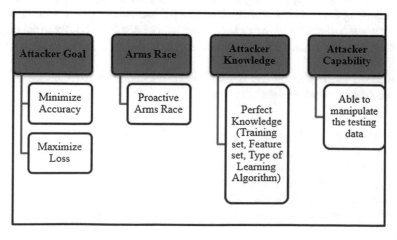

Fig. 5 Security model

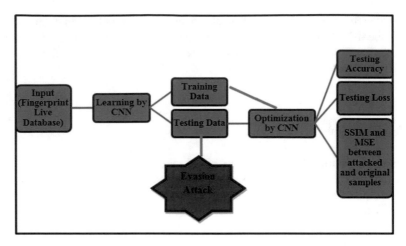

Fig. 6 Overall process flow of proposed framework

Convolution neural networks misclassify the attacked sample as the genuine one. The Gradient descent optimization technique is applied so as to maintain the performance of the system. So, a system needs to be run on a very large number of epochs such that CNN's may learn the genuine database more precisely and perfectly. In this we calculate the performance by accuracy, loss and deterioration rate between the two samples that are attacked and the genuine one so as to determine the rate of attack.

Attack Surface

In the attack scenario, A is the any input sample from the testing data (n) where A^0 is initial attack point and A^* is final attack point. Here, $g(A)$ is discriminative function. Attacker's goal is to minimize the $g(A)$ of the model. A best attacker strategy is to misclassify the model by creating a sample with high confidence, i.e., it can be done by minimizing the $g(A)$ [5]. The attacker has to mutate attack points in the testing data but needs to persist within a defined parameter of distance d_{max} from the original attack input sample. d_{max} provides the attacker more freedom to manipulate the data [5].

Following is the step by step description of methodology used, which explains the complete approach in detail.

Step 1: for $i = 0$ to n
$A[i] = \text{argmin } g(A)$ {$g(A)$ minimize the distance between pixels}
Step 2: If $d(A^*, A^0) > d_{max}$
Then place A^n in the boundary of attack region
Step 3: Applying RGB to gray conversion, Gaussian Blur Function, Fourier Transform and fftshift function which shifts the input image array on input A.
Step 4: Reset the database to the original database by providing any random input to the system.

Countering the Evasion Attack Using CNN

Step 1: Segregate Dataset in Training and Testing (n). Feed as input where each input is of size 240×300 to the Convolutional Neural Network having four convolution layers. Perform Feedforward pass with from L_n activation layers [5].

Step 2: Output layer

$$\delta^{(n)} = -(A - A^*) \cdot f'(z^{(n)})$$

If error is encountered again retrain the model and update the weights

$$\delta^{(l)} = ((w^{(l)})T)\delta^{(l+1)})$$

where $\delta^{(n)}$ = Error or Sudden increase of loss function, A = Normal input A, A^* = Initial Attack point, $f'(z^{(n)})$ = Learning Parameter, $w^{(l)}$ = Updated Weights.

Step 3: Optimize the function $g(A)$ using gradient descent optimization technique built-in CNN to maximize the value of $g(A)$. As a security designer the $g(A)$ should be maximized.

$$\arg \max F(A) = g(A) - \frac{\text{No. of epochs}}{n} \sum w^T w$$

$$\text{St. } g(A)d(A, A^0) \leq d_{\max}$$

An attacker goal is always to increase the loss function and decrease the accuracy, where the accuracy of a model is usually evaluated after the model parameters are learned and fixed and no learning is taking place. Unlike accuracy, loss is not a percentage. It is a calculation of the errors made for each example in training or testing sets.

In the evasion attack surface, the attacker alters the testing data and makes the system to misclassify the genuine fingerprint of a user. The evasion attack mostly relates itself to denial of service. Hence, for a successful attack we need a sample that is to be targeted by the attacker. The attacker always seeks to create fingerprint sample that misclassifies with high confidence, i.e., attacker always tries to minimize the $g(A)$ which is the discriminant function of the model. The attacker as discussed has perfect knowledge of the target classifier, etc. $d_{\max}$ is the maximum distance that permits the maximum alteration of sample under the attack. The distance is important in controlling the safeguarding the structure and originality of the sample [5]. As the $d_{\max}$ value increases, there is high change in the features. Data of fingerprint is segregated into training and testing and fed as input to CNN. CNN performs feedforward pass by providing its all samples to CNN for training. $\delta^{(n)}$ is evaluated as it depicts the error rate or sudden increase in loss. If there is any change then this function needs to be retrained or learnt from its experience by updating the weights of the model [5]. While classifying the result of the model the $g(A)$ needs to be maximized so as to counter the attack. Hence by using gradient descent optimization technique the $g(A)$ is maximized and optimized by CNN. The epochs of the model should be in a huge amount so as to run the optimization technique efficiently. Figure 7 describes the Proactive Architecture of the proposed methodology.

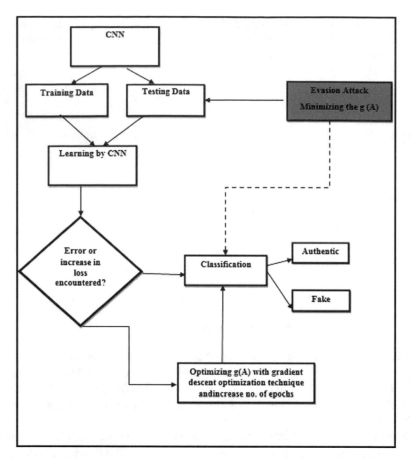

Fig. 7 Proposed methodology

4　Results and Inferences

The results are categorized as without attack, i.e., normal system, after attack and simultaneously countermeasure. A normal system refers to a system which has no attack simulated in it. The Experiment has been done on 3 particular epochs that are at epoch 10, epoch 25, and epoch 75. The graphs are plotted between training loss versus testing loss and training accuracy versus testing accuracy.

- **Database Creation**

The performance of the system is done with the live databases that are acquired in controlled environment. The samples of fingerprint are acquired using Verifier 300 LC. The database is consists of Fingerprints of 53 users (Table 1). Each user-provided 30 samples of left hand with different variations. Data is segregated into 80:20 ratio,

Table 1 Summary of database used

Parameters	Database
Number of users	1590
Number of samples	30
Number of samples for training data	1060
Number of samples for testing data	530

i.e., 1060 fingerprint samples in Training data and 530 samples in Testing data for Convolution neural network.

In a normal system, the system works in a normal way where there is a gradual increase of loss in the system. But when attack is simulated at a particular epoch, the loss increases suddenly (Table 2). When this type of behavior is observed in the system then we can encounter and detect that attack has been simulated into the system. Again after the attack is stopped and the images are restored CNN optimization technique is activated (Table 2). In normal system, when the authentic sample is provided to the system to predict whether it is authentic or fake. The result was authentic, i.e., the provided sample was Authentic. But after attack scenario, when the authentic sample is provided to the system to predict whether it is authentic or fake. The result was Fake, i.e., the provided sample was attacked and system was being misclassified. When the attacked and countermeasure was run simultaneously then, the authentic sample was provided to the system to predict whether it was Authentic or Fake. The result was authentic, i.e., the provided sample was Authentic. Table 2 describes the results of After Attack and simultaneously with countermeasure.

The graphs plotted for epoch 25 have been discussed below. The Normal system has been discussed. In which it is observed that Loss is decreasing at training phase and at validation the Loss has increased but then immediately decreased. Also, it is understandable that loss has increased in small units. Hence, it proves that as the

Table 2 Results of testing in various epochs

Parameter of testing	Epochs	Epoch at which attack has been simulated	Parameter value before attack	Parameter value at attack	Parameter value after attack
Less	10	2	0.032	5.307	0.0333
Accuracy			0.987	0.667	0.972
Less	25	4	0.0323	5.632	0.0434
Accuracy			0.998	0.689	0.986
Less	75	24	0.0145	5.014	0.0528
Accuracy			0.996	0.675	0.984

epochs are increased we encounter ups and downs in loss but not in a greater number (Fig. 8).

In Fig. 9, it is observed that testing accuracy is constant and training accuracy is increasing. Hence we can deduce that small unit ups and downs does not represent any major assault in the system.

In Fig. 10, The Attack and Countermeasure system has been discussed. In which it is observed that at epoch 4 the attack was launched and then the attack was shut after 2 epochs, i.e., at 6. There is a sudden very high increase in the loss of testing that indicates that the attacked was done. But after that CNN intelligently handled the issue and optimized the loss.

Also in Fig. 11, we can observe that our attack has been done from epoch 4 and was shut down at 6 as there is a sudden decrease in the accuracy. The sudden downfall in the accuracy denotes the false happening in the testing, i.e., the simulated attack.

- **Deterioration Rate of Original Data Versus Attacked Data**

The deterioration rate of original data vs attacked data is measured by two parameters that are Mean Squared Error and Structural Similarity Index. In Mean Squared Error, the averaged squared differences between the images are evaluated. In Structural Similarity Index, the main difference between two images is calculated by measuring the similarity of two images.

The MSE and SSIM of the same images are always 0 and 1. If MSE increases from 0 and if SSIM decreases from 0, it means the rate of error between the images is prevailing as in Fig. 12.

The fingerprints of same person after attack deterioration rate was MSE = 32,033.48 and SSIM = 0. Hence, it is deduced that there was a good destruction of fingerprint to fool the system.

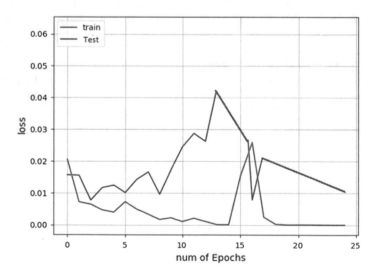

Fig. 8 Normal system, graph between number of epochs versus training loss and testing loss

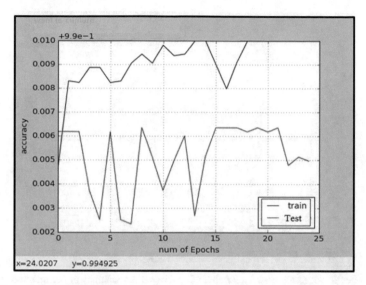

Fig. 9 Normal system, graph between training accuracy and testing

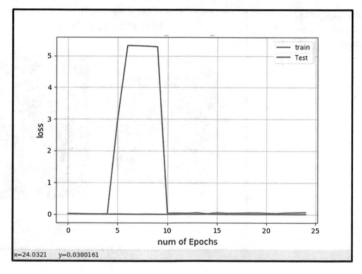

Fig. 10 After attack, graph between training loss and testing

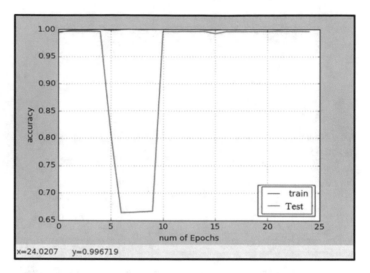

Fig. 11 After attack, graph between number of epochs versus accuracy of training and testing

Fig. 12 MSE and SSIM between the attacked and genuine image

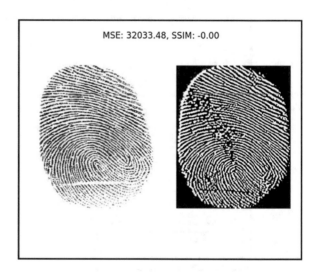

The attack was computed over 228 samples of data, the average deterioration rate

$$\text{MSE} = 25,038.17 \text{ and}$$
$$\text{SSIM} = -0.0056$$

5 Conclusion

The experimental results show that CNN with gradient descent optimization have improved the robustness and reliability of the system against evasion attack. From the obtained results, we can conclude that the attack has been done extensively to deteriorate the fingerprint samples. However, to counter it we need to have more number of epochs so that the CNN has enough test time to optimize the value of discriminant function to maximum. As future work, it is recommended to choose other methods of securing the system such as encryption of testing data, sanitization of data, etc. Also, IDS and IPS system can be used to detect the attack effectively when the loss of the testing increases.

Acknowledgements We would like to express our deep and sincere gratitude to Dr. Manvjeet Kaur for her invaluable encouragement, suggestions, and support in this study and research. We would also like to acknowledge all the cited authors in this study. We would like to express our thanks to all those who contributed in many ways to the success of this study. With the best of our knowledge, all the information provided is authentic and revised.

References

1. Bhattacharyya D et al (2009) Biometric authentication: a review. Int J u- e-Serv Sci Technol 2(3):13–28
2. Khorshidpour Z, Hashemi S, Hamzeh A (2016) Learning a secure classifier against evasion attack. In: 2016 IEEE 16th international conference on data mining workshops (ICDMW). IEEE
3. Biggio B, Fumera G, Roli F (2014) Security evaluation of pattern classifiers under attack. IEEE Trans Knowl Data Eng 26(4):984–996
4. Biggio B et al (2015) Adversarial biometric recognition: a review on biometric system security from the adversarial machine-learning perspective. IEEE Signal Process Mag 32(5):31–41
5. Biggio B, Corona I, Maiorca D, Nelson B, Šrndić N, Laskov P, Giacinto G, Roli F (2013) Evasion attacks against machine learning at test time. In: Machine learning and knowledge discovery in databases. Springer, pp 387–402
6. Barreno M, Nelson B, Sears R, Joseph AD, Tygar JD (2010) Can machine learning be secure? In: ASIACCS'06: Proceedings of the 2006 ACM symposium on information, computer and communication security, Cagliari, Cagliari (Italy), 2010 (cited at p xiv, 4) 111. ACM, New York, pp 16–25 (2006)
7. Biggio B, Fumera G, Roli F (2014) Pattern recognition systems under attack: design issues and research challenges. Int J Pattern Recognit Artif Intell 28(07):1460002
8. Huang L et al (2011) Adversarial machine learning. In: Proceedings of the 4th ACM workshop on security and artificial intelligence. ACM
9. Biggio B, Fumera G, Roli F (2010) Multiple classifier systems for robust classifier design in adversarial environments. Int J Mach Learn Cybern 1(1):27–41
10. Biggio B, Fumera G, Roli F (2011) Design of robust classifiers for adversarial environments. In: IEEE international conference on systems, man, and cybernetics (SMC), pp 977–982
11. Zhang F, Chan PP, Biggio B, Yeung DS, Roli F (2016) Adversarial feature selection against evasion attacks. IEEE Trans Cybern 46(3):766–777
12. Demontis A et al (2017) Yes, machine learning can be more secure! A case study on android malware detection. IEEE Trans Dependable Secur Comput

13. Biggio B, Adversarial pattern classification. Diss. Ph.D. thesis, University of Cagliari, Piazza d'Armi, 09123 Cagliari, Italy
14. Sadeghi K et al (2016) Toward parametric security analysis of machine learning based cyber forensic biometric systems. In: 2016 15th IEEE international conference on machine learning and applications (ICMLA). IEEE
15. Roberts C (2007) Biometric attack vectors and defences. Comput Secur 26(1):14–25
16. Ratha NK, Connell JH, Bolle RM (2003) Biometrics break-ins and band-aids. Pattern Recogn Lett 24(13):2105–2113

Current-Phase Synthesis of Linear Antenna Arrays Using Particle Swarm Optimization Variants

Sudipta Das, Pragnan Chakravorty and Durbadal Mandal

Abstract In this work, the problem of low sidelobe phased array synthesis is taken up, and variants of particle swarm optimization (PSO), like Grey PSO and Novel PSO, are adopted for dealing with this problem. For simplicity, periodic linear array geometries are considered. Effect of position regulation and inertia control strategies on the convergence of PSO variants is studied in this regard. Results reflect the impacts of position regulation and inertia control strategies on the convergence of the algorithms for the problem instance considered. Without the influence of position regulation Grey PSO and Novel PSO have been able to suppress interference levels for a 20-element linear array to -21.31 and -31.23 dB, respectively. Under the influence of the position regulation, their respective values got improved to -27.90 and -43.35 dB.

1 Introduction

Pencil beam pattern with low sidelobe level is the required for peer-to-peer communication. Phased arrays have the advantage of controlling the pattern characteristics without physically rotating or reshaping the entire geometry.

Element feed currents of an array has crucial impact on its radiation characteristics. Hence, synthesizing arrays via adjusting the terminal currents has attracted

---

S. Das (✉)
Department of Electronics Engineering, University Departments,
Rajasthan Technical University, Kota, Rajasthan, India
e-mail: sudipta.sit59@gmail.com

P. Chakravorty
Department of Electronics and Telecommunication Engineering,
Ramrao Adik Institute of Technology, Nerul, India
e-mail: pciitkgp@ieee.org

D. Mandal
Department of Electronics and Communication Engineering, National Institute
of Technology Durgapur, Durgapur, West Bengal, India
e-mail: durbadal.bittu@gmail.com

© Springer Nature Singapore Pte Ltd. 2020
A. Khanna et al. (eds.), *International Conference on Innovative Computing and Communications*, Advances in Intelligent Systems and Computing 1059,
https://doi.org/10.1007/978-981-15-0324-5_8

many researchers. Consequently, many methods are developed so as to meet the prescribed pattern characteristics for a given array geometry. For example, Shcelkounoff's polynomial method [1] is proposed to synthesize patterns with minor lobes interspaced by deep nulls from periodic linear arrays. Proper adjustment of nulls can yield desired radiation characteristics. Dolph's method [2] yields pencil beam low sidelobe radiation pattern from periodic linear arrays that have all sidelobes at a constant level and that maintain the best trade-off between the sidelobe level and the first null beamwidth. Taylor's method [3, 4] for line sources on the other hand can yield a class of steerable pencil beam patterns that maintain prescribed number of near-in sidelobe peaks at a pre-specified level and decaying far-out sidelobe levels. It is interesting that Taylor's distribution can be sampled. Dolph's method is extended to periodic planar arrays for obtaining steerable pencil beam pattern with equiripple nature in the sidelobe region in [5]. On the other hand, Taylor's method is extended to planar circular apertures in [6]. In [7], an iterative method to obtain the desired pattern characteristics from line sources is outlined. Also, prescribed radiation characteristics can be obtained by adopting fast Fourier transform method [8].

Nature-inspired algorithms are very effective for various real-life problem-solving [9, 10]. Swarm intelligence is a family of nature-inspired algorithms which models the collective behaviour of living entities. They retain the memory of the member-wise historically obtained the best experience and rush towards the member which is presently having the best experience. Particle swarm optimization (PSO) is a swarm intelligent algorithm of this type. Owing to the no free lunch theorem [11], there is always a scope to design or adopt algorithms that outperform others on a set of problems. These are called the *specialist* algorithms. Boundary dynamics play crucial role in determining the convergence properties of any metaheuristic algorithm. Recently, position regulation-based controlling strategy for PSO variants is introduced in [12]. In this work, two variants of PSO are chosen, and their boundary dynamics are controlled using one position regulation strategy. These modified PSO variants are then tested on the array design problem. The summary of this work can be given as follows:

- The study on the variation of current-phase variable sets for low sidelobe pattern synthesis problems reveals that currents tend to take symmetric distribution while phase tend to take anti-symmetric distribution about the centre of the array
- The position regulation boundary condition is very effective for different PSO variants.

Hereafter, the work is described as follows: Problem definition is given in Sect. 2; Sect. 3 gives the description of PSO variants adopted and the condition of simulation runs; In Sect. 4, the obtained results are tabulated and discussed, and Sect. 5 concludes on the work and its findings.

2 Problem Definition

2.1 Array Factor and Design Parameters of Linear Arrays

In this work, periodic linear arrays comprising of isotropic elements are considered for simplicity. The array factor of such a linear array comprising of N elements having inter-element distance d and lying along Cartesian z-axis is given in (1).

$$AF_{f_0} = \sum_{n=1}^{N} I_n \exp\{j((n-1)kd \cos\theta + \beta_n)\} \qquad (1)$$

where k is the wavenumber corresponding to the frequency of operation; θ is the angular coordinate of the far-field $\{0 \le \theta \le \pi\}$; $I_n \exp(j\beta_n)$ is the complex excitation coefficient of the nth element of the array. In this work, the set of $I_n \exp(j\beta_n)$ is optimized for obtaining the desired radiation characteristics.

2.2 Definition of the Objective Function

The objective is to obtain pencil beam patterns with as low peak sidelobe level as possible. Hence, the key objective parameter, here, is the relative peak sidelobe level SLL. The parameter SLL may be expressed as given in (2).

$$SLL_{f_i} = \max\{AF_{f_i}|_{\theta \notin \Omega_\tau}\} \qquad (2)$$

where represents the angular region subtended by the target beam. It is important to mention here that the definition of sidelobe level given in (2) does not only apply to sidelobes, but it includes grating lobes also (if appears). While adjusting the phases, frequently beam deviation takes place. Hence, putting a restriction on the beam deviation is necessary to avoid undesired outcome after simulation runs. Accordingly, the objective function should be a constrained objective function. Here, the objective function is designed as given in (3).

$$CF = \text{minimize}\ \{SLL\}$$
$$g = |\theta_\tau - \theta_0| < \varepsilon \qquad (3)$$

where, g represents the constraint function, θ_0 and θ_τ represent the angular locations of the desired and the present main beam peak values, respectively, and ε represents the tolerance level of the beam deviation.

Here, SLL is expressed in dB scale, and θ is expressed in degrees.

3   Adopted PSO Variants and Conditions of Simulation Run

PSO is a population-based swarm intelligent algorithm that models the group-based food searching behaviour of living agents to stochastic searching algorithm. Each agent represents one trial solution in multidimensional solution space. In PSO, each agent is assumed to have knowledge about historically obtained personal best position, as well as the position of the presently outperforming agent in the group. Various assumptions are possible about living agents and consequently, and versions of PSO were proposed to address certain problems more efficiently. In this work, two competitive PSO variants, namely Grey PSO [13] and Novel PSO [14], are adopted for studying the impact of the position regulation. For conciseness, the descriptions of these algorithms are skipped to this end.

It is important to mention that the search is always guided by the decision space limits; otherwise, search efforts will be futile. In order to restrict the search within the limits of the solution space, several velocity and position regulation strategies are proposed in the literature. In this work, a recently proposed position regulation strategy [12], named *mirror boundary condition*, is adopted for modifying the search of these PSO variants.

For this work, the constraint handling technique proposed in [15] is adopted. In this approach, each newly generated solution goes in a tournament with its respective target vector. A newly evolved solution is to replace the older solution if it is either less infeasible (if at least one of them is infeasible) or better fit (if both the solutions are feasible). If a solution has infeasibility less than 0.1, one of these two solutions is randomly selected.

3.1   Conditions of Simulation Run

Decision space dimensions (number of design variables) play crucial role in the convergence properties of any algorithm. Search space increases exponentially with the increase in decision space dimension. This is termed as the curse of dimensionality in evolutionary algorithm community. In such cases, huge effort is required in order to obtain the desired solution quality. Sometimes, the significant modifications are needed in the algorithm definition so as to obtain desired quality solutions within affordable effort limits. Effort, to this end, is referred to as the number of function evaluations.

For the present case, decision space dimension for the present case is defined by the quantity N. Here, N is set as 20; hence, the decision space dimension is 40. For such cases, a limit of 2×10^4 function evaluations are considered sufficient. For sake of comparison, the population size (N_p) is set as 200 and the number of iteration cycles is set as 100, which together match the desired function evaluation limit.

Program is written and simulated in MATLAB software using MATLAB 8.1.0.604 (R2013a) on an Intel Core™ i5-4690CPU with processor speed 3.50 GHz and 4.00 GB RAM and operating system Microsoft Windows 8.1 Professional.

4 Simulation Run Statistics

4.1 Results

Simulation is run for 20-element linear array structures having inter-element separation of $\frac{\lambda}{2}$.

Table 1 shows the details of the optimized patterns such as the peak SLL and the first null beamwidth ($BWFN$) values for the respective PSO variants. Table 1 also records DRR for all these cases.

Figures depict the optimized radiation patterns and their respective optimized currents. Figure 1 compares the obtained radiation pattern and Figs. 2, 3, 4 and 5 plot the respective excitation coefficients.

4.2 Discussions

From Fig. 1, it is visible that the sidelobes are separated by deep nulls. It is seen that the obtained patterns have no beam steering. Figures 2, 3, 4 and 5 reveal that the optimized currents have obtained symmetry about the centre, while the phases turned out to be anti-symmetric. Also, it can be seen that the tapering decreases towards the far-end elements of the array. Also, broadening of the main beam is noted for all these cases of sidelobe level suppression.

Table 1 shows the growth of DRR and $BWFN$ as the SLL values are suppressed. These parameters represent the trade-off between the sidelobe level, and beam broadening, and DRR, and the evolved patterns are non-dominated in this respect. However,

Table 1 PSO variants and the optimized parameter values

PSO variants	Parameter value		
	SLL (dB)	$BWFN$ degrees	DRR (Normalized)
Grey PSO	−21.31	17.15	3.88
NPSO	−31.23	21.25	8.57
Grey PSO with position regulation	−27.90	21.45	6.86
NPSO with position regulation	−43.35	24.55	29.10

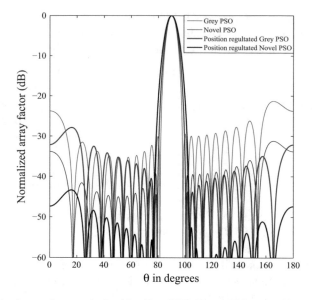

Fig. 1 Optimized array factors obtained by Grey PSO, Novel PSO and their position regulated versions

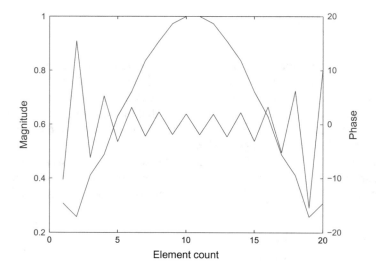

Fig. 2 Optimized excitation coefficients obtained by Grey PSO

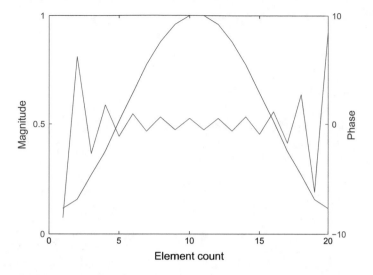

Fig. 3 Optimized excitation coefficients obtained by NPSO

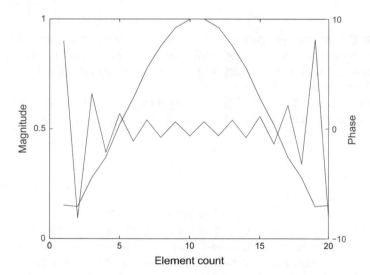

Fig. 4 Optimized excitation coefficients obtained by position regulated Grey PSO

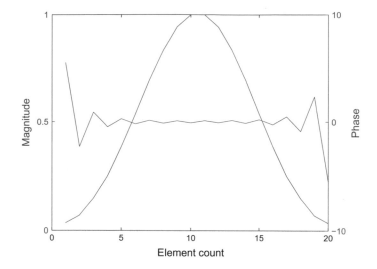

Fig. 5 Optimized excitation coefficients obtained by position regulated NPSO

since the *BWFN* and *DRR* parameters were not included in the objective function (3), the algorithms consider the suppression of *SLL* value with small beam deviation as the main target. Thus, in this respect, position regulation with Novel PSO may be declared the winner in this case.

Mirror boundary condition of position regulation is seen to be very effective to overcome the susceptibility of PSO variants for local traps.

5 Conclusions

In this work, a problem of synthesizing low sidelobe phased linear arrays is taken up, and PSO variants like Grey PSO and Novel PSO are adopted for this purpose. The problem of beam deviation is significant for such cases, and hence, the design problem is formulated as a constrained optimization problem, the amount of beam deviation being treated as the constraint. It is seen that all optimized results are feasible, and the interference suppression level is also good.

It is seen that the obtained excitation coefficients have symmetric current and anti-symmetric phase values about the centre of the array, and the tapering of the currents decrease towards the end elements for almost all evolved solutions.

Position regulation is adopted for modifying the PSO variants, and it is found that the results are significantly improved for the present case study. These approaches may be adopted for other optimization problems as well.

References

1. Schelkunoff SA (1953) A mathematical theory of linear arrays. Bell Syst Tech J 22:80–107
2. Dolph CL (1946) A current distribution for broadside linear arrays which optimizes the relationship between beamwidth and sidelobe level. In: Proceedings of IRE waves and electrons
3. Taylor TT (1955) Design of line-source antennas for narrow beamwidth and low side lobes. IRE Transa Antennas Propag 3(1):16–28
4. Taylor TT (1953) One parameter family of line-sources producing modifed $sin(\pi u)/\pi u$ patterns. Hughes Aircraft Co. Tech. Mem, 324, Culver City, Calif., Contract AF 19(604)-262-F-14
5. Tseng F-I, Cheng DK (1968) Optimum scannable planar arrays with an invariant sidelobe level. Proc IEEE 56(11):1771–1778
6. Hansen RC (1976) A one-parameter circular aperture distribution with narrow beamwidth and low sidelobes. IEEE Trans Antennas Propag 24(4):477–480
7. Elliott RS (1976) Design of line source antennas for sum patterns with sidelobes of individually arbitrary heights. IEEE Trans Antennas Propag 24(4):AP-76–83
8. Keizer WPMN (2009) Low-sidelobe pattern synthesis using iterative fourier techniques coded in matlab, [EM programmer's notebook]. IEEE Antennas Propag Mag 51(2):137–150
9. Roy B, Sen AK (2018) Meta-heuristic technique to solve resource-constrained project scheduling problem. In: Proceedings of international conference on innovative computing and communications, vol 2. Lecture notes in networks and systems, vol 56, pp 93–99
10. Kumar S, Nayyar A, Kumari R (2018) Arrhenius artificial bee colony algorithm. In: Proceedings of international conference on innovative computing and communications , vol 2. Lecture notes in networks and systems, vol 56, pp 187–195
11. Wolpert DH, Macready WG (1997) No free lunch theorems for optimization. IEEE Trans Evol Comput 1(1):67–82
12. Chakravorty P, Mandal D (2015) Role of boundary dynamics in improving efficiency of particle swarm optimization on antenna problems (invited paper). In: 2015 IEEE symposium series on computational intelligence, Cape Town, South Africa, pp 628–634 (2015)
13. Leu M-S, Yeh M-F (2012) Grey particle swarm optimization. Appl Soft Comput 12:2985–2996
14. Mandal D, Ghoshal SP, Bhattacharjee AK (2011) Wide null control of symmetric linear antenna array using novel particle swarm optimization. Int J RF Microw Comput Aided Eng 21(4):376–382
15. Santana-Quintero LV, Coello CAC (2005) An algorithm based on differential evolution for multi-objective problems. Int J Comput Intell Res 1(2):151–169

Fair Channel Distribution-Based Congestion Control in Vehicular Ad Hoc Networks

Swati Sharma, Manisha Chahal and Sandeep Harit

Abstract Vehicular ad hoc network (VANET) serves as one of the significant enabling technologies in intelligent transportation system (ITS). Accurate and up-to-date information received by vehicle-to-vehicle (V2V) communication prevents road accidents. The most critical matter in IEEE 802.11p-based V2V communications is channel congestion, as it results in unreliable safety applications. In this paper, two types of safety messages—beacon and event-driven—are used to reduce hazards. Beacon is disseminated periodically, providing the necessary information about their neighbor status. Event-driven is sent whenever a danger has been detected. As the number of vehicle increases, the number of safety messages disseminated by vehicle also increases, which results in congestion in the communication channel. As a countermeasure, we have proposed the most prominent decentralized congestion control (DCC) algorithm based on transmit rate. The effects of congestion on vehicular safety can be controlled by designing a DCC algorithm including priority model and transmission rate of the messages, which provides more reliable and timely reception of safety messages. DCC algorithm controls the congestion by fairly distribution of channel to each vehicle. A novel approach is proposed regarding congestion control, and performance is analyzed with respect to some indexes such as packet-delivery rate (PDR), end-to-end (E2E) delay, throughput, and transmit frequency.

Keywords VANET · Congestion control · Congestion detection · Channel distribution

S. Sharma (✉) · M. Chahal · S. Harit
Department of Computer Science and Engineering, Punjab Engineering College,
Chandigarh160012, India
e-mail: swati.17.swaift@gmail.com

M. Chahal
e-mail: manisha.chahal2@gmail.com

S. Harit
e-mail: sandeepharit@gmail.com

© Springer Nature Singapore Pte Ltd. 2020
A. Khanna et al. (eds.), *International Conference on Innovative Computing
and Communications*, Advances in Intelligent Systems and Computing 1059,
https://doi.org/10.1007/978-981-15-0324-5_9

1 Introduction

VANET involves many safety applications that can prevent from collision and accidents. The United State Federal Communications Commission (FCC) has assigned a separate spectrum with 75 MHz bandwidth for dedicated short-range communication (DSRC) among VANET communication [2, 5]. It is a special category of mobile ad hoc networks (MANET), which consists of a number of vehicles with the ability of communication to each other without any fixed infrastructure [3, 10]. It is an important ad hoc network because of a key component of intelligent transportation system (ITS). The ITS helps to develop the infrastructure and communication among the vehicles and is an ongoing research area in smart cities. VANET also supports the vehicle-to-vehicle communication and inter-vehicle communication. Nodes in the network move only in predetermined roads, and they do not have the problem of resource limitation regarding data storage and power. VANET enables numerous applications, including the prevention of collisions, accidents, and delay. The basic requirement of safety applications is reliability, collision-free, and fast delivery of messages. MAC protocols in VANET are mainly used to reduce the delay in the transfer of information [9]. For safety applications, the vehicle transmits beacons also known as basic safety messages (BSM), and periodic messages are also known as event-driven. The beacon includes necessary information of the vehicle such as velocity, direction, location, and path history. Event-driven messages are disseminated in the case of emergency in the form of hazard warnings. MAC protocol is mainly used for providing efficient delivery of data packets and applicable to the data link layer.

DSRC uses carrier-sense multiple access/collision avoidance (CSMA/CA) protocol for accessing the channel. Using this protocol, the probability of collision in the MAC frame reduces, however, remains nonzero. As the density of the vehicle increases, the competition for accessing the channel increases, which results in congestion. In order to control the congestion, it is essential to adopt a congestion control algorithm.

In the field of decentralized congestion control (DCC) algorithm [1, 6, 12, 14], most of the paper have used transmit power, transmit rate or either hybrid of both as transmission parameters for controlling congestion. However, bandwidth utilization is not appropriate. To this purpose, we have handled the packets according to the priority queue (PQ) model. This reduces the congestion by categorization of packets according to the priority. Thereafter, an adaptive control algorithm is used to adapt the transmit rate of prioritized messages in order to keep the channel load (CL) below the threshold. This allows all the vehicles to utilize the bandwidth under strict fairness.

This paper contributes to research as follows:

- We have designed a congestion control algorithm by distributing the channel among the vehicles. Fair channel allocation enables proper utilization of bandwidth and solves congestion problem.
- Mathematical equations for message rate, total rate of vehicles are given.

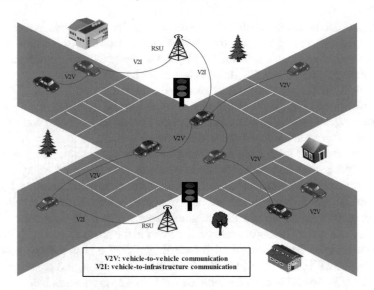

Fig. 1 VANET architecture

– The results show the effectiveness of the proposed algorithm when we compare this with the similar DCC_plain approach.

The rest of the paper is organized as follows. Section 2 discussed the related work. Section 3 described the problem formulation and the proposed algorithm. Section 4 shows the simulation results. Finally, Sect. 5 concludes the paper (Fig. 1).

2　Related Work

This section illustrates the decentralized congestion control (DCC) scheme available in literature.

Schmidt et al. [11] have considered carrier sensing as a parameter for clear channel assessment (CCA). The proposed scheme uses stepwise CCA threshold adaptation (CTA), which measures vehicle waiting time for medium access and sense received power signal. The result discusses the average as well as maximum medium access delay and message queue drops. Authors proposed rate control technique for collaborative road safety applications [4]. He proposed two adaptive message rate control algorithms. One is for low priority and another one for high priority safety messages. The technique continuously calculates the channel load and adjusts the message rate. Proposed provide better utilization of channel for low priority messages and availability of a channel for high-priority messages.

Subramanian et al. [13] designed a TDM overlay on the top of the MAC layer. This design mitigates congestion by controlling transmit power, packet transmission interval, and carrier-sense threshold based on channel load. The authors discuss DCC using the three-state machine and their transitions. According to the state machine, congestion is stated when the channel load (CL) is higher than CL_up for last second and lower congestion when CL_down is lower for five seconds.

Bansal et al. [1] proposed linear message rate integrated control (LIMERIC) algorithm, which controls the message rate to reduce the congestion. LIMERIC is adaptive congestion control that uses binary control variables. It also analyzes noise in the input signal and delays in the update process.

Zemouri et al. [16] proposed a hybrid solution using transmit rate and transmit power called SuRPA algorithm. It has improved requirement of safety messages with high reliability by controlling the transmission parameters of beacons. Proposed algorithm achieves high awareness area and results in better utilization of bandwidth. The authors proposed a new concept of gatekeeper which is inserted above EDCA-based access layer and below the networking layer in the protocol stack [7]. Proposed algorithm adopted contention-based forwarding (CBF) and compared with other protocols like DCF, EDCA, and DCC-gatekeeper. After that, the message sent by CBF to DCC queue is checked for redundancy through duplication list.

Yao et al. [15] proposed loss differentiation rate adaptation scheme (LORA) algorithm, which controls the transmission rate to decrease the congestion. It estimates packet loss rate from each sender and separates the cause of losses from each other like separation of interference losses from fading losses. For the rate adaptation algorithm, the cause of loss helps in selecting the optimal rate, which is very crucial. However, very high data rate decreases collision probability, and low data rate leads to poor channel conditions. Kuhlmorgen et al. [8] compared some basic algorithms like EDCA, DCC plain, robust overhearing recovered algorithm (RORA) and CBF. In a previous paper, it was proved that the limitation of CBF is recovered by introducing control flow w known as DCC advanced. Authors of this paper have optimized the previous result in terms of latency and reliability.

3   Proposed Scheme

In this paper, we assume that there are $N = v_1, v_2, \ldots, v_n$ number of vehicles equipped with global positioning system (GPS) receivers and digital maps. Assume each vehicle generates safety application messages and other application messages along with them. Furthermore, we assume that DCC_{queue} is distributed into three queues, namely $queue_{event}$, $queue_{beacon}$ and $queue_{othermsg}$. All these queues are maintained at MAC sublayer and dequeuing of messages follows the priority queuing model (PQ) and single message is processed in each cycle. In which $queue_{event}$ has assigned highest and $queue_{othermsg}$ has given lowest priority.

The proposed algorithm measures the congestion control by measuring channel load (CL), number of vehicle sharing the same channel. In this scheme, CL is sharing in equal amount by each vehicle and vehicle a chooses its message rate (MR_a) in such a manner that total CL converges below a threshold.

$$MR_a(t) = (1 - c_1)MR_a(t - 1) + c_2(TCL - TMR(t - 1)) \tag{1}$$

$$TMR(t) = \sum_{a=1}^{m} MR_a(t) \tag{2}$$

where c_1 and c_2 are the convergence controlling adaptation parameters, TMR is total rate of m vehicles, and TCL is target channel load. After that, TMR converges to Eq. 3.

$$TMR(t) = \frac{mc_2TCL}{c_1 + mc_2} \tag{3}$$

The TCL is calculated at every t time and message rate is measured according to Eq. 2

Algorithm 1 The proposed algorithm

Input: $queue_{beacon}$, $queue_{othermsg}$, $queue_{event}$
Output: Message dissemination with high transmission rate
1: Upon receiving messages scheduler groups all the messages in queue ($queue_{beacon}$, $queue_{othermsg}$, $queue_{event}$) of same type according to the priority.
2: $queue_{beacon} \leq$ GetMsgNodeBeacon (M,N)
3: $queue_{event} \leq$ GetMsgNodeEvent(M,N)
4: $queue_{othermsg} \leq$ GetMsgNodeOthermsg(M,N)
5: Select $queue_{beacon}$ and $queue_{event}$
6: Message is dequeued according to the priority queuing model
7: Measure CL to calculate the MR
8: **if** $CL \leq TCL$ **then**
9:   apply equation 1 to calculate the message rate of each vehicle.
10:   apply equation 2 to calculate the total message rate of all vehicles.
11: **end if**

---

4 Simulation Results

To illustrate the performance of our proposed algorithm, we have implemented the algorithm in NS2.35 (Network Simulator) and SUMO (traffic simulator). The topology of real vehicle scenario is generated by SUMO by taking .osm file from Openstreetmap of a part of Chandigarh. We have simulated algorithm under bi-directional road scenario. Vehicle travel at the speed of 30–80 Km/h and transmission range is 200 m. The considered simulation network parameters are described in Table 1.

The following metrics are used to check the performance of the proposed scheme.

PDR: It is the ratio of the total messages received to the total messages delivered.

E2E delay: It is the total time taken by message received at destination.

Table 1 Network parameters

Parameters	Values
Area of simulation	2000 m × 2000 m
Time of simulation	200 s
Vehicle density	50–250
Number of lanes	2
Velocity of vehicles	30–80 kmph
Transmission range	200 m
Packet size	256 bytes
DSRC rate	2 Mbps
Radio propagation model	Nakagami-m
Mobility model	Freeway

Transmit frequency: It indicates the frequency by which message will disseminate.

Throughput: The rate by which a message is delivered. It is measured in kbps and calculated by using Eq. 4.

$$Avg.throughput = \frac{received\_packet\_size}{total\_simulationtime} \times \frac{8}{1000} \tag{4}$$

Figure 2 shows the PDR versus density of vehicles. Under low density, the PDR is high because each vehicle take large amount of channel distributions. With the increase in density, vehicle starts gaining small amount of channel distribution and results in chances of congestion. The proposed algorithm improves the PDR as compared to DCC_plain because of the priority model and fair channel distribution to all nodes in the network. Similarly, throughput decreases with increase in number of vehicles and Fig. 5 shows that the proposed scheme attains better throughput than the DCC_plain.

Figure 3 illustrates that the delay increases as the density of vehicles increases. The proposed scheme outperformed with low E2E delay because message transmitted with fast rate. Therefore in the proposed scheme, message reaches at the receiver earlier. Figure 4 describes the transmit frequency along density of vehicles, which decreases as number of vehicle increases because channel distribution will occur among high number of nodes. However, the transmit frequency of the proposed scheme is better than DCC_plain (Fig. 5).

Fig. 2 PDR versus density of vehicles

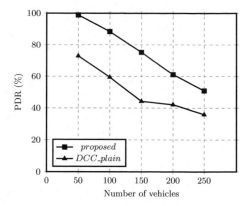

Fig. 3 E2E delay versus density of vehicles

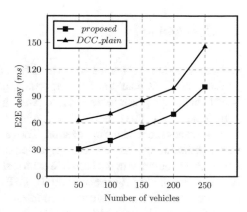

Fig. 4 Transmit frequency versus density of vehicles

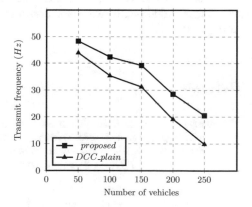

Fig. 5 Throughput versus
density of vehicles

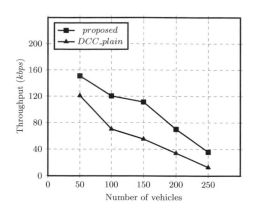

5   Conclusion

Congestion control schemes are needed for efficient V2V communication for safety
applications. DCC is a prominent approach of congestion control at the MAC layer.
It is a challenging task to design a reliable and undelayed DCC scheme under high
vehicle density. In this paper, DCC scheme adapts its transmission parameters such as
transmit rate to control the congestion. The proposed scheme with priority model and
fair distribution of channel load improves congestion control. Proposed algorithm
has been presented with the help of numerical results under high vehicle density. The
results show the improvement in PDR, E2ED, throughput, and transmit frequency
by the proposed scheme as compared to DCC_plain.

In future work, we would enhance the algorithm by adding the meta-heuristic
techniques. Additionally, we would consider more network parameters for analysis
of the work.

References

1. Bansal G, Kenney JB, Rohrs CE (2013b) Limeric: a linear adaptive message rate algorithm for
   DSRC congestion control. IEEE Trans Veh Tech 62(9):4182–4197
2. Chahal M, Harit S (2017) Towards software-defined vehicular communication: architecture
   and use cases. In: 2017 International conference on computing, communication and automation
   (ICCCA). IEEE, pp 534–538
3. Chahal M, Harit S, Mishra KK, Sangaiah AK, Zheng Z (2017) A survey on software-defined
   networking in vehicular ad hoc networks: challenges, applications and use cases. Sustain Cities
   Soc 35:830–840
4. Guan W, He J, Bai L, Tang Z (2011) Adaptive rate control of dedicated short range communications based vehicle networks for road safety applications. In: 2011 IEEE 73rd vehicular
   technology conference (VTC Spring). IEEE, pp 1–5
5. Jiang X, Du DH (2015) Bus-vanet: a bus vehicular network integrated with traffic infrastructure.
   IEEE Intell Transp Syst Mag 7(2):47–57

6. Kenney JB, Bansal G, Rohrs CE (2011) Limeric: a linear message rate control algorithm for vehicular DSRC systems. In: Proceedings of the eighth ACM international workshop on vehicular inter-networking. ACM, pp 21–30
7. Kühlmorgen S, Festag A, Fettweis G (2016) Impact of decentralized congestion control on contention-based forwarding in vanets. In: 2016 IEEE 17th International symposium on a world of wireless, mobile and multimedia networks (WoWMoM). IEEE, pp 1–7
8. Kuhlmorgen S, Festag A, Fettweis G (2017) Evaluation of multi-hop packet prioritization for decentralized congestion control in vanets. In: 2017 IEEE Wireless communications and networking conference (WCNC). IEEE, pp 1–6
9. Menouar H, Filali F, Lenardi M (2018) An extensive survey and taxonomy of MAC protocols for vehicular wireless networks. In: Adaptation and cross layer design in wireless networks, p 6
10. Sakthipriya N, Sathyanarayanan P (2014) A reliable communication scheme for vanet communication environments. Indian J Sci Technol 7(S5):31–36
11. Schmidt RK, Brakemeier A, Leinmüller T, Kargl F, Schäfer G (2011) Advanced carrier sensing to resolve local channel congestion. In: Proceedings of the eighth ACM international workshop on vehicular inter-networking. ACM, pp 11–20
12. Sepulcre M, Gozalvez J, Altintas O, Kremo H (2014) Adaptive beaconing for congestion and awareness control in vehicular networks. In: 2014 IEEE vehicular networking conference (VNC). IEEE, pp 81–88
13. Subramanian S, Werner M, Liu S, Jose J, Lupoaie R, Wu X (2012) Congestion control for vehicular safety: synchronous and asynchronous MAC algorithms. In: Proceedings of the ninth ACM international workshop on vehicular inter-networking, systems, and applications. ACM, pp 63–72
14. Torrent-Moreno M, Mittag J, Santi P, Hartenstein H (2009) Vehicle-to-vehicle communication: fair transmit power control for safety-critical information. IEEE Trans Veh Technol 58(7):3684–3703
15. Yao Y, Chen X, Rao L, Liu X, Zhou X (2017) Lora: loss differentiation rate adaptation scheme for vehicle-to-vehicle safety communications. IEEE Trans Veh Technol 66(3):2499–2512
16. Zemouri S, Djahel S, Murphy J (2014) Smart adaptation of beacons transmission rate and power for enhanced vehicular awareness in vanets. In: 2014 IEEE 17th international conference on intelligent transportation systems (ITSC). IEEE, pp 739–746

Comparative Study of TDMA-Based MAC Protocols in VANET: A Mirror Review

Ranbir Singh Batth, Monisha Gupta, Kulwinder Singh Mann, Sahil Verma and Atul Malhotra

Abstract In recent years, Vehicular ad hoc networks emerge as the promising applications of Mobile ad hoc network. It is specially designed for road safety and comforts to people. It assists vehicles to communicate among themselves and to perceive the road situation such as accidents or traffic jams in their vicinity. This goal can be achieved by using safety applications which can broadcast the warning messages wirelessly between neighboring vehicles informing drivers of any dangerous situation nearby. Vehicles use transmission channels, which is shared medium and neighboring nodes are not allowed to transmit simultaneously because a transmission collision may occur. Therefore, Medium Access Control (MAC) protocols are required to proficiently share the medium and dependable conveyance of messages. Sharing the medium efficiently in VANET is a difficult task due to the special characteristics like high node mobility, frequently changing topology, etc. Various Time Division Multiple Access (TDMA) based MAC protocols have been proposed as it provides bounded delay with less packet loss ratio than other multiple access schemes and experiences no interference from concurrent transmission. In this paper, the comparative analysis of different classification of TDMA-based protocols has presented. The protocols are compared on the basis of different performance metrics such as access collision, merging collision, synchronization, safety services, scalability, etc. Some of the execution measurements are highlighted in the introduction section. In

R. S. Batth · M. Gupta (✉) · S. Verma (✉) · A. Malhotra
Lovely Professional University, Phagwara, India
e-mail: monishagupta2705@gmail.com

S. Verma
e-mail: sahil.21915@lpu.co.in

R. S. Batth
e-mail: ranbir.21123@lpu.co.in

A. Malhotra
e-mail: atul.18011@lpu.co.in

K. S. Mann
GNDEC Ludhiana, Ludhiana, India
e-mail: mannkulvinder@yahoo.com

© Springer Nature Singapore Pte Ltd. 2020
A. Khanna et al. (eds.), *International Conference on Innovative Computing and Communications*, Advances in Intelligent Systems and Computing 1059,
https://doi.org/10.1007/978-981-15-0324-5_10

107

addition to, comparative analysis their design issues, advantages and drawbacks are also discussed.

Keywords MAC · DSRC · V2V · V2I · TDV · TCBT · TCT

1 Introduction

MAC protocols are mainly developed for inter-vehicle communications. Depending on how channels are accessed, MAC protocols are classified into three types—(1) Contention-based medium access method, e.g.,—IEEE 802.11p. (2) Contention-free medium access method. (3) Hybrid method combining the features of above two methods. There is no fixed mechanism used to provide channel access to vehicles, randomly they are assigned slots to transmit which leads to transmission collisions, when density of vehicles increases. Figure 1 delineates the scientific categorization of TDMA-based MAC convention. Contention-based method, i.e., IEEE 802.11p

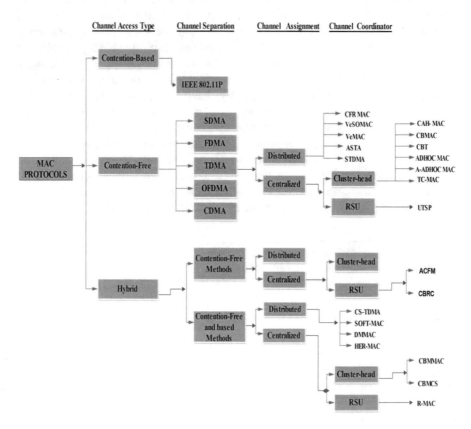

Fig. 1 Taxonomy of TDMA-based MAC protocol [1]

cannot satisfy all the requirements of safety applications. Various mechanisms are proposed to increase the QoS requirements of contention-based access method. In contrast to contention-based, contention-free access method uses a scheduling mechanism for assignment of slots to vehicles which helps them to communicate with other neighboring vehicles. Hybrid protocols associate the advantages of both contention-based and contention-free MAC protocols.

1.1 VANET MAC Protocol Configuration Issues

MAC protocols for VANET must be efficient, reliable and should consider all the key technical challenges:

- **Dynamically varying network topology**—As the nodes are vehicles in VANET, they have high speed and can change their group of vehicles at any time. Due to this, network topology changes frequently.
- **No central coordination**—VANET has no fixed infrastructure, therefore, it has no centralized coordinator and control is equally divided between the vehicles. To avoid collision during channel access, neighboring vehicles transmit control messages. The protocols must ensure that the precious bandwidth should not be consumed by this overhead.
- **Scalability**—MAC protocols are designed in such a way that different traffic load situations, whether it is dense or low are provided with efficient channel utilization.
- **Broadcast support**—MAC protocols must provide broadcast services to announce traffic-related messages over a regional scope.
- **Hidden and exposed node problems**—Broadcast nature of VANET is the primary cause of hidden node problem and control messages like RTS/CTS cannot prevent collisions for broadcast messages.
- **Different QoS requirements**—VANET applications require different QoS requirement. Safety applications require bounded delay as well as having higher priority than any other data messages when broadcasted and infotainment applications require high throughput.
- **Time synchronization**—Vehicles in the network are mostly clock synchronized using a positioning system called as Global Positioning System (GPS), which doesn't operate correctly in all situations like the presence of tunnels, high buildings, etc.

1.2 Performance Metrics

All the existing MAC protocols that are designed for VANETs are assessed based on certain execution measurements. Most commonly used are access delay, packet loss ratio, throughput, stability, support for safety applications and user-oriented applications.

- **Access delay**—It is the average time taken by the vehicle when it begins sending the data until the start of fruitful transmission. The access delay is dependent on both the MAC protocol and the traffic generated by the neighboring node sharing the same channel.
- **Packet loss ratio**—It is characterized as the proportion of number of lost packets to the aggregate number of packets sent. It happens when either control or information packets are neglected to be transmitted effectively.
- **Throughput**—Throughput is the maximum number of successful packets sent over a communication channel, usually measured in bits per second. The main aim of MAC protocol is to increase the throughput of user-oriented applications and decrease the access delay of safety applications.
- **Fairness**—MAC protocols are considered to be fair if during fixed interval, all vehicles gain equal medium access.
- **Stability**—MAC protocols are considered as stable if it can work under different vehicular traffic conditions.

The rest of the paper is stated as follows. Section 2 presents the literature review of the protocols proposed. Section 3 focuses on the different classifications of TDMA-based MAC protocols along with highlighting the differences between each protocol. Section 4 presents the conclusion achieved.

2 Literature Survey

Cao et al. [2] proposed a new protocol for VANETs which is named as VAT-MAC. For transmitting the messages safely, efficient broadcast service is provided by this proposed protocol. Every time frame length is optimized in accurate and adaptive manner by implementing VAT-MAC. The scalability and throughput of the networks are improved as per the experimental results achieved when proposed protocol is implemented.

Ma et al. [3] proposed an enhanced IC-MAC technique for VANETs using which the idle service channels can be utilized. An enhanced approach is proposed to reserve service channel and transmit the service messages on the basis of standard IEEE protocols. In order to transmit the service messages for guaranteeing the QoS of safety messages, the service messages are transmitted using idle service channel of reservation time by the proposed approach. The derivation of throughput and delay of service is done by the proposed approach such that its performance can be evaluated. It is seen that when implementing the proposed approach, it is possible to enhance the throughput and delay performance.

Luo et al. [4] proposed a new MAC protocol for VANETs which is named as sdnMAC. The function of RSU is included by the new node that is designed and named as ROFS. In the form of openflow switches and controllers, the abstraction of

ROFSes is done. MA-ROFS and MA-VEH are the two-tiers amongst which sdnMAC is divided. The cooperative sharing of time slot information is scheduled by the controller and for the vehicles, the transmission range is decided through MA-ROFS. For the allocation of slots, each ROFS acts as a controller within MA-VEH. It is seen here that in comparison to the existing protocols, the performance of proposed approach is better. The merging collisions are eliminated and the packet loss ratio and rate of access collision are reduced as per the simulation results. There is a perfect challenging of vehicular environments by this proposed approach as well. The requirements of cooperative safety are met by the efficient cooperation provided by proposed approach and decoupling of control and data plane.

Lyu et al. [5] proposed a novel approach through which the diverse beacon rates present within the safety applications can be supported and it is named as SS-MAC. For perceiving the occupancy states of time slots online, a circular recording queue is proposed initially. Further, for sharing a particular time slot, the DTSS is designed. RIFF is the random index first fit algorithm that is proposed here such that an appropriate time slot can be chosen such that the maximum resource utilization can be shared. It is seen through the simulations that it is possible to solve all problems observed previously in existing techniques by applying proposed technique. Also, the results achieved are highly accurate and the delay has been reduced so that the overall efficiency of networks can be enhanced. Tianjiao and Qi [6] proposed a GAH-MAC protocol through which the parts of the existing slots are reserved as special slots. For deciding whether the original slot can be reserved or a new one, and utilizing the special slots such that the messages can be broadcasted, the games are played by the colliding nodes. A new modular waiting counter is provided here to maximize the success rate for reservation and ensure fairness within the node reservations. The packet drop rate is minimized and the network throughput is maximized by the proposed protocol. Upadhyay et al. [7] proposed an algorithm to be applied in VANETs to establish a cooperative clustering-based MAC. This is a high priority algorithm through which the transmission delay of safety messages is minimized when there are low and medium traffic scenarios. By choosing the cluster head without causing any delay, a successful packet is received with the cooperative node by the nodes that fail to receive packet on broadcasting. In order to generate clusters and broadcast the packets within mobility conditions, a lane model is proposed in this research. The QoS of networks in the presence of dense traffic conditions will be maintained in future when the high priority algorithm is tried to be implemented.

Khan et al. [8] proposed a distributed hybrid MAC approach which can be applied within VANETs in case of emergency situations. The usage of Control Channel (CCH) is enhanced and the load present on Service Channels (SCHs) is distributed in a uniform manner by the proposed mechanism. The mode is switched from general to emergency for using common service channels by the proposed technique. Therefore, the probability that the messages are being received within the time is increased here. The rate of transmission collisions and latency are minimized by the proposed approach. The probability of delivering messages is high. The simulation of proposed MAC technique is done and for the providing evaluation study, comparisons with the results of existing approaches are made. It is seen that the performance of proposed approach is better.

Nguyen et al. [9] proposed a novel technique for VANETs which is to be applied on the hybrid TDMA/CSMA multichannel MAC. This approach provides efficient time slot acquisition on this hybrid protocol. On the control channel, the throughput is increased along with the removal of excessive control overhead through this proposed protocol. Simulations are performed which show that with respect to the average number of nodes that need a time slot, the performance of proposed protocol is better as compared to the already existing approaches. In case when there is high node density, the information of neighbors is broadcasted with the help of large ANC's payload size needed by the proposed approach.

Priya et al. [10] presented the optimization of MAC layer work's performance by considering the size of queue as an important parameter. With increment in the mobility of vehicle, there is degradation in performance of VANET. It is more challenging to optimize the MAC layer when its mobility is increased. Therefore, for various sizes of queues at different speed levels, the simulative investigations are performed. The AODV routing protocol is considered to be the best as per the simulations performed and results achieved. For this protocol, the queue size of 10 packets has been considered as optimum. In case when the speed of vehicles is high, it is important to consider the longer length of queue sizes as per the inferences drawn in this research. Therefore, for compensating the loss in performance because of the variations in levels of vehicular speed, it is important to include adaptive queue size algorithm.

3   MAC Protocols for VANETs

VANETs are specially intended to serve communication between neighboring vehicles. MAC protocols proposed for VANETs are subdivided into 3 types-contention-based, contention-free and hybrid based on channel access method.

3.1   Classification of TDMA-Based MAC Protocols

- **Protocols working on a completely distributed VANET**: These protocols aim in coordinating access to the channel in a distributed manner.
- **Protocols working in a cluster-based topology**: These protocols elect head from each cluster acting as a coordinator of channel access.
- **Protocols working on a centralized topology**: In this type of protocols, roadside units act as a local coordinator by providing channel access to vehicles within their coverage area.

3.2 TDMA-Based MAC Protocols in a Fully Distributed VANET (TDV)

In a distributed network, we require time-multiplexed protocol; the effectiveness of such protocols is reduced by some issues or collisions which must be addressed, two types of collisions occur mainly access collision and merging collisions. Access collision is caused when vehicles attempt to access the same accessible time slot within the same set of two-hop neighborhood. Merging collision is caused when vehicles from different two-hop sets try to access the same slot and become members of same set.

3.2.1 TDV Protocols

Protocols of fully distributed VANET are listed below where every protocol targets on specific problems with distinct mobility scenarios.

Space-Orthogonal Frequency-Time Multiple Access Control (SOFT-MAC)

It is a protocol for VANET combining all the medium access techniques like FDMA, TDMA, OFDMA, SDMA, and CSMA. The IEEE standards failed to provide higher QoS when we need to exchange, safety-related messages between vehicles and also when the traffic is high in some scenarios like city areas. This problem is mitigated by SOFT-MAC [11]. It provides a higher throughput than IEEE 802.11 but it is quite expensive and complex mechanically due to the use of SDMA, TDMA, and OFDMA mechanism. While using SOFT-MAC various rules and guidelines need to be followed. Vehicles are assumed to be equipped with digital road maps. Thus, these protocols lack efficiency in environment where there is absence of digital road maps.

Self-organizing Time Division Multiple Access (STDMA)

It supports real-time applications. It is currently used in automatic identification systems. Real-time communication has its own requirements like messages should be delivered before deadline, i.e. before accidents with limited error and the communication must be reliable. When the density of node increases in a highway, then CSMA spends a lot of time-solving the requests of channel access, but STDMA being decentralized achieves finite delay and lesser loss of packets. STDMA sometimes results in a packet collision which is very less as compared to that of CSMA. If the sender is using STDMA technique, then in the worst case no packets will be dropped.

Self-organizing MAC Protocol for DSRC Based Vehicular Ad Hoc Networks (VeSOMAC)

VeSOMAC is a contention-free TDMA protocol particularly designed for transmission of safety-related messages between vehicles. It provided efficient communication without the help of the infrastructure designed particularly to relay messages between the vehicles only in the highways. This protocol considers the frequent changes of the topology by the mobile nodes by using a control mechanism named as in-band to exchange slot information [12]. VeSOMAC can be operated in both the modes like-when all the vehicles are time-synchronized with each other using the GPS mechanism and share the equal slot and frame boundaries and also when they are not synchronized having different slot boundaries. It is a fast convergence protocol in dynamically varying topology.

Hybrid Efficient and Reliable MAC (HER-MAC) for Vehicular Ad Hoc Networks

HER-MAC acts as the dynamic slot allocation procedure for VANET. It is designed to provide safety messages to vehicles without any collision in the control channel as well as transmit non-safety messages by efficiently using resources of service channels [13]. It is efficient as well as reliable for transporting messages as compared to the IEEE standard 1609.4. This protocol assumes that the vehicle consists of only one transceiver i.e. half-duplex which can either be transmitting or receiving but not simultaneously. All vehicles are synchronized by time using Global Positioning System. Its architecture and operation are similar to DMMAC. It increases the throughput rate as well as delivery of packets in a large network.

CSMA and Self-organizing TDMA MAC (CS-TDMA)

It represents a multichannel MAC protocol associating CSMA with TDMA and SDMA, solves hidden/exposed terminal problem and a broadcast service which is reliable as well as the ratio between SCH and CCH channels are dynamically adjusted according to density of nodes. As CS-TDMA is self-organized it doesn't require any management from roadside unit. Nodes are synchronized with GPS. It also addressed switching of channel problem concurrently. The utilization of wireless resources is done efficiently and provides guaranteed QoS for safety messages. But the throughput decreases rapidly when the traffic increases, so it is limited to medium dense network.

Near Collision-Free Reservation Based MAC (CFR MAC)

It is a new reservation-based and collision-free based MAC protocol abbreviated as CFR MAC (Collision-Free Reservation based MAC) developed to avoid hidden terminal problem and allow channel access without any collision [14]. It follows the scheduling mechanism same as that of VeMAC protocol, but the slot reservation of the vehicle depends on traffic flow and driving nature of the nodes. CFR MAC provides a reduction in reservation delay and probability of collision compared to VeMAC and IEEE 802.11p.

Adaptive TDMA Slot Assignment (ATSA)

A new TDMA slot assignment protocol called as ATSA (Adaptive TDMA Slot Assignment) proposed for VANETs to improve the efficiency of VeMAC particularly when vehicles move in opposite directions and when densities of vehicles are also not equal [15]. The nodes at a time can access only one slot of a particular frame and frame length can vary according to the neighbor nodes present. In ATSA the nodes acquire slots much faster than VeMAC and ADHOC. It is restricted to only two-lane highways.

Vehicular Ad Hoc Networks MAC (VeMAC)

It represents TDMA-based multichannel MAC protocol providing efficient broadcast services. VeMAC employs new techniques to detect transmission collision and free time slots. It is developed to remove the limitation of ADHOC MAC, as it was not suitable to be used for DSRC architecture due to the use of single-channel [16]. This protocol has certain preliminaries which need to be followed by the node. It minimizes the possibility of access collisions and merging collisions.

Dedicated Multichannel MAC with Adaptive Broadcasting (DMMAC)

A multichannel MAC protocol called as DMMAC (Dedicated Multichannel MAC with Adaptive Broadcasting) developed to provide transmission with less collisions and minimum transmission delay of safety messages for different traffic conditions. Vehicles in DMMAC have half-duplex radio transceiver [17]. It follows the same channel coordination scheme as that of WAVE MAC. The DMMAC adaptive broadcasting mechanism performs well in different traffic conditions. It is highly efficient in transmitting safety data.

3.2.2 Summary of TDV Protocols

Nine TDV types of protocols have been designed. Table 1 presents the comparison and features of these nine protocols. Most of the distributed protocols use the medium which is a single channel and this single channel is shared among all the

Table 1 Comparison of TDMA-based MAC protocols in flat-based network topology [1]

Parameters	VeMAC	HER-MAC	ATSA	SOFT-MAC	DMMAC	VeSOMAC
Access collision	Resolved	Available	Available	Available	Available	Available
Merging collision	Resolved	Available	Resolved	Available	Available	Available
GPS system	Present	Present	Present	Present	Absent	Absent
Synchronization	Yes	Yes	Yes	Yes	No	Yes/No
Channel	Single/multiple	Multiple	Single	Single	Multiple	Multiple
Safety services	Yes	Yes	Yes	Yes	Yes	No
Comfort services	Yes	Yes	No	Yes	N/A	Yes

vehicles for all transmissions, i.e., data, control, and safety transmissions. The protocols using multichannel operation combine various MAC approaches such as TDMA, FDMA, CDMA and SDMA which provides delay-bounded transmission, reliability and higher throughput value which was not possible in single-channel protocol. VeMAC scales well in large network and high load traffic conditions.

3.3 TDMA-Based MAC Protocols in a Cluster-Based Topology (TCBT)

As VANETs lack centralized management; responsible for managing bandwidth and transmission of information, a flat network topology needs to hierarchical network topology, called as clusters. Cluster is a gathering of nodes with comparable attributes; a clustering plan is utilized to partition the vehicles in the gathering. Every cluster has a local management entity known as cluster head (CH) executing intra-cluster communication. Other members of the cluster are called as cluster members (CM). The primary inconvenience of utilizing cluster-based technique include overhead created for selecting the cluster head of entire cluster and managing the members of cluster in highly dynamic varying topology.

3.3.1 TCBT Protocols

Various cluster-based MAC protocols are presented below providing inter-vehicle communications minimizing both intra- and inter-cluster transmission collisions.

AD HOC Medium Access Control (ADHOC MAC)

ADHOC MAC particularly used for inter-vehicle communications. This protocol grouped terminals into clusters, and cluster terminals are internally connected by broadcast radio communication.

Cluster-Based Medium Access Control Protocol (CBMAC)

It is a single channel protocol called as CBMAC (Clustering-based MAC) where the head of the cluster holds the responsibility of assigning bandwidth to all its members and providing fair access to the medium to transmit packets [18]. As bandwidth is already assigned, collisions are greatly reduced which ultimately increases the reliability. It minimizes the hidden and exposed terminal node problem. It provides stability of clusters only when traffic densities are low and medium, but still able to send data reliably in high traffic densities. It is not suitable for DSRC architecture as it has only one channel. It failed to provide communication between vehicle and roadside units.

Clustering-Based Multichannel MAC (CBMMAC)

CBMMAC (Clustering-Based Multichannel MAC) is used for transmission of both safety and warning messages [19]. The three basic functions of CBMMAC are—(1) cluster configuration (2) intra-cluster deployment (3) inter-cluster communication. Clusters are formed only between the vehicles which move in the same direction and each cluster has one head performing all the major functions. It is suited for real-time data transfer only.

A Clustering-Based Multi-channel Vehicle-to-Vehicle Communication System (CBMCS)

It aims to provide protection against accidents and has single data and multiple control channels. Control Channel uses the mechanism of CSMA/CA and TDMA mechanism is implemented in the data channel to provide low transmission delay and prevent collision of packets. This protocol uses a special mechanism known as VAAM (Vehicle Accident Avoidance Mechanism) to inform all the vehicles that are in close proximity to each other about any dangerous situation or accidents.

Adaptive Real-Time Distributed MAC (A-ADHOC)

It is similar to ADHOC MAC. It is mainly designed to be used in large-scale wireless vehicular networks to provide real-time applications [20]. It is highly efficient than ADHOC MAC in utilization of resources of the channel and its response time. It also avoids network failure, even in higher traffic density.

TDMA Cluster-Based MAC (TC-MAC)

It has multiple channels, based on the TDMA mechanism to be used in a cluster-based topology [21]. It is developed to mitigate the issues that are centered in a VANET while exchanging of messages between vehicles such as high mobility of nodes, hidden/exposed terminal problem, a limitation of data rates, etc. It decreases the collision rate and dropping of packets in the channel. It is used in unidirectional traffic but fails in bidirectional traffic.

Cooperative ADHOC MAC (CAH-MAC) for Vehicular Networks

It is mainly designed to maximize the throughput of non-safety applications. Its operation is similar to that of ADHOC MAC, where the access medium consists of frames and frames, have different time slots [22]. It is used to overcome the problem of transmission of packets between sender and receiver which is caused due to the poor channel bandwidth. It can be overcome by a helper node which helps in transmitting the failed packet to the receiver during any idle time slot. But this solution sometimes results in access collision whenever vehicle tries to access the free slot, but that slot is being used by the helper node to relay the lost packet.

Cluster-Based TDMA System for Inter-vehicular Communications (CBT)

It provides efficient communication between inter vehicles. This protocol is used to maintain a reliable and collision-free communication between intra-clusters and inter-cluster. It elects a VANET Coordinator (VC) [18]. Vehicles are synchronized using Global Positioning System. VC decides which slot will be allocated to which member of the cluster. VC is randomly selected and its lifetime is too short which makes the cluster unstable degrading the performance of CBT.

3.3.2 Summary of TCBT Protocols

Eight TCBT types of protocols have been discussed. Table 2 presents the comparison and features of these eight protocols. All these protocols are designed for only highway scenarios but not able to satisfy requirements of urban scenarios. In highways, where vehicles move in opposite directions, CBMMAC is efficient. It is not suitable for higher density as it creates collision due to inter-cluster interference problem. Multichannel protocols (e.g.—TC-MAC, CBMMAC, CBMCS) are preferred than single-channel protocols as it supports wide range of applications, higher density of vehicles and higher throughput performance in different traffic conditions.

Table 2 Comparison of TDMA-based MAC protocols in cluster-based network topology [1]

Parameters	CBMAC	CBT	A-ADHOC	TC-MAC	CAH-MAC	CBMMAC
Access collision	Resolved	Resolved	Available	Resolved	Available	Resolved
GPS system	N/A	Present	Present	Present	Present	Absent
Inter-cluster interference	Available	Available	Available	Resolved	Available	Resolved
Pure TDMA	Yes	Yes	Yes	No	Yes	No
Channel	Single	Single	Single	Multiple	Single	Multiple
Scalability	Low	Very Low	High	Very high	Medium	Low

3.4 TDMA-Based MAC Protocols in Centralized Topology (TCT)

MAC protocols ought to be structured in such a way that it must exploit the variable characteristics of VANET like the presence of infrastructure like RSU, variable speed of vehicles and large transmission range so that messages can be delivered efficiently. Centralized MAC protocols exploit the existence of RSUs for assigning time slots and disseminating control data to lessen the overhead of delay incurred due to channel allocation.

3.4.1 TCT Protocols

Protocols based on centralized topology have been developed providing reliable delivery of messages in VANETs by removing the problem named as access collision caused due to simultaneous access of equivalent time slot.

Adaptive Collision-Free MAC (ACFM)

ACFM assigns slots to vehicles dynamically. It ensures that the node utilizes the slots efficiently. The dynamic assignment of the slots is done by RSUs to vehicles under its coverage. Every frame has two segments—(1) Roadside Unit (RSU) segment (2) Vehicle segment. Roadside Unit segment is used by RSU and vehicle segment consists of 36 data slots which are utilized to send beacon messages. Once a slot is assigned to a vehicle, it cannot be released until it is reallocated. It has limited access delay and less packet loss ratio. It is not efficient to be used in DSRC architecture due to the presence of only one channel and failed to provide QoS for non-safety applications.

Risk-Aware Dynamic MAC (R-MAC)

It is an enhanced version of ACFM for Vehicular Cooperative Collision Avoidance (CCA) system. CCA system is mainly used to deliver warning or safety messages along with beacon messages between vehicles efficiently and reliably [23]. Traditional applications of CCA system has many shortcomings like message collisions due to a large number of vehicles, accidental collisions occur as vehicle become invisible to others due to absence of information about their current position. When the traffic is overloaded, transmission delay is slightly larger and is suitable only for one-lane highway.

Cluster-Based RSU Centric Channel Access (CBRC)

CBRC employs contention-free access method with RSU as the centralized coordinator minimizing channel allocation overhead. Both RSUs and vehicles are operated by CBRC [24]. RSU supports two queues of connection request, for both safety and non-safety application. Queues of safety application are having more priority than non-safety application. It minimizes the hidden and exposed node problem. CBRC do not consider inter-cluster interference. When vehicles are in high speed, it leads to frequent disconnection from one RSU to another which leads to disruption in transmission of messages.

Unified TDMA-Based Scheduling Protocol for V2I Communications (UTSP)

It is intended for transmission of messages between vehicles and RSU. It provides higher throughput value than other protocols. Applications which are throughput-sensitive are supported by UTSP [25]. The protocol performance was evaluated on the basis of single RSU only; therefore, vehicles in the overlapping region suffer from interference problem due to several RSU coordinating access to the channel. It resolves road safety issues.

3.4.2 Summary of TCT Protocols

Four TCT types of protocols have been discussed. Table 3 presents the comparison and features of these four protocols. As known, TCT protocols should consider two issues like inter-RSU interference and short-stay period in RSU region but protocols such as R-MAC and UTSP failed to do so and ACFM and CBRC use different orthogonal frequencies for neighboring RSUs. ACFM and CBRC use both TDMA and FDMA scheme. Even after assignment of fixed frequencies interference doesn't occur. Frequencies are assigned in such that no two neighboring RSU are there. These help in increasing throughput and decreasing the access delay.

Table 3 Comparison of TDMA-based MAC protocols in centralized network topology [1]

Parameters	CBRC	ACFM	UTSP	R-MAC
Access collision	Resolved	Resolved	Resolved	Resolved
GPS system	Present	Present	Present	Present
Inter-RSU interference	Resolved	Resolved	Available	Resolved
Pure TDMA	No	No	Yes	Yes
Channel	Single	Single	Multiple	Single
Scalability	Low	High	Low	High

4 Conclusion

It is a difficult task to define the exact comparison between different classifications of TDMA-based MAC protocols as all the proposed protocols are designed to satisfy different applications of VANET. They use single channel as earlier multichannel concept was not developed. In TDV protocols, there exists three different class of protocols, the first class is efficient in providing real-time applications (e.g.,— STDMA, DMMAC, and ATSA), second class is suitable for transfer of voice and video data (e.g.,—VeSOMAC) and the third class is suitable for safety as well as multimedia applications (e.g.,—VeMAC, HER-MAC). Clustering schemes are judged by their ability to provide better cluster stability. While comparing the transmission power, TC-MAC and CBMAC have medium power and others have negligible power. Large transmission power reduces the count of clusters in the network which ultimately decreases the interference between clusters. All these protocols proposed are designed only for simple highway scenarios but failed for urban scenarios. CBRC and UTSP are enhanced version of centralized MAC protocols. The slot allocation mechanism of R-MAC is most efficient making it scalable. CBRC maintains two queues and provides priorities based on accessibility. TCT protocols require high cost of deployment due to the installation of RSUs, TDV protocols use complex mechanisms compared to other two protocols. TCT protocols of VANETs are given significant consideration over ongoing years but this has limited our research for TDV protocols which expect the topology to be as flat and cannot address the problem in fully distributed VANET due to high speed of vehicles in opposite directions. To lessen interference between covering territories, few protocols make utilization of different access procedures like CDMA, FDMA, etc. which makes them complex protocol. Settling these issues will require more prominent endeavors in the near future. In spite of the impressive research pointing to enhance the execution of MAC protocols in VANETs, no perfect solution has yet been identified that can meet the QoS requirement at the MAC layer and resolve every one of the issues caused by the unique attributes of VANETs.

References

1. https://hal.archives-ouvertes.fr/hal-01211437v1/document
2. Cao S, Lee VCS (2018) A novel adaptive TDMA-based MAC protocol for VANETs. IEEE Commun Lett 22(3)
3. Ma Y, Yang L, Fan P, Fang S, Hu Y (2018) An improved coordinated multichannel MAC scheme by efficient use of idle service channels for VANETs. In: IEEE 87th vehicular technology conference (VTC), June 2018
4. Luo G, Li J, Zhang L, Yuan Q, Liu Z, Yang F (2018) sdnMAC: a software-defined network inspired MAC protocol for cooperative safety in VANETs. IEEE Trans Intell Transp Syst
5. Lyu F, Zhu H, Zhou H, Xu W, Zhang N, Li M, Shen X (2017) SS-MAC: a novel time slot-sharing MAC for safety messages broadcasting in VANETs. IEEE Trans Veh Technol 67(4)
6. Tianjiao Z, Qi Z (2017) Game-based TDMA MAC protocol for vehicular network. J Commun Netw 19(3)
7. Upadhyay A, Sindhwani M, Arora SK (2016) Cluster head selection for CCB-MAC protocol by implementing high priority algorithm in VANET. In: 3rd international conference on electronic design (ICED)
8. Khan S, Alam M, Müllner N, Fränzle M (2016) A hybrid MAC scheme for emergency systems in urban VANETs environment. In: IEEE vehicular networking conference (VNC)
9. Nguyen V, Oo TZ, Chuan P, Hong CS (2016) An efficient time slot acquisition on the hybrid TDMA/CSMA multi-channel MAC in VANETs. IEEE Commun Lett 20(5)
10. Priya K, Malhotra J (2016) Optimisation of MAC layer to mitigate the effect of increased Mobility in VANET. In: 2nd international conference on contemporary computing and informatics (ICCCI), Dec 2016
11. Abdalla GM, Abu-Rgheff MA, Senouci SM (2009) Space-orthogonal frequency-time medium access control (SOFT MAC) for VANET. In: Global information infrastructure symposium (GIIS), vol 1
12. Yu F, Biswas S (2007) A self reorganizing MAC protocol for inter-vehicle data transfer applications in vehicular ad hoc networks. In: 10th international conference on information technology (ICIT), pp 110–115
13. Dang DNM, Dang HN, Nguyen V, Htike Z, Hong CS (2014) HERMAC: a hybrid efficient and reliable MAC for vehicular Ad Hoc networks. In: Proceedings—international conference on advanced information networking applications (AINA), pp 186–193
14. Zou R, Liu Z, Zhang L, Kamil M (2014) A near collision free reservation based MAC protocol for VANETs. In: IEEE wireless communication networking conference (WCNC), vol 2, pp 1538–1543
15. Yang WD, Li P, Liu Y, Zhu HS (2013) Adaptive TDMA slot assignment protocol for vehicular ad-hoc networks. J China Univ Posts Telecommun 20(1):11–18
16. Omar HA, Zhuang W, Li L (2013) VeMAC: a TDMA-based MAC protocol for reliable broadcast in VANETs. IEEE Trans Mob Comput 12(9):1724–1736
17. Lu N, Ji Y, Liu F, Wang X (2010) A dedicated multi-channel MAC protocol design for VANET with adaptive broadcasting. In: IEEE wireless communication networking conference (WCNC), May 2010
18. Günter Y, Wiegel B, Großmann HP (2007) Cluster-based medium access scheme for VANETs. In: IEEE conference on intelligent transportation systems proceedings (ITSC), pp 343–348
19. Hang Su, Zhang Xi (2007) Clustering-based multichannel MAC protocols for QoS provisionings over vehicular Ad Hoc networks. IEEE Trans Veh Technol 56(6):3309–3323
20. Miao L, Ren F, Lin C, Luo A (2009) A-ADHOC: an adaptive real-time distributed vehicular Ad-hoc Networks. In: International conference on communications and networking (ChinaCOM), Xi'an, China, Aug 2009, pp 1–6
21. Almalag MS, Olariu S, Weigle MC (2012) TDMA cluster-based MAC for VANETs (TC-MAC). In: IEEE international symposium on a world of wireless, mobile & multimedia networks (WoWMoM) digital proceedings

22. Bharati S, Zhuang W (2013) CAH-MAC: cooperative ADHOC MAC for vehicular networks. IEEE J Sel Areas Commun 31(9):470–479
23. Guo W, Huang L, Chen L, Xu H, Miao C (2013) R-MAC: risk-aware dynamic MAC protocol for vehicular cooperative collision avoidance system. Int J Distrib Sens Netw 2013
24. Tomar RS, Verma S, Tomar GS (2013) Cluster based RSU centric channel access for vehicular ad hoc networks. Trans Comput Sci XVII 7420:150–171
25. Zhang R, Lee J, Shen X, Cheng X, Yang L, Jiao B (2013) A unified TDMA-based scheduling protocol for vehicle-to-infrastructure communications. In: International conference wireless communications and signal processing (WCSP), Hangzhou, pp 1–6

Node Authentication in IoT-Enabled Sensor Network Using Middleware

Deepak Prashar, Ranbir Singh Batth, Atul Malhotra, Kavita, Varam Sudhakar and Bhupinder Kaur

Abstract Internet of Things (IoT) always has various issues because of its heterogeneous nature and thus is not very effective while accommodating the authentication mechanisms. To resolve this issue of authentication of the nodes involved in a particular application, a middleware has been proposed in this paper. This proposed middleware acts as a gateway between IoT devices and provides the user the ability of node authentication in wireless sensor network (WSN). When a node or a user accesses the network at any time, it is verified by the middleware. It is also capable to handle the heterogeneous messages which are the main challenge in IoT. By executing proposed framework, interoperability issue between IoT networks can be resolved.

Keywords Wireless sensor network · Authentication · Security · Internet of Things

D. Prashar · R. S. Batth · A. Malhotra (✉) · Kavita (✉) · V. Sudhakar · B. Kaur
Lovely Professional University, Phagwara, India
e-mail: atul.18011@lpu.co.in

Kavita
e-mail: dhullkavita@yahoo.in

D. Prashar
e-mail: deepak.prashar@lpu.co.in

R. S. Batth
e-mail: ranbir.21123@lpu.co.in

V. Sudhakar
e-mail: vsudha006@gmail.com

B. Kaur
e-mail: bhupinder.23626@lpu.co.in

© Springer Nature Singapore Pte Ltd. 2020   125
A. Khanna et al. (eds.), *International Conference on Innovative Computing
and Communications*, Advances in Intelligent Systems and Computing 1059,
https://doi.org/10.1007/978-981-15-0324-5_11

1 Introduction

WSN generally consists of nodes having the limited power, computation speed, and memory that disperse the basic information to the main station in the deployed environment. There are some constraints pertaining to security and deployment in WSN which further leads to its limited coverage in the domain of IoT [1]. There is a need of lightweight security protocols for the effective management of the IoT services and devices. Authentication in sensor systems is a challenging task. In conventional systems, that have no constraints with the arrangements of nodes in the area of interest, the public key cryptography (PKC) is more appreciable. The sending node signs the message with its private key and the receiving node checks the genuineness of the message with the public key of the sender shared earlier. But in sensor systems, it is normally expected that PKC cannot be utilized, in light of the asset requirements. Moreover, in the ad hoc network, there is no fixed framework that can be taken into account and even some of the devices may leave the network without any notification. These issues are covered from simple wireless networks to biometrics, cellular telephony, and many other applications. But the major challenge that needs to be addressed in these applications is the authentication of nodes which are involved in the process in hand. So, there is always a requirement of some framework for better authentication so that the applications can be implementing with full confidence and security. Here, first of all, we address the work that has been done in this direction and along with the proposed middleware framework that has been developed in the course of time with its implementation.

2 Background

In this part of the paper, the literature study pertaining to the security mechanisms that are adopted in the domain of IoT has been done.

Jung et al. [2] proposed an improved unknown client validation and key-based method in light of a symmetric cryptosystem in WSNs to address the vulnerabilities that are present in the network. Xiao et al. introduced a biometric verification model to upgrade data security in specially appointed systems. In this approach, there is no requirement for a devoted server as clients can powerfully access the data at any place and at any time. Wong et al. [3] proposed productive client verification technique. It depends on client's secret key and utilizes cryptographic hash. Jiang et al. [4] proposed an appropriated client confirmation approach in WSN. The drawback of this approach is that every node needs to take the permission for the pairwise key and thus cause the additional workload.

Tseng et al. [5] proposed a strict client validation method for remote sensor network based on the elliptic crypto-framework. In this method, key distribution center (KDC) is used in order to authenticate every client. The issue is that clients and nodes need to store a great deal of parameters. Kumar et al. [6] proposed a proficient two-factor client validation system for sensor network, which depends on secret word and utilizes one-way hash function. It additionally experiences synchronization issue as it utilizes time stamp for maintaining a strategic distance from replay attack. Kumar et al. [7] demonstrate technique that has security defects. Client confirmation is a very important step as it guarantees that authentic clients can only be the part of the network. Vaidya et al. [8] propose a user authentication protocol in WSNs. Cheikhrouhou et al. [9] presented a lightweight client validation method in WSNs that gives shared verification and session key utility. The method is executed at two sides; the customer side that controls client's cell phone and server side. Abduvaliev et al. [10] propose straightforward hash-based message verification and trustworthiness code calculation for sensor systems. Benenson et al. [11] develop a client confirmation technique based on cryptography. This method keeps the invalid nodes from getting the information gathered by sensor nodes.

3 Proposed Technique

In order to mitigate the challenges related to the authentication of the nodes involved in the WSN and IoT, we have provided an approach based on middleware deployment for the authentication purpose. This approach is more suited as it covers the main factors that are required for the better validation of the nodes who are involved in the IoT deployment using the WSN. The whole process is divided into various steps like deployment phase, middleware deployment phase, and authentication phase followed by the analysis phase. The simulation is carried in the MATLAB environment.

Deployment Phase
First of all, the deployment parameters are set in terms of the width and height of the network and also the nodes are set up according to the requirements of the application to be run. Once it is done, the nodes are categorized based on their functionality in the given environment.

Deployment Idea

Provide height of deployment area = 1000m
Provide width of deployment area = 1000m
    Set height = 1000m
    Set width = 1000m
    Set N as total node number
    For i = 1 to N
        Draw_node(i)=position(X_i, Y_i)
        Set node_name = N(i)
        Main_Node = random(N)
        End_Node = random(N)
        If Main_Node == End_Node
          Main_Node = rand(N)
          End_Node = rand(N)
        Else
        Main_Node = Main_Node
        End_Node = End_Node
        End
        Set Main_Node as main
        Set End_Node as end
    End

Middleware Deployment Phase

In this phase, we make ensure that each node follows the authentication process set for the application in hand. Once the authentication process is completed, the cover page setup is initiated followed by the calculation of node distance. Later on, the path is finalized using the approach mentioned below.

Area Creation Algorithm

For i = 1 to N
    Ar_set(i) = Area_set(N)
    Ar_list(N, i) = Ar_set(i) **End For**
    i = 1 to N Path(i) = Main_Node
    Path(i) = Main_Node(Ar_se(N))
    If Ar_set(Main_Node) == null Next_node
            = random **End**
    Repeat while end is not found
    Path(last) = End_Node
End

Authentication Phase

Using the IoT system authentication process after route generation, network middleware checks the authentication of the source node and destination node. In this way, each node in the network is validated and the nodes that are not validated are not going to be the part of the process anymore.

Analysis Phase

If attackers are analyzed, we analyze the network and calculate on the basis of iteration the QoS performance metrics such as throughput, delay, bit error rate (BER), execution time, and storage consumption (Fig. 1).

Pseudo Code for the Proposed Middleware:

Define simulation condition and parameters

Rounds   =   R

Nodes =   N

If Source node==Destination node

Warning = '(Please select different source and destination)','communication error')

Else

Set the parameters values for throughput,

delay, error rate and execution time

For Itr = 1: Rounds

x= [200 200 800 800]

y= [200 800 200 800]

plot(x,y)

x-axis =Width of network

y-axis =Height of network

For i=1: numel(x)

Define heterogeneous network sub middleware

xc = x(i) yc = y(i)

radius = 200

theta = linspace(0,2*pi) xr = radis*cos(theta)+xc yr = radis*sin(theta)+yc

plot(xr,yr)

Generate nodes in heterogeneous network

For i1=1: Nodes

radius = 200

theta = rand*(2*pi)

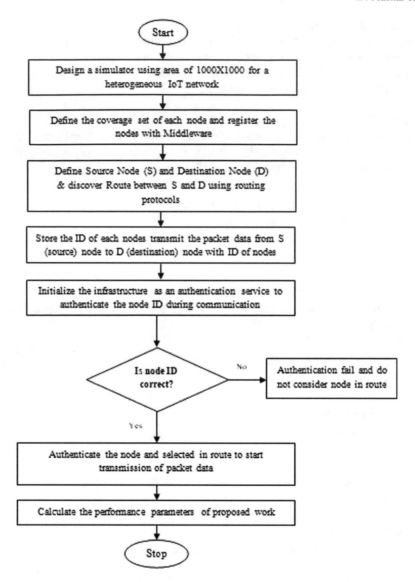

Fig. 1 Work flow of the proposed methodology based on middleware

```
    r = sqrt(rand)*radius  xnn(count) = xc +
    r.*cos(theta)  ynn(count) = yc + r.*sin(theta)
    NodesID= Registration Data
  End End
```

Define main middleware
Bx=500 ,By=500
plot(Bx,By)
Set Possible connection between main middleware and sub middleware

For k=1:4
```
    xx1=[x(k) Bx]
    yy1=[y(k)By]
    Create route
```

End
Check Authentication of Nodes (Unregister Nodes)
For i1=1:10
```
xnn(count) = 1000*rand
ynn(count)   =1000*rand
        plot(xnn(count),ynn(count))
```
End
Create source and destination and define route between source and destination
Authentication registration=0

For k2=1:numel(NodesID)/3
```
        NodeID=find(Source node==str2num(NodesID{k2,2}))
```
        If isempty(NodeID)
```
        NodeID=logical(1)
```
         End
```
        NodeKey=find(Source node==str2num(NodesID{k2,3}))
```
        If isempty(NodeKey)
```
        NodeKey=logical(1);
```
   End End

Return: Simulation Parameters with and without middleware

End

4   Experimental Results

This section presents the proposed algorithm based on middleware in IOT results. Different parameters are used to implement the proposed technique which is mentioned above. Also, the comparison factors taken from the other related work are also mentioned. Here, a graphical user interface (GUI) is developed for the analysis and the execution of the proposed technique.

Figures 2, 3, 4, 5, and 6 show simulation of the proposed technique and its com-

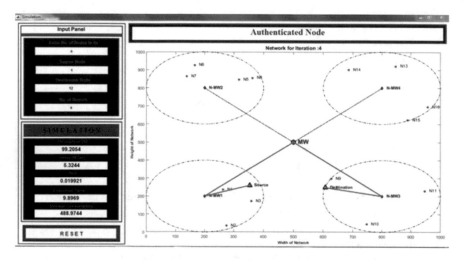

Fig. 2   Simulation of the proposed technique based on middleware

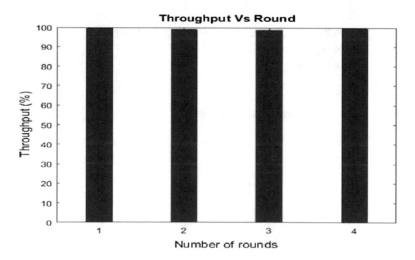

Fig. 3   Throughput comparison with number of rounds

Fig. 4 Delay comparison
with number of rounds

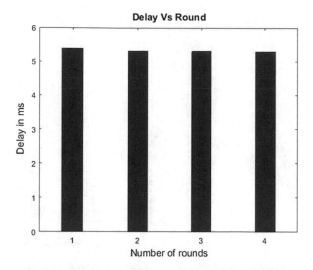

Fig. 5 BER comparison
with number of rounds

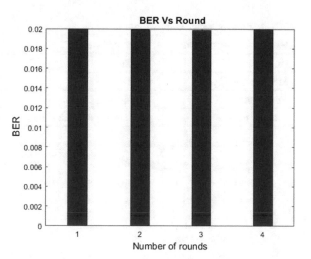

parison in terms of throughput, delay, bit error rate, and storage consumption.

Table 1 and Fig. 7 represent the memory storage comparison that has been done between the proposed work and existing work as mentioned in the paper. It is also evident from the simulation results that the memory storage requirement is more efficient in the proposed work as compared to the other related works and the maximum memory that is required for storage is 487 bytes in case of the proposed work. The proposed work has an average storage of 462 bytes of memory and 132 as compared to the other work as indicated from the results above. This indicates its efficiency and advantage to use for various applications.

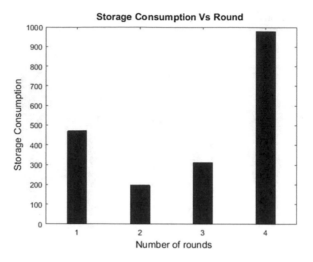

Fig. 6 Storage consumption comparison with number of rounds

Table 1 Metrics for the comparison

Metrics for comparison	Proposed work	Related paper work
Packets count	1000	700
Memory size (bytes)	462	132
Node number	20–100	50
Security triad	Four-way authentication	Four-way authentication

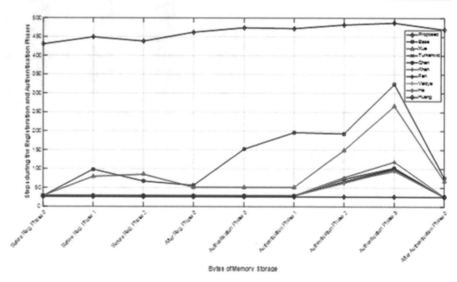

Fig. 7 Parameter comparison of proposed technique

5 Conclusion

Security is a critical network challenges that exists in the present network deployments whether in WSN or in IoT, and thus, the security mechanisms that can be applied on them is in great demand for the easy and efficient implementation of the services. When the IoT comes in the picture, the need of security methods are more required on account of the wide range of applications that are being offered by the IoT. In this paper, a lightweight secure mechanism is being proposed and a framework has been developed on which the applications can be run in smoother way through the deployment of middleware. In this way, the communication is possible in more secure and easier way. The proposed technique is validated through the comparisons along with already existing methods based on the factors like throughput, delay, BER, and storage consumption. The simulation results confirm the efficiency of the proposed approach. The future work will be to integrate the interpretability issues related to the applications by deploying them on the proposed middleware.

References

1. Salman O, Abdallah S, Elhajj IH, Chehab A, Kayssi A (2016) Identity-based authentication scheme for the internet of things. In: IEEE symposium on computers and communication
2. Jung J, Kim J, Choi Y, Won D (2016) An anonymous user authentication and key agreement scheme based on a symmetric cryptosystem in wireless sensor networks. 1–30
3. Wong KH, Zheng Y, Cao J, Wang S (2006) A dynamic user authentication scheme for wireless sensor networks. In: IEEE international conference on sensor networks, ubiquitous, and trustworthy computing, vol 1, p 8
4. Jiang C, Li B, Xu H (2007) An efficient scheme for user authentication in wireless sensor networks. In: 21st international conference on advanced information networking and applications workshops, AINAW'07, vol 1. IEEE, pp 438–442
5. Tseng H-R, Jan R-H, Yang W (2011) A robust user authentication scheme with self-certificates for wireless sensor networks. Secur Commun Netw 4(8):815–824
6. Kumar P, Sain M, Lee HJ (2011) An efficient two-factor user authentication framework for wireless sensor networks. In: 13th international conference on advanced communication technology (ICACT), IEEE, pp 574–578
7. Kumar P, Gurtov A, Ylianttila M, Lee S-G, Lee HJ (2013) A strong authentication scheme with user privacy for wireless sensor networks. ETRI J 35(5):889–899
8. Vaidya B, Sá Silva J, Rodrigues JJ (2010) Robust dynamic user authentication scheme for wireless sensor networks
9. Cheikhrouhou O, Koubaa A, Boujelben M, Abid M (2010) A lightweight user authentication scheme for wireless sensor networks
10. Abduvaliev A, Lee S, Lee YK (2009) Simple hash based message authentication scheme for wireless sensor networks
11. Manivannan D, Vijayalakshmi B, Neelamegam P (2011) An efficient authentication protocol based on congruence for wireless sensor networks. In: International conference on recent trends in information technology, Chennai, India, pp 549–553
12. Xiao Q (2004) A biometric authentication approach for high security ad-hoc networks. IEEE, pp 250–256

An Efficient Data Aggregation Approach for Prolonging Lifetime of Wireless Sensor Network

Bhupesh Gupta, Sanjeev Rana and Avinash Sharma

Abstract In today's environment the aim of wireless sensor network is not restricted to data gathering. But also focus on extraction of useful information. Data aggregation is the term used for extraction of useful information. Data aggregation helps in gathering and aggregating data in energy-efficient way so that network lifetime is heightened. This paper presents a data aggregation approach Mutual Exclusive Sleep Awake Distributed Data Aggregation (MESA2DA). This approach merges with our previous work Mutual Exclusive Sleep Awake Distributed Clustering (MESADC). MESA2DA approach selects cluster head on the basis of MESADC protocol. After that, data aggregation is done by cluster head on its cluster members to remove redundancy, so that the packets delivered to base station are reduced which helps in prolonging lifetime of wireless sensor network. The results obtained with MESA2DA approach are compared with HEED protocols in terms of average energy, delay, throughput, and packet delivery ratio and one finds that the proposed approach is efficient in prolonging lifetime of wireless sensor network.

Keywords Sleep awake · Distributive · Clustering · Sensor · Network

1 Introduction

Wireless sensor networks (WSNs) are composed of networks of lightweight bitsy and slashed devices [1]. These bitsy and slashed devices are known as sensors nodes, which mainly used for monitoring purposes. Sensor nodes preserve path of the substantial criteria such as temperature and humidity. They are in the process of sending

B. Gupta (✉) · S. Rana · A. Sharma
CSE Department, MM (Deemed to be University), Mullana, India
e-mail: bhupeshgupta81@gmail.com

S. Rana
e-mail: dr.sanjeevrana@yahoo.com

A. Sharma
e-mail: sh_avinash@yahoo.com

© Springer Nature Singapore Pte Ltd. 2020
A. Khanna et al. (eds.), *International Conference on Innovative Computing and Communications*, Advances in Intelligent Systems and Computing 1059,
https://doi.org/10.1007/978-981-15-0324-5_12

sensed data to the sink (so-called base station) regularly. A successful wireless sensor network must possess finite and non-rechargeable endurance arrangement of sensors, developing endurance capability, gaining more and more network lifetime, and adjusting the endurance utilization of all nodes [2, 3]. The unadjustable energy utilization is the main problem which arises due to consumption of energy by sensor node via computing unit, communication unit, and sensing unit which need to be figured out to make longer network lifetime [4]. Also, the large portion of energy is used by transceiver. For attaining high energy efficiency in wireless sensor network, one can prefer clustering approach for data aggregation [5]. The reduction in information transformation must be done by following data aggregation [6]. It significantly reduces battery drainage of individual sensor nodes and also simplifies the network management. The topology structure for data aggregation may be classified into flat structure and hierarchical structure. The hierarchical structure has many advantages such as scalability, load balancing, and high efficiency of data aggregation [6–9].

2 Related Work

In order to heighten network lifetime, contemporary mechanisms are being suggested via clustering and data aggregation findings. One of the elementary strategies for prolonging network lifetime via saving energy or endurance consumption of sensor nodes is data aggregation [10]. Redundancy can be removed with the help of data aggregation so that the energy of the sensor nodes is saved and network lifetime can be improved. The major cause for energy consumption is transmission. For that purpose, data aggregation is helpful in reducing battery consumption of the sensor nodes. Clustering can be done in any manner like one can use query clustering [11] for optimization purpose, for making network into connected hierarchy [12], etc. Two categories of clusters are there. Even size clusters and odd size clusters, out of which even size is more efficient due to its energy consumption rate. Energy consumption rate for even is 28% and for odd it is 34% [13]. Hybrid energy-efficient distributed clustering (HEED) [14] comes under the category of distributed clustering protocol. This protocol comes up with two factors: a hybrid of energy and communication cost, for the selection of perfect cluster head. In this distributive protocol, the choice of cluster head is bestow to a merger of two clustering parameters. The initial one is the residual energy of all sensor nodes and the second one is the intra-cluster communication cost. A node that is having high residual energy has better chance of becoming cluster head. For breaking ties, intra-cluster communication cost comes into picture that is nodes that are common to more than one CH. HEED was proposed because it is capable of prolonging network lifetime because of distributing energy consumption, clustering process eliminates within perpetual iterations.

MESADC [15] protocol uses mutual exclusion algorithm in sleep awake mode to fetch cluster head over communication range. Message passing, sensor succession, Sleep awake mode, coeval endurance, and range of communication are the bases of MESADC. Cluster head formed with this protocol helps in prolonging lifetime of

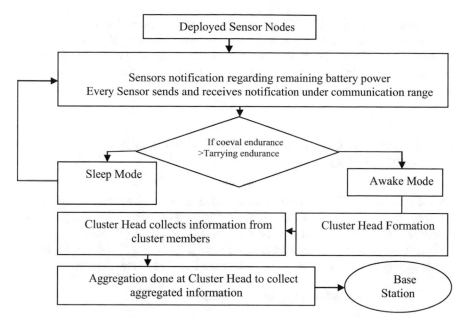

Fig. 1 Working of MESADC

WSN. Criteria for sensor node division is if coeval endurance of any node is greater than all tarrying nodes endurance at that succession, then that sensor node is used for cluster head formation. Figure 1 shows the working procedure of MESADC protocol. For simulation purpose, MATLAB is used. Experimental results of MESADC protocol show enhancement in network lifetime. The basis for results in MESADC protocol is a number of nodes becoming dead versus the number of rounds taken at different range of communications.

3 Methodology

Data aggregation is all about how to combine data which are collected from different sources so that it can be carried out in a number of ways. Eradication and illustration of useful data is the primary key for data aggregation. Proposed MESA2DA approach is capable of removing indeterminateness of data. The indeterminateness arises due to emulous nature of sensors in wireless sensor network, due to which they can sense the same phenomena at the same time. In our proposed MESA2DA approach, cluster head will be chosen on the basis of our previous MESADC work in which cluster head knows how many sensors are under his belt from which it is getting data. For eradication and illustration of useful data, cluster head will calculate standard deviation. The blended data packet is committed up to the base station; so, the base station is awake of the values being generated by the sensor nodes. This access provides decreasing data redundancy at the base station such that only demanding data are delivered along the network which further diminish energy utilization, dispose of time and adaptability of the network. Given below are criteria for our proposed approach MESA2DA.

3.1 Criteria for Cluster Head Selection in Proposed MESA2DA Approach

For the proposed MESA2DA approach at initial stage, each sensor node possesses the same energy in a network. Now, cluster head is selected by using our previously proposed MESADC protocol as shown in example. Beginning of track

Let randomly deployed sensor nodes = 08, let it be $S1, S2, S3, S4, S5, S6, S7, S8$.

Let endurance of each node is 0.4 at the time of deployment. Therefore,

$$S1 = S2 = S3 = S4 = S5 = S6 = S7 = S8 = 0.4$$

For identification of nodes, all of them have unique number. Let unique number for $S1 = UN1$, $S2 = UN2$, $S3 = UN3$, $S4 = UN4$, $S5 = UN5$, $S6 = UN6$, $S7 = UN7$, $S8 = UN8$.

Now, under range of communication, Initially, $UN1 = 0.4$, $UN2 = 0.4$, $UN3 = 0.4$, $UN4 = 0.4$, $UN5 = 0.4$, $UN6 = 0.4$, $UN7 = 0.4$, $UN8 = 0.4$.

Now, under range of communication sensor nodes starts sending notifications regarding coeval endurance to each other sensor. This process continues until every sensor under range of communication sends and receives notification. During this process, there is some loss in endurance for each sensor node.

Let it be $UN1 = 0.1$, $UN2 = 0.2$, $UN3 = 0.11$, $UN4 = 0.25$, $UN5 = 0.35$, $UN6 = 0.32$, $UN7 = 0.22$, $UN8 = 0.19$.

Now, every node has a succession of coeval endurances.

For UN1 (0.1), succession of remaining nodes is

$UN2 = 0.2$ $UN3 = 0.11$ $UN4 = 0.25$ $UN5 = 0.35$ $UN6 = 0.32$ $UN7 = 0.22$ $UN8 = 0.19$

For UN2 (0.2), succession of remaining nodes is

$UN1 = 0.1$ $UN3 = 0.11$ $UN4 = 0.25$ $UN5 = 0.35$ $UN6 = 0.32$ $UN7 = 0.22$ $UN8 = 0.19$

For UN3 (0.11), succession of remaining nodes is

$UN1 = 0.1$ $UN2 = 0.2$ $UN4 = 0.25$ $UN5 = 0.35$ $UN6 = 0.32$ $UN7 = 0.22$ $UN8 = 0.19$

For UN4 (0.25), succession of remaining nodes is

UN1 = 0.1 UN2 = 0.2 UN3 = 0.11 UN5 = 0.35 UN6 = 0.32 UN7 = 0.22 UN8 = 0.19

For UN5 (0.35), succession of remaining nodes is

UN1 = 0.1 UN2 = 0.2 UN3 = 0.11 UN4 = 0.25 UN6 = 0.32 UN7 = 0.22 UN8 = 0.19

For UN6 (0.32), succession of remaining nodes is

UN1 = 0.1 UN2 = 0.2 UN3 = 0.11 UN4 = 0.25 UN5 = 0.35 UN7 = 0.22 UN8 = 0.19

For UN7 (0.22), succession of remaining nodes is

UN1 = 0.1 UN2 = 0.2 UN3 = 0.11 UN4 = 0.25 UN5 = 0.35 UN6 = 0.32 UN8 = 0.19

For UN8 (0.19), succession of remaining nodes is

UN1 = 0.1 UN2 = 0.2 UN3 = 0.11 UN4 = 0.25 UN5 = 0.35 UN6 = 0.32 UN7 = 0.22

Now, apply criteria for sensor division in sleep awake mode:

If coeval endurance of any node > all tarrying nodes endurance at that succession

For UN1, UN1 < all tarrying nodes endurance at that succession; for UN2, UN2 < all tarrying nodes endurance at that succession; for UN3, UN3 < all tarrying nodes endurance at that succession; for UN4, UN4 < all tarrying nodes endurance at that succession; for UN5, UN5 > all tarrying nodes endurance at that succession; for UN6, UN6 < all tarrying nodes endurance at that succession; for UN7, UN7 < all tarrying nodes endurance at that succession; and for UN8, UN8 < all tarrying nodes endurance at that succession.

After applying sleep awake criteria on all successions, we find only UN5 satisfies the condition. Therefore, UN5 comes under the category of awake mode.

Operations perform in awake mode:

Node becomes cluster head, collects information from cluster members, performs aggregation, and sends aggregated information to base station.

In above example, node S5 comes under the category of awake mode and performs all operations of awake mode.

3.2   *Life Road of MESA2DA Approach*

See Fig. 2.

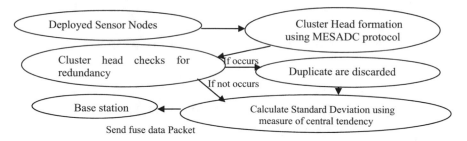

Fig. 2 Life road of MESA2DA

3.3 Criteria for Aggregation and Cluster Head Data Collection in Proposed MESA2DA Approach

Cluster head assigns each node a time slot via TDMA schedule so that node can transmit sensed data. Now, after collection of sensed data, cluster head performs the following operation. Firstly, check for repetition of data until timeout occurs, if found must be removed. After that, perform descriptive statistics on collected data by using the measure of central tendency. The measures of central tendency have twofold values. First, it is an "average" which represents all of the scores made by group and gives the concise description of the performance of the group as a whole and secondly, it enables us to compare two or more groups in terms of some typical performance. Calculate variance with the help of average data and perform the standard deviation to know how much data deviate from average. Now, data are sent to the base station via pre-limit and post-limit. Pre-limit is a difference of calculated mean and standard deviation and post-limit is sum of calculated mean and standard deviation. Figure 2 is the life road of MESA2DA approach.

3.4 MESA2DA Algorithm for Cluster Head Formation and Data Collection

Number of sensors $= n$
All sensors have unique number so
$UN_i =$ Unique number of node I, $T_a =$ Nodes in Awake mode, $T_s =$ Nodes in Sleep mode, $T_t =$ Nodes in Transmitting mode, $T_r =$ Nodes in Receiving mode, $E_t =$ Node energy during transmission, $E_r =$ Node energy during receiving
$R_f =$ Frequency radius, $S_i = i$th sensor's succession
Track 1: Layout Cluster formation (n), For each next track, For each UN_i
Counter $= 0$: For each UN_j of T_t and T_r within R_f of UN_i

   If $E_r \leq E_t$ Notify E_i
   For each UN_i, Put all incoming notifications from sensor j into S_i

For each UN_i, While S_i is not empty
If $E_i \geq E_j$ Put nodes in T_s

Counter $= 1$ Else Put nodes in T_a

For each UN_i, If Counter $= 0$
T_a send Cluster head declaration to UN_j under R_f

Track 2: For each d_i, If$(d_i = d_{i+1})$ Add d_i to $A[]$

Else
Add d_i and d_{i+1} to $A[]$

3.5 Data Aggregation by Proposed MESA2DA Approach

The proposed MESA2DA approach performs aggregation by firstly checking for reiteration of information as far as meantime occurs, if raise must be evacuated and after that using measures of standard tendency (standard deviation and mean) so that information packet size will decrease. Hence, in proposed approach, the cluster head in lieu of dispatching all the sensed information of sensor nodes, it will just send the post-limit and pre-limit of the information sensed by the sensor nodes to the base station as shown below

$$\text{Calculate } \mu = 1/n \sum_{i=1}^{n} d_i, \quad \text{calculate } v = 1/n \sum_{k=0}^{n} (d_i - \mu)^2 \text{ and}$$
$$\text{calculate } \sigma = \sqrt{v}$$

Calculates the post-limit and pre-limit by using $\mu + \sigma$ and $\mu - \sigma$, respectively. Where $d_i =$ sensed value, $\mu =$ mean, and $\sigma =$ standard deviation.

3.6 Communication Energy

Communication energy plays a vital role in our proposed protocol. Energy/bit is used for communication. Transmission energy E_{tx}, electronics energy E_{elect}, transmission amplification energy E_{tx_amp}, receiving energy E_{rx}, receiving energy dissipation, and transmission energy dissipation must be taken into consideration. Table 1 shows usage of communication energy. If k bits are there having distance d, then

Transmission energy $= E_{tx}(k)$,
Transmission amplification energy $= E_{tx\_amp}$
Transmission energy dissipation : $E_{tx}(k) = k * E_{elect} + E_{tx\_amp}(k, d)$

Table 1 Parameters for performing MESADC

Parameters	Symbol	Value
Number of nodes	n	100
X-value for plot area	X_m	100
Y-value for plot area	Y_m	100
Initial energy of sensor	E_o	0.05
Transmission energy	$E_{tx}()$	50 * 0.000000000001
Data aggregation energy	EDA	05 * 0.000000000001
Energy used for advertisement	E_{adv}	50 * 0.000000000001
Amplification energy	E_{fs}	10 * 0.000000000001
Receiving energy	$E_{rx}()$	50 * 0.000000000001

$$E_{tx\_amp}(k, d) = k * d2 * E_{mp}, \; E_{tx\_amp}(k, d) = k * d4 * E_{fs}$$
$$\text{Receiving Energy dissipation}: E_{rx}(k) = k * E_{elect}$$

4　Experimental Results

For executing proposed MESA2DA approach, MATLAB is used. The following parameters are taken into consideration:

In this segment, we present the simulation results of our proposed approach MESA2DA. The results are obtained by simulation using MATLAB. Hundred nodes are randomly deployed in a $100 \times 100 \text{ m}^2$ area. A node becomes lifeless and is not able to forward or receive data when it runs out of energy. Figure 3 shows how clusters

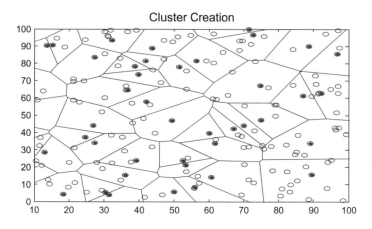

Fig. 3 Creation of clusters

are created in MESA2DA approach. For obtaining results of MESA2DA, following set of performance menstruations are suggested on the basis of delay, packet delivery ratio, throughput, and average energy. Results of proposed MESA2DA are compared with well-known HEED protocol. Average energy deals with endurance consumed by the network. Figure 4 shows average energy for both the protocols: well-known HEED and proposed MESA2DA protocol. The comparison graph is shown which is on the basis of average energy versus each round. Comparison graph shows average energy in case of MESA2DA is higher as compared with existing HEED protocol. Delay or we can say standard discontinuation copes with average lag time of a packet between broadcast from origin and acquiring at the terminal. Figure 5 shows delay for both the protocols: well-known HEED and proposed MESA2DA protocol. The comparison graph is shown which is on the basis of delay versus each round. Comparison graph shows delay in case of MESA2DA is less as compared with existing HEED protocol. Packet delivery ratio deals with amount among sum of data packets received by terminal and the sum sent by origin. It describes the percentage of the packets that grasp the destination. Figure 6 shows packet delivery ratio for both the protocols: well-known HEED and proposed MESA2DA protocol. The comparison graph is shown which is on the basis of packet delivery ratio versus each round.

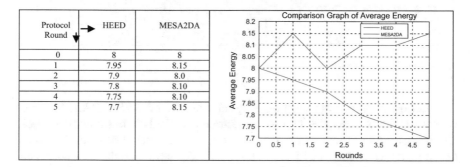

Protocol Round	HEED	MESA2DA
0	8	8
1	7.95	8.15
2	7.9	8.0
3	7.8	8.10
4	7.75	8.10
5	7.7	8.15

Fig. 4 Average energy for each round

Protocol Round	HEED	MESA2DA
0	2.3	0
1	3	1
2	3.3	1.2
3	3.8	2.4
4	4	2.8
5	4.5	3.1

Fig. 5 Delay for each round

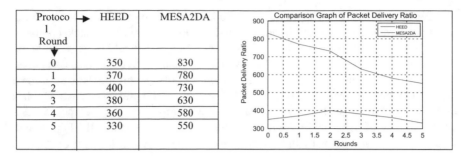

Protocol Round ➡	HEED	MESA2DA
0	350	830
1	370	780
2	400	730
3	380	630
4	360	580
5	330	550

Fig. 6 Packet delivery ratio for each round

Protocol Round ➡	HEED	MESA2DA
0	1.1	1.9
1	1	1.7
2	1.05	1.5
3	1	1.5
4	0.95	1.3
5	0.9	1.1

Fig. 7 Throughput for each round

Comparison graph shows packet delivery ratio in case of MESA2DA is higher as compared with existing HEED protocol. Throughput deals with sum of data packets sent per entity of time. Also, it deals with average amount of fruitful message delivery over a link channel. Figure 7 shows throughput for both the protocols: well-known HEED and proposed MESA2DA protocol. The comparison graph is shown which is on the basis of throughput versus each round. Comparison graph shows that throughput in case of MESA2DA is higher as correlated with existing HEED protocol.

5 Conclusion and Future Scope

This paper reported with a new approach Mutual Exclusive Sleep Awake Distributed Data Aggregation (MESA2DA). MESA2DA approach is capable of gathering and exemplary data into an erect with reduced packet size and data recurrence. Also, it is expected to crop gratifying results for the excellent use of the battery power of the sensor nodes deployed in a region. The proposed technique is modeled on the basis of MESADC protocol, which was our previously proposed protocol. Then, our proposed approach is successfully simulated and compared with HEED protocol. We discuss regarding the energy constraint in wireless sensor network. Replenishment of endurance for sensor node is a big challenge in WSN as these nodes once deployed may not be physically usable. Huge data packet transmission increases overhead and absorbs more energy. Simulation result shows our proposed approach is more energy-efficient than HEED in prolonging network lifetime.

References

1. Gupta B, Rana S (2018) Energy conservation protocols in wireless sensor network: a review. IJFRCSCE 4(3):16–20
2. Heinzelman WR, Chandrakasan AP, Balakrishnan H (2000) Energy-efficient communication protocol for wireless micro sensor networks. In: Proceedings of the 33rd annual Hawaii international conference on system sciences, pp 1–10
3. Ahlawat A, Malik V (2013) An extended vice cluster selection approach to improve V-leach protocol in WSN. In: 3rd international conference on advanced computing and communication technologies. IEEE
4. Liao Y, Qi H, Li W (2013) Load-balanced clustering algorithm with distributed self-organization for wireless sensor networks. IEEE Sens J 13(5)
5. Villas LA, Boukerche A, Ramos HS (2013) DRINA: a lightweight and reliable routing approach for in-network aggregation in wireless sensor networks. IEEE Trans Comput 62(4)
6. Mantri D, Prasad NR (2013) Grouping of clusters for efficient data aggregation (GCEDA) in wireless sensor network. In: 3rd IEEE international advance computing conference
7. Li G, Wang Y (2013) Automatic ARIMA modeling-based data aggregation scheme in wireless sensor networks. EURASIP Wirel Commun Netw
8. Abirami S (2019) A complete study on the security aspects of wireless sensor networks. In: Bhattacharyya S, Hassanien A, Gupta D, Khanna A, Pan I (eds) International conference on innovative computing and communications. Lecture Notes in networks and systems, vol 55. Springer, Singapore
9. Devi DC, Vidya K (2019) A survey on cross-layer design approach for secure wireless sensor networks. In: Bhattacharyya S, Hassanien A, Gupta D, Khanna A, Pan I (eds) International conference on innovative computing and communications. Lecture Notes in networks and systems, vol 55. Springer, Singapore
10. Ji P, Li Y, Jiang J, Wang T (2012) A clustering protocol for data aggregation in wireless sensor network. In: International conference on control engineering and communication technology
11. Gupta B et al (2012) A review on query clustering algorithm for search engine optimization. IJARCSSE 2(2)
12. Younis O, Krunz M, Ramasubramanian S (2006) Node clustering in wireless sensor networks: recent developments and deployment challenges. IEEE
13. Ruperee A, Nema S, Pawar S (2014) Achieving energy efficiency and increasing network life in wireless sensor network. IEEE
14. Younis O, Fahmy S (2004) HEED: a hybrid, energy-efficient, distributed clustering approach for ad hoc sensor network. IEEE Trans Mob Comput 3(4)
15. Gupta B, Rana S (2018) Mutual exclusive sleep awake distributive clustering (MESADC): an energy efficient protocol for prolonging lifetime of wireless sensor network. Int J Comput Sci Eng 1–7

A Comprehensive Review of Keystroke Dynamics-Based Authentication Mechanism

Nataasha Raul, Radha Shankarmani and Padmaja Joshi

Abstract Keystroke dynamics, also called keystroke biometrics or typing dynamics, is a biometric-based on typing style. Typists have unique typing patterns that can be analyzed to confirm the authenticity of the user. Keystroke dynamics is most often applied in situations where the authenticity of a user must be ascertained with extreme confidence. It could be used as an additional degree of security for password-protected applications. If user's password is compromised, and the keystroke dynamics of the real user is known, the application may be able to reject the impostor despite having received valid credentials. Different types of keyboards and remote access are major problems of keystroke dynamics authentication technique. In this paper, a comprehensive analysis of contemporary work on keystroke dynamic authentication mechanisms is summarized to analyze the effectiveness of various methodologies in present. Also, various statistical-based and machine learning-based algorithms are analyzed with their strengths and weaknesses. From this survey, it was observed that there is a need to strengthen the keystroke dynamics dataset which has all essential features. Also an efficient algorithm is required to obtain high accuracy to make authentication effective, as the performance of biometric keystroke authentication is still an open research.

Keywords Keystroke dynamics · Machine learning · Template update · Feature extraction · Classification

N. Raul (✉) · R. Shankarmani
Sardar Patel Institute of Technology, Mumbai, India
e-mail: nataasharaul@spit.ac.in

R. Shankarmani
e-mail: radha_shankarmani@spit.ac.in

P. Joshi
C-DAC, Mumbai, India
e-mail: padmaja.cdac@gmail.com

© Springer Nature Singapore Pte Ltd. 2020   149
A. Khanna et al. (eds.), *International Conference on Innovative Computing and Communications*, Advances in Intelligent Systems and Computing 1059,
https://doi.org/10.1007/978-981-15-0324-5_13

1   Introduction

Generally, the biometric-based authentication system is categorized into two types. First type of authentication works based on physiological characteristics such as fingerprint, iris, voice, and face recognition. This authentication mechanism suffers from concerns such as real-world attacks that may affect their biometric security deployment [1–3], cost for additional device requirement [4] and misinformation user acceptance that makes the user to be aware always with their changes properties [5]. These drawbacks enforce us to move to another authentication system which is based on behavioral characteristics that may not be same for everyone such as typing speed, mouse dynamics, swipe gestures, etc. This new keyboard-based passwordless authentication system is known as keystroke dynamics or typing dynamics. Since it doesn't require any additional hardware devices. This could be deployed in the system by simply installing and upgrading a software.

The features of the keystroke dynamics are extracted based on the timing information of the KeyPress/Release/Hold events. The keystroke dynamics is defined as the process of analyzing the way of user typing. The keystroke dynamics recognizes the activities of keyboard for identifying the unauthenticated access, which depends on the typing way of different people. Naturally, it is used for the computer security, which analyzes the latency between the keystrokes, typing durations, finger placement, pressure on keys, etc. Authentication is performed based on the typing rhythm of a person. Typically, the keystroke verification technique can be classified into static and dynamic. In static verification, the analysis takes place only at particular incidences such as login. In dynamic verification, the user's typing behavior is continuously monitored and analyzed. In order to handle different types of keyboards and remote access, an enhanced dynamic keystroke authentication system is required. When the user frequently uses the same password, there will be the variation in the typing rhythm and speed [6]. The major benefit of using keystroke dynamics is, it detects the imposter's based on the analysis of typing rhythms. Keystroke dynamics suffers from user vulnerability to exhaustion, vigorous variation in typing forms, injury, typing ability of the user, variation of keyboard hardware, etc.

This paper is organized as follows. Section 2 analyzes the background and various process steps of keystroke dynamics authentication system. Section 3 exhibits the data collection and feature extraction methods which keep record of typing pattern based on acknowledged keyword, idiom or some other predetermined text and feature selection. Section 4 describes various classification approaches that have been used in analyzing keystroke dynamics. Section 5 delivers the metrics and evaluation carried out for keystroke dynamics authentication system with various datasets and Sect. 6 concludes the survey with overall observation.

2 Background of Keystroke Dynamics

The keystroke dynamics authentication system is characterized into two varieties called static and dynamic authentication system. Dynamic authentication system is further categorized as periodic and continuous authentication system.

The static keystroke dynamics works with a single known phrase such as password or username. The degree of resemblance between typing estimation of the registered static password and the current typing rhythm of the same password is compared and based on the comparison the acceptance or rejection of the user is determined. The profile could be built rigidly around the timing of the training attempts of the same phrase [7]. Combining username and password helps in strengthening the static authentication mechanism [8]. M. S. Obaidat et al. provide a comprehensive evaluation related to static keystroke recognition on a personal data [9]. The user may be rejected by this approach due to inconsistent typing speed and will require multiple login attempts to get authenticated.

Periodic dynamic authentication system is introduced in order to overcome multiple login attempts. The technique also helps to overcome the limitation of requiring specific or known text for authentication [10]. Periodic authentication can be applied to varied inputs. This method is not reliant on accessing of definite text and is capable to accomplish authentication on any input.

Continuous keystroke analysis is an enhancement to periodic keystroke analysis. It captures keystroke features during the entire duration of the login session. In this method, the imposter can be detected at the earlier stage than under a periodically monitored implementation [11, 12]. The major disadvantage of this approach is additional processing. It makes the approach costlier and affects the performance of the system. Continuous keystroke analysis will be useful when the user uses the keyboard after the login procedure for different activities such as to use the internet, to type the document, to chat, etc. whereas periodic analysis is preferred when the time to monitor the user is very short like the time period when the user enters username and password.

A. *Performance Parameters*:

Keystroke dynamic approaches are evaluated based on False Acceptance Rate (FAR) and False Rejection Rate (FRR). False Acceptance Rate (FAR) is defined as percentage of identification instances in which the biometric security system falsely accept an access attempt by an unauthorized user. False Rejection Rate (FRR) is defined as percentage of identification instances the biometric security system will falsely reject an access attempt by an authorized user.

B. *Process Flow of Keystroke Dynamic*:

The process flow followed in keystroke dynamics-based authentication system is described ahead. The process follows the steps mentioned in Fig. 1 and is explained further in this section. Once feature vectors are extracted to represent the typing characteristics, they are then classified for authentication and identification purposes.

Fig. 1 General process flow
of keystroke dynamics
authentication system

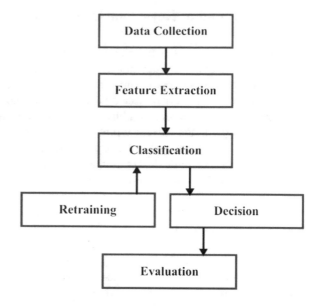

Ali et al. studied various verification approaches where the extracted features are to make an authentication decision [13].

- **Data collection** involves capturing of Keystroke dynamics features of an individual. These features are stored in the profile database at the time of registration. This collected dataset helps train the system. However, before doing the data collection it is important to identify the features to be captured. In Fig. 1, it is a pre-requisite and is considered as a default.
- **Feature extraction** is about creating a user profile based on the feature set collected for an individual in the data collection phase.
- **Classification** phase is about comparing the captured keystroke features with the stored profile and generate the decision. For keystroke dynamics-based authentication, the classification will be in binary form means either authenticated or not.
- **Evaluation** could be done for both authentication, i.e., verification and identification.

This section focused on understanding various aspects of authenticating users based on keystroke dynamics. Various aspects of keystroke dynamics using machine learning are explored in this survey. Different existing approaches and their virtues and imperfections are also analyzed.

C. *Template Update Mechanism*:

Keystroke dynamics eliminates the requirement of an additional device; however, the approach has its own challenges. As mentioned earlier section, the user profile or template is created during the training phase. User typing speed being inconsistent, the template needs to be updated. The template is defined during the training and is dependent on multiple factors such as user's comfort in using the keyboard, familiarity with password.

As the typing skill of the user improves or changes, the data collected at the time of registration may be different from the initial data template computed in the earlier or recent stage. Change in the typing rhythm of the user results in authentication failures which degrade the performance of the system. To overcome this issue template update mechanism is recommended. In this mechanism, the classifier automatically adapts user's current data through learning methodologies like supervised or semi-supervised. This helps in decreasing deviation caused due to template aging [14, 15]. Thus to improve the system performance template update mechanism should be adopted in the process. Machine learning techniques are used for updating template. The machine learning techniques suffer due to the retraining of classifiers, at the time of template updating [16]. So, it is essential to check whether there is a probability to measure the user's typing pattern deprived of entirely retraining the whole classifiers.

During the user profile generation, some of the rhythms will outfall and become noise, which degrades the verification performance [17]. These outliers should be removed based on the global keystroke rhythms or some other valid approaches like pairwise user coupling [18]. The pattern profile size for the pattern formation has a great influence on the social biometric investigation [19].

3 Keystroke Dynamics Features

Keystroke dynamics of users is collected initially for creating user profiles for every individual as explained earlier. The keystroke dynamics evaluates the typing pattern of users and matches it with the relevant profile record. In order to achieve these different features are extracted from a user's particular rhythm. In this section, we summarized various features researchers have explored and studied.

The keystroke dynamics is an analysis of a sequence of key events and the relevant time that everyone befalls. The event could be either a KeyPress or KeyRelease. Some of the most commonly extracted features are based on KeyPress and KeyRelease. In Fig. 2, KeyPress event is shown as P and KeyRelease events are shown as R. These features are explained ahead.

- Dwell time is estimated based on the latency between the KeyPressing (P) and KeyReleasing (R) events. The feature is represented by D1, D2 and D3 in Fig. 2.
- Flight time is the difference between the current KeyRelease (R) and the next KeyPress (P). The same is shown by F1 and F2 in Fig. 2.
- Wordgraphs is the distance from the first KeyPress to the last KeyRelease of events on a single word. Thus, if the word is WELCOME, then the time difference between P of W and R of E.

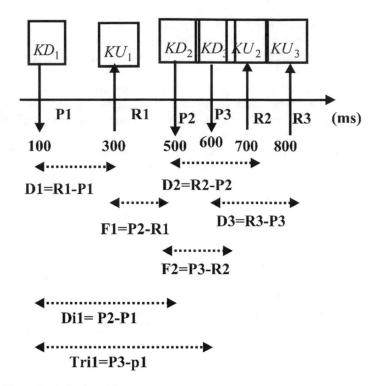

Fig. 2 Keystroke timing based features

- Digraph is estimated based on the time in the middle of pair of events. The same is shown by Di1 in Fig. 2.
- Trigraph is estimated based on the time of latencies between every three successive keys. The same is shown by Tri1 in Fig. 2.
- n-graphs is estimated based on the delay between the n number of keystroke dynamics events.

Different authors used different feature sets for authenticating users using keystroke dynamics. The approaches are summarized ahead for understanding the contribution of various features in keystroke dynamics.

Typically, digraph indicates the sequence of two characters including punctuation, numerals, letters and space. Digraphs are used along with other features for various purposes. Gentner et al. [20] used trigraph with digraphs for identifying the types of errors that the user makes while typing. Roth et al. [21] applied digraph features to the keystroke sound generated when a user types using keyboard. One of their main assumptions was, when a key is pressed, each key will emit a slightly different acoustic signal dependent upon the user. This inspired them to cluster sample keystroke sounds to learn a virtual alphabet. The digraph latencies within the pairs of the virtual letters were then used to generate the score. These features, i.e., digraphs, trigraphs

Table 1 Most common feature set for Keystroke Dynamics authentication

Features used	Description	Performance value
KeyPress, Dwell and Flight time [23]	The authors have used Artificial Bee Colony (ABC) algorithm for feature extraction and applied to digraph. It is observed that it has given the best performance as compared to KeyPress and Dwell time	Error rate: 0.0445% Accuracy: 95.5%
KeyPress and KeyRelease time [24]	The authors have used Partially Observable Hidden Markov Model (POHMM) to form feature vector. POHMM being a sequential model can only be applied in continuous verification process	Error rate: 0.042% Accuracy: 78.8%
Digraph features [25]	The Gaussian Mixture Model (GMM) is used to select the features used for keystroke dynamics. Digraphs do not show pure Gaussian but a mixture of Gaussians. The best result was seen with 2 components Gaussian	FAR: 0.09% FRR: 2.94%
Timing features such as Dwell Time, latency time and non-conventional feature set [26]	The fusion of using timing and non-conventional feature set give better accuracy as compared to individual feature set	FAR: 0.09% FRR: 0.21% Accuracy: 80%
Dwell time and latency time [27]	Dwell time and latency time attributes in the dataset used separately for authentication led to higher performance	EER: 0.07%
Dwell time and latency time [28]	The heterogeneous vectors (i.e., combination of Dwell time and latency time) yields better result as compared to aggregate vectors (i.e., KeyPress and KeyRelease Time)	EER: 1.72%

and n-graphs depend only on the word context [22]. To overcome these dependencies Dowland and Furnell [7] used digraphs, trigraphs and n-graphs features along with the keyword latencies i.e., AutoID, Left character, Right character, Latency and Timestamp. They achieved the most promising results when using digraph latencies.

The most commonly used features for the keystroke dynamics authentication systems are represented in tabular form in Table 1 along with its performance value. It has been observed from the studied that not much work has been carried out using non-conventional features such as Shift Key usage, Caps Lock Key usage, Number Key usage and use of Left or Right Shift Key for authenticating user.

4 Classification Methods

Identification of features to create a unique profile for every individual is the most important aspect in machine learning-based keystroke dynamics. Classification algorithms in machine learning is a supervised learning method in which the system designed learns from the input data given to it during training phase and then uses the trained data learning to classify the given test data.

User profiling and classification to authenticate the user are the next important step in keystroke dynamics for user verification. Classification algorithms require that the user undergoes a training period, during which the algorithm learns the typing pattern of every user. The types of classifiers in machine learning are Linear Classifier (Logistic classifier and Naive Bayes), Support Vector Machines (SVM), Decision Trees, Boosted Trees, Random Forest, Nearest Neighbor (NN) and Neural Network. The features selection is an important area of research. Once this is done the model is developed and the classification is performed. This section enumerates some of the research efforts done in utilizing classification algorithms for authenticating user.

Keystroke dynamics authentication method reduces the weakness of passwords guessing attacks by breaking down the typing patterns of the client. In keystroke analysis, it is typical to distinguish users by running a classifier over the whole pool of users that produces a particular classification accuracy. Different classification algorithms used in keystroke dynamics with different objectives are summarized ahead. Out of these classifiers, neural networks are popularly being used in keystroke dynamics in the past [23, 24, 29–31]. Deep learning is the recent advancement in neural network, which has achieved the state of art recognition performance for voice biometrics and many other object recognition applications. Deng and Zhong [32] applied the deep learning method to keystroke dynamics user authentication. Their study shows deep learning method significantly outperforms other algorithms on the CMU keystroke dynamics dataset [33].

Support vector machine (SVM) is the most prevalent machine learning method which processes choice limits by maximizing the boundary to reduce the generalization error. SVM is used as a classification algorithm by [26, 34–40] in keystroke dynamics analysis.

Hidden Markov Model (HMM) is a model of a sequence of hidden variables with generative or conditional dependency. HMM can be an addressing model for keystroke dynamics as it naturally represents the interconnections between consecutive features and noise in the data. Jiang et al. [41] and Monaco et al. [24] have, respectively, used the concept of Hidden Markov Models to learn the keystroke dynamics in typing rhythms. Typically for machine learning-based methodologies larger training set helps in improving the performance of the system. However, increasing the number of samples in the training set becomes a tedious task.

To reduce the possible number of samples utilized in the learning phase, K-Nearest Neighbor (KNN) classification technique used by [42–48] to develop a keystroke dynamic authentication model. Giot et al. [39] used SVM to reduce the number of entries required during the registration step while maintaining performance of EER value 15.28%. Similarly, Ahmed and Traore [49] used neural network architecture to predict the missing digraphs by analyzing the relation between the keystrokes. The enrollment of the user was done solely on the samples that were collected from users at the time of registration, so that prediction can be started with only one enrolled user. The experimental results were FAR equal to 0.0152% and FRR equal to 4.82%. Different studies have announced an extensive change in performance for comparable and similar algorithms, in light of the fact that most investigations utilized their very own dataset. To address this issue, Killourhy and Maxion collected a keystroke dynamics benchmark dataset and also published it [33]. They evaluated this dataset on fourteen existing keystroke dynamics algorithms which include Manhattan, Nearest Neighbor, SVM (one-class), Mahalanobis, Neural Network, Euclidean, Euclidean, Fuzzy Logic and K-Means. This keystroke dataset with the assessment approach and the performances of the algorithms give a decent benchmark to dispassionately survey advances of new keystroke biometric algorithms.

5 Metrics and Datasets for Evaluation

Once the data is classified, one needs to evaluate it for its correctness. For which it is essential to provide a standardized testing methods and performance metrics. The datasets used also play a very important role in this testing as it is known fact that the training dataset if used for testing it can not comment on the performance of the classification. Hence, a separate more appropriate dataset should be chosen for testing. Various existing performance metrics and keystroke dynamics datasets are evaluated in this section. The metrics chosen to test the feasibility of the system by various researches to evaluate their methods are summarized in this section.

A. *Metrics*

FAR and FRR explained in Section II are the core performance metrics applied to assess the performance of keystroke dynamics systems [22, 36, 50]. The lower values for both these metrics mean better performing system. Accuracy and Equal Error Rate (EER) are the other two performance metrics that are commonly used for evaluating systems [36]. EER is the point where the False Acceptance and False Rejection rate are minimal and optimal. The lower the equal error rate value, the higher is the accuracy of the biometric system. Accuracy is one metric for evaluating classification algorithms. It is a fraction of number of correct predictions made by the total number of predictions made. Based on these metrics, a system verifies whether the level of trust and the genuineness of the user attains global (common to all user) or personal (different for every user) threshold value. The metrics are used quite effectively for finding the performance of the system.

Table 2 Details of existing Datasets

Dataset name	Description
GREYC Keystroke Benchmark [51] and GREYC12 Static Keystroke Dynamics Benchmark Dataset [52]	KeyPress, KeyRelease and Flight Time information
Si6 Labs Keystroke rhythm Dataset [53]	Digraphs details for the fixed text which is in Spanish
Beihang University Static Keystroke Dynamics Benchmark Dataset [30]	Dwell time and Flight time of 117 subjects in which 2057 are test samples and 556 are training samples
GREYC-NISLAB Keystroke Dynamics Soft Biometrics Dataset [54]	Timing information of (i) code of the key, (ii) the type of event (press or release), and (iii) the time of the event
The Sapientia University dataset [55]	Timing information of Key Hold time, Flight Time and Average hold time
Clarkson University Dataset [56]	Keystroke data for short pass-phrases, fixed text and free text
Rhu Keystroke dataset [57]	Timing information of each KeyPress and Release, timing between two key press and release and date of trial
CMU dataset [33]	Keystroke-timing information from 51 users, each typing a password .tie5Roanl 400 times

B. *Datasets*

Recently, many datasets were introduced in this field to encourage researchers to compare the performance of their algorithms with other existing algorithms is represented in Table 2. GREYC Keystroke Benchmark [51] and GREYC12 Static Keystroke Dynamics Benchmark Dataset [52] include KeyPress, KeyRelease and Flight Time information. Si6 Labs Keystroke rhythm Dataset [53] contains digraphs details for the fixed text which is in Spanish, Beihang University Static Keystroke Dynamics Benchmark Dataset [30] contains dwell time and flight time of 117 subjects in which 2057 are test samples and 556 are training samples, and GREYC-NISLAB Keystroke Dynamics Soft Biometrics Dataset [54] contains timing information of (i) code of the key, (ii) the type of event (press or release) and (iii) the time of the event. The Sapientia University dataset [55] contains timing information of Key Hold time, Flight Time and Average hold time. Clarkson University's Dataset [56] includes keystroke data for short pass-phrases, fixed text and free text. Rhu Keystroke dataset [57] includes timing information of each KeyPress and Release, timing between two KeyPress and KeyRelease and date of trial. CMU dataset [33] consists of keystroke-timing information from 51 users, each typing a password .tie5Roanl 400 times. Though these datasets help to have the standard benchmark to compare with other algorithms, when any new feature is introduced, it is essential that the data must be captured.

6 Conclusion and Future Work

The paper has focused on the work done in the area of keystroke dynamics using machine learning. It is observed during this study that majority of the work focuses on time-based features for static analysis. The impact of non-conventional features on the performance of keystroke dynamics is not much explored. The knowledge may help improve the performance of keystroke dynamics. As the area is not much explored, datasets with non-conventional features are not available for testing. Another important aspect of machine learning is classification. The objective was to identify if there is any classification algorithm that is more suited for keystroke dynamics and helps to get better accuracy. After studying many papers, we realized that it is very difficult to conclude about the classification approach as some authors used existing dataset and some used their very own datasets. It was also observed from the survey that it is necessary to travel a long path for providing an effective keystroke dynamics authentication system. In order to analyze the efficacy values of the system, a proper selection of features set and benchmark test methods are needed. It is also observed that whenever a new feature is proposed a new dataset needs to be created for testing it. The paper helps identify some of the white spaces such as use of non-conventional features, having a common dataset for testing these features and identifying the most suitable classification approach for machine learning-based keystroke dynamics-based authentication.

References

1. Uludag U et al (2004) Biometric cryptosystems: issues and challenges. Proc IEEE 92:948–960
2. Uludag U, Jain AK (2004) Attacks on biometric systems: a case study in fingerprints. Proc IEEE 92:948–960
3. Uludag U et al (2004) Biometric template selection and update: a case study in fingerprints. Pattern Recognit 37:1533–1542
4. Jain AK et al (2005) Biometric template security: challenges and solutions. In: Signal processing conference, pp 1–4
5. Moody J (2004) Public perceptions of biometric devices: the effect of misinformation on acceptance and use. In: Issues in informing science & information technology
6. Umphress D, Williams G (1985) Identity verification through keyboard characteristics. Int J Man-Mach Stud 23:263–273
7. Dowland PS et al (2004) A long-term trial of keystroke profiling using digraph, trigraph and keyword latencies. In: Security and protection in information processing systems. Springer, Berlin, pp 275–289

8. Mondal S et al (2016) Combining keystroke and mouse dynamics for continuous user authentication and identification. In: IEEE international conference on identity, security and behavior analysis (ISBA), pp 1–8

9. Obaidat MS et al (1999) Estimation of pitch period of speech signal using a new dyadic wavelet algorithm. Inf Sci 119:21–39

10. Gunetti D et al (2005) Keystroke analysis of free text. ACM Trans Inf Syst Secur (TISSEC) 8:312–347

11. Mondal S, Bours P (2015) A computational approach to the continuous authentication biometric system. Inf Sci 304:28–53

12. Mondal S et al (2015) Continuous authentication in a real world settings. In: Eighth international conference on advances in pattern recognition, pp 1–6

13. Ali ML et al (2015) Authentication and identification methods used in keystroke biometric systems. In: High performance computing and communications (HPCC), pp 1424–1429

14. Mhenni A et al (2016) Keystroke template update with adapted thresholds. In: 2nd international conference on advanced technologies for signal and image processing (ATSIP), pp 483–488

15. Pisani PH et al (2016) Enhanced template update: application to keystroke dynamics. Comput Secur 60:134–153

16. Çeker H et al (2016) Adaptive techniques for intra-user variability in keystroke dynamics. In: Biometrics theory, applications and systems (BTAS). IEEE, New York, pp 1–6

17. Rahman KA et al (2013) Snoop-forge-replay attacks on continuous verification with keystrokes. IEEE Trans Inf Forensics Secur 8:528–541

18. Koakowska A et al (2018) Usefulness of keystroke dynamics features in user authentication and emotion recognition. In: Human-computer systems interaction. Springer, Berlin, pp 42–52

19. Kim S et al (2014) A correlation method for handling infrequent data in keystroke biometric systems. In: International workshop on biometrics and forensics (IWBF), pp 1–6

20. Gentner DR et al (1983) A glossary of terms including a classification of typing errors. In: Cognitive aspects of skilled typewriting. Springer, Berlin, pp 39–43

21. Roth J et al (2015) Investigating the discriminative power of keystroke sound. IEEE Trans Inf Forensics Secur 10:333–345

22. Zhong Y, Deng Y (2015) A survey on keystroke dynamics biometrics: approaches, advances, and evaluations. In: Recent advances in user authentication using keystroke dynamics biometrics. Science Gate Publishing, pp 1–22

23. Akila M et al (2012) A novel feature subset selection algorithm using artificial bee colony in keystroke dynamics. In: Proceedings of the international conference on soft computing for problem solving (SocProS), pp 813–820

24. Monaco JV, Tappert CC (2018) The partially observable hidden markov model and its application to keystroke dynamics. Pattern Recognit 76:449–462

25. Ceker H, Upadhyaya S (2015) Enhanced recognition of keystroke dynamics using Gaussian mixture models. In: Military communications conference, MILCOM IEEE, pp 1305–1310

26. Alsultan A et al (2017) Improving the performance of free-text keystroke dynamics authentication by fusion. Appl Soft Comput

27. Ho J, Kang D-K (2018) One-class nave bayes with duration feature ranking for accurate user authentication using keystroke dynamics. Appl Intell 48:1547–1564

28. Balagani et al (2011) On the discriminability of keystroke feature vectors used in fixed text keystroke authentication. Pattern Recognit Lett 32(7):1070–1080

29. Loy CC et al (2005) Pressure-based typing biometrics user authentication using the fuzzy artmap neural network. In: Proceedings of the twelfth international conference on neural information processing (ICONIP), pp 647–652

30. Li Y et al (2011) Study on the Beihang keystroke dynamics database. In: International joint conference biometricson, pp 1–5

31. Sridhar M et al (2013) Intrusion detection using keystroke dynamics. In: Proceedings of the third international conference on trends in information, telecommunication and computing. Springer, Berlin. https://doi.org/10.1007/978-1-4614-3363-7_16

32. Deng Y, Zhong Y (2013) eystroke dynamics user authentication based on Gaussian mixture model and deep belief nets. In: ISRN signal processing. ArticleID 565183. https://doi.org/10.1155/2013/565183

33. Killourhy KS, Maxion RA (2009) Comparing anomaly-detection algorithms for keystroke dynamics. In: Proceedings of IEEE/IFIP international conference on dependable systems & networks (DSN), pp 125–134

34. Kim J et al (2018) Keystroke dynamics-based user authentication using freely typed text based on user-adaptive feature extraction and novelty detection. Appl Soft Comput 62:1077–1087

35. Lau S, Maxion R (2014) Clusters and markers for keystroke typing rhythms. In: The LASER workshop: learning from authoritative security experiment results

36. Alsultan A et al (2017) Non-conventional keystroke dynamics for user authentication. Pattern Recognit Lett 89:53–59

37. Ho G et al (2014) Tapdynamics: strengthening user authentication on mobile phones with keystroke dynamics. Technical report, Stanford University

38. Li BZY et al (2011) Study on the Beihang keystroke dynamics database. In: Proceedings of the international joint conference on biometrics (IJCB), pp 1–5

39. Giot R et al (2011) Unconstrained keystroke dynamics authentication with shared secret. Comput Secur 30(6–7):427–445

40. Martono W et al (2007) Keystroke pressure-based typing biometrics authentication system using support vector machines. In: Proceedings of computational science and its applications ICCSA 2007: international conference, Part II, pp 85–93

41. Jiang SSC, Liu J (2007) Keystroke statistical learning model for web authentication. In: 2nd ACM symposium on information, computer and communications security (ASIACCS), pp 359–361

42. Wesoowski TE et al (2015) User verification based on the analysis of keystrokes while using various software. J Med Inf Technol 24:13–22

43. Longi K et al (2015) Identification of programmers from typing patterns. In: Proceedings of the 15th Koli Calling conference on computing education research, pp 60–67

44. Mhenni A et al (2018) Towards a secured authentication based on an online double serial adaptive mechanism of users' keystroke dynamics. In: International conference on digital society and eGovernments

45. Kambourakis G et al (2016) Introducing touchstroke: keystroke based authentication system for smartphones. Secur Commun Netw 9:542–554

46. Zhong Y et al (2012) Keystroke dynamics for user authentication. In: IEEE computer society conference on computer vision and pattern recognition workshops (CVPRW), pp 117–123

47. Ivannikova E et al (2017) Anomaly detection approach to keystroke dynamics based user authentication. In: IEEE symposium on computers and communications (ISCC), pp 885–889

48. Dholi PR, Chaudhari K (2013) Typing pattern recognition using keystroke dynamics. In: Mobile communication and power engineering. Springer, Berlin, pp 275–280

49. Ahmed AA, Traore I (2014) Biometric recognition based on free-text keystroke dynamics. IEEE Trans Cybern 44:458–472

50. Alghamdi SJ et al (2015) Dynamic user verification using touch keystroke based on medians vector proximity. In: 7th international conference on computational intelligence, communication systems and networks, pp 121–126

51. Giot R et al (2009) Greyc keystroke: a benchmark for keystroke dynamics biometric systems. In: IEEE 3rd international conference biometrics: theory, applications, and systems, pp 1–69

52. Giot R, El-Abed M, Rosenberger C (2012) Web-based benchmark for keystroke dynamics biometric systems: a statistical analysis. In: Eighth international conference on intelligent information hiding and multimedia signal processing (IIH-MSP), pp 11–15

53. Bello L et al (2010) Collection and publication of a fixed text keystroke dynamics dataset. In: XVI Congreso Argentino de Ciencias de la Computacin

54. Idrus SZS et al (2013) Soft biometrics database: a benchmark for keystroke dynamics biometric systems. In: IEEE conference on biometrics special interest group (BIOSIG), pp 1–8

55. Antal M et al (2015) Keystroke dynamics on android platform. Proc Technol 19:820–826

56. Vural E et al (2014) Shared research dataset to support development of keystroke authentication. In: IEEE International joint conference on biometrics, pp 1–8
57. El-Abed M et al (2014) Rhu keystroke: a mobile-based benchmark for keystroke dynamics systems. In: International Carnahan conference security technology (ICCST), pp 1–4

Performance Analysis of Routing Protocols in Wireless Sensor Networks

Jameel Ahamed and Waseem Ahmed Mir

Abstract Wireless Sensor Networks (WSNs) are interconnected sensor devices that can detect process and communicate data over large distances in large geographical areas. Today sensor devices have found their use in defense services, notably military services, disaster management, medical services, wildlife monitoring and precision agriculture, and in habitat monitoring and logistics. Since the sensor devices are small, they have limited energy resources that reduce their service life. Because of this restriction, energy management is the primary research area in these networks. This study determines the performance of five routing protocols: Ad hoc On-Demand Distance Vector (AODV), Temporally Ordered Routing Algorithm (TORA), Optimized Link State Routing Protocol (OLSR), Geographical Routing Protocol (GRP), Distance Routing Effect Algorithm for Mobility (DREAM) and their comparison. The results are shown in three scenarios whereby DREAM is considered as one of the best routing protocols based on four performance metrics.

Keywords Routing protocols · AODV · DREAM · TORA · OLSR · GRP

1 Introduction

A Wireless Sensor Network comprises self-organized, autonomous, low power sensor nodes which operate in cooperation with complex physical conditions. They sense the change in physical variables like motion, temperature, and sound, translate the minute changes into data packets and transmit these packets of useful information over the networks [1]. The energy limitations and requirement of reduced transmission of excessive information lead to existence of algorithms to be distributed [2]. The power source often comprises a battery with a limited energy budget. Military

J. Ahamed (✉) · W. A. Mir
Department of Computer Science and Information Technology, MANUU Hyderabad, Hyderabad, India
e-mail: jameel.shaad@gmail.com

W. A. Mir
e-mail: waseemmir78177@gmail.com

© Springer Nature Singapore Pte Ltd. 2020
A. Khanna et al. (eds.), *International Conference on Innovative Computing and Communications*, Advances in Intelligent Systems and Computing 1059,
https://doi.org/10.1007/978-981-15-0324-5_14

applications like battlefield surveillance originally motivated the development of wireless sensor network. However, WSNs are now employed in many civilian application areas comprising the environment and habitat monitoring. Because of distinct limitations originating from their reasonable nature, limited size, weight, and ad hoc approach of distribution each sensor faces the challenge of limited energy resources [3]. Various approaches have been adopted to optimize the energy consumption both at the hardware and software level in the Wireless Sensor Networks [4]. Different routing schemes have been devised to optimize the performance parameters. These routing protocols are mostly operation-specific and not applicable in the wide range of applications. This paper has selected five routing protocols TORA, AODV, DREAM, OLSR, and GRP to analyze the performance parameters of these protocols.

2   Routing Protocols in WSNs

There are numerous routing protocols devised for WSNs having resource management capability mainly in terms of energy efficiency, deployment capability in diverse terrains. Broadly, the energy-efficient protocols in WSNs can be divided into four categories and are further subdivided into many types which are given in Fig. 1.

From the enormous number of routing protocols, we selected the following five routing protocols for the simulation purposes.

(a)   Temporally ordered routing algorithm (TORA): It is a reactive, extremely adaptive, efficient and scalable distributed routing algorithm based on the notion of link reversal [5].

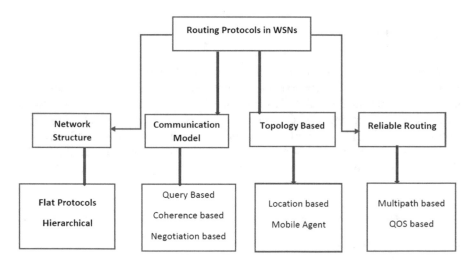

Fig. 1 Routing protocols in WSNs

(b) Ad Hoc On-Demand Distance Vector (AODV): It is a reactive routing protocol which establishes route, when a node requires sending data packets. AODV is capable of both unicast and multi-cast routing. The operation of the protocol is divided into two functions: route discovery and route maintenance [6–8].

(c) Distance Routing Effect Algorithm for Mobility (DREAM): DREAM a location-based routing protocol framework where each node maintaining a position database are the part of the network [9]. In DREAM a GPS system is installed which enables each node to know its geographical coordinates that make DREAM protocol more reliable [10].

(d) Optimized Link State Routing Protocol (OLSR): It is a table-driven routing protocol meant for mobile adhoc networks and optimization of pure link state algorithm [11, 12].

(e) Geographical Routing Protocol (GRP): Geographical Routing Protocol is a proactive routing protocol [13]. It uses Global Positioning System (GPS) for finding the location of node to collect network information at a source node with a very small amount of control overheads.

3 Simulation Setup and Results

Opnet Modeller 17.5 Academic edition is used for simulation of the study and the sensor networks used are wireless LAN workstations (mobile nodes) as they can act same as that of any wireless sensor node. The entire work in this paper is divided into three modules and four scenarios; each scenario runs on the same set of sensor nodes while routing protocols are changed. Light traffic in module one and heavy traffic in module two has been incorporated. In module 1 we simulated AODV and TORA protocols and in Module 2 we simulated DREAM and TORA. In Module 3 under scenario 1, two more routing protocols namely OLSR and GRP have been added and simulation results have been recorded for light traffic. The scenario 2 of this module has recorded simulation results for different parameters for routing protocols namely AODV, TORA, OLSR and DREAM under heavy traffic. As shown in Fig. 2a,

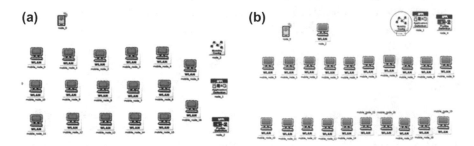

Fig. 2 Light traffic and heavy traffic incorporation

each scenario thereby is made of WSN comprising of wireless sensor nodes (wireless LAN workstations), a wireless LAN server, Application configuration object, Profile configuration object, and Mobility configuration object. Then simulation results of selected algorithms in reference to the light traffic and heavy traffic are compared separately. The description of the modules and their corresponding scenarios are given below.

Module 1.
Scenario 1: AODV with light traffic
Scenario 2: TORA with light traffic

Module 2.
Scenario 3: DREAM with heavy traffic
Scenario 4: TORA with heavy traffic

Module 3.
Scenario 1: Comparison of AODV-TORA-OLSR-GRP.

With the help of application and profile configuration objects, network traffic is incorporated. A profile named PROF1 was created using profile configuration object which was configured to support only the file transfer protocols (FTP) packets and with application configuration object, the application definition was set on FTP, so that the nodes could be configured for only FTP traffic. The sensor nodes will now be able to access the FTP traffic from the wireless LAN server. The paths followed by the sensor nodes to receive and send the traffic/packets were defined using the mobility configuration object.

Performance Metrics:

(a) **Throughput**: Throughput can be described as the data packets a node can process in a fixed interval of time. In terms of the communication network, it is the rate at which data packets reached the terminal node successful and is measured in bits per second.
(b) **Delay**: Delay is described as the time taken by a bit of data to go from transmission node to receiver node in a network. Delay depends upon the routing protocols and location of nodes in a network.
(c) **Traffic Received**: (Data Packet Received): Data packets transmitted across the network and arrived at the terminal node in a network is called traffic received.
(d) **Load**: It is defined as the total traffic across the network.

Module 1

In first scenario of module one, we configured the wireless sensor nodes for light traffic with the help of application and profile configuration object, the server, and the nodes were configured to run only on the AODV routing protocol. The second scenario was also configured using the WSNs for light traffic with the help of application and profile configuration object where the WLAN server and the sensor nodes

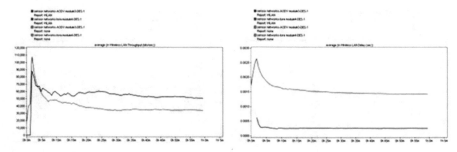

Fig. 3 Throughput and delay of AODV and TORA

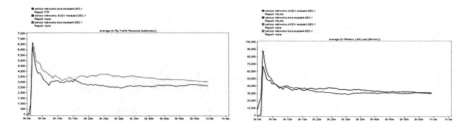

Fig. 4 Traffic received and Load of AODV and TORA

were configured to run only on the TORA routing protocol. Then for the entire four parameters throughput, delay, traffic received and load, the system was simulated which is shown in Figs. 3 and 4. In Fig. 3a, throughput is shown for both TORA (represented by red curve) and AODV by blue curve. Delay for both protocols is shown in Fig. 3b where TORA is represented by red curve and AODV by blue curve. In Fig. 4a, traffic received in bytes per second is shown for both the protocols where red curve shows TORA and Blue AODV while in Fig. 4b; load is shown as red curve for TORA and Blue for AODV. From the Figs. 3 and 4, it is evident that TORA has edge over the AODV in terms of throughput, delay, traffic received and load. All the parameters are measured in bits per second.

Module 2

In first scenario of module two, configuration of the wireless sensor nodes for heavy traffic with the help of application and profile configuration object was done, where the server and the nodes were configured to run only the TORA routing protocol. The second scenario was also configured using the wireless sensor nodes for heavy traffic with the help of application and profile configuration object, where the WLAN server and the sensor nodes were configured to run only the DREAM routing protocol. Then for the entire four parameters throughput, delay, traffic received and load, system was simulated and results shown in Figs. 5 and 6. Figure 5a shows the throughput for TORA represented by green curve and DREAM by blue curve.

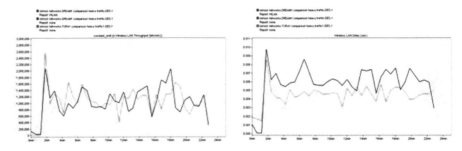

Fig. 5 Throughput and delay of TORA and DREAM

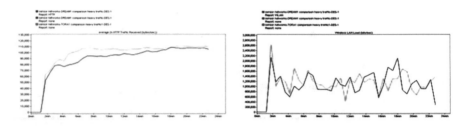

Fig. 6 Traffic received and load of TORA and DREAM

Figure 5b shows the delay for TORA represented by green curve and DREAM by blue curve. Figure 6a shows the traffic received in bytes per second for TORA represented by green curve and DREAM by blue curve whereas in Fig. 6b, load is shown where green curve represents TORA and blue curve represents DREAM. In this way, the results are compared for both TORA and DREAM routing protocols through the four parameters. Figures 5 and 6 draw the comparison between TORA and DREAM. It is evident from the graphs that DREAM is much better than the TORA routing protocol. Although the throughput of TORA shows an instant steep increase at the beginning session with time it is as good as the DREAM routing protocol. In all other parameters namely delay, traffic received and load, DREAM routing protocol has more advantages over the TORA routing protocol.

Module 3: Scenario 1

In this scenario, the comparison of AODV, TORA, OLSR and GRP based on four performance factors is shown below.

Throughput: Figure 7a shows the comparison of throughput for all the routing protocols and it can be observed that out of the four protocols OLSR has the maximum throughput at an average of 10,000 bits/s. The fluctuations are very little in OLSR and the graph runs smooth throughout the simulation. GRP protocol has an average throughput of almost 4000 bits/s and so is the case with AODV and the lowest among the four is TORA with an average throughput of almost 2000 bits/s.

Delay: From Fig. 7b, we found that among all the protocols used, OLSR is the most efficient in terms of delay; this protocol has lowest delay among all the other

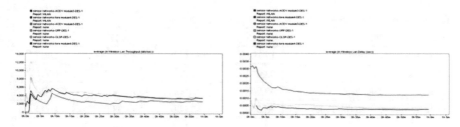

Fig. 7 Comparison of throughput and delay of AODV-TORA-OLSR-GRP in bits/sec and in seconds

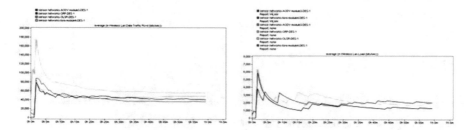

Fig. 8 Comparison of traffic received and Load of AODV-TORA-OLSR-GRP in bits/sec and in seconds

protocols used with almost 0% delay. TORA is having maximum delay of an average of 0.0020 s.

Traffic Received: From Fig. 8a, it is clearly visible that at the beginning of the simulation, there is a sharp increase in the traffic received by all the networks, wherein TORA implementation receives maximum traffic. Later there is a shift from maximum value to a more constant trend of traffic received in all the networks. TORA remains constant at 60,000 bits/s followed by OLSR with a value of 50,000 bits/s. This shows the best efficiency of TORA routing protocol in case of traffic received.

Load: The graph in Fig. 8b, presents a comparison of Load (bits/sec). On an average load received by TORA is maximum than all the other routing protocols at an average value of 2000 bits/s. The load value shows the efficiency of TORA to withstand and carry a heavy traffic by ease and efficient way. The next efficient routing protocol is AODV with a value close to TORA at almost 1800 bits/s. The fluctuations in all the protocols are considerably negligible. OLSR protocol has also a comparably efficient load capacity (Table 1).

Module 3: Scenario2

In this scenario, the comparison of AODV, TORA, OLSR and DREAM based on four performance factors is shown below.

Throughput: The graph in Fig. 9a shows the comparison of throughput for all the routing protocols. From the study of the graph obtained, it is observed that the average throughput is maximum in case of DREAM routing protocol, followed by TORA and AODV. Moreover, the graph obtained also shows that the throughput of

Table 1 Values of parameters on light traffic under simulation

Parameter studied	TORA	OLSR	AODV	GRP
Throughput (bits/sec)	170,000	50,000	60,000	59,000
Delay (seconds)	0.0015	0.0001	0.0003	0.0004
Load (bits/sec) average	90,000	72,000	68,000	64,000
Data traffic received (bits/sec)	60,000	50,000	40,000	38,000

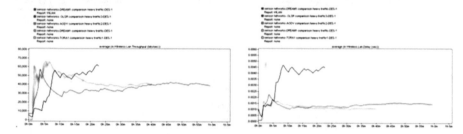

Fig. 9 Comparison of throughput and Delay of AODV-TORA-OLSR-DREAM in bits/sec and in seconds

DREAM is almost linear and ends at a maximum value of 60,000 bits/s at the end of simulation. TORA protocol starts at a high value but is not able to keep a track of its progress and ends at a low value of almost 42,000 bits/s. OLSR routing protocol maintains its value of 30,000–40,000 bits/s throughout the simulation time of 1 h.

Delay: From the study of the graph in Fig. 9b, we could easily conclude that delay is less in case of AODV at a value of 0.0010 s, delay in case of DREAM is at a high value of 0.0045 s but the peculiar feature is that it starts at a low value of 0.0004 s and ends at the high delay value. OLSR remains constant at a value of 0.0015 s.

Traffic Received: The comparison of routing protocols for Data traffic received is shown in Fig. 10a. It is observed that TORA has gained the highest value of traffic received but after almost 20 min of simulation, it has stopped receiving further data, same is the case with AODV and OLSR, but only DREAM protocol maintained

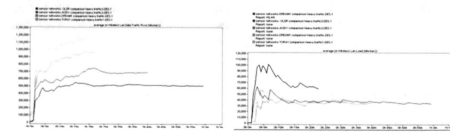

Fig. 10 Comparison of traffic received and Load of AODV-TORA-OLSR-DREAM in bits/sec and in seconds

Table 2 Values of parameters on heavy traffic under simulation

Parameter studied	TORA	OLSR	AODV	DREAM
Throughput (bits/sec)	45,000	40,000	40,000	60,000
Delay (seconds)	0.0038	0.0015	0.0010	0.0045
Load (bits/sec) average	600,000	400,000	300,000	800,000
Data traffic received (bits/sec)	800,000	700,000	600,000	1,000,000

its constant value of data reception throughout the simulation process. The DREAM protocol has maintained a constant value of 50,000 bits/s for a whole 1 h of simulation time.

Load: The graph in Fig. 10b presents a comparison of routing protocols on the selected parameter of Load (bits/sec). DREAM is seen as having the maximum load while AODV is having lowest as compared to other routing protocols (Table 2).

4 Discussion

For module one, it is seen that TORA with an average throughput of almost 2000 bits/s is ahead of AODV protocol and fairs well in this parameter. In terms of delay, TORA is having maximum delay of an average 0.0020 s, which is more than AODV protocol. In terms of traffic received TORA implementation receives maximum traffic but there is a shift from maximum value to a more constant trend of traffic received in all the networks. TORA remains constant at 60,000 bits/s followed by AODV with a value of 50,000 bits/s. This shows the enhanced efficiency of TORA routing protocol in case of traffic received and in terms of delay, on an average load received by TORA is maximum than all the other routing protocols at an average value of 2000 bits/s. The load value shows the efficiency of TORA to withstand and carry a heavy traffic by easy and efficient way. The next efficient routing protocol is AODV with a value close to TORA at almost 1800 bits/s. The fluctuations in all the protocols are considerably negligible.

For the second module, the average throughput is maximum in case of DREAM routing protocol, followed by TORA. Moreover, the graph obtained shows that the throughput of DREAM is almost linear and ends at a maximum value of 60,000 bits/s at the end of simulation. TORA protocol starts at a high value but is not able to keep a track of its progress and ends at a low value of almost 42,000 bits/s. In terms of delay, it is concluded that the delay is less in case of TORA at a value of 0.0010 s, delay in case of DREAM is at a high value of 0.0045 s but the peculiar feature is that it starts at a low value of 0.0004 s and ends at the high delay value. In terms of traffic received TORA has gained the highest value of traffic received but after almost 20 min of simulation it has stopped receiving further data, but only DREAM protocol has maintained its constant value of data reception throughout the simulation process. The DREAM protocol has maintained a constant value of 50,000 bits/s for a whole

1 h of simulation time and in terms of load it is clear that DREAM routing protocol has the maximum load value of 60,000 bits/s followed by TORA with 52,000 bits/s. DREAM protocol has reached a maximum value of 100,000 bits/s among all the other protocols.

For module three, it is again proved that DREAM is the best routing protocol among all others discussed and simulated for various performance metrics.

5 Conclusion

This paper describes the comparison of different routing protocols on four parameters for light traffic and heavy traffic. In case of light traffic incorporation, TORA proven to be an efficient routing protocol while in heavy traffic incorporation, DREAM routing protocol is proven as the efficient routing protocol among all. Overall graphs and results discussed both for light and heavy traffic depicts that DREAM is considered as the efficient routing protocol followed by TORA and AODV.

References

1. Jamatia A et al (2015) Performance analysis of hierarchical and flat network routing protocols in wireless sensor network using NS-2. Int J Model Optim 5(1)
2. Rezaei Z, Mobininejad S (2012) Energy saving in wireless sensor networks. Int J Comput Sci Eng Surv 3(1)
3. Mir W, Ahmad J (2017) Energy management in wireless sensor networks—a review. Int J Adv Res Comput Sci 8(4)
4. Khiavi MV, Jamali S, Gudakahriz SJ (2012) Performance comparison of AODV, DSDV, DSR and TORA routing protocols in MANETs. Int Res J Appl Basic Sci 3(7):1429–1436
5. Kumar S, Saket RK (2011) Performance metric of AODV and DSDV routing protocols in MANETs using NS-2. Int J Recent Res Appl Stud 7(3)
6. Mauve M, Widmer J, Hartenstein H (2001) A survey on position-based routing in mobile ad hoc networks. IEEE Network 1(6):30–39
7. Mohapatra S, Kanungo P (2012) Performance analysis of AODV, DSR, OLSR and DSDV routing protocols using NS2 simulator, vol 30. In: International conference on communication technology and system design, pp 69–76
8. Clausen T, Philippe J (2003) Optimized link state routing protocol (OLSR). No. RFC 3626
9. Mittal P, Singh P, Rani S (2013) Performance analysis of AODV, OLSR, GRP and DSR routing protocols with database load in MANET. Int J Res Eng Technol 2(9):412–420
10. Yassine M, Ezzati A (2014) Performance analysis of routing protocols for wireless sensor networks. In: 2014 Third IEEE international Colloquium in Information Science and Technology (CIST), IEEE, pp 420–424
11. Yan J, Zhou M, Ding Z (2016) Recent advances in energy-efficient routing protocols for wireless sensor networks: a review. IEEE Access 4:5673–5686
12. Luo RC, Chen O (2012) Mobile sensor node deployment and asynchronous power management for wireless sensor networks. IEEE Trans Industr Electron 59(5):2377–2385
13. Khan JA, Qureshi HK, Iqbal A (2015) Energy management in wireless sensor networks: a survey. Comput Electr Eng 41:159–176

Controlled Access Energy Coding (CAEC) for Wireless Adhoc Network

Bharati Wukkadada and G. T. Thampi

Abstract Currently, the mobile ad hoc network is outstanding and developing in an outstanding manner and the research on this topic is rigorous research area globally. The evolving wireless networking has an incorporation of cellular technology, personal computing internet, etc. is due to global communication and computing interactions. Ad-hoc networks are very beneficial to the circumstances where an arrangement is unobtainable. Power-Protocols are required to assign separately to multiple terminals, where this problem in the minor networks, it becomes dominative among bigger networks, where the number of terminals to have power and the reuse of powers. This paper discusses the issues of power allocation in communication in MAI in wireless network. It discusses the MAI-multiple access interference and the limiting near-far problems. It will be addressing the decreasing throughput performance of mobile ad hoc network. This paper will discuss the communication minimizing MAI impact with the power assignment scheme for the proper utilization of power in MANET.

Keywords Mobile adhoc network · Multiple access interference-MAI · RTS-CTS

1 Introduction

The wireless and mobile ad hoc are constructed to work in co-ordination without the usage any practice of centered admission point or any infrastructure having fixed setup for this purpose. Expecting that nodes must work and communicate in a bi-directional way small distance as well as for emergency. If they want to communicate

B. Wukkadada (✉)
K. J. Somaiya Institute of Management Studies and Research, Vidyavihar, Mumbai 400077, India
e-mail: wbharati@rediffmail.com; wbharati@somaiya.edu

G. T. Thampi
Thadomal Sahani Engineering College, Banadra (W), Mumbai 400050, India
e-mail: gtthampi@gmail.com; gtthampi@yahoo.com

B. Wukkadada · G. T. Thampi
University of Mumbai, Mumbai, India

© Springer Nature Singapore Pte Ltd. 2020   173
A. Khanna et al. (eds.), *International Conference on Innovative Computing and Communications*, Advances in Intelligent Systems and Computing 1059,
https://doi.org/10.1007/978-981-15-0324-5_15

from the long distance, it has to use multi-hop concept using multi-hop routing method so that it can interact with other in-between nodes without fail. In this case, there is the constant option of change in the MANET, as the nodes are in full actions in nonstop and endless mode. Because of the uniqueness of this MANET, there are n number of proposed routing protocols are available till today. The growth of wireless and mobile ad hoc network communication are on demand for its compatibility and useful for the coming decades for the coming generation. The improved and incorporated alternative wireless request will be proved for the coming generation, with a wonderful service of compatibility and greater support and service because the present is failing with the designs of the wireless communication.

The chief application, MANET is primarily used in the professional domain, where huge calculations and important concept is game theory to introduce and for decision making for communication purposes. With the attention of wireless mobile ad hoc network has got the attention on the ability to supply the capability in a situation where there is a problem of fixed infrastructure and to miss with a too. So it is much costly to workout. The ad hoc network design is a diverse topic, and it proposes the many mixing wireless concept of better performances. There are upcoming many communication-based approaches for the packet transmission in the MANET.

In mobile ad hoc network, there are various coding—based aces protocols are available in the past as well as in the present, and various research is also going on. Random-access channel of design primarily used in the protocols, when the transaction accepts the transmission packets are transmitted to the destination without the channel exist in the network. Studied various random-access methods for cryptographic MANET, the mobile ad-hoc network, highly confidential, integrated, authenticity is used for some data used by cryptographic algorithm. In the cryptography protocol has been used for the confidence and safe for the unrestricted crash of the node because it is main key element so the power assignment and power-spreading schemes will use. In multiple accesses, the transmitters can send data simultaneously over single-channel communication, where various nodes can share a frequency, which is exactly same bandwidth. Moreover, there is interference between the nodes without the undue permission. It is nothing but the coding scheme, where each node will assign a code (CDMA). If the source node and other nodes have different desired signals in the transmission then it is performing best. The correlating of the different signals sends and received with a normally generated code of the desired user. If it is matching instead of different signals then, the function of correlation is high and it will extract the signal. A cross-correlation occurs when a desired user's code is different with signals the correlation must be near to zero as far possible. If it is not zero then, the correlation will be close to zero as possible, is mentioned as auto-correlation, and used for rejecting the multi-path correlation.

The non-zero cross-correlation among the completely of not likely cryptography powers brings the multi-access interference (MAI). It will lead to subordinate crash of the system on the destination node side, i.e., different cryptography will be used by both the sender and receiver nodes. Near-far problem is one of the disadvantages of this type, and this happens because major reduction in the fall of the network.

These disadvantages will be correct with the good planned coding—based access protocol. The problem of near-far in the cryptography is mobile ad hoc network has a further problem for the causes of ruins the crashing of the network. For this reason, the method is used to minimize the problem and to get good performance for mobile ad hoc network a MAI used. Our research will be reducing the MAI problem of cryptography, for the mobile ad-hoc network it will be very cost-effective performance. For this research of MAI reducing cryptography, a self-motivated power allocation technique will be used for the good effects for the network.

In [41], the researcher has a plan for the protocol for powered nature sense protocol, where the K is the number of tones which are busy, cooperated with the spreading power. During the reception of packets at particular energy at the receiving end. Transmissions are busy in nature. Researcher of [42], the standard power packets send by RTS-CTS by all terminals. Whereas the information packets area unit sent employing transmitter/receiver end approach, slightly common methods having slight resemblance will be planned for all protocols, authors are predicting good orthogonally across spreading powers that it will ignore the issue of adjoining to near far.

Plan a theme based on reservation [41], whereby only a few packets are adopted to slots for management are information packets. The researcher [45] applied the MAI in the direction to avoid employment of FHSS. They suggested approach, will not be applied to DSS which is the methodology for selection of wireless standards applicable to present situation. In [31], the researcher planned new algorithms for assignment of distributed channel of multi-hop networks. Protocol, still may not yield any MAI, so for the completely not like powers it cannot support overlapping transmissions of signals. For power management in powered networks by the clusters as planned by another eye-catching approach. It makes things easier than the supply performance of many terminals. Also, additionally result in the creation of bottlenecks will be on cost for the network will be reduced operation ought to bear meanwhile entire communications the cluster heads.

The near-far issue will be planned to mitigate by the employment in a multiuser detection at the receiving end in MANET. This planned theme additionally needs the employment of GPS receivers to produce correct position and temporary order data. Moreover, though it is possible to set up multi-user GPS receiving is thinkable at the dominant central position. Currently the above-mentioned is not practical, has expensive implementation; such as receivers inside mobile nodes, etc. lately, a motivating way to joint programming and management of power in impromptu networks has been planned. In order to mitigate MAI physical layer techniques are heavily relied upon. Moreover, it will not put effort for MAI on the access layer. However, a central controller is needed to execute the programming formula, so it is not a very distributed resolution. The existed feedback channel allows receivers to transmit SNR measurements by conflict-free manner of their several transmitters. For this research to attain integration of cryptography theme over an ad hoc network to target, cut back the MAI impacted.

2   Access Interference Problem

The most significant access interference problems are crucial issues for ad hoc networks and will be solved, and are required for the efficient use, capability of Quality of service, and widely applicable. There are various consistent techniques for permitting for EIM protocols to generate a series of adaptive and optimized various necessities, situations, and developing technologies.

It is vital that differentiate in the middle of the spreading power protocol and therefore the access protocol. The previous resolves that the power be employed to unfold the signal, however, don't solve the competition in the medium. It implies the access protocol that though a terminal associated has proposed spreading power, it is going to be not allowed to transmit. The objectives are: The protocol should be asynchronous, disseminated, plus climbable in lieu of big networks. This one should conjointly require involve minimal data exchange, should remain appropriate intended period implementation. Frequently vigorous as an outcome of it is sometimes tough to ensure proper power assignment in any respect times once topology is incessantly dynamical.

3   Objectives

The design of our Controlled Access protocol, define thoroughly in consequent sections. The followed objectives will be guided subsequently:

1. To asynchronous the protocol for the distributed as well as for the climbable for the huge networks.
2. To conjointly implicate marginal exchange of data and should appropriate time for implementation.
3. The receiving electronic equipment must not to a fault advanced within the sense where it must not needed just observe the complete power set.
4. The protocol ought to channel to adapt changes as quality patterns.
5. Though we tend to assume that an influence assignment protocol is in a row at the next layer, so access protocol should minimize as eliminate crashes, though the facility is not assigned as "correct". Frequently and dynamically as an outcome, it is sometimes dangerous to ensure proper power assignment.

4   Controlled Access Energy Coding (CAEC)

Practical explanation controlled access energy coding which is proposed as an architecture name for the concurrent as mobile ad hoc network. The RTS AND CTS packets will be transmitting on the controlled channel with the regular power at a fixed power altogether potentially in an IEEE system, retain these data packets Following are the two constant parameters used for transmission by sender and receiver:

- The power for spreading
- The power for transmission

According to any power assignment scheme, the power selection can be done. Most critical is the choice of an energy level, which depicts or shows a MAI a link quality tradeoff. It explains further precisely, due to the increment of the transmission power, the bits and the minute error in the receiving intended station descends, that's improvement of the quality of the link. However, simultaneously for other ongoing parallel receptions, MAI that the receptions quality depreciation as observed. As an addition to the accounts of above 2 factors, this rule (protocol) assimilates into the computation of power an interference margin. Now the said margin as discussed as let's the terminals so some certain interfering space from the anticipated receiving station to surprised new broadcasts will be for the future. There were 2 types of frequency data channels used data & control, respectively, as partitioning as frequency division multiplexing. So frequent distribution power was been utilized by all the nodes over the governing channel, disseminating power above the steering channel terminal-specific powers for the several may be above the channel for data. These distinct powers utilized above the channel for data remain not effortlessly orthogonal. That shows, as of these partition of the frequency. A signal is seen completely orthogonal over the control channel to the power over the data rate or any signal. The portioning of the bandwidth available into 2 which are not overlapping frequency bands is essentially required headed for receive the control and to run the transformation to terminal simultaneously, where data channels are notwithstanding the directed power of signal.

This approach is a combination node for minimum MAI coding architecture for power allocation in MANET and the efficiency improvement during transmission as unintended enhanced non-zero MAI throughput at the receiver while dispreading. The severe and the main significant consequence of MAI is the near-far issue, whereas the transmitters i th signal will be very close like ample receiver will receive, so the trying is to detect the duration of the variance to. And around, so transmitter of J, will then I thru the trans spreader. By this way as soon as all communication powers which are equitable so the signals from a trans sender J arrive in a comparatively larger power than that of transmitter I at the receiver. This results in the transmissions at I t be decoded incorrectly, i.e., subordinate interference. The margin of interference supposed to be needed for the terminals to allow the space. Hence, in order to initiate, fresh transmissions in the upcoming interference margin is required for terminals at some space of distance from the receiver.

Consider a random receiving I, suppose $\mu *$ be the Eb/N$_0$, at a movement of time where the ratio of the receiver must achieve the given the target bit error rate. In this situation, the paths after that are to attain the objective error rate, and for this, it is essential that this calculation will be demonstrated below.

$$\frac{P_0^{(i)}}{P_{thermal} + P_{MAI}^{(i)}} \geq \mu^*$$

In the above formula, $P_{thermal}$ is power for the noise thermal, where the MAI of $p(i)$ as the entire recipient I of MAI. Subsequently the lowest prescribed accepted $(P(I)\ 0)$ power is min $= \mu * (P_{thermal} + P(i) * MAI)$. Interfering verge powerfully impends weight of the network. The aforementioned will transmit expression, so appeal to rise of the noise as $(\xi(I))$, determined succeed:

First, with a maximum transmission power, the $(P(i)0)$ receiver power min essentially $\mu *$ and $P_{thermal\ in}$ increased stay constants by this noise growth is enhanced. Moreover, hence, there is a decrease, this also lowers the max coverage range for reliable communication.

Second, used to generally transmit packet which leads to increase in power, and increases the consumption of energy for a very infrequent resources in MANETs. Hence, energy throughput tradeoff is undesirable. The transmitter's interference margin is sitting at the max planned noise growth (ξ max),. Thus takes into account the 2 on $\xi(i)$ restrictions mentioned above.

Allowance plan permits merely trans missions reason either original or ancillary collisions headed for go on untidily. Here the packets like CTS and RTS are usage towards contributing 3 performances.

1. The RTS data packet format falls in the line with the IEEE 802. 11 and apart from the two-byte will include the $p(j)$ field in diagram, it is shown in the full diagram (Fig. 1).

The above-said packets permit nodes to the assessment of the channel improvements among the pairs of the transmitter and the receiver.

1. The CTS packet is used by a receiver i to notify and advice its neighboring nodes of the added sound power will be denoted as noise $P(i)$. this adds each neighbor to I terminal without impacting is reception. Hence, a set of possible potential terminals interfering has been constituted by these neighbors.

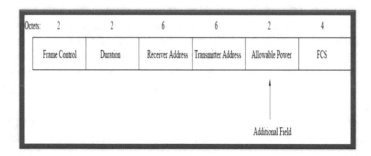

Fig. 1 CEAC protocol for RTS packet-layout

2. As a final point, for each terminal to attention on retain channel control irrespective of the destination of signal in instruction to record the average active handless respective neighborhoods.

Technically the networking practice over which the packet transfer is done is actually termed and the j terminal to communicate the packet. The packets of RTS sends packet done to the control P max channel. Moreover, it will comprise the above-said packet with extreme permissible ($P(j)$ map) power level. The J terminal will use undisturbed somewhat continuing permanent reception in j's neighborhood. The proposed receiver, when receives the RTS packet, roughly and nearly said as terminal i, and it uses the prearranged the value of P_{max}. The power of $P(ji)$ received signal of the estimated the $Gji = P(ji)$ channel gain acknowledged/P_{max} among j and I terminals on that movement. I terminal will power the data packets will be correctly unknowingly get the transmission at a power $P(ji)$ min specified through:

$$P^{(ji)}_{\min} = \frac{\mu^* P_{\text{thermal}} + P^{(i)}_{\text{MAI-current}}}{G_{ji}}$$

In the formula power, the efficient and common MAI of all in-progress transmissions and communications into e network is $p(i)$ MAI current. Note that for of the supposed for the stable situation for channel suitable in excess of short measure intervals, Gji is roughly fixed throughout process the transmissions of control data packets. With this $P(ji)$ min power for the min strength, j terminal has been used the transmission of the data in such a method for I terminal. Exactly packet for the data power for the interfering level currently.

The $P(ji)$ min power, on the other hand, permit will not be for any interference endurance at i terminal, all adjoining of I terminal have towards communication delay during i's terminal continuing retaking no contemporaneous transmissions has yield a position in the i neighborhood. That terminal j will be assigned to an application and that will grant i has been stated:

$$P^{(ji)}_{\text{allowed}} = \frac{\zeta \max \mu^* P_{\text{thermal}}}{G_{ji}}$$

Here the $P(ji)$ permit $< P(ji)$ min, at that juncture the MAI neighborhood of i terminal is better as if the single permit means of the attach accumulation. These coincidences, replied by i an indirectly with CTS as j's instruct cannot continue transmit the data. This has been anticipated, transmissions begin links over the endow the MAI gallant. Consequently extends numerousness lively links for the network (liable for usable limit of power). Added to this, a condition as $P(ji)$ assign $>P(ji)$ min, next it has practicable to i terminal to accept signal of j's, nevertheless only condition of $P(ji)$ assign fewer than $P(j)$ sketch (RTS enclosed). Latest requisite stays essential for the j's transmission sort out not to trouble at all progress transmissions neighborhood. Popularly this event, i terminal compute the power interference $P(i)$ MAI-prospective endurance as

$$P_{\text{MAI-current}}^{(i)} = \frac{3W\ G_{ji}}{2\ \mu^*}\left(P_{\text{allowed}}^{(ji)} - P_{\min}^{(ji)}\right)$$

In the formula, 3 w/2 spreading gain factor comes from the step for succeeding is justifiably as well as defensibly transmit tolerance of power for the future potential users for interfering as the i district avoid the total $p(I)$ MAI—future consumption by anyone neighbor only as the justification and explanation.

The propagation of the above said power acceptance is assumed that $K(i)$ is the transmission for the neighborhood tracking. Then it will be monitored by the exchanges CTS/RTS above the channel regulation. The accrual to i retains the notation $K(i)$ Avg for $K(i)$ a typically besides average concluded by a window specified. After this later, $K(i)$ will be designed and as calculated:

$$K^{(i)} = \begin{cases} \beta\left(K_{\text{avg}}^{(i)} - K_{\text{inst}}^{(i)}\right) & \text{if } K_{\text{avg}}^{(i)} > K_{\text{inst}}^{(i)} \\ \beta & \text{otherwise} \end{cases}$$

whereas the safety margin will be $\beta > 1$ though the communication in the system, observation is done wherein the neighbor interference is less compared to the interference with the system. Then this eventually leads to the next observation where its level of effect which was ultimate decreases (Fig. 2).

The CTS packet linked the power limit, next will be transmitted to the control channel to back for acceptance. In these circumstances, the invited power trans out to be the supplementary additional to already existing limiting power. This leading to the dropping of the request made. A coding scheme which is bean proposed due to the above problem for randomly distributed wireless ad hoc network hence certain results as well as opinions are been made.

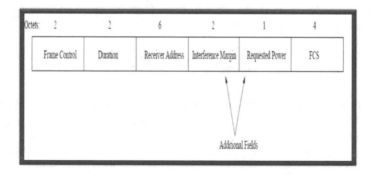

Fig. 2 Layout of CTS packet-proposed protocol

5   Results and Observations

The energy coding called "controlled Energy Access coding" (CEAC) for MANET are measured in the indoor environment using Matlab. Initially, the research has been interfaced with 5 laptops and desktops that have Matlab software installed.

5.1   Simulation

An arbitrarily disseminated wireless ad hoc network has been well-thought-out and the simulation of controlled Access coding protocol is used to modify RTS-CTS reservation mechanism and the based of dispute. The CTS and RTS packets are communicated above the control channel at a common power and at a stable Pmax max power. All the nodes potentially interfere in standard IEEE 802.11, so scheme receives these packets. Though, IEEE 802. 11 in contrast scheme as coding protocols, interference of nodes supposed to allow transmitting the data concomitantly. The nodes interfering are permitted to concurrently transmit data, which implies its dependency on certain criteria. On behalf of this certifying the data container receiver and sender will have an agreement on two Factors:

- Spreading power
- Power for transmission

In a case where the wireless network is scattered randomly, any power assignment scheme can be applied for power selection. And having been considered that, a codeless scheme simulation but yet with protocol will follow different topology conditions.

The aimed simulation performed as;

Structure-network: Delta Topology
Number of nodes: 10
Simultaneous offered load: 5

5.2   The Effect of CAEC of Generated Scattered Network

The effect of CAEC shown in Fig. 3 describes a graph having randomly generated scattered network with the count our points as nodes in a random region on the graph.

5.3   The Effect of CAEC of Throughput Plot

The effect of CAEC shown in Fig. 4 illustrate the result throughput for the above

Fig. 3 Generated scattered
network

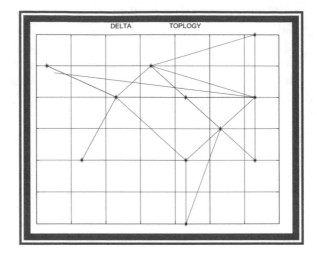

Fig. 4 Throughput plot over
variable offered load

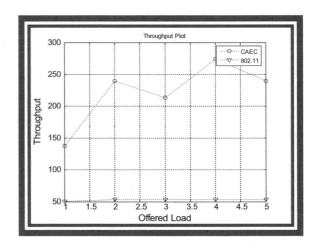

system mentioned for the conventional way according to 802.11. The observation
illustrates about the marginal interference which has been nodes shared in the process
of their communication. this is done by obtaining the throughput which is ideally
higher than any conventional system in this context.

5.4 The Effect of CAEC of Energy Consumption Plot

The effect of CAEC shown in Fig. 5, where there has been an evaluation done on
the quiet dropping the energy consumption, in given data load. There has been an
observation done where the total consumed energy is reduced done to the prior

Fig. 5 Energy consumption plot

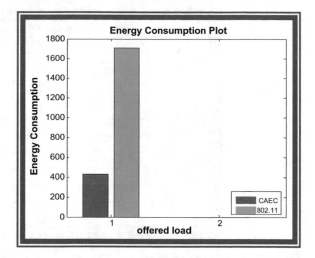

knowledge of channel having power generated by interference. This further helps in reduction of the total packet generation based on the various channel connection.

5.5 The Effect of CAEC of Simultaneous Transmission Probability

The effect of CAEC shown in Fig. 6 that there is probability of data transfer at the same time due to which there can be two different methods in order to study the effect with reference to the load offered. In the channel leads to increment when compared to the other conventional method.

Fig. 6 Simultaneous transmission probability

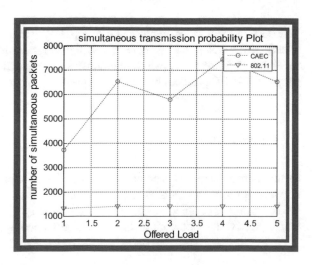

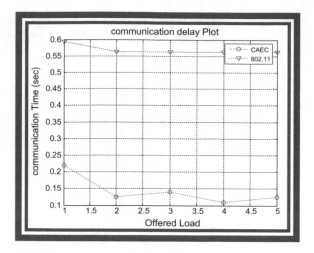

Fig. 7 Delay-plot for communication

5.6   The Effect of CAEC of Communication Delay Plot

The effect of CAEC shown in Fig. 7 the total data packets for the communication where the load capacity has been used as differently is detected. Here the interruption has been observed and it has been proposed in approach to reduce the conventional approach. There is channel having lower congestion for the no generation for the packet interfering. It is observed that scattered topology having total 6 nodes and offered load also 6.

5.7   Real-Time Implementation

The development scenario presented in Fig. 8 illustrate the creation of our research tested with wireless ad-hoc network, as the diagram depicts the working environment tested. Some of our pictures demonstrating the real-time implementation of the proposed CAEC protocol in the below figure.

6   Conclusion

In this paper, there has been an execution of access protocol of a wireless ad hoc network protocol and which is power controlled. Discussed the formula and the CAEC protocol of the above described the system which reduces the near-far issue which MANET's undermines the throughput accomplishment by amounting for

Fig. 8 Working environment snapshots illustrating of protocol

multi-access interference. Whereas CAEC practices channel gain with the CTS and RTS packets above a widely known control range. As a result of this simulation, it describes that the coding is done which is based on CAEC can make the network throughput even better. Moreover, specifically in this movement the accomplishment of fifty percent diminution for consumption of energy to distributing a delivery of packet satisfactorily to the end from the starting. (source–to-destination).

References

1. Chalmatac I, Conti M, Liu JJN (2003) Mobile adhoc Networking: Imperatives and challenges. Ad-hoc Networks, Elsevier, pp 13–64
2. Sun J-Z (2004) Mobile ad hoc networking: an essential technology for pervasive computing media team. Machine Vision and Media Processing Unit, Infotech, Oulu
3. Sharp BA, Grind rod EA, Camm DA (1995) Hybrid TDMA/CSMA protocol for self-managing packet radio networks. In: 4th IEEE ICUPC, Tokyo, pp 929–933
4. Sunil Kumar A, Vineet S, Raghvan B, Deng Jing (2006) Medium access control protocols for adhoc wireless networks. Surv Adhoc Netw 4:326–358
5. Clementi AEF, Penna P, Silvestri R (2000) The power range assignment problem in radio networks on the plane. Berlin, Heidelberg
6. Andreas T Mobile ad-hoc networks, Pearson, 2nd edn
7. Chen B, Jamieson K, Balakrishnan H, Morris R (2008) Span: an energy efficient coordination algorithm for topology maintenance in ad hoc wireless networks
8. Yu C, Lee B, Yong Youn H (1999) Energy efficient routing protocols for mobile ad hoc networks
9. Hwang C-H, Wu ACH (2000) A predictive system shutdown method for energy saving of event-driven computation. ACM Trans Des Autom Electron Syst 5(2):226–241
10. Jones CE, Siva Lingam KM, Agrawal P, Chen JC (2001) A survey of energy efficient network protocols for wireless networks. Wireless Netw 7:343–358
11. Estrin D, Govindan R, Heidemann J, Kumar S (1999) Next century challenges: scalable coordination in sensor networks, USA Copyright ACM
12. Krevat E, Shahdadi A (2001) Energy-efficient dynamic source routing Environment, IEEE infocom
13. Jung E-S, Vaidya NH (2002) A power control MAC protocol for ad hoc networks, ACM
14. Zussman G, Segall A (2003) Energy efficient routing in ad hoc disaster recovery networks, IEEE

15. IEEE Computer Society (1999) Wireless LAN Medium Access Control (MAC) and Physical Layer (PHY) Specifications, ANSI/IEEE Std 802.11, Edition. In Ad-Hoc Wireless Networks
16. Chang J-H, Tassiulas L (2000) Energy conserving routing in wireless ad-hoc networks, IEEE INFOCOM, IEEE
17. Raju J, Garcia-Luna-Aceves JJ (2003) Scenario-based comparison of source-tracing and dynamic source routing protocols for ad-hoc networks
18. Feeney LM (1999) An energy consumption model for performance analysis of routing protocols for mobile ad hoc networks. IETF MANET WG, July 14
19. Feeney LM, Nilsson M (2006) Investigating the energy consumption of a wireless network interface in an ad hoc networking environment
20. Tamilarasi M, Chandramathi S, Palanivelu TG (2010) Overhead reduction and energy management in DSR for MANETs
21. Čagalj M, Hubaux JP, Enz C (2002) Minimum energy broadcast in all wireless networks: NP completeness and distribution issues, USA
22. Karayiannis NB, Nadella S (2004) Power-conserving routing of ad hoc mobile wireless networks based on entropy-constrained algorithms
23. Johansson P, Larsson T, Hedman N (2007) Scenario-based performance analysis of routing protocols for mobile ad-hoc networks, SE-412 96 Göteborg, Sweden
24. Bergamo P, Giovanardi A, Travasoni A, Maniezzo D, Mazzini G, Zorzi M Dipartimento di Ingegneria, Universitá di Ferrara, Distributed Power Control for Energy Efficient Routing in Ad Hoc Networks
25. Quick Installation Guide (2009) Wave LAN/IEEE Turbo 11 Mb PC Card
26. Gallager RG, Humblet PA, Spira P (1983) A distributed algorithm for spanning trees minimum-weight
27. Sivakumar R, Sinha P, Bharghavan V (1999) CEDAR: a core-extraction distributed ad hoc routing algorithm. IEEE J Sel Areas Commun 17(8)
28. Ramanathan R, Rosales-Hain R (2000) Topology control of multihop wireless networks using transmit power adjustment. IEEE INFOCOM
29. Lindsey S, Raghavendra CS (2001) Energy efficient broadcasting for situation awareness in ad hoc networks, IEEE
30. Guha S, Khuller S (1996) Approximation algorithms for connected domain sets ESA
31. ElBatt T, Ephremides A (2000) Joint scheduling and power control for wireless ad-hoc networks. IEEE INFOCOM 2002, IEEE
32. Xu Y, Bien S, Mori Y, Heidemann J, Estrin D (2003) Topology control protocols to conserve energy in wireless ad hoc network
33. Xu Y, Mori Y, Heidemann J, Estrin D (2001) Geography informed energy conservation for ad hoc routing, July 1621, Rome, Italy
34. Eu ZA, Tan H-P, Seah WK (2009) Routing and relay node placement in wireless sensor networks powered by ambient energy harvesting. In WCNC '09: proceedings of the IEEE wireless communications and networking conference, pp 1–6
35. The ns-2 Network Simulator, [online] available http://www.isi.edu/nsnam/ns
36. Subramanian S, Ramachandran B (2012) QOS Assertion in MANET routing based on trusted AODV (ST – AODV)
37. Heinzelman WR, Chandrakasan A, Balakrishnan H (2000) Energy-efficient communication protocol for wireless microsensor networks, IEEE, pp 7695–0493
38. Chen B, Jamieson K, Balakrishnan H, Morris R Span: an energy efficient coordination algorithm for topology maintenance in ad hoc wireless networks
39. Agarwal M, Cho JH, Gao L, Wu J Energy efficient broadcast in wireless ad hoc networks with hitch-hiking
40. Sycard Technology 1994–96, PCCextend 140 CardBus Extender User's Manual
41. Chenl J-C, Sivalingam KM, Agrawal P, Kishore S (1998) A comparison of MAC protocols for wireless local networks based on battery power consumption, IEEE

IoT-Based HelpAgeSensor Device for Senior Citizens

Ravi Shekhar Pandey, Rishikesh Upadhyay, Mukesh Kumar, Pushpa Singh and Shubhankit Shukla

Abstract Recent trends and reports show the cases of undetected death of senior citizens in India due to negligence of family member and society. This issue became a major social issue and also pulls back the development of any country and nation. The phenomenal growth in terms of science and technology can play a key role in the direction of safety of the senior citizens. Technology like IoT can be utilized in order to keep safe and secure senior citizens. The purpose of this work is to develop a IoT-based *HelpAgeSensor device* which is in a ring form and easily wearable. This device would sense the pulse rate and send alert messages along with location information of the device holder to connected entity. These entities could help device holder by providing medical treatment or dignified cremation ceremony in case of death. The main components of this device are pulse monitor, GPS system, and threshold generator. This device is based on threshold values of pulse rate. This HelpAgeSensor device would be able to help the society to combat unnoticed death of people behind the closed doors.

Keywords IoT · Threshold · Sensor · Senior citizens · Pulse rate

R. S. Pandey · R. Upadhyay · M. Kumar · P. Singh (✉) · S. Shukla
Accurate Institute of Management and Technology Greater Noida, Noida, India
e-mail: pushpa.gla@gmail.com

R. S. Pandey
e-mail: ravishekhar322@gmail.com

R. Upadhyay
e-mail: rishikeshupadhyay1997@gmail.com

M. Kumar
e-mail: mk586440@gmail.com

S. Shukla
e-mail: shubhankitshukla9@gmail.com

© Springer Nature Singapore Pte Ltd. 2020
A. Khanna et al. (eds.), *International Conference on Innovative Computing and Communications*, Advances in Intelligent Systems and Computing 1059,
https://doi.org/10.1007/978-981-15-0324-5_16

1 Introduction

Internet of things (IoT) is the modern concept of remotely connecting and monitoring the any real-world objects or things with the help of the Internet [1]. IoT is proven not only effective in home automation, smart city, but also admiring in social issue. IoT provides advanced connectivity of different types of equipment, various services, numerous protocols, and different applications; moreover, IoT is characterized with the vision of heterogeneity [2]. In the concept of IoT, the devices are able to collect important information with the help of heterogeneous technologies and then ensure flow the data between the devices. Rapid development and social change in India have led to a large pro-portion of senior citizens living alone. There are around 15 million Indian senior citizens are living alone. According to a 2015 Global AgeWatch Index [3], India does not care for its senior citizens. It ranked 71st in a list of 96 countries and according to the Proceedings of the National Academy of Sciences, both social isolation and loneliness are associated with a higher risk of mortality in adults aged 52 and older.

According to Agewell foundation[1] 43 out of every 100 senior citizens in India are sufferers of psychological difficulties due to loneliness and other relationship problems. Sometimes, the death of senior citizens is only noticed when a malodor spread to the environment or maggots leaves a house through the doorsill. This is very horrible and undignified situation for any society and nation. Every person ever lived must have the right to dignified cremation ceremony.

Last few years, cases of people died unwanted and unloved behind closed doors, their bodies lying in a "state of shocking decomposition" are expeditiously increasing. It is also alarming because India has more than 100 million senior citizens at present and the number is only increasing. By 2050, India's 60 and older population is expected to reach 323 million peoples.

Till date, no such device is available to address this type of social problem. This motivates us to create a IoT-based sensor device and referred as HelpAgeSensor a device for the older citizens who are living alone and isolated. This device is in the form of a ring which could be easily wearable. The proposed device is based on IoT which senses the pulse rate and would send an alert and location information of a deceased person to family members, society's RWA representatives, or NGO/organizations in case of low or lost pulse. This device consists of some important parts such as pulse monitor, GPS system, and threshold generator. IoT is the main baseline to connect these components with server, and Arduino is the medium to connect software and hardware.

---

[1] https://www.ndtv.com/india-news/study-says-43-per-cent-of-elderly-in-india-face-psychological-problems-1731428.

2 Related Works

Recently, IoT gained a great attention of researchers and developers. It allows communications between objects, machines, and everything together with peoples. The state of the art of the IoT application requirements and their related communication technologies are surveyed by Akpakwu et al. [4]. Karimi [5] contended that the world is entering at the age of IoT computing technology with great prospective to permit the communication between different machines, infrastructure and machine, and machines and the environment. Zeinab and Elmustafa [6] reviewed the concept of many IoT-based applications and future possibilities for new related technologies and challenges of facing the implementation of the IoT.

Moubarak [7] presented a method for smart home automation by using Internet of things (IoT) integrated with computer vision, Web services, and cross-platform mobile services. IoT-based smart home automation system for elderly and handicapped people was discussed and proposed by Ghazal and Al-Khatib [8]. The system was recognized by using XBee transceivers that maintain RF wireless communication between the remote control and the master control panel board to provide ubiquitous access.

Kodali et al. [9] represented the application of an IoT in hospital healthcare system by using Zigbee protocol. The healthcare system could periodically observe the physiological parameters of the patients. IoT-enabled devices increased the quality of care at minimum cost and also collected data for analysis. Lakshminarasimhan [10] proposed an advanced traffic management system in heavily populated cities like Los Angeles and Amsterdam. The system is implemented using Internet of things (IoT). Most of IoT applications related to smart home and healthcare system are discussed and reported in the literature. However, IoT can also be important to solve social problems. Lee [11] proposed a system related to social issues. Their system could identify the schedules of pedestrian, unlike the current security cameras, because its camera tracks the schedule of pedestrian. The system was related to actively preventing crime and actively monitored crime scenes. Similarly, we can have many IoT-based systems and these research areas are open and need to be explored.

The objective of this paper is to design an IoT-based sensor device for those senior citizens living alone or neglected by their families and society. We proposed a HelpAgeSensor device which is in a ring form and easily wearable. This device would work as interface by sending an alert message and GPS location of connected NGO/organization or family member, in case of any causality to provide medical help or cremation rituals.

3 System Description and Components

In this paper, we are proposing a design of HelpAgeSensor device which can be easily wearable by any senior citizens at any time and any place. This device is used to monitor pulse rate of senior citizens continuously. When the pulse rate is below or equal to the threshold value, then it started to send alert messages to connected NGO/organizations or family members. This HelpAgeSensor device would be able to help the society to combat unnoticed death of people behind the closed doors. This device could be extended for any kind of people living alone, isolated or in medical field. Through this device, one can provide help in various ways to the device holder. The main components of this device are given below.

Pulse Monitor:

The pulse monitor is used for observing the pulse rate of the person. These devices are sensor-based and can monitor and record real-time information about the pulse condition and motion activities. A sensor is able to measure a physical phenomenon like temperature, pulse rate, and blood pressure and convert it into an electric signal. Wearable sensor-based devises may contain different types of flexible sensors that can be integrated into textile fiber, clothes, and elastic bands or directly attached to the human body [12]. Here, we supposed this device in the form of ring which can be worn by person any time and any place.

Arduino:

Arduino consists of programmable circuit board also called microcontroller and a software module or IDE that is used to create electronic projects. Arduino is used to collect data from sensors and then transfer to the cloud. Here, we are proposing the device based on Arduino because of low cost: Arduino is microcontroller and can run one program at a time, over and over again, while a Raspberry Pi is a general-purpose computer and has the ability to run multiple programs.

Bluetooth System:

Bluetooth is the communication support between sensor-based pulse monitor device and smartphone. Presently, more than one-third IoT devices are planned to connect with Bluetooth standards. Bluetooth is a short-range technology of connectivity. Various types of connections such as RFID, Wi-fi, Bluetooth, and Zigbee are used to connect GSM, GPRS, 3G, and LTE.

Cloud Technologies:

Cloud technologies have been extensively researched due to their effectiveness in big data management [13]. Cloud is the backbone of IoT revolution. The cloud provides data collection and data processing and sharing with third parties. Here, pulse information about a person is continuously collected and processed by cloud. If this rate decreases as compared to the stored threshold value, then the alert message is sent to connected device of the NGO/organization and family member.

Table 1 List of hardware and software requirement

Hardware requirements	Software requirements
Rechargeable battery	Arduino IDE
Microcontroller	GPS navigation
GPS	Android
LTE modem	
Sensors	
Arduino	
Relay	

GPS System:

GPS location is sent to the registered members and the NGOs. GPS receiver will receive the location information and give to the microcontroller. The microcontroller will send an alert message to the family member to inform about the emergency situation with location information so that they can take necessary action to help the senior citizens.

List of important hardware and software that is required to design the proposed device is represented in Table 1.

4 Proposed Framework of HelpAgeSensor Device

Here, we have designed a *HelpAgeSensor* device for senior citizens, especially those are living alone. The objective of this project is to provide help to citizens living alone or neglected by their families and society. This device would work by sending an alert message along with location to connect NGO/organization or family member, in case of any causality to provide medical help or cremation rituals. The working of the proposed work is given in the following steps:

Step-1: Initialized threshold value of pulse rate, IP address or phone number of main connected NGO/organization or family members.
Step-2: Continuously monitoring the pulse rate of device holder.
Step-3: If the pulse rate becomes low or equal to threshold value, it sends alert message as "Immediate Help" with location information to connected entity.
Step-4: If the pulse rate becomes zero, then it sends alert message as "No Pulse" with location information to connected entity.
Step-5: Based on step-3 and 4, the entity provides the device holder either medical facility or cremation rituals.

The proposed design architecture of *HelpAgeSensor* device is represented in Fig. 1.

According to the proposed framework, sensors are used to sense the pulse rate of senior citizens. These sensors are placed on the human body, which help to monitor

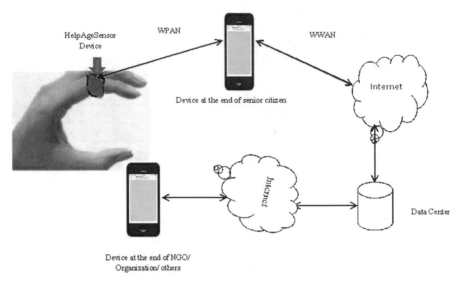

Fig. 1 Design architecture of HelpAgeSensor device

the pulse rate condition without disturbing the daily routine of the device holder. Hel-pAgeSensor device will use WPAN protocol Bluetooth 4.0 or also known Bluetooth Low Energy (BLE) module that connects sensor devices to smartphones, and tablets [14]. The smartphone device uses long-range wireless technology such as WWAN and communicated to the cloud (data server). This data server is holding threshold generator which will send appropriate alert messages like "Immediate help" or "No Pulse" to the NGO/Organization or their family member. The proposed framework is used for improving society by investigating social issue solutions and technologies centered on social organizations and systems.

Future Scope

This device could be extended in the health sector. In future, this device can be used to monitor various deceases like anaphylactic Shocks, tachycardia, heart attack by sensing the pulse rate. The proposed work could be used by any kind of people who is living alone, isolated, or in remote locations.

5 Conclusion

The HelpAgeSensor device is a revolution in the field of communication and health care. This device would give timely information to family members or NGO/organization about any severe casualty happened to device holder. The main issue solved by this invention is diminishing the cases of unnoticed death of people living alone. This device would also help in medical emergent situations by sending timely alerts to concerned person. In this way, this device is helping the society to combat unnoticed death of people behind the closed doors and enabling people to have dignified death rituals.

References

1. Kodali RK, Jain V, Bose S, Boppana L (2016) IoT based smart security and home automation system. In: 2016 International Conference on Computing, Communication and Automation (ICCCA), IEEE, pp 1286–1289
2. Singh P, Agrawal R (2018) A customer centric best connected channel model for heterogeneous and IoT networks. J Organ End User Comput JOEUC) 30(4):32–50
3. HelpAge International (2015) Global age watch index 2015', Insight report. Available online at http://www.helpage.org/global-agewatch/population-ageing-data/global-rankings-table/
4. Akpakwu GA, Silva BJ, Hancke GP, Abu-Mahfouz AM (2018) A survey on 5G networks for the Internet of Things: communication technologies and challenges. IEEE Access 6:3619–3647
5. Karimi K (2014) What the Internet of Things (IoT) needs to become a reality. Available http://www.freescale.com/files/32bit/doc/white_paper/INTOTHNGSWP.pdf. Accessed 24 Nov 2018
6. Zeinab KAM, Elmustafa SAA (2017) Internet of Things applications, challenges and related future technologies. World Sci News 2(67):126–148
7. Moubarak MH (2016) Internet of Things for home automation. Media Engineering and Technology Faculty German University in Cairo, vol 15
8. Ghazal B, Al-Khatib K (2015) Smart home automation system for elderly, and handicapped people using XBee. Int J Smart Home 9(4):203–210
9. Kodali RK, Swamy G, Lakshmi B (2015) An implementation of IoT for healthcare. In: 2015 IEEE Recent Advances in Intelligent Computational Systems (RAICS), IEEE, pp 411–416
10. Lakshminarasimhan M (2014) Advanced traffic management system using Internet of Things
11. Lee HJ (2015) A study on social issue solutions using the "Internet of Things" (focusing on a crime prevention camera system). Int J Distrib Sens Netw 11(9):747593
12. Majumder S, Mondal T, Deen MJ (2017) Wearable sensors for remote health monitoring. Sensors 17(1):130
13. Baker SB, Xiang W, Atkinson I (2017) Internet of Things for smart healthcare: technologies, challenges, and opportunities. IEEE Access 5:26521–26544
14. Aliev K, Rugiano F, Pasero E (2016). The use of bluetooth low energy smart sensor for mobile devices yields an efficient level of power consumption. In: Proceedings of the 1st international conference on Advances in Sensors, Actuators, Metering and Sensing (ALLSENSORS'16), pp 5–9

Optimization of LEACH for Developing Effective Energy-Efficient Protocol in WSN

Avinash Bhagat and G. Geetha

Abstract In the present research paper, the author is trying to improve LEACH for increasing energy efficiency of protocol within wireless sensor network by optimizing Low Energy Adaptive Clustering Hierarchy (LEACH). Optimizing LEACH protocol has certain limitations like routing, aggregation, selection of cluster head (CH) and transmission of data to cluster head which tends to uneven distribution of Energy. Taking this problem into consideration, we are focusing to enhance energy efficiency protocol. We have proposed a solution to improve LEACH by implementing new scheme for selection of cluster head and changing the routing techniques.

Keywords Cluster head · Hopfield neural network · LEACH · Mobile sink · Rendezvous node · Residual energy · WSN

1 Introduction

Wireless sensor network (WSN) contains a battery function and small sensor nodes which are positioned over a wide geographical area to monitor the events and to accumulate the collected data to a distant centralized location called as base station. Organization of nodes is done in such a way that entire area is covered with wireless nodes. Deployed nodes sense the data from its neighbour and transmit the collected data for further processing. Difference between ad hoc networks and wireless sensor networks is their distinct area of applications. Ad hoc networks mainly focus on communication aspects, whereas WSN focuses more on monitoring and information collection. Wireless nodes are bound by several resource restrictions such as the memory availability, battery power, bandwidth requirement, and the data rate. These tiny nodes may work for a longer duration of time from few months to many years depending upon application requirements. To enhance the network lifetime, many suggestions have been proposed. The power consumption of wireless sensors can be reduced by creating clusters of sensors. Designing a clustering algorithm remains

A. Bhagat (✉) · G. Geetha
Lovely Professional University, Phagwara, India
e-mail: avinash.bhagat@lpu.co.in

© Springer Nature Singapore Pte Ltd. 2020   195
A. Khanna et al. (eds.), *International Conference on Innovative Computing and Communications*, Advances in Intelligent Systems and Computing 1059,
https://doi.org/10.1007/978-981-15-0324-5_17

a challenging issue [1]. In order to improve network lifetime, use of battery power needs to be employed proficiently. Sleep mode operation is one of the effective methods to increase network lifetime; when needed, they can switch to wake-up mode if node needs to sense environment. Sensor node should communicate through special routing technique like Hopfield neural networks to maximum battery life [2]. Sensor nodes reduce the redundancy of data transmission by collecting data from the neighbouring nodes through cluster heads (CH). Data is collected and then transmitted to mobile sink. Further in research, we have proposed to use Hopfield neural networks routing technique to improve LEACH in which nodes are grouped in the form of clusters using the least distance benchmarks [3]. The routing is done by the ACO-PSO hybrid approach to find optimum routing that connects the elected cluster heads to base station. Nodes are homogeneous and have limited energy. The results based on certain parameters are compared with existing basic LEACH and optimized LEACH [4].

The paper is organized in six sections as below:

- Brief review of the literature presented in Sect. 2.
- The network and radio model of our improved model is discussed in Sect. 3.
- Improved LEACH protocol analysis and simulation is discussed in Sect. 4.
- Simulation parameters, simulation model and results are discussed in Sect. 5.
- Conclusion is done in Sect. 6.

2   Related Work

Singh et al. [4] investigated and concluded that even after 16 years of existence of Low Energy Adaptive Clustering Hierarchy (LEACH) protocol, very few measured the utilization of energy during CH selection and cluster formation during simulation.

Arumugam and Ponnuchamy [5] suggested energy-efficient LEACH according to which an optimized energy-aware routing protocol is obligatory for gathering data. As seen, almost all sensor nodes have related impact and equal competences which inspires the need for refining lifespan of the sensor nodes and sensor network. According to the author, the proposed objective of EE-LEACH protocol is to reduce the energy consumption and increase the network longevity.

Mottaghi and Zahabi [4] discussed that the construction of small and smart sensors with a very little cost is a result of wireless sensor network developed from recent development in the area of wireless sensor network, communication and microelectronic mechanical systems (MEMS). WSN is a decentralized, self-configurable network which is made by large number of interconnected sensor nodes and sink in predesigned sample framework. Sensor nodes send its data to the sink for processing. Sensor nodes use single-hop transmission technique for short distance [6].

The network efficiency can be improved by decreasing distance of transmission, as energy consumption is more in case of long-distance communication. Reduction in distance also improves the network lifetime. The distance can be reduced by using multihop method where each node transfers data through other nodes into sink. All intermediate nodes act as routers, this method reduces the distance of the transmission, and hence, total energy consumption can be reduced. A suitable protocol must be used to select an optimized route for transmitting the data from node to sink. Only drawback of this technique is energy drop rate is very high for the nodes which are located near mobile sink.

Another solution for optimizing energy utilization will be use of clustering techniques where cluster members transmit data to cluster head and collected data by cluster heads is transferred to mobile sink [4].

Heizelman et al. [7] suggested an effective clustering algorithm in which nodes located in a cluster transmit data to CH; this algorithm is also popularly known as LEACH. In LEACH algorithm, the signals received by the cluster head from the nodes of a cluster and then cluster transmit the effective information to the sink after decreasing the total number of bits. The communication ability of the cluster head ends when it uses all its energy and dies, if the cluster head operations are fixed for some specific nodes within the cluster. Therefore, cluster head is not able to communicate to the rest of the cluster members. To distribute energy consumption evenly in considered area, a random rotation structure is followed through all the nodes used by the LEACH. In LEACH, the cluster heads are selected randomly and a node can select the cluster where it belongs. The decision for transmission from each node is done according to the time-division multiple access (TDMA) policy [4, 6, 7]. In LEACH, we follow adaptive clustering and random CH selection on rotation basis allows distributed energy which makes LEACH a better protocol than its counterpart conventional protocols in terms of energy consumption. Data aggregation by CH in LEACH helps for effective utilization of power consumption during data transmission; further, it also helps to reduce energy consumption by implementation of mobile sink. MS gathers data from all the nodes while moving in or around the environment. Sink movement can be controlled or uncontrolled, but the path is controlled and predefined and movement of the mobile sink is uncontrolled and random in specified environment. Mobile sink movement in close proximity of normal nodes decreases the transmission distance and time. Implementation of mobile sink was introduced to increase lifetime of wireless network, increasing distance between two movements and minimizing stopover time and location.

According to study, mobile sink could not be very close from all nodes for gathering data. An idea of Rendezvous Points (RPs) is developed [5] in which RP near trajectory of sink and a node, i.e. rendezvous node located nearby which transmits data to the mobile sink as it passes nearby [4]. Previous studies [4, 6, 8] demonstrated the effectiveness of RPs on the performance of mobile sink. [4, 9, 10], studied the effect of MS movement on a predetermined path which uses RN for data collection. The MS sends beacon signals to notify the RNs of the MS arrival. The advantage of using LEACH algorithm is its self-organization property, each node is free to join any of the clusters, and it decides whether to be CH or not. MS is not a part of CH selection.

According to the present study, cluster head selection is done using Hopfield neural network method and data aggregation is done using hybrid ACO-PSO algorithm to get the optimized results. The present study is sought to preserve the property of self-organization in LEACH algorithm and improve CH selection. The proposed improved LEACH decreases energy consumption in WSNs as compared to traditional LEACH, particularly when the network is large [4, 11, 12].

3 Network Model Evolution

Various assumptions made while designing improved LEACH are as follows:

- BS is located far from the sensor network and moving along Y-axis.
- Nodes are homogeneous and have limited energy.
- Data is gathered periodically from the given environment.
- The energy needed for transmission from node i to node j is the same as energy needed from node j to node i, for a given signal-to-noise ratio.

Energies required for transmitting and receiving data packets are $E_{elec} = E_{TX} = E_{RX} = 50$ nJ/bit. Energy model to calculate energy consumption during transmission or receipt of data used in present study is given by Eq. (1) [4, 5]. In the model, the energy dissipated to transmit l bits of message to another location at distance d metre is given by $E_{Tx} = E_{Tx\_Elec(l)} + E_{TX\_amp(l; d)}$ as shown in Eq. 1

$$E_{TX} = \begin{cases} l * E_{TX} + l * d^2, & \text{if } d < d_0 \\ l * E_{TX} + l * d^4, & \text{if } d \geq d_0 \end{cases} \qquad (1)$$

where

$$d_0 = \sqrt{\frac{E_{mp}}{E_{fs}}}$$

and E_{TX} is the energy consumed by the radio electronic circuit for l bit transmission, E_{fs} is the energy consumed by power amplifier on the free space model, and E_{mp} is the energy consumed by power amplifier in the multipath model. If the transmission distance is less than d_0, the energy consumed is proportional to d^2, and if the transmission distance is more than d_0, the energy consumed is proportional to d^4. The total energy consumed to transmit and receive data is given by Eq. 2:

$$E_{Tx} = E_{Tx}(l, d) + E_{Rx}(l) \qquad (2)$$

where E_{Rx} is the energy consumed by a node in receiving mode.

Fig. 1 Initial setup with
fixed network area of
150 m × 150 m

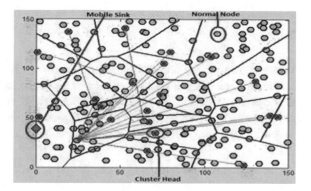

4 Improved LEACH

Improved LEACH is the routing algorithm over the optimized LEACH. It works
in four main phases: network construction phase, cluster head selection phase, task
ordination phase and data transmission phase.

4.1 Network Construction Phase

In this phase, fixed network area of 150 m × 150 m is taken; in this area, initially 100
nodes are placed. Node scalability is taken from 100, 120, 140, up to 400 nodes. It is
assumed that position of BS is moving along the Y-axis as in Fig. 1. The procedure
for the selection of cluster head is discussed in the next section.

4.2 Network Construction Phase

Limitations on the parameters of the network are initialized with number of nodes,
initial energy and sink location. Following are the steps to find cluster head using
Hopfield neural network.

- Initialize weights, $T_{XY} = \sum_{c=0}^{M-1} i_x^c i_y^c$,
- where $x \neq y$, i_x^c is element x of class c exemplar and i_y^c is element y of class c
  exemplar.
- Apply input on the outputs $z = i$.
- Iterate until the network converges.

$$z_y^+ = f_h \sum_{x=0}^{N-1} T_x^y x^z z^x \tag{3}$$

where f_h is hard limiter. The output converges to the best matching exemplar. A single Hopfield, is a popular method of Hopfield neural network is implemented to monitor group of nodes to check eligibility for becoming cluster head. Hopfield network also helps for pattern recall. In this method, we assign weights to every node according to certain decided parameters by taking the value of -1 or $+1$ [13]

$$T = [+1, +1, +1, +1, \ldots, -1, -1, -1, -1]$$

As per above illustration, first parameter indicates energy consumption (minimum), second parameter indicates distance (minimum), third parameter highlights surrounding nodes in the clusters (which should be maximum), and fourth constraint highlights ratio of consumed energy to allotted energy (which should be maximum). As per one-hop-field method, best case obtained value will be $[-1 \ -1 \ +1 \ +1]$, whereas in worst case obtained values will be $[+1 \ +1 \ -1 \ -1]$. All nodes with best and average set of values will be considered for cluster head selection [13].

Computation of average energy consumption by all eligible set of nodes is called a threshold value d_0. As per condition, only those nodes whose energy level is $\leq$ threshold value d_0 will participate in CH selection and also generate random number between 0 and 1, whereas nodes which do not fulfil required condition lead to wait for $1/p$ rounds. In order to become eligible as cluster head, the number should be less than the threshold $Th(\text{rnd})$ [4, 13]. The threshold in our case is given below in Eq. 4.

$$Th_{\text{rnd}} = \left\{ \begin{array}{ll} \frac{p}{1-p*(\text{rmod})\frac{1}{p}}, & \text{if } n \, \varepsilon \, G \\ 0, & \text{otherwise} \end{array} \right\} \tag{4}$$

As per above equation, 'p' represents total percentage of nodes eligible to be cluster head, r stands for current round, 'G' signifies subset of nodes which is selected as cluster head, and all leftover nodes which are not CH and RN will serve as normal nodes [13].

4.3   Network Construction Phase

Let us assume all are normal nodes; each of them decides whether RN condition is satisfied or not. To become RN, a prerequisite is to be placed and all necessary conditions must be satisfied. Distance of nodes is equated with MS trajectory [13]. Nodes which fulfil all necessary conditions will be labelled as RN, and necessary conditions are given below in Eq. 5.

$$\frac{y_w}{2}(1 + R_x) \le Y_y \ge \frac{Y_w}{2}(1 - R_x) \tag{5}$$

where Y_W stands for sampling region width, Y_Y signifies position of node in y-direction, and constant R_x needs to have value less than one [13].

4.4 Data Transmission

To transmit the data, shortest route search is necessary which also helps to optimize use of energy consumption only after creating and establishing schedule of data transmission. Once shortest routes identified, data transmission begins. Let us assume all rounds nodes hold some data, and at the same they also generate data at the same rate. In order to increase the energy efficiency and utilization, all the normal nodes communicate and send data to cluster head in its allotted time frame, and it is also important to keep radio on of normal node. It is also mandatory for all receiving CHs and RNs to be kept ON during data transmission. Once data transmission completes, all gathered data from all the nodes then CH sends data either sent to MS or closest RN [13].

5 Simulation Model and Results

Simulation is performed on MATLAB version R2014a considering fixed area of 150 m × 150 m. Initially, 100 nodes are taken as sample population sink is moving along y-axis as shown in Fig. 1. In the present work, we have taken static nodes and mobile sink taking nodes scalability from 100, 120, 140, increasing by 20 nodes and maximum up to 400 nodes. Initial parameters used in simulation model are described in Table 1.

Nodes are placed randomly in the given area. In the proposed work, the performance of improved LEACH is compared with LEACH [7] and optimized LEACH [4] observing number of rounds consumed for first (Fig. 2a), teenth (25%) (Fig. 2b) and all nodes dead (Fig. 2c).

Other parameters observed are residual energy, packets sent to base station and cluster head after round numbers 100, 120, 140,... 400. Initially, all the nodes are at same energy level, i.e. 0:3 J, and by scaling the nodes, we compare residual energy of the nodes shown in Fig. 3f. It is noticed from the graphs Fig. 3 and Tables 2, and 3 that improved LEACH is having better results in terms of number of rounds for first, tenth and all nodes dead for the given population of interest. Further, residual energy is marginally more, and a number of packets sent to base station and cluster head are more in improved LEACH. All these values lead to reduced energy consumption and hence increased network lifetime. A number of packets sent to base station and cluster head are significant performance evaluator. Simulation environment is designed to

Table 1 Simulation parameters

Simulation parameter		
Parameters	Values	Description
E_0	0.3 J	Initial energy
E_{amp}	0:0013 pJ/bit/m^4	Energy consumed by the power amplifier on multipath model
E_{fs}	10 pJ/bit/m^2	Energy consumed by the power amplifier on the free space model
E_{RX}	50 nJ/bit	Energy consumed by radio electronics in receiving mode
E_{TX}	50 nJ/bit	Energy consumed by radio electronics in transmit mode
l	4000 bits	Size of data packet
R_X	16% bits	Constant related to the width of region where RN is chosen
$X_m; Y_m$	100	Length and width of the region

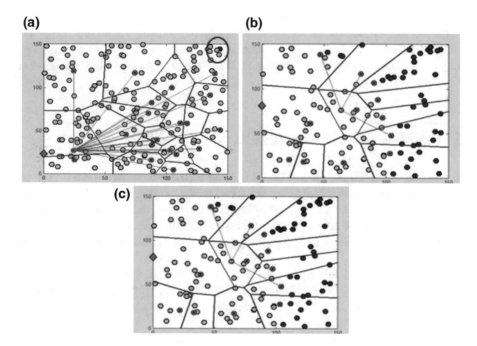

Fig. 2 Simulation for (**a**) first node dead; (**b**) teenth or 25% nodes dead and (**c**) all nodes dead

send data packets of fixed size of 4000 bits are sent to cluster head and base station, and number of packets sent till all the nodes are dead are counted and compared as shown in Fig. 3d, e shows number of packets sent to cluster head and Fig. 3f shows residual energy.

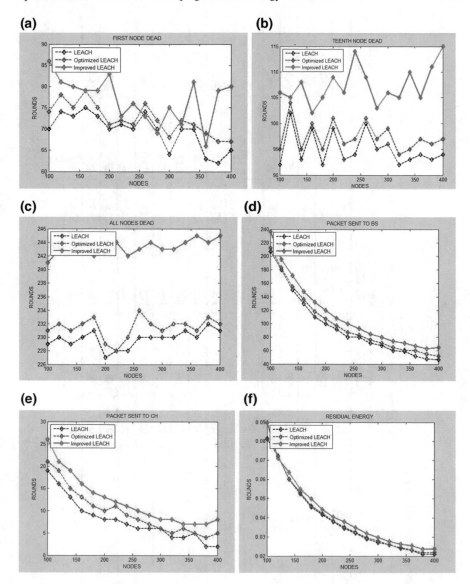

Fig. 3 Comparison among LEACH, optimized and improved LEACH first node dead (**a**); 25% nodes dead (**b**); all nodes dead (**c**); packets sent to BS (**d**); packets sent to CH (**e**); and residual energy (**f**)

Table 2 Simulation results for first, tenth and all nodes dead

No. of nodes	No. of rounds for first node dead			No. of rounds for tenth 25% nodes dead			No. of rounds for all nodes dead		
	LEACH	Optimized LEACH	Improved LEACH	LEACH	Optimized LEACH	Improved LEACH	LEACH	Optimized LEACH	Improved LEACH
100	70	74	86	92	95	106	222	232	241
120	73	78	81	100	104	105	222	233	243
140	70	75	80	90	95	108	220	233	244
160	73	79	79	96	100	102	224	234	243
180	69	75	79	92	95	105	224	234	242
200	65	71	83	99	101	109	223	233	244
220	68	72	73	93	96	106	224	233	244
240	67	71	76	93	97	114	222	232	242
260	72	76	73	95	101	109	223	235	243
280	67	72	69	94	97	103	224	235	244
300	63	68	75	95	99	106	222	234	243
320	67	72	71	91	94	105	223	233	243
340	67	71	81	91	95	110	224	235	244
360	64	69	66	94	97	105	225	234	245
380	62	67	79	94	96	111	225	234	244
400	62	67	80	94	97	115	224	235	245

Table 3 Results for residual energy packets received by cluster head and packets received by base station

No. of nodes	Residual energy			Packets received by cluster head			Packets received by base station		
	LEACH	Optimized LEACH	Improved LEACH	LEACH	Optimized LEACH	Improved LEACH	LEACH	Optimized LEACH	Improved LEACH
100	0.0815	0.0817	0.0896	212	214	236	21	23	26
120	0.0725	0.0726	0.071	181	185	195	18	20	21
140	0.0603	0.0604	0.0636	155	156	171	15	17	19
160	0.0532	0.0531	0.055	135	138	148	13	15	16
180	0.0460	0.0462	0.0498	118	120	132	11	13	14
200	0.0421	0.0423	0.0443	107	109	120	11	12	13
220	0.0383	0.0385	0.0403	96	99	108	10	11	12
240	0.0350	0.0352	0.0379	88	90	100	9	10	11
260	0.0325	0.0325	0.0348	82	85	93	8	9	10
280	0.0294	0.0296	0.0317	75	78	85	8	8	9
300	0.0026	0.0028	0.0297	71	73	80	7	8	8
320	0.025	0.026	0.0277	64	67	74	7	7	8
340	0.0241	0.0244	0.0263	61	64	71	6	7	7
360	0.0231	0.0233	0.0254	57	61	67	6	6	7
380	0.0216	0.0218	0.0236	57	57	63	5	6	7
400	0.020	0.022	0.0238	59	57	65	6	7	8

6 Conclusion

The present paper proposes an improved LEACH protocol which is an enhancement over the optimized LEACH; this protocol adopts the novel approach for cluster head selection based on the parameter energy consumed, distance (minimum), neighbours and ratio of current energy by remaining energy. The overall improvement in case of improved LEACH is 9–15% in terms of packets sent to base station and 7% to 9% in terms of residual energy. Further, the present work can be extended using SWARM optimization techniques to get better results.

References

1. Ari AAA, Labraoui N, Yenké BO, Gueroui A (2018) Clustering algorithm for wireless sensor networks: the honeybee swarms nest-sites selection process-based approach. Int J Sens Netw 27(1)
2. Nitesh K, Jana PK (2015) Grid based adaptive sleep for prolonging network lifetime in wireless sensor network. Procedia-Procedia Comput Sci 46:1140–1147
3. Bozorgi SM, Bidgoli AM (2018) HEEC: a hybrid unequal energy efficient clustering for wireless sensor networks HEEC: a hybrid unequal energy efficient clustering for wireless sensor networks. Wirel Netw
4. Mottaghi S, Zahabi MR (2015) Optimizing LEACH clustering algorithm with mobile sink and rendezvous nodes. AEU-Int J Electron Commun 69(2):507–514
5. Singh SK, Kumar P, Singh J (2017) A survey on successors of LEACH protocol. IEEE Access 5:4298–4328
6. Arumugam GS, Ponnuchamy T (2015) EE-LEACH: development of energy-efficient LEACH Protocol for data gathering in WSN. EURASIP J Wirel Commun Netw 2015(1):76
7. Kumar GS, Vinu MV, Jacob KP (2008) Mobility metric based LEACH-mobile protocol. In: Proceedings of the 16th international conference on advanced computing and communication. pp 248–253. https://doi.org/10.1109/ADCOM.2008.4760456
8. Heinzelman WR, Chandrakasan A, Balakrishnan H (2000) Energy-efficient communication protocol for wireless microsensor networks. System sciences. In: Proceedings of the 33rd annual Hawaii international conference on IEEE
9. Xing G, Wang T, Xie Z, Jia W (2007) Rendezvous planning in mobility-assisted wireless sensor networks. In: Proceedings real-time systems symposium. pp 311–320
10. Liu Q, Zhang K, Liu X, Linge N (2016) Grid routing: an energy-efficient routing protocol for WSNs with single mobile sink. Lect Notes Comput Sci (including Subser Lect Notes Artif Intell Lect Notes Bioinf) 10040:232–243
11. Singh M, Kaur R (2019) LP-ACO technique for link stability in VANETs. In: International conference on innovative computing and communications, vol 55. Springer, Singapore, pp 111–121
12. Jain S, Agrawal S, Paruthi A, Trivedi A, Soni U (2019) Neural networks for mobile data usage prediction in Singapore. In: International conference on innovative computing and communications. Lecture notes in networks and systems, vol 56. Springer, Singapore, pp 349–357
13. Konstantopoulos C, Pantziou G, Gavalas D, Mpitziopoulos A, Mamalis B (2014) A rendezvous-based approach enabling energy-efficient sensory data collection with mobile sinks. IEEE Trans Parallel Distrib Syst 23(5):809–817

Automated Vehicle Management System Using Wireless Technology

Indranil Sarkar, Jayanta Ghosh, Soumya Suvra Ghosal and Soumyadip Deb

Abstract The research is focused on the intelligent traffic system which is roughly based on the concepts of wireless transmission using radio frequency (RF) transmission and image processing. In the proposed system, FM receivers are fitted on traffic signals and FM transmitters are fixed on high-priority vehicles. The transmitter transmits the GPS location of a vehicle at a constant interval. On receiving the RF signal from the high-priority vehicle, the traffic signal, closest on its route, gets activated. From the received GPS location which is nearly 2 km away from the signal, the system finds out the direction and speed of the oncoming vehicle and releases the traffic on that route. The directions found by RF antenna [1] situated in the receiver. If no FM signals are received from any high-priority vehicle, then the system uses image processing [2] to find the vehicle density on each side. This is done by counting the number of vehicles in the four videos taken by a closed-circuit camera preinstalled in the traffic signal. The cameras have a range of 100 m each. An intense signal would make that side having the highest vehicular density green if and only if the vehicle number exceeds a threshold. Otherwise, the system mimics an ordinary traffic system. The system gives an accuracy of more than 91% while calculating the number of vehicles and in the other cases the error is as low as zero. The novel system simulation is achieved through CircuitMaker 2000 and MATLAB software.

Keywords RF transmission · RF antenna · Image processing · MATLAB · CircuitMaker 2000 software

I. Sarkar (✉) · J. Ghosh · S. S. Ghosal · S. Deb
Department of Electronics & Communication Engineering, National Institute of Technology Durgapur, Durgapur, India
e-mail: is.16u10428@btech.nitdgp.ac.in

J. Ghosh
e-mail: jg.16u10555@btech.nitdgp.ac.in

S. S. Ghosal
e-mail: ssg.16u10436@btech.nitdgp.ac.in

S. Deb
e-mail: sd.16u10394@btech.nitdgp.ac.in

© Springer Nature Singapore Pte Ltd. 2020
A. Khanna et al. (eds.), *International Conference on Innovative Computing and Communications*, Advances in Intelligent Systems and Computing 1059,
https://doi.org/10.1007/978-981-15-0324-5_18

1 Introduction

India grew tremendously post the economic reforms of 1990 in spite of infrastructure not growing commensurately. Presently, it has become extremely commonplace to see an ambulance stuck in metropolitan cities. The Secretary-General of Natural Institute of Emergency Medicine, Anucha Setthasathian, said, "More than 20% of patients needing emergency treatments have died on their way to the hospital because of delays due to traffic jams." Congested roads are decreasing vehicle average speed by 10% per year, leading to a huge loss in revenue and time. The loss incurred due to vehicular mismanagement in the big four metros only—Bangalore, Kolkata, Mumbai, and Chennai—was $22 billion (1536 crore rupees) in the year 2016. To address this problem of vehicle management, we propose an intelligent traffic system based on concepts of wireless transmission and image processing [2]. Direction-sensitive frequency-modulated (FM) antennas are fitted on traffic signals, and FM transmitters are fixed on high-priority vehicles. The transmitter transmits a signal of fixed intensity at a constant interval of time. Receiving the signal from high-priority vehicle, the traffic signal which is directly in front of the vehicle calculates the position of the vehicle based on the intensity of the incoming signal. The direction of the vehicle is also determined with the help of direction-sensitive FM antennas. Thus, the system finds out the direction and speed of incoming vehicles from the received information and releases traffic on that route. If no FM signals are received, then the system uses image processing to find vehicle density on each side. This is done by finding the number of vehicles [3] in each side with the help of videos taken by camera preinstalled in the traffic signal. The camera has a range of 100 m. An intense signal would make that side having the highest vehicular density green, if and only if the entropy exceeds a threshold. Otherwise, the system mimics an ordinary traffic system. The automated system is a viable, practical solution to mitigate the loss due to vehicular mismanagement in India (Fig. 1).

Fig. 1 Loss due to traffic jam in India per year (reference ATLAS)

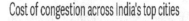

Cost of congestion across India's top cities

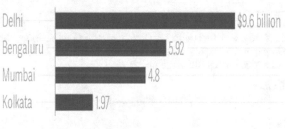

Delhi $9.6 billion

Bengaluru 5.92

Mumbai 4.8

Kolkata 1.97

2 Some Previous Works

Xui and Zhong [4] designed the overall vehicle management systems. The basis of their work was multi-node RFID cards.

Keun and Eung [5] designed a vehicle information management system. The system is able to deal with locations effectively.

Anand [6] resolved the compressive sensing approach to the reconstruct path in wireless sensor networks. As path length is quite smaller compared to network size, such path vectors are sparse. This is quite useful in the field of vehicle management.

Wu and Lu [6] gave the approaches of friendly interfaces and easy operation. The system was controlled through Baidu map positioning vehicle without geographical restrictions.

Daler Kaur and Mrs. Maninder Kaur [7] resolved the approach of the mobile wireless sensor network. The work was based on how we can take the coverage of a target with minimum movement from the present place.

In 2011, an intelligent traffic system [8] was created using RFID technology applying Dijkstra's algorithm in Bangalore City. The system calculates the minimum time paths of the whole city by creating a full map of the city. The system was first experimentally implemented in 35 traffic signals in December 2017. Then, it was implemented in 453 signals.

Lee, Kim [9] described a vehicle management system. The system has two main parts. The first one is tracking of the vehicle, and the second one is LPR. The system has some sensors which are used to detect the vehicles.

Hamad [10] described a general new approach to the fleet management system. His approach was to control and monitor the GPS-based system. There are two parts in this system. The hardware part consists of a tracker which is installed in the vehicle. Along with this, the second part, the Web-based software part, is located on the server side.

3 Simulation Algorithm

The proposed vehicle management system has three major steps to work with. In between those three steps, some minor steps are there. For every time, the system will give the green signal only for 30 s. The steps are described as follows:

1. *Search for receiving frequency of ambulance*: At first, the system will check if any radio frequency (RF) signal is coming from the ambulance or not. The RF receiver antennas are in search of the frequency in the range between 97.3 and 97.7 MHz. If any signal is received within that frequency range, then an energy harvesting antenna converts the signal into DC voltage. Then, the system checks in which direction RF antenna creates highest voltage (i.e., *north*). The system gives the green signal to that direction of the road (*north*) (Table 1, Fig. 2).

2. ***Search for receiving frequency of other priority vehicles***: In the absence of ambulance, the system checks if any radio frequency (RF) signal is coming from other priority vehicles or not. The RF receiver antenna checks for the frequency in the range between 96.6 and 96.9 MHz. If any signal is received within that frequency range, then an energy harvesting antenna converts the signal into DC voltage. Then, the system calculates in which antenna highest voltage is created (i.e., *south*). The system gives a green signal to that direction of the road (*south*).

3. ***The image processing part***: When there is no such high-priority vehicle, then this part of the system works. The following steps will be performed in sequence.

   (i)   The system takes the videos, each of length 1 s, from all the sides of the traffic signals. Those videos are to be taken by the preinstalled closed-circuit camera in the signal.

   (ii)   All the videos are to be processed by Otsu's method (white top method for the day and black top method for the night). The image processing part then calculates the number of vehicles in videos (i.e., in each side of the signal).

   (iii) After getting the number of the vehicle in each side, the system compares which side has the most number of vehicles. If and only if the highest number of vehicles is more than a predecided threshold, then only the system gives a green signal to that side.

   (iv)   If the maximum number of the vehicle is not more than the threshold (varies from city to city), then normal traffic signal continues.

   (v)   The system counts the number of consecutive green signal to a single side. If the consecutive green signal equals to three, then it mimics the normal traffic system.

Table 1 Proposed traffic signal system

IMAGE PROCESSING	PROPOSED SYSTEM
Input: Video signal *Vi*	Input: RF signal & vehicles

While (hVideoSrc != 0)
{ I1 = step(hVideoSrc);
Im1 = imtophat(I1, strel()); (Matrix)
th1 = multithresh(Im1);
th1 = multithresh(Im1);
BW1 = Im1 > th1;
Centroids1 = step(hBlob, BW1);
StapleCount=int32(size(Centroids,1));
Centroids1(:, 2) = Centroids1(1,2);
step(hVideoOut, It1); }
	i = input(RF Signal);
	j = input(Video of each side);
	while TRUE
	{ if (97.3<=i<=97.7)
	       Circuit 1 (Ambulance);
	   if (97.3<=i<=97.7)
	       Circuit 2 (Priority Vehicles);
	   else
	       image processing part (Count vehicles; Output most No of vehicles); }

Output: Most No. of vehicle & its direction if more than Threshold	Output: Green Signal as per the algorithm has defined

(vi) The loop of the image processing part continues until and unless any of the high-priority vehicles come (Fig. 3).

In the system, we are using the frequency range from 96.6 to 97.7 MHz as this frequency range is not used in commercial FM transmission in India. So there is no chance of the receiving RF antenna being affected by the any other commercial frequency or other than the priority vehicles transmitting frequency.

```
        It2 = insertMarker(It2, Centroids2, 'o', 'Size', 6, 'Color', 'r');
        It2 = insertMarker(It2, Centroids2, 'o', 'Size', 5, 'Color', 'r');
        It2 = insertMarker(It2, Centroids2, '+', 'Size', 5, 'Color', 'r');

        step(hVideoOut, It2);
    end

while ~isDone(hVideoSrc3)
        I3 = step(hVideoSrc3);
        Im3 = imtophat(I3, strel('square',18));
        Im3 = imopen(Im3, strel('rect',[15 3]));
        th3 = multithresh(Im3); % Determine threshold using Otsu's method
        BW3 = Im3 > th3;
        Centroids3 = step(hBlob, BW3);              % Blob Analysis

        StaplesCount3 = int32(size(Centroids3,1));

        txt3 = sprintf('Vehicles count: %d', StaplesCount3);
        It3 = insertText(I1,[10 280],txt3,'FontSize',22); % Display staples count

        Centroids3(:, 2) = Centroids3(1,2);          % Align markers horizontally

        It3 = insertMarker(It3, Centroids3, 'o', 'Size', 6, 'Color', 'r');
        It3 = insertMarker(It3, Centroids3, 'o', 'Size', 5, 'Color', 'r');
        It3 = insertMarker(It3, Centroids3, '+', 'Size', 5, 'Color', 'r');

        step(hVideoOut, It3);
    end
```

Fig. 2 MATLAB simulation algorithm

Fig. 3 Analog-to-digital converter

4 Hardware Device Specification

1. **ADC**: In digital electronics, ADC stands for analog-to-digital converter. This is an electronic component which converts the analog signal into digital bits. Here, in our integrated circuit, we used the ADC for conversion of analog RF signal into binary bits (Fig. 4).

   In the proposed system, we are using IC-AD9688BQ as our ADC.

2. **Decoder**: In digital electronics, this decoder is a special combinational circuit. The decoder converts the input information in desired form. If the number of given inputs to a decoder circuit is n, then the number of output lines will be 2^n. And if in some cases the output lines will be less than 2^n, then some output lines will be repeated for more than one times (Fig. 5).

   In our integrated circuit, we propose the use of 2- to 4-line decoder 74HC139.

3. **RF 433 MHz Transmitter/Receiver Module**: In the system, the FS1000A 433 MHz transmitter RF module is used. These RF transmitters transmit the RF signal for sending the message. The RF transmitters [11] transmitted the RF signal from the high-priority vehicles. And the RF receivers are present in each traffic signal. In our integrated system, we use the RF module to detect the presence of an ambulance within the range of 2 km radius (Fig. 6).

Fig. 4 IC 74LS139

Fig. 5 2 * 4 decoder

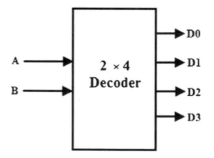

Fig. 6 Transmitter/receiver
pair

The features are

- Modulation technique: amplitude-shift keying (ASK)
- Input voltage: 5 V
- Frequency range: 433.92 MHz.

4. **Microcontroller**: Microcontroller is the integrated device which consists of processing unit, memory (RAM, ROM), and other input–output interfacing ports. Here, we used LPC-2148 ARM-7, i.e., Advanced RISC Machine (originally Acron RISC Machine) architecture-based microcontroller, which is a 32-bit general-purpose microprocessor (Fig. 7).

The main features are as follows:

- The performance of the processor is very efficient as it uses the pipelining method to process more than one block at a time.
- ISP (in system programming) or IAP (in application programming) is using on-chip boot loader software.
- On-chip static RAM is an 8 KB-40 KB, on-chip flash memory is 32 KB-512 KB, the wide interface is 128 bit, and the accelerator allows 60 MHz high-speed operation (Fig. 8).

5. **RF Energy Harvester** [12]: It is an integrated energy management subsystem which integrates a rectifier, which extracts AC power from ambient RF waves simultaneously [13]. These energy harvesters basically convert the RF wave into AC power [14] to simultaneously store energy in a rechargeable element [15] and supply the system with two independent regulated voltages. This allows product designers and engineers to extend battery life and ultimately get rid of the primary energy storage element in a large range of wireless applications [16] like industrial monitoring, home automation, and wearable (Fig. 9).

Fig. 7 ARM7

*ARM7 Based LPC2148
Microcontroller
Architecture*

Fig. 8 LPC2148

Fig. 9 Energy harvester IC
[17]

Component details are:

- TEXAS INSTRUMENTS CC2540F128RHAT IC
- SOC, RF, 128 K, 2.4 GHZ, 40VQFN.

5 Software Specification—Otsu's Method

The system uses Otsu's method [3] to calculate the number of vehicles. The main concept of Otsu's method is to calculate the number of objects by comparing the image pixel with the structuring elements. It finds the optimal value for the global threshold. For this, there is a simple algorithm:

- At first to define the global threshold value by generating a square matrix.
- Then to make the segmentation of the images.
- To compare if the pixel is brighter or darker than the threshold.
- After computing, the comparison threshold should be updated.

Two-Dimensional Otsu's Method: We are using the two-dimensional Otsu's method in counting the number of vehicles. The methods are:

White Top Hat: It takes an image and transforms it. After transforming, it returns an image containing those "objects" or "elements" of that input image that is

- Smaller than the structuring element
- Brighter than their surroundings.

Black Top Hat: It takes an image and transforms it. After transforming, it returns an image containing those "objects" or "elements" of that input image that is

- Smaller than the structuring element
- Darker than their surroundings.

The MATLAB code algorithm:

1. Firstly, the system checks for any type of high-priority vehicles coming toward the signal. In the absence of priority vehicles, the next steps will follow.
2. Four videos have been taken by the preinstalled camera, and then, the system processes those videos. Each of the videos is of one-second length.
3. For each video, the system creates a system object to read the avi file using the *hVideoSrc* function.
4. Then, the system creates a blob analysis system object to count the objects (i.e., vehicles here). And then, it finds the centroids of each object. Those are done by *a hBlob* function.
5. The system creates the system object to find the output display shown for the side having the most number of vehicles using *the hVideoOut* function.
6. Stream Processing Loop:
   The loop is used to count the number of vehicles in each input videos. The loop stops when the iteration reaches the end of the input video file. The steps of this processing loop are:

   (a) The system divides the given video signal into frames using step function, and then, it reads the produced images.
   (b) Using the *imtophat* system makes the top-hat filtering. For the day part, the system uses white top hat, and for the night part, system uses black top hat filtering. It creates an array of 0's and 1's that stands for the size and shape of the structuring element [18].
   (c) Processor opens the images created by the framing of videos earlier. And then, it compares the images with the structuring element and determines the center element.
   (d) Using Otsu's method, the processor determines the threshold of each part of the images.
   (e) Then, the processor compares the structure element and images, and if the images' frames are smaller and brighter (i.e., darker for Black top-hat) than the structure element, then it counts it as a countable object.
   (f) After that, it detects the centroids of the different frames of the video signal given as the input. And it calculates the total number of vehicles present in videos.
   (g) After calculating the number of vehicles in each signal, it compares the output and gives the most number of vehicles as the final output.

7. If the highest number of the vehicle is more than a threshold (i.e., 40), then the system gives green signal to that side, otherwise normal traffic mimics.

The MATLAB programs controls all the steps and the proposed system also. The program has to be loaded in the processor given in the previous part of descriptions.

6 Results and Discussions

The system output has nearly full accuracy (i.e., 100%) in the case of RF signal analysis as the RF energy harvesting antenna has perfect accuracy while detecting the direction. From the back-to-back two signals received in a traffic signal, the average speed of the vehicle can also be calculated. In the given table, there are some of the experimental data we got while testing the system. We experimented for every possible condition that can be faced by any traffic signals.

Here, we have tested the number of vehicle counting accuracy using the Otsu's white top method (for the day) and black top method (for the night). Here, in Table 2, some of the examples of given input and outputs are there.

After checking the vehicle counting output form of the system, we have checked what is the output when the input is 4 videos from the 4 direction of the signal. At this experiment, the system is giving the correct output most of the time (i.e., 97%). If the number of vehicles is the same in more than one side, then the system gives a green signal (i.e., algorithm output) to the side which is processed earlier. The system gives the green signal only when the most number of vehicles is more than that of a predefined threshold (i.e., 45 as an example here); otherwise, the output is "Normal".

From Table 2, we have got almost full accuracy while detecting which part of the proposed system will work in different conditions.

In Table 3, we have got 92% accuracy almost. The table is based on the data of calculating the number of vehicles from different videos.

From Table 4, we got almost accurate output while calculating which side of the

Table 2 Output of the system in each condition

Ambulance arriving	Other high-priority vehicle arriving	Image processing algorithm			
		North	South	West	East
Yes	× (No)	×	×	×	×
No	Yes	×	×	×	×
No	No	On	On	On	On
No	No	On	On	On	On
Yes	× (Yes)	×	×	×	×
No	Yes	×	×	×	×
Yes	× (No)	×	×	×	×

Table 3 Image processing vehicle counting output of the system

Manual counting of dense side	The result of the algorithm	Error percentage (in %)
53	58	9.433
102	94	7.834
88	95	7.954
75	80	6.667
55	50	9.090
73	77	5.479
41	45	9.756
97	92	5.155
56	51	8.928
31	34	9.677

Table 4 Image processing output of the system

Number of vehicles in east	Number of vehicles in west	Number of vehicles in south	Number of vehicles in north	Most dense side	Algorithm output
56	**87**	62	58	West	West
89	20	78	60	East	East
63	125	25	**145**	North	North
50	34	**50**	21	East	East
12	**30**	23	7	West	Normal
43	5	11	6	East	East
8	**41**	**41**	**41**	West	West
5	7	**13**	9	North	Normal
45	34	**122**	100	South	South
32	12	67	**104**	North	North

traffic signal the system is considering as the highest density side after calculating the number of the vehicle in each side.

So, the system is giving a satisfactory output. And we are still trying to achieve more and more accuracy and to overcome the problem of same numbers of vehicle and the problem of coming ambulance from more than one direction at the same time.

7 Conclusion

The problem of vehicular mismanagement has reached a threatening apex and is posing as a deterrent to the operation of smart cities. Though several attempts have already been made to solve this ever-growing problem, most of them are either pragmatic or require high resources for installation and maintenance. In this paper, a novel and cost-effective method is being proposed to solve the problem of vehicular mismanagement. We propose to install modified traffic signals consisting of direction-sensitive antennas and an image processing algorithm linked with an advanced processor to control the traffic flow. Four cameras of range 100 m each are fitted on each side of the signal. The images which are captured by the cameras are processed by the image processing algorithm to measure entropy to define the densest side, i.e., the side with most number of vehicles. On testing the algorithm on 50 images to measure the output, an average minimal error of 9% is obtained. And the output of the model in the presence of ambulance and other high-priority vehicle is almost fully accurate. The proposed model uses direction-sensitive antennas which are cost-effective as they are readily available in the market. Along with this, the total cost of the system is low in comparison with the total loss of revenue and lives.

In the future, we would like to work on improving the accuracy of the model to make it impeccable. Implementing the model in different weather conditions such as fog, night, and rain also needs to be considered in depth. Since real-time execution needs to be performed at a respectable speed, a more advanced and robust processor needs to be used for this model.

References

1. Spooner CM, Brown WA, Yeung GK (2002) Automatic radio-frequency environment analysis, IEEE. https://ieeexplore.ieee.org/document/910700
2. Sathuluri MR, Bathula SK, Yadavalli P, Kandula R IMAGE processing based intelligent traffic controlling and monitoring system using Arduino. https://ieeexplore.ieee.org/document/7987980
3. Hidayah MR, Akhlis I, Sugiharti E Recognition number of the vehicle plate using Otsu method and K-nearest neighbour classification. e-ISSN 2460-0040. http://journal.unnes.ac.id/nju/index.php/sji
4. Lee H, Kim D, Kim D, Bang SY Real-time automatic vehicle management system using vehicle tracking and car plate number identification. In: ICME '03, IEEE. https://ieeexplore.ieee.org/document/1221626/
5. Watson MD, Johnson SB (2007) A theory of vehicle management systems. In: 2007 IEEE aerospace conference, 2007, IEEE. ISSN 1095-323X. https://ieeexplore.ieee.org/document/4161686
6. Wu L, Qiao F, Lu J (2013) Design and implementation of a vehicle management system based on the ubiquitous network. In: 2013 IEEE 4th international conference on electronics information and emergency communication, IEEE. ISBN 978-1-4673-4933-8. https://ieeexplore.ieee.org/document/6835510

7. Wang Y, Ho OKW, Huang GQ, Li D, Huang H (2008) Study on RFID-enabled real-time vehicle management system in logistics. In: Proceedings of ICAL, the IEEE international conference on automation and logistics 2008, pp 2234–2238, IEEE. http://hdl.handle.net/10722/129746
8. Pan Y, Ge N, Dong Z (2008) Mixed-signal modeling and analysis for a digital RF direct sampling mixer. In: 2008 4th IEEE international conference on circuits and systems for communications, China, 26th–28th May 2008, Publisher—IEEE. https://ieeexplore.ieee.org/document/4536830
9. Lee EJ, Ryu KH Design of vehicle information management system for effective retrieving of vehicle location. Part of the Lecture Notes in Computer Science book series (LNCS, volume 3481). Online ISBN - 978-3-540-32044-9, Springer, Berlin. https://link.springer.com/chapter/10.1007/11424826_107
10. Madanayake A, Wijenayake C, Belostotski L, Bruton LT (2015) An overview of multi-dimensional RF signal processing for array receivers. In: 2015 Moratuwa Engineering Research Conference (Mercon), 2015, IEEE, Accession Number: 15180278. https://ieeexplore.ieee.org/document/7112355
11. Chen XM, Wei Z-H (2011) Vehicle management system based on multi-node RFID cards. In: Proceedings of the 30th Chinese control conference, 22-24th July 2011, IEEE. https://ieeexplore.ieee.org/document/6000392
12. Aminov P, Agrawal JP (2014) RF energy harvesting. In: 2014 IEEE 64th Electronic Components and Technology Conference (ECTC), Print ISSN: 0569-5503, IEEE. https://ieeexplore.ieee.org/document/6897549
13. Mrnka M, Vasina P, Kufa M, Hebelka V, Raida Z The RF Energy harvesting antennas operating in commercially deployed frequency bands: a comparative study. https://www.hindawi.com/journals/ijap/2016/7379624/
14. Kumari P, Sahay J (2017) Investigation on RF energy harvesting. In: 2017 Innovations in Power and Advanced Computing Technologies (i-PACT), India, IEEE. https://ieeexplore.ieee.org/document/8245021
15. Uzun Y (2016) Design and implementation of RF energy harvesting system for low-power electronic devices. J Electron Mater 45. https://link.springer.com/article/10.1007/s11664-016-4441-5
16. Mouapi A, Hakem N, Delisle GY (2017) A new approach to design of RF energy harvesting system to enslave wireless sensor networks. ICT, December 18, pp 228–233. https://www.sciencedirect.com/science/article/pii/S2405959517300218
17. Nintanavongsa P A survey on RF energy harvesting: circuits & protocols. https://doi.org/10.1016/j.egypro.2014.07.174
18. Li-Sheng J, Lei T, Rong-ben W, Lie G, Jiang-Wei C An improved Otsu image segmentation algorithm for path mark detection under variable illumination. In: Intelligent vehicles symposium. https://ieeexplore.ieee.org/document/1505209
19. Liu F, Zeng Z, Jiang R A video-based real-time adaptive vehicle-counting system for urban roads. https://doi.org/10.1371/journal.pone.0186098

A Supply Chain Replenishment Inflationary Inventory Model with Trade Credit

Seema Saxena, Vikramjeet Singh, Rajesh Kumar Gupta, Pushpinder Singh and Nitin Kumar Mishra

Abstract This research work considers a problem of obtaining the optimal replenishment schedule in a supply chain with a parameter of credit period rate. The model is designed for time-dependent quadratic demand and deterioration. The model is generalized considering partially backlogged shortages under inflation. However, as a peculiar case, an example is aimed for a model without lost sales. Sensitivity analysis is performed to analyze the mathematical formulation, and numerical examples are examined to study the effect of inflation and the time value of money on the economic order quantity model. The model evaluates the optimal replenishment schedules for the single retailer and single supplier for a single product in the supply chain subject to inflation.

Keywords Inflation · Supply chain · Credit term · Inventory

AMS Subject Classification 90B05 · 90B99

S. Saxena (✉) · R. K. Gupta · N. K. Mishra
Department of Mathematics, Lovely Professional University,
Phagwara, Punjab 144411, India
e-mail: mseemamishra@gmail.com

R. K. Gupta
e-mail: rajesh.gupta@lpu.co.in

N. K. Mishra
e-mail: snitinmishra@gmail.com

V. Singh
Department of Mathematics, IKGPTU Campus,
Batala, Punjab 143506, India
e-mail: vikram31782@gmail.com

P. Singh
Department of Mathematics, Mata Gujri College, Fatehgarh Sahib,
Punjab 140407, India
e-mail: pushpindersnl@gmail.com

© Springer Nature Singapore Pte Ltd. 2020
A. Khanna et al. (eds.), *International Conference on Innovative Computing and Communications*, Advances in Intelligent Systems and Computing 1059,
https://doi.org/10.1007/978-981-15-0324-5_19

1   Introduction

Considering real-life scenarios some goods/commodities, for example, medicines, vegetables, volatile liquids, fruits, electronic goods, etc., deteriorate constantly with respect to time due to spoilage, evaporation, etc. Ghare and Schrader [5] first introduced a model for an exponential deterioration rate of inventory. An extension in the inventory model of [5] was done by [12]. They extended for deterioration in Weibull form. Also, [16] considered a green inventory model under Weibull deterioration. The products after screening were remanufactured and recycled in a finite planning horizon. Chang [3] presented an inventory replenishment model considering deterioration and trade credit where the delay in payment is offered for large purchase quantity under the effects of inflation. Chung and Huang [4] demonstrated an optimal replenishment schedule for deteriorating goods incorporating delay in payment under inflation during the finite planning horizon.

Shortages occur when the inventory does not meet the demand. However, shortages are completely or partially backlogged. For fashionable goods or electronic equipment where the product has a short life cycle, the consumers do not wait for a long period and so switch to the other options available in the market, thus resulting in lost sales. Papachristos and Skouri [11] stated in their inventory model that backlogging is exponentially distributed when the waiting time of customer increases in the system. Palanivel and Uthayakumar [10] studied a linear trend of the demand function for a deterministic EOQ model with two separate warehouses and different deterioration rate under inflation allowing partial backlogging.

To demonstrate sales of the product during the product life cycle, the inventory models are made for the demand varying with respect to time as done by [8]. The research articles which include time quadratic demand are done by [6, 13]. Generally, the inventory models presume that the retailer pays for the products as they are received but realistically this does not happen. The supplier extends to the retailer a discount, either in terms of quantity or permissible delay of payment. Palanivel and Uthayakumar [10] framed an inventory model with permissible delay for time-differing demand with inflation. Inflation has been a key concern for the developing countries like India and China where the rate of inflation is in double digits. Inflation on variables related to total cost is not considered in the traditional inventory model. Singh et al. [15] studied an EOQ model considering credit term with effects of inflation on cost for deteriorating products. Yang et al. [20] have studied a model for spoilable products with the demand that is depending on stock under inflation with credit policy between supplier and retailer.

However, the effect cannot be ignored while framing an EOQ model. Bierman and Thomas [2] studied the model with inflation considering time discount. Yang et al. [19] extended the inventory model to include the effects of inflation and calculating the economic ordering schedule for the time-differing demand under shortages and deterioration. Jaggi et al. [9] derived an inventory model in a planning horizon that is finite for profit maximization, incorporating partially backlogged shortages and deterioration. The subject of [7, 14] is to calculate the optimal ordering point for the model with inflation, and [10] discusses ordering model in a planning horizon that is finite. Later, proceeding with the research paper, Sect. 2 shows the postulates and nomenclature, and Sect. 3 is the model framing for both the situations. Furthermore, Sect. 4 shows the optimality of the total cost. Finally, numerical illustration and sensitivity analysis are shown in the next section followed by a conclusion.

2 Postulates and Terminology

2.1 Postulates

1. Negligible lead time is considered.
2. Supplier offers credit term ρ to retailer for repaying the dues.
3. The demand $z(t)$ is time-dependent and quadratic in nature.
4. The deterioration of goods starts as they arrive in the stock. The rate of deterioration is $\psi(t) = \mu t$, with $(\mu > 0)$ and $(t > 0)$.
5. Deteriorated goods are neither replaced nor repaired.
6. Shortages are countenanced, and the unfulfilled demand is back-ordered. The notation $A(t)$ denotes the unfulfilled demand. The back-ordered shortages are the fraction $1/(1 + \gamma t)$ of demand, where $0 \le \gamma \le 1$ and t is the waiting time for the next replenishment.
7. The inventory model is for the finite planning horizon H.
8. During shortage period, $1 - A(t)$ fraction of demand turns into lost sales.
9. There is the same rate of inflation for all the inventory-related costs.
10. The model is assumed to have zero initial inventory level. It starts with shortages in the beginning until the first replenishment takes place. Also, the last cycle does not have any shortages.

2.2 Terminology

1. The inventory level is R_i^N, deteriorated quantity is D_i^N, shortage level is S_i^N, and lost sales are L_i^N for the time period $[t_i, s_{i+1}]$, $[i = 1, 2, 3, \ldots, n_1]$ in an individual system where there is no coordination between the retailer and the supplier and R_j^J, D_j, S_i^J, and L_i^J are the total inventory level, total quantity deteriorated, the total shortage level, and amount of lost sales during the time period $[t_j, s_{j+1}]$, $[j = 1, 2, 3, \ldots, n_2]$ for the joint system where the retailer is offered trade credit.
2. Capital cost $C_c(\$/unit/year)$ is equal for both the supplier and the retailer.
3. The cost of carrying inventory is $h(\$/unit/year)$, and $h = I_{hr} + C_c$.
4. Setup cost for supplier is S_s ($/order).
5. Purchasing price for retailer is P_o and for the supplier is P_s ($/unit) and $P_s < P_o$.

3 Model Approach and Analysis

The model is devised for the time-varying demand and deterioration with partially backlogged shortages. Figure 1 depicts the inventory theory under review.

Model 1: Individual model

The instantaneous level of inventory $\left(I_{1i}^N(t)\right)$ is given by following differential equation:

$$\frac{d\left(I_{1i}^N(t)\right)}{dt} + \psi(t)I_{1i}^N(t) = -z(t), t_i \leq t \leq s_{i+1}, \quad (i = 1, 2, \ldots, n_1) \quad (1)$$

with the boundary condition is $I_{1i}^N(s_{i+1}) = 0$

$$\frac{d\left(I_{2i}^N(t)\right)}{dt} = z(t)A(t) = \frac{z(t)}{1 + \gamma(t_i - t)}, s_i \leq t \leq t_i, \{i = 1, 2, \ldots, n_1\} \quad (2)$$

Taking $\psi = \mu t, (0 < \mu < 1)$, differential equation (1) is solved as

$$I_{1i}^N(t) = \int_t^{s_{i+1}} e^{(\mu/2)(u^2 - t^2)} f(u)du, t_i \leq t \leq s_{i+1}, \{i = 1, 2, \ldots, n_1\}. \quad (3)$$

Neglecting higher orders of μ, as μ is small. We have,

$$R_i^N = \int_{t_i}^{s_{i+1}} \left[\left(1 + (\mu/2)t^2\right)(t - t_i) - (\mu/6)\left(t^3 - t_i^3\right)\right] z(t)dt, (i = 1, 2, 3, \ldots, n_1). \quad (4)$$

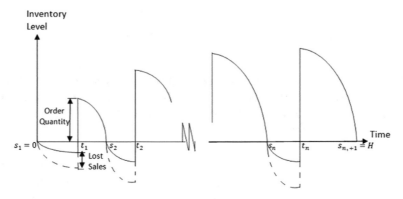

Fig. 1 Inventory model under study

For interval $[s_i, t_i]$, shortages in individual model amounts to

$$S_i^N = \int_{s_i}^{t_i} (((t_i - t)) \div (1 + \gamma \, (t_i - t))) \, z(t) dt, \, (i = 1, 2, 3, \ldots, n_1) \quad (5)$$

Quantity ordered in the horizon is

$$Q^N = \sum_{i=1}^{n_1} Q_i^N = \sum_{i=1}^{n_1} \left(R_i^N + S_i^N \right). \quad (6)$$

Present estimate of the various costs associated is as follows:
Ordering cost C_i is

$$C_i = C_0 * e^{(\alpha - \pi) * t_i} \quad (7)$$

Holding cost H_i is

$$H_i = H_0 * \int_{t_i}^{s_{i+1}} e^{(\alpha - \pi) * t} * \left[\left(1 + \frac{\mu}{2} t^2 \right) (t - t_i) - \frac{\mu}{6} \left(t^3 - t_i^3 \right) \right] z(t) dt \quad (8)$$

Deterioration cost D_i is

$$D_i = D_0 * \int_{t_i}^{s_{i+1}} e^{(\alpha - \pi) * t} * (\mu * t) \, (t - t_i) \, z(t) dt \quad (9)$$

Shortage cost S_i is

$$S_i = S_0 * \int_{s_i}^{t_i} e^{(\alpha - \pi) * t} * (((t_i - t)) \div (1 + \gamma \, (t_i - t))) * z(t) dt \quad (10)$$

Cost of lost sales L_i is

$$L_i = L_0 * \int_{s_i}^{t_i} e^{(\alpha - \pi) * t} * \frac{\gamma \, (t_i - t) \, z(t)}{1 + \gamma \, (t_i - t)} \, dt \quad (11)$$

Purchase cost P_i is $P_i =$

$$P_0 * e^{(\alpha - \pi) * t_{i-1}} * \left(\int_{t_i}^{s_{i+1}} \left[1 + (\mu/2) \left(t^2 - t_i^2 \right) \right] z(t) dt \right.$$
$$\left. + \int_{s_i}^{t_i} (1 \div (1 + \gamma \, (t_i - t))) \, z(t) dt \right) \quad (12)$$

Present estimate of the total cost for the retailer is $PWTC_r^N$ which is summing the cost of purchasing, holding cost, ordering cost, cost of deterioration, shortage, and lost sale cost.

$$PWTC_r^N \left(n_1, s_1, t_1, \ldots, s_{n_1+1}\right) = \sum_{i=1}^{i=n_1} \left(C_i + D_i + S_i + L_i + P_i + H_i\right)$$

$$PWTC_r^D \left(n_1, s_1, t_1, \ldots, s_{n_1+1}\right) = \sum_{i=1}^{i=n_1} \left(C_0 * e^{(\alpha-\pi)*t_i} + D_0 * \int_{t_i}^{s_{i+1}} e^{(\alpha-\pi)*t} * (\mu * t)\right.$$

$$(t - t_i) z(t)dt + S_0 * \int_{s_i}^{t_i} e^{(\alpha-\pi)*t} * \frac{(t_i - t) z(t)dt}{(1 + \gamma (t_i - t))}$$

$$+ L_0 * \int_{s_i}^{t_i} e^{(\alpha-\pi)*t} * \frac{\gamma (t_i - t) z(t)dt}{(1 + \gamma (t_i - t))}$$

$$+ P_0 * e^{(\alpha-\pi)*t_{i-1}} * \left(\int_{t_i}^{s_{i+1}} \left[1 + (\mu/2) \left(t^2 - t_i^2\right)\right] z(t)dt + \int_{s_i}^{t_i} \frac{z(t)dt}{(1 + \gamma (t_i - t))}\right)$$

$$+ H_0 * \int_{t_i}^{s_{i+1}} e^{(\alpha-\pi)*t} * \left[\left(1 + (\mu/2)t^2\right) (t - t_i)\right.$$

$$\left. - (\mu/6) \left(t^3 - t_i^3\right)\right] z(t)dt, \quad \{i = 1, 2, \ldots, n_1\} \tag{13}$$

The inventory system is designed to optimize the present estimate of total cost by determining the values of s_i, t_i, and n_1. $PWTC_r^N$ is minimum when

$$\frac{\partial PWTC_r^N (t_i, s_i; n_1)}{\partial t_i} = 0 \tag{14}$$

and

$$\frac{\partial PWTC_r^N (t_i, s_i; n_1)}{\partial s_i} = 0 \tag{15}$$

which is further solved to obtain the optimum values for the double derivative which minimizes the total cost.

Equations (14) and (15) are solved for the optimal solution for the minimum value of $PWTC_r^N \left(n_1, s_1, t_1, \ldots, s_{n_1+1}\right)$. Let the optimal solution be n_1^{N*}, s_1, t_1^{N*}, s_2^{N*} ..., $s_{n_1+1}^{N*} = H$.

Supplier also calculates the present estimate of total cost for the individual system by the equation:

$$PWTC_s^N \left(n_1^{N*}, s_1, t_1^{N*}, s_2^{N*}, \ldots, s_{n_1+1}^{N*} = H\right)$$
$$= n_1^{N*} * e^{(\alpha-\pi)*t_i} * S_s + n_1^{N*} * e^{(\alpha-\pi)*t_i} * P_s * Q_{OPT}^*$$
$$= n_1^{N*} * e^{(\alpha-\pi)*t_i} * S_s + n_1^{N*} * e^{(\alpha-\pi)*t_i} * P_s * \left(R_i^{N*} + S_i^{N*}\right). \tag{16}$$

The optimum quantity ordered is

$$Q^*_{OPT} = \sum_{i=1}^{n_1^{N*}} Q_i^{N*} = \sum_{i=1}^{n_1^{N*}} \left(R_i^{N*} + S_i^{N*} \right). \tag{17}$$

Model 2: Joint Model

The supplier expects the retailer to agree to this new replenishment schedule. The increase in the holding cost tends to increase the net cost of the retailer in the joint model, yet the increase is not only compensated by the supplier, but also the savings realized are shared by him in the coordinated system.

For reordering cycle n_2, the present estimate of the overall cost is

$$PWTC_r^J (n_2, s_1, t_1^J, \ldots, s_{n_2+1}^J) = \sum_{j=1}^{j=n_2} (C_j + D_j + S_j + L_j + P_j + H_j) \tag{18}$$

$$= \sum_{j=1}^{j=n_2} \left(C_0 * e^{(\alpha-\pi)*t_j^J} \right.$$

$$+ D_0 * \int_{t_j^J}^{s_{j+1}^J} e^{(\alpha-\pi)*t} * (\mu * t) \left(t - t_j^J \right) z(t) dt$$

$$+ S_0 * \int_{s_j^J}^{t_j^J} e^{(\alpha-\pi)*t} * \left(\left(\left(t_j^J - t \right) \right) \div \left(1 + \gamma \left(t_j^J - t \right) \right) \right) z(t) dt$$

$$+ L_0 * \int_{s_j^J}^{t_j^J} e^{(\alpha-\pi)*t} * \left(\left(\gamma \left(t_j^J - t \right) \right) \div \left(1 + \gamma \left(t_j^J - t \right) \right) \right) z(t) dt$$

$$+ P_0 * e^{(\alpha-\pi)*t_{j-1}^J} * \left(\int_{t_j^J}^{s_{j+1}^J} \left[1 + (\mu/2) \left(t^2 - \left(t_j^J \right)^2 \right) \right] z(t) dt \right.$$

$$\left. + \int_{s_j^J}^{t_j^J} \left(1 \div \left(1 + \gamma \left(t_j^J - t \right) \right) \right) z(t) dt \right)$$

$$+ H_0 * \int_{t_j^J}^{s_{j+1}^J} e^{(\alpha-\pi)*t} * \left[\left(1 + (\mu/2)t^2 \right) \left(t - t_j^J \right) \right.$$

$$\left. - (\mu/6) \left(t^3 - \left(t_j^J \right)^3 \right) \right] z(t) dt \tag{19}$$

The supplier compensates the retailer's increase in his/her cost in the joint model.

$$PWTC_s^J \left(n_2, s_1, t_1^J, \ldots, s_{n_2+1}^J \right) = n_2 * e^{(\alpha-\pi)*t_j} * S_s + n_2 * e^{(\alpha-\pi)*t_j} * P_s * Q_j^J +$$
$$PWTC_r^J \left(n_2, s_1, t_1^J, \ldots, s_{n_2+1}^J \right) - PWTC_r^N \left(n_1^{N*}, s_1, t_1^{N*}, s_2^{N*}, \ldots, s_{n_1+1}^{N*} = H \right) \tag{20}$$

$$= n_2 * (S_s + C_0) * e^{(\alpha - \pi)*t_j} + n_2 * (P_s + P_0) * e^{(\alpha - \pi)*t_j} * Q_j^J$$

$$+ H_0 * \int_{t_j^J}^{s_{j+1}^J} e^{(\alpha - \pi)*t} * \left[\left(1 + (\mu/2)t^2 \right) \left(t - t_j \right) - (\mu/6) \left(t^3 - t_j^3 \right) \right] z(t) dt$$

$$+ D_0 * \int_{t_j^J}^{s_{j+1}^J} e^{(\alpha - \pi)*t} * (\mu * t) \left(t - t_j \right) z(t) dt$$

$$+ (S_0 + L_0 * \gamma) * \int_{s_j^J}^{t_j^J} e^{(\alpha - \pi)*t} * \left(\left((t_j - t) \right) \div \left(1 + \gamma \left(t_j - t \right) \right) \right) z(t) dt \quad (21)$$

The optimum result of the present estimate of the total cost for the supplier $PWTC_s^J \left(n_2, s_1, t_1^J, \ldots, s_{n_2+1}^J = H \right)$ is $n_2^{J*}, s_1, t_1^{J*}, s_2^{J*}, \ldots, s_{n_2+1}^{J*} = H$.

The optimum quantity to be ordered in the joint model is $Q^J = \sum_{j=1}^{n_2^{J*}} Q_j^{J*} = \sum_{j=1}^{n_2^{J*}} \left(R_j^{J*} + S_j^{J*} \right)$.

The new ordering plan is accepted by the retailer only if

$$PWTC_r^N \left(n_1^{N*}, s_1, t_1^{N*}, s_2^{N*}, \ldots, s_{n_1+1}^{N*} = H \right)$$

$$\geq PWTC_r^J \left(n_2^{J*}, s_1, t_1^{J*}, s_2^{J*}, \ldots, s_{n_2+1}^{J*} = H \right) - \sum_{j=1}^{n_2^{J*}} e^{(\alpha - \pi)*t_j} C_c.\rho \left(s_{j+1}^{J*} - s_j^{J*} \right) Q_j^{J*} \quad (22)$$

where credit period duration is M_j^J and an equal trade credit rate ρ is followed for every ordering cycle.

$$M_j^J = \rho \left(s_{j+1}^{J*} - s_j^{J*} \right)$$

Credit term is given for a minimum time to retailer and so the present estimate of the cost for the retailer with minimum credit rate $\rho_{\min}$ is

$$PWTC_r^J \left(n_2^{J*}, s_1, t_1^{J*}, s_2^{J*}, \ldots, s_{n_2+1}^{J*} = H \right) - \sum_{j=1}^{n_2^{J*}} e^{(\alpha - \pi)*t_j} C_c.\rho_{\min} \left(s_{j+1}^{J*} - s_j^{J*} \right) Q_j^{J*}$$

$$= PWTC_r^N \left(n_1^{N*}, s_1, t_1^{N*}, s_2^{N*}, \ldots, s_{n_1+1}^{N*} = H \right) \quad (23)$$

Also for the supplier, the rate of credit period is maximum, $\rho_{\max}$.

$$n_2^{J*} S_s.e^{(\alpha - \pi)*t_j} + \sum_{j=1}^{n_2^{J*}} P_s.e^{(\alpha - \pi)*t_j}.Q_j^{J*} + \sum_{j=1}^{n_2^{J*}} .e^{(\alpha - \pi)*t_j}.C_c.\rho_{\max} \left(s_{j+1}^{J*} - s_j^{J*} \right) Q_j^{J*}$$

$$= PWTC_s^N \left(n_1^{N*}, s_1, t_1^{N*}, s_2^{N*}, \ldots, s_{n_1+1}^{N*} = H \right) \quad (24)$$

The extra cost saving in joint model is divided equally by $\bar{\rho}$, i.e., average of $\rho_{\max}$ and $\rho_{\min}$. Present estimate of the net cost for retailer and supplier is evaluated using $\bar{\rho}$:

$$PWTC_r^{J*\rho}\left(n_2^{J*}, s_1, t_1^{J*}, s_2^{J*}, \ldots, s_{n_2+1}^{J*} = H\right)$$

$$= PWTC_r^J\left(n_2^{J*}, s_1, t_1^{J*}, s_2^{J*}, \ldots, s_{n_2+1}^{J*} = H\right) - \sum_{j=1}^{n_2^{J*}} .e^{(\alpha-\pi)*t_j}.C_c\bar{\rho}\left(s_{j+1}^{J*} - s_j^{J*}\right)Q_j^{J*}$$

(25)

and

$$PWTC_s^{J*\rho}\left(n_2^{J*}, s_1, t_1^{J*}, s_2^{J*}, \ldots, s_{n_2+1}^{J*} = H\right)$$

$$= n_2^{J*}.e^{(\alpha-\pi)*t_j}.S_s + \sum_{j=1}^{n_2^{J*}} P_s.e^{(\alpha-\pi)*t_j}.Q_j^{J*} + \sum_{j=1}^{n_2^{J*}} .e^{(\alpha-\pi)*t_j}.C_c.\bar{\rho}\left(s_{j+1}^{J*} - s_j^{J*}\right)Q_j^{J*} \quad (26)$$

4 Optimality Condition for $PWTC_r^N$ and $PWTC_s^J$

Theorem 1 $PWTC_r^N\left(n_1^{N*}, s_1, t_1^{N*}, s_2^{N*}, \ldots, s_{n_1+1}^{N*} = H\right)$ *is convex in* n_1.

Total cost follows convex nature which is presented in Figs. 2 and 3 for $a_1 = 225$ in individual and joint system. The same can be used to show that $PWTC_s^J$ is convex in nature and $PWTC_s^j(n_2, s_1, t_1^J, s_2^J, \ldots, s_{n+1}^J)$ reaches a minimal.

4.1 Numerical Exercise

A set of values for the parameters is taken to support the inventory model with an optimum result.

Example 1 Taking $a_1 = \{225\}$ products/year, $b_1 = 110$ products/year, $c_1 = 5$ products/year, $P_o = 0.3$ \$/product, $S_o = 18$ \$/product, $L_o = 22$ \$/product, $\gamma = 0.8$, $\mu = 0.01$, $s_1 = 0$, $C_o = 300$ \$/ order, $S_s = 320$ \$/setup, $H = 4$, $P_s = 0.3$ \$/product,

Fig. 2 Convexity of total cost for $a_1 = 225$ in individual system

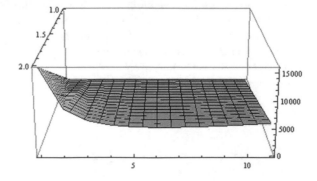

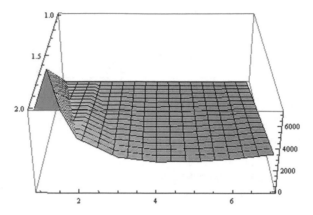

Fig. 3 Convexity of total cost for $a_1 = 225$ in joint system

Table 1 For $a_1 = \{225\}$ inflated cost of retailer in an individual system

$\downarrow \quad \rightarrow n_1$ a_1	1	2	3	4	5
225	15639.3	8383.29	6267.48	5457.99	5129.8
$\downarrow \quad \rightarrow n_1$ a_1	6	7	8	9	10
225	5037.26	5077.04	5186.41	5367.41	5582.42

$C_c = 2.5$ \$/product/year , $H_o = 3$ \$/product/year, $\alpha = 0.115$, $\pi = 0.05$, $DC = 12$ \$/product/year. Use of Mathematica(version 8.0) is done to calculate the solution of the nonlinear equations in individual model for different replenishment cycles $n_1 = \{1, 2, \ldots, 7\}$.

Table 1 shows the present estimate of the overall retailer's cost for $a_1 = \{225\}$. The optimized total cost for retailer under inflation for the values of $a_1 = \{225\}$ is 5037.26 for replenishment cycle 6. The decrement in the inflated cost from $n_1 = 1$, till it reaches a minimum or optimum value at $n_1 = 6$ after which the value again gradually increases for remaining cycles following its convexity for different values of a_1. The graphical representation as in Fig. 2 for the overall cost function shows the convex nature of the cost subject to inflation rate.

The optimum values of time s_i's and t_i's alongwith the optimal cycles are presented in Table 2 for distinct values of a_1.

Table 2 Optimum ordering period for retailer in individual model

a_1	n_1^{N*}	s_1	s_2	s_3	s_4	s_5
225	6	0	0.662171	1.3173	1.97399	2.6376
a_1	n_1^{N*}	s_6	s_7			
225	6	3.31183	3.99944			
a_1	n_1^{N*}	t_1	t_2	t_3	t_4	
225	6	0.0551381	0.717079	1.37316	2.03152	
a_1	n_1^{N*}	t_5	t_6			
225	6	2.6973	3.37412			

Table 3 For $a_1 = \{225\}$, inflated cost for supplier in joint system

$\rightarrow n_2$							
a_1	1	2	3	4	5	6	7
225	11542.	4719.51	2996.54	2564.47	2608.93	2867.12	3287.23

Table 4 Optimal ordering period for supplier in joint model

a_1	n_2^{J*}	s_1	s_2	s_3	s_4	s_5
225	4	0	1.08096	2.08518	3.05131	3.99858
a_1	n_2^{J*}	t_1	t_2	t_3	t_4	
225	4	0.10232	1.17385	2.17539	3.14158	

Table 5 Cost saving percentage for $a_1 = \{225\}$

Individual system

a_1	$PWTC_r^{N*}$	$PWTC_s^{N*}$	n_1^{N*}	Q^{N*}	ρ_{min}	ρ_{max}	$\bar{\rho}$
225	5037.26	2859.29	6	1895.32	0.020	0.034	0.27

Joint system

a_1	$PWTC_r^{J*\rho}$	$PWTC_s^{J*\rho}$	n_2^{J*}	Q^{J*}
225	4889.85	2711.88	4	1897.92

% Cost saving

a	$\dfrac{\Delta PWTC}{PWTC_r^{N*}}$	$\dfrac{\Delta PWTC}{PWTC_s^{N*}}$
225	2.92	5.15

Table 3 displays the optimized total cost for supplier under inflation in a joint model for different values of $a_1 = \{225\}$ and for different replenishment schedules n_2 going from 1 to 7. The optimum value of the total cost for $a_1 = \{225\}$ is \$2564.47. The optimal refilling schedule is 4. Table 4 comprises of the optimal replenishment cycles and the optimal time s_i's and t_i's. Table 5 shows the cost-saving percentage for the retailer as well as for the supplier when they both agree for the new replenishment schedule in the joint system. The convex nature of the final cost under inflation is graphically representation for $a_1 = 225$ in Fig. 3.

5   Sensitivity Analysis

Sensitivity analysis is done for the inventory model analyzing whether the formulated model is influenced by the alterations in the input parameters.

The following table demonstrates the result when every parameter is altered by $-50, -25, 25$, and 50%, of the initial cost taken in Example 1.

1. $PWTC_r^{J*\rho}$ increases for parameters $a_1, b_1, 1, H_o, P_o, C_o, d$, and H. It decreases for the parameters P_s, S_s, π, and γ. Also, $PWTC_r^{J*\rho}$ less sensitive to $c_1, \mu, s, l, P_o, P_s, S_s, DC, \alpha$, and π.
2. $PWTC_s^{J*\rho}$ increases for the parameters $a_1, b_1, c_1, \mu, H_o, s, P_s, S_s, DC, \alpha$, and H. It decreases for the parameters l, P_o, C_o, π, and γ. Also, $PWTC_s^{J*\rho}$ is extremely responsive to the parameters H and S_s, fairly responsive to $a_1, H_o, P_s, C_o, \alpha$, and γ. $PWTC_s^{J*\rho}$ is unresponsive to the parameters π, DC, P_o, l, s, b_1, c_1, μ.

6   Conclusions

The financial state of an organization is substantially affected by the inflation. Every cost associated with the inventory model is greatly influenced as there is an increase in the rate of inflation due to which buying index of money decreases. The present paper demonstrates an inferred inventory joint framework where supplier and the retailer enters in a common agreement for the replenishment schedule and quantity ordered with permissible delay. Present estimate of the total cost is determined for the retailer as well as for the supplier. Also, numerical examples validate the model for any set of values. Lastly sensitivity analysis is done for the change in every factor to ascertain their impact on the overall cost. The present model can be extended by adding fuzzy logic as in [1, 17, 18].

	Sensitivity analysis for the change in each parameter	
% Change in parameter	$\dfrac{\Delta PWTC_r^{J*\rho}*100\%}{PWTC_{r*}^{J*\rho}}$	$\dfrac{\Delta PWTC_s^{J*\rho}*100\%}{PWTC_{s*}^{J*\rho}}$
a_1 −50	−18.7886	−24.1073
−25	−9.046	−11.9912
25	11.0582	7.01499
50	21.7364	13.4031
b_1 −50	−12.7777	−7.65148
−25	−6.38878	−3.82572
25	3.58144	9.09425
50	9.67571	13.1159
c_1 −50	1.31821	−5.61953
−25	2.2008	−5.40583
25	0.698756	0.516567
50	1.39578	1.03274
μ −50	0.435129	−5.5699
−25	1.60703	−5.11528
25	1.10117	0.530871
50	2.19322	1.05689
H_0 −50	−21.2359	−17.6589
−25	−9.43805	−8.03695
25	10.689	3.37463
50	20.324	5.57955
s −50	2.73164	−6.09767
−25	2.80691	−5.43692
25	0.0693557	0.507132
50	0.158604	0.904935
l −50	−2.92412	0.170032
−25	−1.316	0.0915276
25	1.08737	−0.0904029
50	4.73686	−4.79181
P_o −50	−8.66318	0.0735532
−25	−4.33156	0.0367678
25	4.3315	−0.0367505
50	8.66295	−0.0734837
P_s −50	0.0368718	−14.5863
−25	0.0184476	−7.29314
25	−0.0184711	7.2931
50	−0.0369657	14.5862
S_s −50	3.4603	−34.0989
−25	2.04299	−16.5223
25	−2.18021	16.2911
50	−4.3617	32.58
C_o −50	−25.877	17.2343
−25	−9.30514	3.5487
25	11.26	−7.14465
50	22.2856	−13.8979
DC −50	0.591263	−5.39095
−25	1.68416	−5.02671
25	1.02882	0.438901
50	2.05003	0.874761
α −50	−10.2732	−16.4019
−25	−6.70085	−6.27349
25	7.29999	6.8106
50	15.2526	14.2153
β −50	6.31203	5.88979
−25	3.09734	2.89179
25	−2.98423	−2.79019
50	−5.85919	−5.48356
γ −50	23.6259	−15.8831
−25	9.40406	−7.67832
25	−1.99311	1.9056
50	3.12699	−3.00943

References

1. Asopa P, Asopa S, Mathur I, Joshi N (2019) A model of fuzzy intelligent tutoring system. In: International conference on innovative computing and communications. Springer, Berlin, pp 303–311
2. Bierman H, Thomas J (1977) Inventory decisions under inflationary conditions. Decis Sci 8(1):151–155
3. Chang C-T (2004) An EOQ model with deteriorating items under inflation when supplier credits linked to order quantity. Int J Prod Econ 88(3):307–316
4. Chung K-J, Huang C-K (2009) An ordering policy with allowable shortage and permissible delay in payments. Appl Math Modell 33(5):2518–2525
5. Ghare PM, Schrader GF (1963) A model for an exponential decaying inventory. J Ind Eng 14:238–243
6. Ghosh S, Chaudhuri K (2006) An EOQ model with a quadratic demand, time-proportional deterioration and shortages in all cycles. Int J Syst Sci 37(10):663–672
7. Gilding BH (2014) Inflation and the optimal inventory replenishment schedule within a finite planning horizon. Euro J Oper Res 234(3):683–693
8. Hariga M (1995) An EOQ model for deteriorating items with shortages and time-varying demand. J Oper Res Soc 46(3):398–404
9. Jaggi C, Khanna A, Nidhi N (2016) Effects of inflation and time value of money on an inventory system with deteriorating items and partially backlogged shortages. Int J Ind Eng Comput 7(2):267–282
10. Palanivel M, Uthayakumar R (2015) Finite horizon EOQ model for non-instantaneous deteriorating items with price and advertisement dependent demand and partial backlogging under inflation. Int J Syst Sci 46(10):1762–1773
11. Papachristos S, Skouri K (2000) An optimal replenishment policy for deteriorating items with time-varying demand and partial-exponential type-backlogging. Oper Res Lett 27(4):175–184
12. Philip GC (1974) A generalized EOQ model for items with weibull distribution deterioration. AIIE Trans 6(2):159–162
13. Singh P, Mishra NK, Singh V, Saxena S (2017) An EOQ model of time quadratic and inventory dependent demand for deteriorated items with partially backlogged shortages under trade credit. In: AIP conference proceedings, vol 1860. AIP Publishing, p 020037
14. Singh V, Saxena S, Singh P, Mishra NK (2017) Replenishment policy for an inventory model under inflation. In: AIP conference proceedings, vol 1860. AIP Publishing, p 020035
15. Singh V, Saxena S, Gupta RK, Mishra NK, Singh P (2018) A supply chain model with deteriorating items under inflation. In: 2018 4th international conference on computing sciences (ICCS). IEEE, New York, pp 119–125
16. Singh V, Mishra NK, Mishra S, Singh P, Saxena S (2019) A green supply chain model for time quadratic inventory dependent demand and partially backlogging with Weibull deterioration under the finite horizon. In: AIP conference proceedings, vol 2080. AIP Publishing, p 060002
17. Srivastava P, Sharma N (2019) Fuzzy risk assessment information system for coronary heart disease. In: International conference on innovative computing and communications. Springer, Berlin, pp 159–170
18. Thorat S, Mahender CN (2019) Domain-specific fuzzy rule-based opinion mining. In: International conference on innovative computing and communications. Springer, Berlin, pp 287–294
19. Yang H-L, Teng J-T, Chern M-S (2010) An inventory model under inflation for deteriorating items with stock-dependent consumption rate and partial backlogging shortages. Int J Prod Econ 123(1):8–19
20. Yang S, Lee C, Zhang A (2013) An inventory model for perishable products with stock-dependent demand and trade credit under inflation. Math Prob Eng 2013

Customer Churn Prediction in Telecommunications Using Gradient Boosted Trees

Tanu Sharma, Prachi Gupta, Veni Nigam and Mohit Goel

Abstract Customer churn is a critical problem faced by many industries these days. It is 5–10 times more valuable to keep a long-term customer than acquiring a new one. This paper addresses the problem of customer churn with respect to telecommunication industry as churn rate is quite high in this industry (ranging from 10 to 60%) in comparison to others. Predicting customer churn in advance can help these companies in retaining their customers. The paper proposes XGBoost algorithm as a model with the best performance among other state-of-the-art algorithms. The previously used models focus more on the accurate prediction of churners as compared to non-churners, whereas the proposed model classifies churners among the total churners correctly and is able to achieve the highest True positive rate of 81% and AUC score of 0.85. Also, concepts of data transformation, feature selection, and data balancing using oversampling are applied for the same.

Keywords Boosting · Churn prediction · Telecommunications · XGBoost

1 Introduction

Customer churn refers to the likeliness of a customer to stop using services from a service provider. Customer churn prediction is the way of recognizing those customers who could move away from the present service provider. It also includes

T. Sharma (✉) · P. Gupta · V. Nigam · M. Goel
Department of Computer Science, Bhagwan Parshuram Institute of Technology, GGSIPU, New Delhi, India
e-mail: tanusharma1217@gmail.com

P. Gupta
e-mail: prachigupta360@yahoo.in

V. Nigam
e-mail: veninigam18@gmail.com

M. Goel
e-mail: mohitgoel188@gmail.com

© Springer Nature Singapore Pte Ltd. 2020   235
A. Khanna et al. (eds.), *International Conference on Innovative Computing and Communications*, Advances in Intelligent Systems and Computing 1059,
https://doi.org/10.1007/978-981-15-0324-5_20

those customers who may not switch to other service provider but terminate their relationship with the existing company.

Customer churn prediction is a well-known issue in various fields such as sales, Internet providers, telecommunications, satellite TV, and banking services. A large number of companies today are facing the challenge of retaining their existing customers as they are the major source of revenue. It is more commonly seen in telecommunications industry which experiences the highest annual churn rate ranging from 10 to 60% and is significantly increasing month by month. An organization loses its customers due to various reasons such as bad customer service, poor onboarding, and ineffective relationship building. Also, due to high competition between the companies and saturated markets, it is very easy for a customer to get attracted toward price and quality of other services and switch from one company to another. This affects the reputation and overall growth of the company for long term. Also, it costs about 5 to 10 times more to acquire new customer which incurs heavy loss to the revenue generated and resources invested by the company on the previous ones. Therefore, it is very important for the telecom companies to retain existing subscribers than adding new ones.

In order to retain subscribers, the company should predict those customers who are at a higher risk of churning in a given time period and take appropriate marketing strategies which will have greatest retention impact on them. The ability to predict and timely measures can eliminate a large proportion of customer churn and would represent a huge additional potential revenue source for the company.

However, the various attempts to predict customer churn that were made can only identify a certain percentage of customers at risk. Further, sometimes it provides incentives to happy and active customers which lead to loss in revenues for no reason. These attempts based on old statistical techniques and data mining [1] methods are relatively inaccurate and offer a little value.

A better and more accurate way is to implement supervised machine learning models [2] which are investigated as the best way to tackle the highly increasing churn rate. Supervised machine learning models learn from a training dataset and approximate a mapping function so that it can make predictions based on the new input data. This paper proposes a churn prediction model using extreme gradient boosting which is considered to be robust and successful algorithm for churn prediction. The proposed XGBoost-based model has also been compared with other state-of-the-art algorithms.

The remaining paper is organized as following: next section briefs about the related work in customer churn prediction. Section 3 specifies our methodology which describes the algorithm XGBoost and underlying algorithms. Experiments are provided in Sect. 4 followed by evaluation and results in Sect. 5. Lastly, the conclusion and future scope are discussed in Sects. 6 and 7, respectively.

2 Related Work

Extensive research has been done on predicting customer churn in different industries over the years. This section presents the recent and eminent publications on customer churn prediction in recent years.

In [3], authors provide a statistical analysis tool for prediction of customer churn in telecommunication industry by data mining and machine learning models, namely logistic regression and decision trees. The system is built on three phases—development of a Web interface model, a feature extraction model, and prediction model. Information gain is used by decision tree for calculation of each feature and logistic regression uses maximum likelihood estimation for transformation of dependent variables into a logistic variable. Therefore, model suggested data mining to be a reliable way for customer churn prediction.

In [4], Bayesian belief network is used for identification of customer behavior and the factors of paramount importance that effect the customer churn in telecommunication industry. CHAID algorithm is used to convert continuous behavior of variables into discrete variable on a dataset consisting of 2000 customers. The paper identifies that average minutes of calls, billing amount, and calls frequency to people are the paramount factors for prediction of customer churn. Lastly, three different scenarios are analyzed on the conclusions. Bayesian belief network is the most efficient way to deal with casual relationship between different variables.

In [5], SVM algorithm comprising of four kernel functions is used for assessment of customer churn on the dataset of mobile telecommunication. Gain measure is used as an evaluation tool. On result evaluation, it was observed that for both the predictions, i.e., for churn and non-churn, 'Polynomial kernel' gave the best performance with an accuracy of 88.56%(on an average).

In [6], the authors have used boosting to improve the performance on model of customer churn. In contrast to other researchers who used boosting to improve the accuracy of a given base model directly by using boosting algorithm, in their study, they separated the training data into two groups based on the weight designated using boosting and they have built a separate model for every cluster. They compared the results with another machine learning model—logistic regression—on entire training dataset and observed that boosting provided a good separation and defined a higher-risk customer group. They proposed that the results regarding performance can be further improved by other classification methods.

In this study [7], the authors worked on the exploration of the multilayer perceptron neural network along with back-propagation learning for churn prediction in a telecommunication company. It used two developed MLP-based approaches; the typical change on error method and ANN weights and then compared them to rank most influencing factors in customers churn. The study discovered and inferred common attributes by both methods in order to plan customer churn.

In [8], authors developed a new set of features and then applied seven models (LR, LC, DT, MLP, SVM, DMEL) for prediction of customer churn. They compared the new features set and current features and the effectiveness of each of them.

They found out the new set of features with seven models yield much improved results for prediction as compared to the current set for customer churn prediction in telecommunications. They proposed the limitations with the new set of features and also the methods to be focused in the future to improve the results.

In [9], SVM has been compared with four machine learning methods—ANN, decision tree, LR, and Naive Bayesian classifier—and it was observed that it has the best accuracy rate, covering rate, lift coefficient, hit rate. It concluded that SVM have simple classification plane, strong generation ability, and good fitting precision. Different traits of customers having high churn rate have been found.

In [10], distance factor is used for classifier certainty prediction for presenting customer churn approach. The dataset was grouped into two parts one with high certainty and other with low certainty. They presented an approach based on distance factor and its effect in different zones to estimate the expected certainty of the decision.

In [11], a proposal for customer churn prediction for a telecom company located in South Asia is based on fuzzy classifiers. The model utilized FuzzyNN, DWANN, and VQNN to predict the churners from the customer records accurately. Several classification models like neural network, C4.5, SVM, Adaboost, gradient boosting, random forest, and linear regression in contrast with fuzzy classifiers were used to determine the highest accuracy.

3   Methodology

In this section, a brief overview of underlying algorithm and some of the features of XGBoost [12] are presented. XGBoost (extreme gradient boosting) is a fast, distributed, and scalable machine learning system introduced by Chen [12] and Guestrin in 2016. This section explains the various concepts involved in the proposed methodology which includes ensemble learning, bagging, and boosting.

3.1   Ensemble Learning

XGBoost is an ensemble (i.e., Meta) learning method. Sometimes, the results of a single machine learning model are not reliable due to errors such as bias and variance. In such case, we need to combine the power of multiple learning models to provide a stable and more accurate solution. Ensemble learning provides a single model that gives combined output from various models also called as base learners derived from same or different learning algorithm. The two most used ensemble learners are bagging and boosting. Although they can be implemented with various different models, the most widely used model is decision tree since they are the easiest to interpret and have high variance.

3.2   Bagging and Boosting

Bagging and boosting [13] generate additional data by random sampling the original dataset. The new N training datasets are obtained with some repeated observations in each set. In bagging, each observation has equal chances of occurring in new set, whereas in boosting, the elements are weighted and then put into new sets.

The same learner is trained on these multiple sets and thus produces different classifiers. In bagging, classifiers are built independent of each other; however, boosting builds the classifier sequentially, i.e., after building each classifier, the weights of the observations are redistributed such that the weight of data which is more misclassified is increased so that the subsequent learner can focus more on them during their training.

After training, the learners are applied to new observations and result is obtained. In bagging, average of all results is taken, whereas in boosting, weighted average of the classifiers is taken to estimate the result. The learner with more errors is given less weight thus keeping the track of classifier errors.

Based on different conditions and methods of calculating weights to be used in next step, several techniques exist like AdaBoost [14], LPBoost, XGBoost, GradientBoost, and BrownBoost. In the next section, we will discuss in brief about the GradientBoost and XGBoost only.

3.3   Gradient Boosting

Gradient boosting [15] is a method for both regression and classification. It ensembles the weak prediction models (usually decision trees) to produce a single strong prediction model. Every subsequent tree is built sequentially and learns from the errors of previous tree. Each model commits some mistakes in predicting the values called residuals. The algorithm finds some pattern in the residuals and then fits a model to that pattern to update the value of predictions using gradient descent. The process is repeated until there is no pattern that can be modeled and the sum of residuals becomes close to zero and the predicted values come close to actual values.

3.4   XGBoost

XGBoost is the fast, portable, and distributed implementation of gradient boosting.

- It is fast as it is built in C++ and higher-level languages like Python and R.
- It is portable as it can be used on different architectures.
- It is distributed as it can run on multiple cores of single machine or multiple servers.

XGBoost [12, 16] lessens the search space by considering the feature distribution over all the data points and therefore decreasing the possible splits. It overcomes the inefficiency of gradient boosting in which further splitting of a node is stopped once it comes across a negative loss, whereas XGBoost first goes up to the maximum possible depth that is defined and then it discards those splits in which there occurs any negative loss by going backward [16].

Mathematically, it can be expressed as [17] (1):

$$f_t(x) = w_{q(x)}, w \in R^T, q : R^d T \rightarrow \{1, 2, \ldots, T\} \tag{1}$$

Here, "w" refers to the vector of scores on leaves, "q" refers to the function that assigns each data point to its corresponding leaf, and the total numbers of leaves are represented by "T." The complexity of the model is described as below (2):

$$\Omega(f) = \gamma T + \frac{1}{2}\lambda \sum_{j=1}^{T} w^2 j \tag{2}$$

For any iteration t, the objective function can be given as (3):

$$\text{Obj}^{(t)} = \sum_{j=1}^{T} \left[G_j w_j + \frac{1}{2}(H + \lambda)w_j^2 \right] + \gamma T \tag{3}$$

On evaluating the above equation, the best objective reduction and the best W_j for a given structure $q(x)$ can be formalized as (4) and (5):

$$w_j^* = -\frac{G_j}{H_j + \lambda} \tag{4}$$

$$\text{Obj}^{(t)} = -\frac{1}{2} \sum_{j=1}^{T} \frac{\text{sqr}(G_j)}{H_j + \lambda} + \gamma T \tag{5}$$

Splitting a leaf into two gains a score of (6):

$$\text{Gain} = -\frac{1}{2}\left[\frac{\text{sqr}(GL)}{HL + \lambda} + \frac{\text{sqr}(GR)}{HR + \lambda} - \frac{\text{sqr}(GL + GR)}{HL + HR + \lambda} \right] - \gamma \tag{6}$$

where $G_J = \sum_{i \in Ij} g_i$ and $H_J = \sum_{i \in Ij} h_i$; the definitions of which are as per [12, 17].

4 Experiments

This section describes about the dataset, preprocessing [18] the dataset and building of models. The proposed work is depicted in Fig. 1.

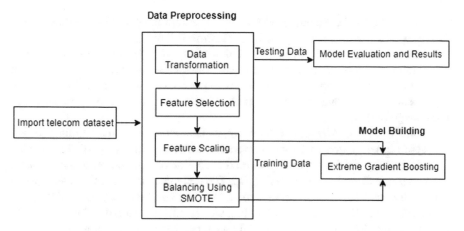

Fig. 1 Flowchart for the proposed model

4.1 About the Dataset

The dataset used in this work is publicly available on kaggle [19]. The raw data contains 3333 rows and 21 columns. The "Churn" column is the target variable. For each subscriber, their total ingoing and outgoing calls count, SMS count, and voice mail count are provided.

4.2 Data Preprocessing

Noise removal: The data is collected in raw format which is not feasible for the analysis and thus needs to be preprocessed before analyzing. On exploring the data, it was found that there are no null or missing values.

Conversion of categorical data into numeric data: The categorical data converted to numeric data enhances the data so that the classification models can make meaningful predictions. The categorical values of state column are converted into numeric data using label encoding and the binary data (such as "yes/no" or "true/false") is replaced with "1" and "0," respectively.

Feature selection: Feature selection [20] is the process of reducing the inputs or of finding the meaningful features for the analysis. It helps in training the model faster and reduces the complexity. Feature selection is done on basis of correlation between variables. It is observed that the minutes and charge of calls perfectly correlate with each other, which means they have repeated information. Keeping both the minutes and charge of calls would be redundant. We will, therefore, remove all the minutes. Also, voice_mail_plan and number_vmail_messages are highly correlated. We can remove voice_mail_plan since number_vmail_messages = 0 essentially means voice_mail_plan = 0. All the numbers of calls and account_length are irrelevant since they have little correlation with churn. The area code and phone number are also dropped since they are only for identification and not relevant.

Feature scaling: It is a technique used so that every feature must have equal opportunity to influence the target variable. It involves standardizing the different range of independent variables. It is also called as data normalization and is usually implemented during data preprocessing. All the numeric data has been transformed into a range of 0–1.

Balancing using SMOTE [21] *technique*: The churners in the training set are 483, approximately 14% and non-churners are 2850, approximately 86%. Since the classification algorithms require balanced distribution of both the yes and no classes, we oversample the non-churners using SMOTE technique. In this technique, replicas of minority class are created using a subset of data and then they are added to the original dataset. Therefore, after balancing, we obtain a balanced distribution of both the classes with 2271 rows each and 4542 total records in the training set.

4.3   Model Building

It is the process of choosing the learning models and providing them data to learn from called as training data. The dataset is separated into training and test data in the random ratio of 80:20 consisting of 2666 rows in training set and 667 in the testing set. The test data is kept aside for evaluation and training data is used to fit and build the various classification models. Classification techniques include logistic regression, SVM, decision tree, random forest, and XGBoost.

5   Evaluation and Results

For the evaluation purpose of the proposed model, two evaluation metrics are used.

The churn prediction model is evaluated by its potential to find maximum number of customers who may churn. The performance of the classification models can be visually compared by observing the confusion matrix in Table 1. The diagonal values depict the correct classifications, whereas the off-diagonal values show the incorrect classifications by the corresponding model in the leftmost column.

It can be observed that XGBoost has the highest percentage, i.e., 81.81% of correctly predicting customers as churners, whereas SVM and decision trees have the lowest percentage, i.e., 71.59% among all.

Another measure that can be used for evaluation is the receiver operating characteristic (ROC) curve where the value of area under curve (AUC) is used to measure the effectiveness of the model. AUC [22] is the graph of true positive versus false positive and is more adequate measure than accuracy since it is not affected by class imbalance and evaluation data.

Table 1 Confusion matrix for various models

Model name	Observed	Predicted		
		No	Yes	% Correct
Logistic regression	No	429	150	74.09%
	Yes	21	67	76.13%
	Overall %	67.46%	32.53%	74.36%
Support vector machine	No	530	49	91.53%
	Yes	25	63	71.59%
	Overall %	83.20%	16.79%	88.90%
Decision trees	No	503	76	86.87%
	Yes	25	63	71.59%
	Overall %	79.16%	20.83%	84.85%
Random forests	No	537	42	92.74%
	Yes	23	65	73.86%
	Overall %	83.95%	16.04%	90.25%
XGBoost	No	516	60	89.11%
	Yes	16	72	**81.81%**
	Overall %	79.76%	19.79%	88.15%

Experimental results show that the area under curve for class churner has been increased after balancing the dataset since recall rate shows significant improvement for all classifiers as shown in Tables 2 and 3.

Figures 2 and 3 show the area under curve of different classification models and it is observed that XGBoost performed better with respect to AUC score in contrast to

Table 2 AUC score for unbalanced dataset

Classifier	FP rate	TP rate	AUC
XGBoost	0.0189983	0.647727	0.814364
Random forest	0.0224525	0.670455	0.824001
Decision trees	0.0811744	0.75	0.834413
Logistic regression	0.0224525	0.170455	0.574001
Support vector machine	0.0172712	0.602273	0.792501

Table 3 AUC score for balanced dataset

Classifier	FP rate	TP rate	AUC
XGBoost	0.103627	**0.818182**	**0.857277**
Random forest	0.058722	0.727273	0.834275
Decision trees	0.136442	0.75	0.806779
Logistic regression	0.259067	0.761364	0.751148
Support vector machine	0.084629	0.715909	0.81564

Fig. 2 AUC curve for
unbalanced dataset

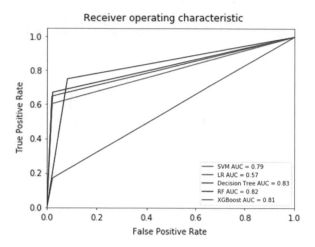

Fig. 3 AUC curve for
balanced dataset

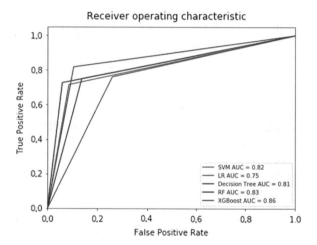

other classifiers after balancing the dataset. Table 3 clearly depicts that the maximum
number of churners have been identified by XGBoost since it has highest TP rate of
approximately 81%.

6 Conclusion

In this paper, XGBoost-based model has been proposed for customer churn prediction. The acquired data from an anonymous telecom provider could not be directly
applied to the churn prediction models. Data cleaning along with feature selection
and balancing were implemented before building the model.

From the experiments, it can be clearly seen that the proposed model has been capable of achieving a TP rate of 81% and an AUC score of 0.85. A comparison between several classifiers, specifically logistic regression, support vector machines, decision tree, and random forest, has been done with XGBoost to emphasize the dominance of the proposed model in correctly predicting the largest number of churners.

7 Future Scope

An ample amount of work can be done to further enhance the accuracy. Although various classification models like logistic regression, support vector machines, decision trees, and random forests have been used, finding other methods that can improve the performance is also essential. Models based on artificial intelligence and neural networks can be investigated. Also, combination of classifiers can also be implemented to tackle imbalanced data.

In this application, we have only reduced the dimensions of input features by using feature selection. We can add new features by using feature extraction. Inclusion of more data samples to third data source might increase confidence in quantitative results.

In spite of determining the accuracy of prediction models which can only help in short-listing and prioritizing certain customers. It is also important to identify the actual factors affecting the churn rate in any business under different circumstances and then finding appropriate solutions and strategies for business growth.

References

1. Hand D (2007) Principles of data mining. Drug Saf 30:621–622
2. Osisanwo FY, Akinsola JE, Awodele O et al (2017) Supervised machine learning algorithms: classification and comparison. Inter J Comput Trends Technol 48:128–138
3. Dalvi PK, Khandge SK, Deomore A et al (2016) Analysis of customer churn prediction in telecom industry using decision trees and logistic regression. In: Symposium on colossal data analysis and networking (CDAN)
4. Kisioglu P, Topcu YI (2011) Applying bayesian belief network approach to customer churn analysis: a case study on the telecom industry of Turkey. Expert Syst Appl 38:7151–7157
5. Brandusoiu I, Toderean G (2013) Churn prediction in the telecommunications sector using support vector machines. Margin 1:x1
6. Lu N, Lin H, Lu J, Zhang G (2014) A customer churn prediction model in telecom industry using boosting. IEEE Trans Ind Inform 10:1659–1665
7. Adwan O, Faris H, Jaradat K et al (2014) Predicting customer churn in telecom industry using multilayer preceptron neural networks: modeling and analysis. Life Sci J 11(2):75–81
8. Huang B, Kechadi MT, Buckley B (2012) Customer churn prediction in telecommunications. Expert Syst Appl 39:1414–1425
9. Xia G-E, Jin W-D (2008) Model of customer churn prediction on support vector machine. Syst Eng Theor Pract 28:71–77

10. Amin A, Al-Obeidat F, Shah B et al (2019) Customer churn prediction in telecommunication industry using data certainty. J Bus Res 94:290–301
11. Azeem M, Usman M, Fong ACM (2017) A churn prediction model for prepaid customers in telecom using fuzzy classifiers. Telecommun Syst 66:603–614
12. Chen T, Guestrin C (2016) Xgboost: a scalable tree boosting system. In: Proceedings of the 22nd ACM SIGKDD international conference on knowledge discovery and data mining, pp 785–794
13. Maclin R, Opitz D (1997) An empirical evaluation of bagging and boosting. AAAI/IAAI, pp 546–551
14. Ying C, Qi-Guang M, Jia-Chen L, Lin G (2014) Advance and prospects of adaboost algorithm. Acta Automatica Sinica 39:745–758
15. Friedman JH (2001) Greedy function approximation: a gradient boosting machine. Ann Stat 1:189–232
16. Pohjalainen V (2017) Predicting service contract churn with decision tree models
17. Ajit P (2016) Prediction of employee turnover in organizations using machine learning algorithms. Algorithms 4(5):C5
18. Zhao J, Wang W, Sheng C (2018) Data preprocessing techniques. In: Data driven prediction for industrial process and their applications. Springer, Berlin
19. Churn in telecom's dataset. https://www.kaggle.com/becksddf/churn-in-telecoms-dataset. Accessed 20 Sep 2018
20. Dash M, Liu H (1997) Feature selection for classification. Intell Data Anal 1(3):131–156
21. Chawla NV, Bowyer KW, Hall LO et al (2002) SMOTE: synthetic minority over-sampling technique. J Artif Intell Res 16:321–357
22. Jiménez-Valverde A (2012) Insights into the area under the receiver operating characteristic curve (AUC) as a discrimination measure in species distribution modeling. Glob Ecol Biogeogr 2(4):498–507

Efficient Evolutionary Approach for Virtual Machine Placement in Cloud Data Center

Geetika Mudali, K. Hemant Kumar Reddy and Diptendu Sinha Roy

Abstract Administering energy and resource management are two vital managing components of cloud data centers. From last two decades, most of cloud data centers (CDC) are suffering from these two; the former has become a serious issue nowadays. In this paper, we focused on effective virtual machine placement (VMP). Evolutionary approach is applied to place the virtual machine in an effective way which properly utilizes the underutilized resources and reduced the active physical servers. After experiencing the performance of particle swam optimization (PSO) algorithm for combinatorial problems, a distributed PSO approach is modeled to minimize energy consumption of CDCs. The proposed PSO and DPSO algorithms are applied on VMP over large distributed cloud data centers. Experimental results of PSO and distributed PSO algorithms are presented. The model is applied with variety of placement problems with varying data center network topology. The performance of the model outperforms the traditional heuristic and several optimizations approaches.

Keywords Virtual machine placement · PSO · Distributed PSO · Energy efficient · Cloud computing · Evolutionary approach

1 Introduction

Cloud computing has set a new trend for enacting massive scale distributed computing [1]. At present cloud computing serves, there are a number of services over Internet, but the major services are IaaS, PaaS, and SaaS for a runtime environment. Cloud

G. Mudali · K. H. K. Reddy (✉)
Department of Computer Science and Engineering, National Institute of Science and Technology, Berhampur, India
e-mail: khemant.reddy@gmail.com

D. S. Roy
Department of Computer Science and Engineering, National Institute of Technology, Shillong, Meghalaya, India
e-mail: diptendu.sr@gmail.com

© Springer Nature Singapore Pte Ltd. 2020   247
A. Khanna et al. (eds.), *International Conference on Innovative Computing and Communications*, Advances in Intelligent Systems and Computing 1059, https://doi.org/10.1007/978-981-15-0324-5_21

providers are able to serve these seamless services to end user with the help of virtualization [2]. The existing physical servers are virtualized to form a uniform virtual machines (VM) and these VMs are used efficiently for parallel and distributed computing [3, 4]. In order to handle million of users' request simultaneously, limited physical machines are virtualized to create the number of virtual machine of different computation, memory, and disk capacity with required operating systems as per request. These virtual machines have to run over these physical hosts and fulfill customers' requirement [5, 6]. It also supports execution of multiple VMs on the same physical machine and shares hardware resources among them if required. As users' demands are always dynamic in nature in terms of number of VMs and its computation capability, it is very difficult to predict. Frequent variation in users' demand and maintaining QoS in such scenario is very difficult [7]. Even it is more difficult to manage in large scale which leads to maximum activation of physical machine at instance.

There has been rapid growth in establishing cloud data centers to maintain the users' demand that leads to increase in energy consumption by millions of physical servers and as well as other equipment like cooling, network switches, and others [8]. Nowadays, energy consumption is a prime issue in cloud data centers. In our earlier paper, we focused on this energy issue and proposed a scheduling of resources on class constraint basis [9]. As per the current survey, around 11–50% of cloud servers are utilized most of time [10], whereas active servers consume around 50–70% in idle cases [11]. Therefore, an efficient VM placement can greatly minimize the power dissipation by placing underutilized servers' VMs to other servers. We propose an evolutionary scheduling scheme for allocating VMs within fewer physical machines in order to reduce energy consumption cloud data centers [12]. Nowadays, most of data centers are build on hierarchical structure using FAT tree topology with three levels of switches that is presented in Fig. 1. An ample amount work has been devoted to energy management in FAT structures [13], and in addition to FAT structure, a coordinated provisioning approach is also proposed in our earlier work [12]. Most of evolutionary approaches are focus on VM placement within data centers or within the one POD that is within the different servers under aggregation level or access level and many efficient allocations approaches suffers migration latency problems when allocated VMs need to migrate to other servers for sudden increase in demand. These migrations occur within the switch levels which is limited space. Due to limited space, there is a possibility of increase in number of migrations and frequent migrations for single VM will increase the latency. So, we aligned our work to develop a distributed evolutionary approach that considers the geographical distribution with related parameters. Virtual machine consolidation is a NP-complete problem and solution used for this class of problem evolutionary algorithms is genetic algorithm (GA), [14] ant colony optimization [15], particle swarm optimization (PSO) [16], and others. Few of them achieves very good performance in case of small-scale simulation, whereas others shown the good performance in larger scale also. In this paper, we considered classical PSO and distributed PSO approach for VM consolidation. The rest of the paper is organized as follows. The VM consolidation problem and the outline of the model of evolutionary approach are

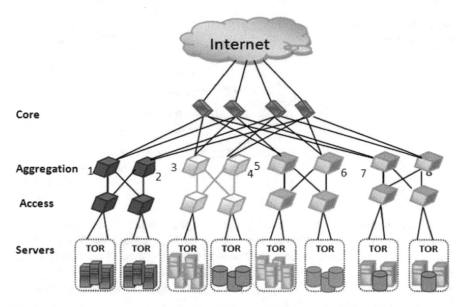

Fig. 1 Cloud data center network topology (FAT tree) with four pods

presented in Sect. 2. Section 3 presents the details of mapping of VM consolidation problem with distributed evolutionary approach. It is then applied in Sect. 4 that presents the simulation details and experimental results and undertaken validation is compared with heuristic approaches such as first-fit decreasing approach [17], EC-based approaches such as reordering grouping GA [18], ACO-based method [13], and a hybrids ACO and PSO [19] are discussed in Sect. 5. Finally, Sect. 6 presents the summary as conclusion.

2 Cloud Data Center Hierarchical Network Model

Cloud data center is build with set of computational resources, storage resources, and network resources. In order to access these pool of computational and storage resources in an efficient way, a set of network resources are used and by interconnecting these resources, a hierarchical network structure models are designed. The entire data center network is build with computational, storage, and network nodes and these nodes are interconnected with bidirectional communications links. Fat topological model is converted into a graph, where nodes indicate the network switches or Tor switch that is connected to physical servers. Each network switch consumes certain amount of energy and time delay starting from core-level to Tor-level switches. The lines that connect these switches called bidirectional links, which also consumes certain amount of energy and time depending on the bandwidth available. The cloud network could be represented by undirected weighted graph $G =$

Fig. 2 Graph representation
of cloud data center FAT
topology

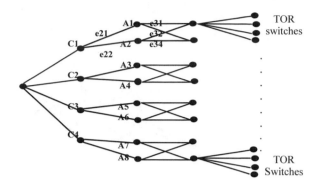

(V, E) with vertices (V) and edges (E). The weight of vertex v_i, $(w(v))$ depends on the amount energy and delay elapsed at vertex v_i during transmission. The weight of the vertex is inferred by the number of flows connected to it and their load flow. For example, there are three communication links at A_1, and then the weight of switch A_1 is: $A_1 = \omega(e_{21}) + \omega(e_{31}) + \omega(e_{32}) + \tau$. Figure 1 would be transformed into the coarsened graph shown in Fig. 2.

3 PSO Algorithm

Particle swarm optimization is a voluntary algorithm and it works on the collective intelligence of the populations. PSO algorithm is developed by Kennedy and Eberhart [20] which is based on the common behavior of animals that move in groups for searching food or shifting seasonwise locations. A population, the so-called swarm, consists of particles which move through multidimensional space and change their positions, depending on their own experience and the experience of the other particles. At first, the algorithm was developed for continual systems, and later, algorithms for discrete systems were introduced: binary PSO [21] and integer PSO [22, 23]. With the binary algorithm, each particle from the population can take binary values (0 or 1). In more general cases, such as integer PSO, an optimal value is achieved by rounding the real optimum to the nearest integer.

3.1 Particle Representation

Supposing the given graph $G = (V, E)$ with a set of vertices (roots) $V = \{v_1, v_2, ..., v\}$ should be divided into p partitions. By adapting PSO algorithm to the graph partitioning problem, the dimension of particles is equal to the number of vertices in the graph. Particle positions are updated in a defined number of iteration, in accordance to optimization criteria, and the best particle position is searched. The change of particle i position is achieved by moving the particle from the previous position,

$$x_i^{k+1} = x_i^k + v_i^{k+1}, \tag{1}$$

where x_i^k is the position of ith particle in iteration k, and v_i^{k+1} is the velocity of ith particle in iteration $k + 1$

$$v_i^{k+1} = \omega \cdot v_i^k + c_1 \cdot r_1 \cdot (b_i - x_i^k) + c_2 \cdot r_2 \cdot (b_g - x_i^k) \tag{2}$$

where ω is the inertia factor, c_1, c_2 are constants, $b_i - x_i^k$ is the individual component (b_i—best position of particle 'i'), $b_g - x_i^k$ is the social component, and (b_g is the global best position). r_1 and r_2 are random number generated within [0, 1]. The optimization criteria F is calculated and updated particle positions are memorized as the local best (b_i) and the entire population is global best (b_g). Particle movement velocity need to be controlled, otherwise high velocity movement particles change the result significantly. Velocity of the particle (v_m) is limited to:

$$\text{round}\left(v_i^{k+1} = \begin{cases} -v_m, \text{ if } v_i^{k+1} < -v_m \\ v_i^{k+1}, \text{ if } -v_m \leq v_i^{k+1} \leq v_m \\ v_m, \text{ if } v_i^{k+1} > v_m \end{cases} + \partial x_{\min} \cdot r \right) \tag{3}$$

In order to bring the dynamism in velocity movement of the entire set of particles, $\partial x_{\text{avg}} \cdot r$ is added to the updated velocity, where ∂x_{avg} is the average possible change of particle position and r is the random value generated between [0, 1]. As the PSO works on discrete problem and its particle position should be in integer form and it should lie in between [0, p], then modules can be applied to get the range within the defined space. After iterating definite number of times particles achieves b_g position.

4 Distributed PSO Algorithm

Distributed PSO can be applied for searching the available solutions in parallel in the search space [24]. The entire set of population is divided into number of subset of populations and distributed in different processor for searching the possible best solution in parallel. Figure 3 depicts the small scenario of swarm distribution architecture. After certain number of iterations, all these sub-swarms achieve their best positions, and then, the solutions are exchanged in between.

The number of iterations can be controlled by the distributed in order to get the local best solution at each swarm. After completing the definite number of iterations, the global best is communicated to the each sub-swarm to update their velocity using Eqs. 1, 2, and 3. The communicated global best limits the sub-swarm to spread over the search space and time period that controls the number of iterations at sub-swarm

Fig. 3 Smarm distribution
architecture

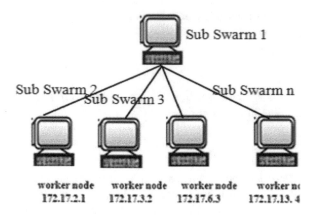

level can be controlled as distributor level. Even the sub-swarms are assigned to different processor for execution, but less parallelization is achieved due to less synchronized time period between the distributor and sub-swarm. Inter communication between swarm for updating the movement velocity is an additional burden, which slow down the algorithm execution. Whereas independence of sub-swarms can be increased with higher value of swarms synchronization period, but it impact the efficacy of the solution. In order to maintain the efficacy of the system, a hierarchical distribution approach is adopted. In [25], the efficacy of the hierarchical distribution is presented. Figure 4 presents the hierarchical swarm distribution architecture. The .hierarchical is completely independent of the physical connection among the processors. The deployed three-level hierarchical architecture is dynamic in nature in the sense that for different instances of scheduling events, the nodes that act as the distributor at different levels are not unique. Moreover, the groups of nodes that fall within a site or a cluster are not the same at different times. In the model shown in Fig. 4, the server or load distributor node that distributes the load to other nodes or second level distributor nodes. The rest of the nodes in the hierarchy are the worker

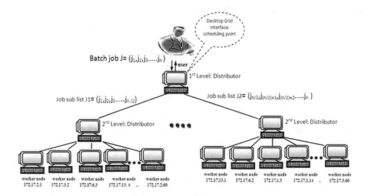

Fig. 4 Hierarchical smarm distribution architecture

nodes. Second-level coordinators are in dynamic and these are selected by the server node and rest of the hierarchy are treated as worker nodes. The number of distributor nodes in second level depends on the total number of swarms available in the population. The model is restricted to three levels, and it depends on the number of nodes available and swarm size to optimize.

5 Experimental Results

The model is tested on simulated three-level fat tree topology with different POD size varying from 10 to 100. Assumption taken in the simulation is static and T is ms. The network properties considered in our simulation are presented in Table 1. The models used are completed on FAT Tree topology based and the numbers of populations are generated based on the number of edges and nodes in the fat tree structure. The delays incurred at different level of switches are considered different but static for entire simulation, similarly delay incurred at different links and link bandwidth also Table 2.

6 Conclusion and Future Work

The approach presented in this paper is applied on VMP over simulated distributed cloud data centers. The efficacy of the proposed PSO and DPSO algorithms presents the delay incurred and the number of physical servers used to process the required service requests. The model is applied with variety of placement problems with varying data center network topology. The performance of the model outperforms the traditional heuristic and several evolutionary approaches. The energy consumption,

Table 1 Fat tree topology architecture with 3-level

Core level switches	No. of nodes	No. of edges	No. of servers
10	90	180	40
20	180	560	80
30	270	1140	120
40	360	1920	160
50	450	2900	200
60	540	4080	240
70	630	5460	280
80	720	7040	320
90	810	8820	360
100	900	10,800	400

Table 2 Result of PSO and DPSO for 300 iterations with fixed request size

Core level switches	Number of particles	PSO		DPSO			
		Execution time in ms	Number of active servers	$T = 10$/execution time in ms	Number of active servers	$T = 50$/execution time in ms	Number of active servers
10	90	3.10	44	3.10	39	3.06	38
20	280	9.18	45	9.18	39	9.00	38
30	570	18.28	45	18.28	40	18.06	40
40	960	30.67	44	30.67	39	30.25	39
50	1450	45.92	46	45.92	39	45.56	38
60	2040	64.36	48	64.36	38	64.00	37
70	2730	87.18	45	87.18	38	85.56	38
80	3520	111.54	46	111.54	39	110.25	39
90	4410	140.91	45	140.91	38	138.06	37
100	5400	172.06	45	172.06	38	169.00	38

server placement mapping procedure, and details of simulated parameters are not included in this paper. It is assumed that reduction in active physical machines leads to energy efficient. In addition to this, resource-aware virtual machine placement with DPSO and energy model is taken as future work.

References

1. Foster Y, Zhao I, Raicu, Lu SY (2008) Cloud computing and grid computing 360-degree compared. In: Proceedings of the IEEE grid computing environments workshop, Austin, TX, pp 1–10
2. Lawey AQ, El-Gorashi TEH, Elmirghani JMH (2014) Distributed energy efficient clouds over core networks. J Lightw Technol 32(7):1261–1281
3. Liu X-F, Zhan Z-H, Lin J-H, Zhang J (2016) Parallel differential evolution based on distributed cloud computing resources for power electronic circuit optimization. In: Proceedings of the genetic and evolutionary computation conference, Denver, CO, pp 117–118
4. Zhan ZH et al (2016) Cloudde: a heterogeneous differential evolution algorithm and its distributed cloud version. IEEE Trans Parallel Distrib Syst. https://doi.org/10.1109/tpds.2016.2597826
5. Chen Z-G et al (2015) Deadline constrained cloud computing resources scheduling through an ant colony system approach. In: Proceeding of the international conference on cloud computing research and innovation, Singapore, pp 112–119
6. Li H-H, Chen Z-G, Zhan Z-H, Du K-J, Zhang J (2015) Renumber coevolutionary multiswarm particle swarm optimization for multi-objective workflow scheduling on cloud computing environment. In: Proceedings of the genetic and evolutionary computation conference, Madrid, Spain, pp 1419–1420
7. Mastroianni C, Meo M, Papuzzo G (2013) Probabilistic consolidation of virtual machines in self-organizing cloud data centers. IEEE Trans Cloud Comput 1(2):215–228

8. Greenpeace (2010) Make it green: cloud computing and its contribution to climate change. Greenpeace International. [Online]. Available http://www.thegreenitreview.com/2010/04/greenpeacereports-on-climate-impact-of.html
9. Reddy K, Mudali G, Roy DS (2016, March) Energy aware Heuristic scheduling of variable class constraint resources in cloud data centres. In: Proceedings of the 2nd international conference on information and communication technology for competitive strategies. ACM, p 13
10. Dasgupta G, Sharma A, Verma A, Neogi A, Kothari R (2011) Workload management for power efficiency in virtualized data centers. Commun ACM 54(7):131–141
11. Greenberg A, Hamilton J, Maltz DA, Patel P (2009) The cost of a cloud: research problems in data center networks. ACM SIGCOMM Comput Commun Rev 39(1):68–73
12. Reddy KHK, Mudali G, Roy DS (2017) A novel coordinated resource provisioning approach for cooperative cloud market. J Cloud Comput 6(1):8
13. Mishra J, Sheetlani J, Reddy KHK, Data center network energy consumption minimization: a hierarchical FAT-tree approach. Inter J Inf Technol, 1–13
14. Bui TN, Moon BR (1996) Genetic algorithm and graph partitioning. IEEE Trans Comput 45(7):841–855
15. Liu X-F, Zhan Z-H, Du K-J, Chen W-N (2014) Energy aware virtual machine placement scheduling in cloud computing based on ant colony optimization approach. In: Proceedings of the ACM genetic evolutionary computation conference, Vancouver, BC, pp 41–48
16. Zhan Z-H et al (2013) Multiple populations for multiple objectives: a coevolutionary technique for solving multiobjective optimization problems. IEEE Trans Cybern 43(2):445–463
17. Ajiro Y, Tanaka A (2007) Improving packing algorithms for server consolidation. In: Proceedings of the international conference Computer Measurement Group, pp 399–406
18. Wilcox D, McNabb A, Seppi K (2011) Solving virtual machine packing with a reordering grouping genetic algorithm. In: Proceedings of the IEEE congress of evolutionary computation, New Orleans, LA, pp 362–369
19. Suseela BBJ, Jeyakrishnan V (2014) A multi-objective hybrid ACOPSO optimization algorithm for virtual machine placement in cloud computing. Int J Res Eng Technol 3(4):474–476
20. Kennedy J, Eberhart R (1995) Particle swarm optimization. In: Proceedings of the 1995 IEEE international conference on neural networks (ICNN), vol 4. IEEE Service Center, Piscataway, New Jersey, pp 1942–1948
21. Kennedy J, Eberhart R (1997) A discrete binary version of the particle swarm algorithm. In: Proceedings of the 1997 IEEE international conference on systems, man, and cybernetics, vol 5. IEEE Service Center, Piscataway, New Jersey, pp 4104–4108
22. Laskari E et al (2002) Particle swarm optimization for integer programming. In: Proceedings of the IEEE congress on evolutionary computation, vol 2. Honolulu, Hawaii, pp 1582–1587
23. Capko D et al (2009) PSO algorithm for graph partitioning. 17th Telecommunication Forum 2009, Belgrade
24. Laguna-Sánchez GA et al (2009) Comparative study of parallel variants for a particle swarm optimization algorithm implemented on a multithreading GPU. J Appl Res Technol 7(3):292–307
25. Reddy KHK, Roy DS (2012, March) A hierarchical load balancing algorithm for efficient job scheduling in a computational grid testbed. In: IEEE 1st international conference on recent advances in information technology (RAIT), pp 363–368

SCiJP: Solving Computing Issues by Java Design Pattern

Hanu Gautam, Rahul Johari and Riya Bhatia

Abstract The types, needs, advantages and disadvantages of Java Design Patterns have been explained, and Java Cryptography Architecture (JCA) case study has been discussed. The broad survey cross the spectrum of high-end programming language shows Java is the only programming language that offers concept and features which can be used extensively to design applications and software to harness the advantages of exascale computing.

1 Introduction

The J2SE (Java 2 Standard Edition) and J2EE (Java 2 Enterprise Edition) offer the concept of JDP (Java Design Patterns) which facilitates the design, develop and deployment of platform independent Java programs that can successfully be deployed both in n-Tier Client Server Architecture Environment and MVC(Model View Controller Architecture) Environment. A pattern language is a technique of describing good design methodologies usually adopted by organization for design and development of state of the art product. The word "JDP" was first coined by technical evangelist Christopher Alexander and popularized by his 1977 book A Pattern Language [1]. A pattern language is also an attempt to use indepth wisdom of what brings aliveness within a particular area of human endeavor, through a set of homogeneous and heterogeneous interconnected patterns.

Coders, Programmers and Developers (CPD) started writing different types and styles of the codes for solving same type of problems. Usually, their are certain set of problems that are repetitive in nature. In fact, their was a pattern in problems

H. Gautam · R. Johari · R. Bhatia (✉)
USICT, GGSIPU, Delhi, India
e-mail: riyabhatia26@gmail.com

H. Gautam
e-mail: hanugautam96@gmail.com

R. Johari
e-mail: rahul.johari.in@ieee.org

© Springer Nature Singapore Pte Ltd. 2020
A. Khanna et al. (eds.), *International Conference on Innovative Computing and Communications*, Advances in Intelligent Systems and Computing 1059,
https://doi.org/10.1007/978-981-15-0324-5_22

encountered, as these problems were faced by the CPD in US, Europe and Asia Pacific Region, and everyone was wasting time to find the solutions to same set of problems, everywhere, everytime but in a different way, of which some were good and optimized solutions and some were awkward and badly written LoC resulting in poorly designed Software. This led to the Technical Evagelist(s) of IBM Apple, Microsoft, Google, SunMicroSystem and major industry vendors like W3C to announce Best Programming Practices (BPP). Parallel to BPP, on the other hand in order to eradicate this anomaly, design pattern (DP) came into being where strong emphasis was on pattern usage and code reusability. Design patterns helped to write effective and efficient object-oriented programming code which when adopted to solve the problems using computing approach helped to achieve fast and accurate results. Of the 23 Java Design Pattern, none is dedicated to the security issues. In the current research work, a Secure Java Design Pattern is hereby proposed, and the simulation of the same has been shown taking DES Algorithm as example using Java Cryptography Architecture.

2 Related Work

In [2], authors have implemented modifications to Java using design choices in Polygot framework. In [3], authors have discussed a innovative model for modularization of design patterns. In [4], authors had introduced jContractor library which uses conventions that facilitates the users to use classes designed in Java Programming Language and implements Design by Contracts construct. In [5], authors have presented problems in functional logic languages and proposed solutions using software design patterns. In [6], authors have developed a class-based macro system Open-Java which uses metaobjects to represent source code of object-oriented programs. In [7], authors have shown that modularizing the implementation of GoF (Gangs of Four) patterns improves using AspectJ. In [8], authors have come up with a tool which analyzes Java programs and detects and classifies potential instances of design pattern. In [9], authors have demonstrated the accuracy and efficiency of their graph algorithm-based design pattern detection algorithm on three open-source projects. In [10], authors have proposed an approach to identify key set of spots and a refactoring strategy to convert it into the design pattern structure. In [11], authors have introduced a toolkit based on XML standard to explore patterns by identifying the characteristics,structural in nature in terms of both weight and matrix. In [12], authors have devised a tool, PINOT, which uses simple analysis techniques to detect design patterns. In [13], authors have come up with an approach object-oriented design patterns based on a multi-stage filtering strategy and also developed a portable environment in Java to assess its effectiveness. In [14], authors have discussed the architecture, mechanism as well as security issues in delay tolerant networks and proposed an approach for secure communication by using Java Cryptography Architecture (JCA) and Oracle.

3 Gang of Four (GoF)

Design patterns: elements of reusable object-oriented software [15], a software engineering book illustrating software design patterns. The book's authors are E. Gamma, R. Helm, R. Johnson and J. Vlissides.

Design pattern defines structure and way how set of common problems should be solved (leaving aside domain specific problem): like specific problems faced by Banking sector, Insurance sector, Aviation Sector Hospitality Sector (Hotels and Hospitals) Educational Sector, etc. They capture the software engineering experiences. Design patterns are independent of Programming Language Syntax and notations. To become a professional software developer, one must be aware of some popular solutions (i.e., design patterns) to the coding problems. Design patterns provide answer to some general set of problems that many software engineers face during development stage of the software. These solutions were obtained by sincere efforts of software architects belonging to many IT Companies by spending lot of man hour(s) working on software. GoF Book talks about two simple things.

3.1 Favor Object Composition over Inheritance

Whenever PDC have two options, to select either 'Is' a Relationship or 'Has' a Relationship, PDC should always prefer 'Has' a Relationship kind of structure in a program (Fig. 1).

For example, suppose user wants to create a class for which only a single object can be instantiated and that single object can be used by all other classes within same package.

3.2 Program to an Interface Not an Implementation

Interface is a aggregation of Abstract Functions. A class usually implements the interface and not vice-versa. An interface usually contains Abstract Functions, Constants and Static methods. An interface usually do not possess instance fields. However,

Fig. 1 Class B 'Is' a Class A

they possess fields that are declared both as static and final. An interface is different from a class in various ways, including

- Interface can not be instantiated.
- Constructors do not exist in interfaces.
- Interface is congregation of Abstract Functions.
- An interface can extend multiple interfaces, a class cannot.

A class, abstract in nature usually have data member, abstract function, method body, constructor and even main() method. It can be inherited but cannot be instantiated.

Algorithm 1 Example showing use of interface
_____
1: interface Fruit
2: {
3:        public void grow();
4:        public void eat();
5: }
6: Class Apple
7: Class Apple implements Fruit   { \\Class Definition }
8: To Create an object in class Apple :-
9: Apple xyz = new Apple();   \\Wrong Method
10: Fruit xyz = new Apple();   \\Correct Method
_____

4 Classification of Design Patterns

Broadly speaking,the design patterns can simplified into the following branches as shown in Table 1.

5 Proposed Design Pattern Approach for Java Cryptography Architecture (JCA)

5.1 DES (Data Encryption Standard)

DES, as literature say is primarily a Shared-Symmetric Key Algorithm that is block cipher in nature—meaning thereby it operates on 64 bit plaintext to produce a 64 bit ciphertext, of the same size. As well known, DES results in a permutation among the 2^{64} arrangements of 64 bits, which are either 0 or 1. Each block comprises of 64 bits which are further sub-divided into two blocks of 32 bits each, wherein a left half block is of L bits and a right half is of R bits.

Table 1 Classification of design patterns

Type of design pattern	Corresponding subtypes
Creational design pattern	• Singleton pattern
	• Factory pattern
	• Abstract factory pattern
	• Builder pattern
	• Prototype pattern
Behavioral design pattern	• Template method pattern
	• Mediator pattern
	• Chain of responsibility pattern
	• Observer pattern
	• Strategy pattern
	• Command pattern
	• State pattern
	• Visitor pattern
	• Interpreter pattern
	• Iterator pattern
	• Memento pattern
Structural design pattern	• Adapter pattern
	• Composite pattern
	• Proxy pattern
	• Flyweight pattern
	• Facade pattern
	• Bridge pattern
	• Decorator pattern

5.2   Design and Implementation

5.2.1   JCA [16] API's

Some of the most popular JCA API are detailed as :

- javax.crypto.Cipher;
- javax.crypto.KeyGenerator;
- javax.crypto.SecretKey;
- java.security.InvalidKeyException;
- java.security.NoSuchAlgorithmException;
- javax.crypto.BadPaddingException;
- javax.crypto.IllegalBlockSizeException;
- javax.crypto.NoSuchPaddingException;

Algorithm 2 DES : JCA lines for encryption using Design Pattern

```
1: KeyGenerator keygenerator = KeyGenerator.getInstance("DES");
2: SecretKey myDesKey = keygenerator.generateKey();
3: \\Create the cipher
4: Cipher desCipher;
5: \\Initialize the cipher for encryption
6: desCipher = Cipher.getInstance("DES/ECB/PKCS5Padding");
7: desCipher.init(Cipher.ENCRYPT_MODE, myDesKey);
8: byte[] textEncrypted = desCipher.doFinal(plaintext);
```

6   Advantages of Design Pattern

The design pattern are highly reusable in nature and, therefore, can be applied to multiple projects. They provide the flexible and easy way to adopt solutions that aid in defining the system design and architecture. They provide fluidity and high transparency to the design of an application. They are well-proven and well tested as they have been built upon the wisdom, vision and experience of expert architects. Having said all,still design patterns do not offer an absolute solution to all the problem(s) which the software engineers face. But indeed, they become handy and provide clarity to the system design and architecture and increases the possibility of a delivery of high-quality software product.

7   Drawbacks of Design Pattern

Usage and adoption of too much of pattern devoid developer(s) of creativity, and it stops software engineer from thinking new. It makes them monotonous and mundane. After certain period of time, sticking to same routing template would produce same set of results. But for the nave and beginners, it is good in the sense that developer(s) are able to obtain optimized results of their program by using pre-defined packages and namespaces.(e.g., DES Programs) For the academicians and researchers, it throws enormous research potential to design and develop new and innovative patterns and templates covering each and every branch of computer science.

8   Conclusion

The program designed using pattern-oriented framework are not only robust, dynamic and scalable, but they since have their roots from pre-designed and pre-defined pattern and framework; they are optimally written in minimum lines of code (LoC) with high-quality output. The megascale computing is easy to achieve nowadays due to the usage of Computing Principles and Algorithm which exploit the features of distributed computing to achieve it.

Acknowledgements We would like to express our sincere thanks and gratitude to our institution Guru Gobind Singh Indraprastha University for providing a great exposure to accomplish research-oriented tasks and achievements.

References

1. Alexander C (1977) A pattern language: towns, buildings, construction. Oxford University Press, New York
2. Nystrom N, Clarkson MR, Myers AC (2003) Polyglot: an extensible compiler framework for Java. In: Hedin G (eds) Compiler construction. Lecture notes in computer science, vol 2622. Springer, Berlin
3. Garcia A, Sant'Anna C, Figueiredo E., Kulesza U., Lucena C., von Staa A. (2006) Modularizing design patterns with aspects: a quantitative study. In: Rashid A, Aksit M (eds) Transactions on aspect-oriented software development I. Lecture notes in computer science, vol 3880. Springer, Berlin
4. Karaorman M, Hlzle U, Bruno J (1999) jContractor: a reflective Java library to support design by contract. In: Cointe P (eds) Meta-level architectures and reflection. Lecture notes in computer science, vol 1616. Springer, Berlin
5. Antoy S, Hanus M (2002) Functional logic design patterns. In: Hu Z, Rodrguez-Artalejo M (eds) Functional and logic programming. FLOPS, Lecture notes in computer science, vol 2441. Springer, Berlin
6. Tatsubori M, Chiba S, Killijian MO, Itano K (2000) OpenJava: a class-based macro system for Java. In: Cazzola W, Stroud RJ, Tisato F (eds) Reflection and software engineering. OORaSE, Lecture notes in computer science, vol 1826. Springer, Berlin
7. Hannemann J, Kiczales G (2002) Design pattern implementation in Java and AspectJ. In: ACM Sigplan, vol 37, no 11. ACM, pp 161–173 (November)
8. Heuzeroth D, Holl T, Hogstrom G, Lowe W (2003) Automatic design pattern detection. In: 11th IEEE international workshop on program comprehension, 2003. IEEE, pp 94–103 (May)
9. Tsantalis N, Chatzigeorgiou A, Stephanides G, Halkidis ST (2006) Design pattern detection using similarity scoring. IEEE Trans Softw Eng 32(11):896–909
10. Jeon SU, Lee JS, Bae DH (2002) An automated refactoring approach to design pattern-based program transformations in Java programs. In: Ninth Asia-Pacific Software Engineering Conference, 2002. IEEE, pp 337–345
11. Dong J, Lad DS, Zhao Y (2007) DP-miner: design pattern discovery using matrix. In: 14th annual IEEE international conference and workshops on the engineering of computer-based systems, 2007. ECBS'07. IEEE, pp 371–380 (March)
12. Shi N, Olsson RA (2006) Reverse engineering of design patterns from java source code. In: 21st IEEE/ACM international conference on automated software engineering, 2006. ASE'06. IEEE, pp 123–134 (September)
13. Antoniol G, Fiutem R, Cristoforetti L (1998) Design pattern recovery in object-oriented software. In: 6th international workshop on program comprehension, 1998. IWPC'98. Proceedings. IEEE, pp 153–160 (June)
14. Johari R, Gupta N (2011) Secure query processing in delay tolerant network using java cryptography architecture. In: 2011 international conference on computational intelligence and communication networks (CICN). IEEE, pp 653–657 (October)

15. Gamma E (1995) Design patterns: elements of reusable object-oriented software. Pearson Education India
16. Gong L, Ellison G (2003) Inside Java (TM) 2 platform security: architecture, api design, and implementation. Pearson Education

Permissioned Blockchain-Based Agriculture Network in Rootnet Protocol

Amal C. Saji, Akshay Vijayan, Ann Jerin Sundar and L. Baby Syla

Abstract The agriculture sector has seen considerate changes in its framework and working methodologies since its inception. Current trend in supply chain management includes the interception of prejudiced middlemen taking advantage of the anonymity in relationships existing between the actual producers and ultimate consumers. This trend puts two main areas in the supply chain at risk, the profit factor of the cultivators and quality of the final product delivered to the consumers. This paper aims at enhancing the supply chain performance by incorporating blockchain technology to solve the aforementioned problems. It includes the creation of a blockchain network encompassing the producers and consumers, and an automated Distributed Digital Ledger mechanism that effectuates two-way product traceability. It will consequently reduce the gap between the market price and the farmer's selling price to a great extent. The consumers on the other hand will receive better and healthier agro products that will eventually eradicate deadly ailments from the ecosystem.

Keywords Blockchain · Agriculture · Supply chain management · DHT

1 Introduction

Agriculture is the main source of daily bread for the whole Homosapien community. In spite of the preponderance of farmers among the whole population, the primary producers continue to be in turmoil due to the lack of transparency involved in

A. C. Saji (✉) · A. Vijayan · A. J. Sundar · L. Baby Syla
Department of Computer Applications, College of Engineering Trivandrum, Trivandrum, Kerala, India
e-mail: amalcs23@gmail.com

A. Vijayan
e-mail: akshayvijayan@cet.ac.in

A. J. Sundar
e-mail: annjerinsunder@cet.ac.in

L. Baby Syla
e-mail: syla@cet.ac.in

© Springer Nature Singapore Pte Ltd. 2020
A. Khanna et al. (eds.), *International Conference on Innovative Computing and Communications*, Advances in Intelligent Systems and Computing 1059,
https://doi.org/10.1007/978-981-15-0324-5_23

the whole business commute. The prevailing system projects several inefficiencies, with the inability of farmers to sell their produce in the regular market bereft of a decent selling price being the potent one. The supply chain lacks transparency [16] in its entirety, with upto 30–40% of the actual produce being deteriorated in the incompetent cold storage [14] facilities. Denied access to the credit system, about 60% of the farmers turn to private money lenders, paying high interest rates between 40 and 60% p.a. Approximately one-third of the food produced worldwide is wasted on a daily basis, in addition to food fraud.

This study proposes a protocol architecture with blockchain technology as its pivotal element to meet the following objectives:

- Enhance the current trend of supply chain performance.
- Minimize the loss of post-harvest produce.
- Eliminate the interception of middlemen compromising quality trade.
- Build transparency among the actual producers in conjunction with the ultimate consumers.

The remaining paper is categorized into six sections as follows. Section 2 reviews related literature pertaining to this study. Section 3 discusses the proposed method. Section 4 highlights the key elements of the technology used. Section 5 describes the system design. Finally, Sect. 6 concludes the paper.

2 Literature Review

Agricultural farming has evolved magnanimously with the influence of novel technologies in today's progressing world. The intervention of IOT and other mechanism has led to smart farming [7], which yields a contrasting difference in the produce [2]. Nevertheless, the focus still continues to be on the manufactured product rather than the accompanying processes and other specific components involved post-manufacture. The farmers often succumb to poor pricing strategies devoid of a network that connects them together.

The prevailing credit rating mechanism reflects a prejudiced perspective for different realms in the society [4]. A drastic variation in the financial status is noticed, concerning the farmers involved in the country's agriculture trade. Denied credit access [8], they often turn to money lenders for basic agricultural inputs, resulting in deteriorated life conditions with no proper farming strategies.

The agricultural supply chain has been reconstructed with a centralized information sharing platform [9], where information regarding the movement of goods can be acquired through a common interface. The collaboration of supply chains with blockchain technology will pave the way to innovative and improved means of agriculture trade.

3 Proposed Model

This study proposes a permissioned blockchain network in rootnet protocol. The producer–consumer blockchain network is created by an initial enrolment procedure for the consumers and producers, respectively. The authorizing components of the rootnet protocol include a root centre and root controller. The root centre authenticates the users based on strict procedures, and the root controller is responsible for the consequent transactions. The root controller appends data regarding the producer along with the product being transacted in a Distributed Digital Ledger.

The consumer gets the transaction history regarding the required product from the digital ledger. It will include the cultivator's credentials, the Kissan score, the time and place where the raw materials for the product were cultivated, the selling price, the cold storage facilities and the distribution centres for the product have passed through and much more.

The producers get to trace the whole transaction of the corresponding product till its delivery. This will eliminate middlemen intervention and creates a broader market space for the farmers, increasing their profit factor.

4 Rootnet Protocol

Rootnet is a global, decentralized protocol for authenticating commercial trade of products, through identity verification and creation of a peer-to-peer network of organizational credit vouching.

4.1 Blockchain

In this study, blockchain [6] is adopted for the conception of a distributed ledger mechanism that gives access to all members to record transactions in a decentralized data log, maintained on a network of computers, which replaces the current physical ledger or a single database [15]. The transactions are secured through the incorporation of cryptography and must be approved through consensus.

The decentralized digital ledger is formed by connecting together a series of blocks, resembling a chain. Each block corresponds to essential information regarding a set of transactions, specifically the timestamp, hash of the previous block, nonce, hash of the Merkle root [13] and body of the block.

Data entered into the ledger is accessible by all members, and any updates to it may be recorded or tracked along independently. Blocks are added on top of one another. Entry once made is recorded permanently and cannot be altered or deleted by any member.

Blockchain has the upper hand over several other technologies due to the following reasons:

Secure: Blockchain ensures highly secure and tamperproof information storage facility, efficient supply of products, fair pricing and improved tracking of products.

Food Safety: Introduction of transparency to the supply chain inhibits bad actors and poor processes as a resultant of the identification of affected nodes from a vantage point. This enables precise tracking of the defective source in the event of a food safety outbreak.

Traceability: Irrespective of the geographical location of the terminal food vendor, one can easily track down the primary source of the product, when it was harvested and processed, and the detailed description of the primary producer.

Opening New Markets: The current market space of the farmers remains bounded to a limited geographical area. Incorporation of blockchain technology shall see the advent of new markets in this developing world [10].

4.2 Permissioned Blockchain

Fabric [12] is an open-source system that exhibits modularity and extensibility for operating and deploying permissioned blockchains. It allows the system to be customized for specific use cases and trust models, since it supports modular consensus protocols [1, 12].

A permissioned blockchain facilitates secure means to interact among groups of entities with minimum trust that share a common goal. Hyperledger Fabric grants the plug-and-play feature for components such as consensus and membership services. The Hyperledger Fabric Network gives access to its members by an enrolment procedure through a trusted Membership Service Provider or MSP.

4.3 Rootnet Architecture

The existing drawbacks in the current trend of agricultural trade are inhibited through the amalgamation of a connected network of primary producers and supply chain, in a blockchain. The components are as follows.

Transaction ID (ID_t): Implements traceability in the food supply chain and streamlines the food products on a farm-to-market basis.

FarmerID (ID_f): Unique key element that identifies each farmer, along with his trade history.

ProductID (ID_p): Movement of agri-products along the whole commute is traced back through this particular identification code.

Kissan score (K_{score}): The producer's creditworthiness based on their productivity, cultivation history and other evaluating features is analysed through this metric.

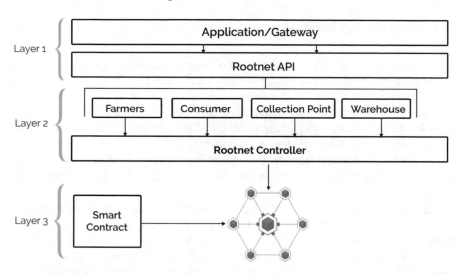

Fig. 1 Rootnet protocol architecture

Figure 1 depicts the architecture of the rootnet protocol. The protocol components are bound together to the rootnet controller. The rootnet controller plays a significant role of issuing node credentials for authentication and authorization purposes. The controller holds and maintains the identities of the system nodes.

The terminal users can make use of an application or a gateway interface to connect with the Rootnet API. The resultant connection will trigger the rootnet controller to perform the authentication verification procedures of individual nodes involved in the network. Upon successful authentication of the latter node, the rootnet controller authorizes the same to take part in the consensus concerned with the required trade in the network.

A smart contract [3, 5, 11] is a digital contract responsible for controlling the digital currency transfer between the parties involved in a transaction. It defines the necessary rules and penalties involved in the agreement, and the participants are enforced to fulfil the terms and conditions automatically by signing the contract. The rootnet controller establishes smart contracts between the participatory users on either ends, for facilitating the digital ledger storage mechanism in the decentralized blockchain network.

5 System Design

The rootnet system is composed of various components and processes. It initially needs the terminal user to provide it with the necessary details of the concerning trade and trader and moves on to the verification and storage of essential information in the digital ledger.

5.1 Rootnet Registration

The blockchain network is an assimilation of heterogeneous actors including peers, orderers, client applications, administrators and much more. Unlike other systems, agri-transactions demand, strict user authentication and accurate data collection mechanisms. The Root centre is responsible for carrying out the authentication process for the agricultural traceability system, to evidently identify the credentials of the participant submitting the data in the network. It effectuates the registration and verification of individual farmers, through online identity proof verification using distributed hash table data stored in the blockchain network [6], or offline onsite verification.

The root centre, on the reception of a particular registration application, undertakes a comprehensive verification process. Consequently, based on the verified information, the user generates a pair of keys through an asymmetric encryption algorithm. The private key is kept classified by the user, while the public key is uploaded to the root centre and connected with the distributed hash table which marks the end of the user's registration process.

5.2 Rootnet Transactions

The selling participant or the primary cultivator uploads potent information regarding the product to be sold, consequent to the registration process. Each piece of data uploaded regarding product traceability and pertaining components needs to be digitally signed with a private key. The root centre verifies the seller and the product, while the root controller verifies the farmer ID and his digitally signed signature. The successful authentication process will result in the root controller digitally signing the aforesaid information with his private key. This verified information is then broadcasted onto the participating nodes in the network and is termed as Transaction Proposal.

The individual permission nodes execute the same smart contract called the endorsement peer, verifying the data format, the digital signature of the root centre and the digital signature of the seller. A successful verification sends back an acknowledgement to the root centre. The root centre computes the minimum percentage of agreement based on the response and reciprocates it to the rootnet controller that maintains a cache pool of unarchived data.

The root controller then commences the process of appending the transaction into the blockchain. It sends the new block data to every node in the blockchain network. The individual nodes verify the received data based on some consensus mechanism and consequently record the transactions into the Distributed Digital Ledger. Distributed ledger technology (DLT) implements synchronized shared ledger mechanism in a distributed system. In order to change the product information or product ownership, the state data or state version of the data is to be changed.

When the farmer sells the product to the root centre, the product is verified by the root centre, which results in the generation of a unique product ID by hashing the farmer ID and the product details. This product ID is used to uniquely identify the product.

Suppose there are two other transaction parties having their own unique identification details interested in buying some products. This transaction is recorded into the blockchain by changing the state of the data. A new transaction hash is generated and latched on to the previous hash, which will help the end-user to track the product history. The creation of a new transaction hash starts by fetching the last transaction hash of the particular product ID. The transaction party details, timestamp and other pertaining data are then fed to the hash function, along with the previously obtained hash value. The new transaction hash thus obtained is appended to the particular product ID.

Algorithm 1 Build Rootnet   // Procedure to building agri-transaction chain

Require: ID_f, ID_p, D_p, D_t
  if exist(ID_f) && **isRightFormat**(D_p) **then**
    $ID_p \leftarrow \{ID_f, D_p\}$
  else
    if exist(D_t) && **exist**(ID_p) **then**
      lastTrHash $\leftarrow$ *getLastTransactionHash*(ID_p)
          newTr $\leftarrow$ {*lastTrHash*, D_t}
          transactionHash $\leftarrow$ *sendTransaction*(*new*Tr)
          updateLastTransactionHash(ID_p, transactionHash)
    end if
  else
    return false
  end if

5.3 Consumers

All transaction details related to a product ID is recorded in the blockchain as key-value pairs. Initially, the transaction hashes related to the consumer's product ID, from the first to the last transaction, is queried from the blockchain. Further, the transaction details of the individual transaction hashes are fetched and stored using an appropriate data structure, which is displayed thereafter. This resultant data structure aids the consumer in verifying and tracking the history of the product.

5.4   Producers

Significant leverage the farmer holds in this system is the efficient tracking of his agri-products and the corresponding selling price of the respective products, using the product ID, which yields a profitable price for the farmer. On that account, the root centre fixes a feasible price for each product, which is unanimously agreed upon by both the parties involved in the transaction. This ensures a standard global rating mechanism. Accordingly, the requisite amount is credited to the farmer's account as wallet balance. The root centre ensures that each farmer is credited with the proper revenue and monitors the entire supply chain. It holds the control over the pricing strategies and their timely regulations accordingly.

6   Conclusion

Permissioned blockchain-based agriculture network in rootnet protocol ensures better credit facilities for the actual primary producers involved in the agri-trade. It eliminates the intervention of middlemen by interconnecting the retailers and the consumers directly. It endorses the financial stability of the cultivators by improving the farming profitability. This will ease the process of availing government aids through the provision of farmer history evaluation credentials.

The introduction of a token-based transaction system guarantees effective supply chain management which helps to enhance food safety and boosts proper delivery mechanisms through the blockchain. The additional benefits encompass a fresh supply of healthy agro products, reduced food wastage and eradication of harmful chemically treated products in the marketplace.

References

1. Androulaki E, Cachin C, Ferris C, Sethi M, Stathakopoulou C (2018) Hyperledger fabric: a distributed operating system for permissioned blockchains. In: Proceedings of the thirteenth EuroSys conference on EuroSys '18. https://doi.org/10.1145/3190508.3190538
2. Caro MP, Ali MS, Vecchio M, Giaffreda R (2018) Blockchain-based traceability in Agri-Food supply chain management: a practical implementation. In: IoT Vertical and Topical Summit on Agriculture—Tuscany (IOT Tuscany). https://doi.org/10.1109/IOT-TUSCANY.2018.8373021
3. Dinh TTA, Liu R, Zhang M, Chen G, Ooi BC, Wang J (2018) Untangling blockchain: a data processing view of blockchain systems. IEEE Trans Knowl Data Eng. https://doi.org/10.1109/TKDE.2017.2781227
4. Hilscher J, Wilson M (2017) Credit ratings and credit risk: is one measure enough? Manage Sci 63(10):3414–3437. https://doi.org/10.1287/mnsc.2016.2514
5. Hinckeldeyn J, Jochen K (2018) (Short Paper) Developing a smart storage container for a blockchain-based supply chain application. In: Crypto valley conference on blockchain technology (CVCBT). https://doi.org/10.1109/CVCBT.2018.00017

6. Hua J, Wang X, Kang M, Wang H, Wang F (2018) Blockchain based provenance for agricultural products: a distributed platform with duplicated and shared bookkeeping. In: IEEE intelligent vehicles symposium (IV). https://doi.org/10.1109/IVS.2018.8500647
7. Kamilaris A, Gao F, Prenafeta-Boldu FX, Ali MI (2016) Agri-IoT: a semantic framework for Internet of Things-enabled smart farming applications. In: IEEE 3rd world forum on internet of things (WF-IoT). https://doi.org/10.1109/WF-IoT.2016.7845467
8. Katchova AL, Barry PJ (2005) Credit risk models and agricultural lending. Am J Agric Econ 87(1):194–205. https://doi.org/10.1111/j.0002-9092.2005.00711.x
9. Ming Q, Jingxu X (2008) The reconstruction of agriculture supply chain based on information sharing. In: ISECS international colloquium on computing, communication, control, and management. https://doi.org/10.1109/CCCM.2008.282
10. Mor RS, Singh S, Bhardwaj A, Singh LP (2015) Technological implications of supply chain practices in agri-food sector–a review. Int J Supply Oper Manage 2(2):720. https://doi.org/10.22034/2015.2.03
11. Parizi RM, Amritaj, Dehghantanha A (2018) Smart contract programming languages on blockchains: an empirical evaluation of usability and security. In: ICBC 2018. LNCS 10974, pp 75-91. https://doi.org/10.1007/978-3-319-94478-4_6
12. Sukhwani H, Martínez JM, Chang X, Trivedi KS, Rindos A (2017) Performance modeling of PBFT consensus process for permissioned blockchain network (hyperledger fabric). In: IEEE 36th symposium on reliable distributed systems (SRDS). https://doi.org/10.1109/SRDS.2017.36
13. Sun H, Hua S, Zhou E, Pi B, Sun J, Yamashitha K (2018) Using ethereum blockchain in internet of things: a solution for electric vehicle battery refueling. In: ICBC 2018. LNCS 10974, pp 3–17. https://doi.org/10.1007/978-3-319-94478-4_1
14. Tian F (2016) An agri-food supply chain traceability system for China based on RFID & blockchain technology. In: 13th international conference on service systems and service management (ICSSSM). https://doi.org/10.1109/ICSSSM.2016.7538424
15. Tse D, Zhang B, Yang Y, Cheng C, Mu H (2017) Blockchain application in food supply information security. In: IEEE international conference on industrial engineering and engineering management (IEEM). https://doi.org/10.1109/IEEM.2017.8290114
16. Wu H, Li Z, King B, Miled ZB, Wassick J, Tazelaar J (2017) A distributed ledger for supply chain physical distribution visibility. Information 8(4):137. https://doi.org/10.3390/info8040137

Load Balancing and Fault Tolerance-Based Routing in Wireless Sensor Networks

Priti Maratha and Kapil

Abstract Energy-efficient data collection from the environment is a critical operation in many applications areas of wireless sensor networks (WSNs). Unprecedented techniques which help in ameliorating the energy efficiency are highly required to elongate the lifetime of the network. In WSNs, sensed data needs to be forwarded to the sink node in an energy-efficient manner. Multi-hop communication helps a lot in reducing the energy consumption if the parent node through which data needs to be transmitted is selected in an efficient manner. Some nodes get overloaded while others are having very less load when parent nodes are selected in a random manner. In this paper, we have proposed a linear optimization-based formulation to balance the load of the nodes by selecting the parent node in an efficient manner. Moreover, when after some interval of sensing, some parent nodes start to get dying, the child nodes change their parent node to avoid the packet loss. So, a fault tolerance strategy is also proposed. Simulation results verify that proposed work is outperforming in terms of packet delivery ratio, network lifetime, and count of dead nodes.

Keywords Load · Linear programming problem · Next hop (parent) · Network lifetime · Dead nodes · Wireless sensor networks

1 Introduction

The popularization of wireless sensor networks (WSNs) has been driven due to the reduction in the cost of wireless communication, information communication technology (ICT), and the Internet of things (IoT) paradigm [1, 2]. WSNs have opened the door to collect, compute, and transmission of the data even from a harsh

P. Maratha (✉) · Kapil
Department of Computer Applications, National Institute
of Technology Kurukshetra, Haryana, India
e-mail: niki.maratha19@gmail.com

Kapil
e-mail: kapil@nitkkr.ac.in

© Springer Nature Singapore Pte Ltd. 2020
A. Khanna et al. (eds.), *International Conference on Innovative Computing
and Communications*, Advances in Intelligent Systems and Computing 1059,
https://doi.org/10.1007/978-981-15-0324-5_24

environment. The application domains are natural disaster relief, health monitoring, and military surveillance, etc. [3].

WSNs are basically a composition of hundreds or thousands of sensors used for monitoring the phenomena's or detecting events, etc. Nodes are generally deployed in a random manner. These nodes communicate with each other in a multi-hop fashion to relay the data to the sink. During this process, the battery of some nodes drains off quickly because they are heavily loaded (as many nodes have selected these as their next hop (or parent)). These heavily loaded nodes are used exhaustively by other nodes to relay the data. Hence in some areas, a fraction of the nodes die earlier. In WSNs, such kind of problem is referred to as *load balancing problem*. These nodes, whose battery level drain off at a faster rate become non-operational, result in the formation of energy consumed areas (dark zones) after a short period of time [4, 5]. This will result in reduced network lifetime. If WSN optimization is considered, an immense number of problems have been discussed in the literature. Dealing with the issue of energy consumption is one of them. Battery recharging is a practically unattainable operation in economic or practical terms. Also, limited energy is an unavoidable question in the field of WSNs as it urges strict restrictions on network operations [6, 7]. As a matter of fact, the energy consumption of the sensor node plays a crucial role in defining the lifetime of the network. So, the main goal of research studies is to optimize the energy by using various innovative techniques to revamp the network performance which includes maximization of the lifetime of the network [8, 9]. This network lifetime can be revamped in many ways. Firstly, once the nodes are deployed or when the parent of a node is about to die, routes are reconstructed to avoid the load imbalance and packet loss respectively. In other words, network protocols should be capable enough for the self-organization of the nodes. Secondly, heavily loaded dying nodes send a notification to the sink about their dying status earlier, and then redeployment is done. But the second approach is more expensive. Hence, maintaining the network quality with route reconstruction is a better approach.

In this paper, a load balancing approach using linear programming problem and fault tolerance approach by next hop selection is proposed for efficient routing (LBFTR). Load balancing can be done at the initial stage, once the network gets established in a temporary manner or at later stages. In this paper, we are focusing on the load balancing of the nodes at the initial stage. The load balancing process starts from the leaf nodes (that are farthest away from the sink). Linear programming problem (LPP)-based formulation is proposed for the selection of better next hops and hence to solve the load balancing problem. The solution of LPP results into load of heavily loaded nodes is shifted toward lightly loaded nodes, and hence, the parameters such as packet delivery ratio get improved. Using simulation results, we will see how it will help in improving the network lifetime and count of dead nodes. Also in LBFTR, the next hop change strategy is proposed for the nodes whose next hops are dying. This helps in maintaining the packet delivery ratio.

Various definitions of the lifetime of the network exist in the literature [10]. For example, a network is considered as dead, when it does not have the capability to transmit the data toward the sink node or when first node death (FND) occurs or a percentage of node dies or existence of one target out of many which cannot be monitored. In this paper, we have considered that a network is said to be alive till we get a 50% packet delivery ratio at the sink.

Motivated by the above shortcomings, a novel mechanism to balance the load of the nodes is proposed to improve the network lifetime. The contributions in this paper are summarized below:

- A generic formulation of load balancing has been provided. LPP has been used to find the optimal solution, i.e., next hops are selected in the best possible manner initially just after the deployment. It is used to set a limit on network lifetime and dead nodes. To prevent the packet loss, when parent node of few nodes is about to dead, child node will change their parent during sensing operation.
- A fault tolerance strategy is also proposed to prevent the packet loss; when parent node of few nodes is about to dead, child node will change their parent during sensing operation.
- A comparison of proposed work (LBFTR) is done with traditional hop-by-hop communication approach [4, 11, 12]. Performance analysis is done using the parameters like network lifetime and number of dead nodes.

The rest of the paper is organized as follows: Sect. 2 reviews the related literature. The proposed work is discussed under Sect. 3. In Sect. 4, performance analysis of LBFTR is done and it is verified that LBFTR is doing far better than the traditional approach. In the end, the conclusion and future scope of the work are discussed in Sect. 5.

2 Related Work

In this section, we review the most admissible energy-efficient algorithms which are based either linear or nonlinear optimization approach. Optimization techniques serve the designers to meet the domain requirements at different layers of the network [13, 14]. An immense amount of work has been done in WSNs in different contexts. Nevertheless, stating optimization problems is a concern. Optimization problems can be classified into linear and nonlinear optimization problems. Many researchers and industrialists have proposed significant multi-objective optimization algorithms to leverage the efficiency of the resource-constrained WSNs effectively [15, 16]. A routing method based on minimum total energy has been suggested by [17]. In this paper, the idea of minimizing the total energy consumption to reach the sink has been proposed. Nevertheless, if the whole traffic load is transmitted via a minimum energy route, the energy of the sensor nodes on this particular routing path will get drained off in a fast manner leading to partitions in the network while other nodes are still having sufficient energy to survive. Authors have not considered residual energy of the nodes, so it cannot elongate the network lifetime in an efficient manner.

Singh et al. [18] have proposed a min–max battery cost routing. Residual energy has been considered as one of the metrics. In this, routing tasks are often handled by the nodes which are having the high residual battery. This algorithm does not guarantee the minimization of the total consumed energy over the selected route. Hence, [19] introduced the minimum drain rate (MDR)-based mechanism. In order to predict the network lifetime as per traffic conditions, a new metric drain rate is introduced along with remaining energy. The objective behind this new metric is that the best routes cannot be established if residual energy is considered as the only metric in the routing protocols. Because if a node accepts all route requests just because of its high residual energy, traffic load will be very high. This could result in an energy drain off in a fast manner and hence the node gets dead. So, for the efficient functioning of the network, traffic load-based characteristic is required. However, the minimization of total energy is not guaranteed. So, a new protocol conditional minimum drain rate (CMDR) has been suggested by the same authors which is a modified version of MDR. The nodes which satisfy a lifetime threshold constitute a set of all possible paths. Lifetime threshold basically represents how long each node can survive by bearing the current load along with its current residual energy and drain rate. Among all the possible paths, a path with minimum transmission energy is given a priority. The performance of the algorithm is greatly affected by the proper choice of threshold.

A shortest path routing algorithm named as flow augmentation (FA) has been proposed by [3] to prolong the network lifetime. It depends on the link costs which is a combination of residual energy and transmission energy. The traffic load of the nodes has not been considered in route selection. Also, the empirical value assigned to the variables greatly affects the performance of the network. Same authors have identified the maximum lifetime problem as a linear programming problem (LPP), and the problem has been extended to a multi-commodity case. Since it is an LPP, so it is solvable in polynomial time. Authors have framed a mixed integer linear programming to resolve the data aggregation problem [20]. Total transmission power has been minimized by considering the parameters of radio resource constraints, co-channel interference constraints.

Ok et al. [21] have suggested a distributed energy balanced routing protocol. This algorithm utilizes a combination of residual and transmission energy to judge the optimal routing paths as used in FA. Before sending the data toward the base station, each sensor node checks how much expensive it is to transmit the data toward one of their neighbors or directly to the base station. In this work, one special thing has been added that all the nodes can transmit the data to the base station in a single hop manner. Another energy-aware cost function named as exponential and sine cost function has been proposed by [22]. This function assists in the mapping of little update in residual energy to a large update in cost. In between, the double cost function-based route accounted for both residual energy and energy consumption rate of the sensor nodes. High consumption rates experienced by the nodes near the sink have been considered by this cost function which in turn improves the energy balance. The shortest path-based distributed routing algorithm has been given to analytically derive the performance of the system [23]. The energy efficiency of existing protocols can be evaluated by considering this as a benchmark. Due to the exponential nature of computational complexity, this is unreasonable for large-scale WSNs. In this paper, the nodes and the topology of the network are static. So, the results are applicable to static networks. In LBFTR, initially the sensor nodes are deployed randomly as in traditional hop-by-hop communication approach and they select their parents to forward their data in a random fashion which is not a good choice. So just after the deployment, an LPP-based formulation is proposed for the better selection of next hops. Then an approach is suggested to circumvent the packet loss of the nodes whose parents are dying. By simulation results, we verify the designed framework is giving far better results than existing approaches in terms of network lifetime and the number of dead nodes.

3 Proposed Work

Energy Model

From the energy model [24], the energy consumed while transmitting data to the BS is given as:

$$ET_x = \begin{cases} E_{elec} \times k + \varepsilon_{fs} \times k \times d^2, & d < d_{threshold} \\ E_{elec} \times k + \varepsilon_{mp} \times k \times d^4, & d \geq d_{threshold} \end{cases}$$

where E_{elec} represents amount of energy required to activate the electronic circuits. ε_{fs} and ε_{mp} refer to the amount of energy required by the amplifier to send a packet of k-bits using free space and multi-path models, respectively. Here d is the distance between sender and receiver.

While the energy consumed during receive is given by:

$$ER_x = E_{elec} \times k$$

Network Model

In the network field, sensor nodes are deployed in a random manner as in traditional approaches (as shown in Fig. 1, sensor nodes are marked as stars while black diamond represents sink node). Sensor nodes are immobile in nature. Two nodes (say $n_i, n_j \in S$) are assumed as neighbors if one node is within the communication range of the other sensor node, then a control packet, i.e., *RREQ MSG* (*sender_id, h*) is broadcasted into the environment to know about neighboring nodes. Here *sender_id* is the node through which this message has come and h is hop count from the sink. A neighbor list (i.e., $Neighbor_i$) corresponding to each ith node is prepared using all nodes from which it has received *RREQ MSG*. All the sensor nodes set their next hop from which it receives *RREQ MSG* on a *first come, first serve* basis. Sink is assumed at hop count 0. Firstly, sink node broadcasts the packet (0,0) into the network field. In this way, initial routes are setup.

Let P_{ij} is the probability that tells n_i will select n_j as the next hop and it is defined as:

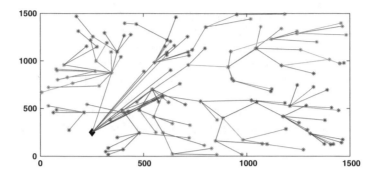

Fig. 1 Network field after random deployment

$$P_{ij} = \begin{cases} 1, & if\ n_i\ selects\ n_j\ as\ next\ hop \\ 0, & otherwise \end{cases}$$

where S is the set of sensor nodes.

Also, each node n_i must select only one node as its next hop toward the BS. Therefore, P_{ij} will be one only if $n_i \in S_h$, $n_j \in S_{h-1}$ such that

$$\sum_{j=0}^{n} P_{ij} = 1,\ \forall i\ |\ 1 \leq i \leq n,\ \forall n_i \in S_h,\ \forall n_j \in S_{h-1}$$

where S_h and S_{h-1} denote the set of sensor nodes at hop count h and $h - 1$, respectively.

Once the deployment is done, all sensor nodes have the information about their next hops and neighboring nodes after the RREQ message has been broadcasted by the sink node. All the nodes send their respective information to the sink node. Then sink node calculates the best possible next hops corresponding to each node on a respective hop count. Load balancing is done just after the deployment and initial setup. In *load balancing phase*, an LPP-based formulation is proposed for the better selection of next hops. Also, the nodes whose parents are about to die, they change their parents to avoid the packet loss. It is discussed under *fault tolerance phase*.

3.1   Load Balancing Phase

In this phase, firstly load of all the nodes is calculated. Load of the nodes having maximum hop count down to 1 hop count is calculated. It is assumed that initially all the nodes are having load value 1 (own sensed data), i.e., $LD(n_i) = 1$, $\forall n_i \in S$. The load corresponding to each jth node is find out using

$$LD(n_j) = 1 + \sum_i P_{ij} \times LD(n_i) \mid 1 \leq i \leq n, \; \forall n_i \in S_h, \; \forall n_j \in S_{h-1}$$

where $LD(n_i)$ is the load of the node n_i, i.e., number of packets transmitted by n_i. $EC(n_i, n_j)$ is the energy consumption of node n_i while transmitting the data toward n_j and is given by

$$EC(n_i, n_j) = \begin{cases} (E_{\text{elec}} \times k + \varepsilon_{fs} \times k \times d^2) \times LD(n_i), & d < d_{\text{threshold}} \\[2ex] (E_{\text{elec}} \times k + \varepsilon_{mp} \times k \times d^4) \times LD(n_i), & d \geq d_{\text{threshold}} \end{cases}$$

Then the lifetime of a node is determined, i.e., how much of the time the node is going to alive. For a node n_i, lifetime is given as

$$L(n_i) = Initial\ Energy / \sum_{n_j} EC(n_i, n_j)$$

Now our objective is to maximize the lifetime of the network. So, to achieve this objective, linear programming problem (LPP) of network lifetime can be formulated as follows:

$$L = min \frac{1}{L(n_i)} \; \forall n_i \in S_h, \; h \in 1 \ldots MaxHop$$

subject to

$$P_{ij} \geq 0, \quad \forall\, i, j \tag{1}$$

$$\sum_j P_{ij} = 1 \tag{2}$$

The constraint in Eq. 1 specifies the probability with which node i transfers its data to node j. The constraint in Eq. 2 specifies that each node can be assigned to one and only one node out of all neighbor nodes.

Once all the P_{ij}'s are found, and the lifetime corresponding lifetime is estimated. Then for a particular node n_i, only that node (say n_j, where $n_j \in Neighbor_i$ s.t. $\forall n_i \in S_h, \forall n_j \in S_{h-1}$) is selected as next hop corresponding to which lifetime is maximum and according to that i and j, value 1 is assigned to P_{ij} and for rest it is set as 0. The *benefit* of keeping only one node as next hop is that the packet coming over a particular node i will not be segmented. It will lessen the time and resources consumed during segmentation and reassembly. It also reduces the delay may get increased due to increasing path length over different routes. Once the P_{ij}'s are decided, this information is transmitted to the nodes and in this way, next hops are decided. Once the routes are reconstructed, the sensing operation starts. Nodes that are at maximum hop from the sink will send their sensed data first while the nodes which are at one hop distance will send the data at the end.

3.2 Fault Tolerance Phase

After some duration of sensing, a time comes when the battery of some nodes (which are the parent for some other nodes) drops below a particular threshold, then these nodes are assumed as *Dead nodes*, and hence, they do not perform any task. At this time, the nodes which are transmitting their data to the BS indirectly through these parent nodes should change their parents using their neighboring nodes. Corresponding to a particular node, the node which is having maximum energy is selected as the new parent out of all its neighboring nodes. Let $NH(n_i)$ current next hop of a node n_i. Out of neighbor list of n_i (i.e., *Neighbor$_i$*), let n_k is the node which is having maximum residual energy (*RE*). Then n_k is assigned as the new parent node of n_i. In other words,

$$NH(n_i) = n_k \mid n_k \in Neighbor_i \cap n_k \in S_{h-1} \ (if \ n_i \in S_h) \cap$$
$$n_k = arg_k(max(RE(Neighbor_i)))$$

4 Simulation Results

We assume that the nodes having maximum hop from the sink send the data first while the nodes at one hop distance send the data at the end. Total of 50 experiments have been done to analyze the average behavior in MATLAB 2017. A comparison is done between traditional hop-by-hop communication (named as *MHC* in the graph), hop-by-hop communication with proposed fault tolerance (named as *MHC-FT* in the graph) and hop-by-hop communication with proposed load balancing and fault tolerance (named as *LBFTR* in the graph) approaches. The behavior of said approaches is same on overlapping lines in the graphs. In this section, we will discuss the performance parameters and simulation results corresponding to them. Simulation parameters are given in Table 1.

Various performance metrics used for analysis are defined as follows:

Network Lifetime: In this paper, we have considered that a network is said to be alive till 50% PDR. Packet Delivery Ratio (PDR) is defined as the ratio of number of packets delivered to the BS to the total number of packets generated.

Number of Dead Nodes: It is defined as the ratio of number of total number of nodes which have got died in a particular sensing wave to the total number of nodes deployed.

We have analyzed the average behavior by considering 50 experiments corresponding to network lifetime and number of dead nodes.

Figure 2 is representing the average behavior of *packet delivery ratio*. A comparison is done between traditional MHC, MHC-FT, and LBFTR. From the figure, it is easily analyzed that

- LBFTR is doing far better than MHC. Because in proposed work (LBFTR), next hops are selected in a better manner after the setup phase, instead of the random selection of parent nodes as in traditional MHC.
- Further in the proposed work, when the parent nodes are about to die, corresponding parent nodes are changed to avoid the packet loss. The green line in the figure represents the impact of the change of next hops when parent nodes of respective nodes are about to die.

From Fig. 2, it can be inferred that proposed work outperforms others in terms of *network lifetime*. We can see that the network lifetime of traditional MHC, MHC-FT, and LBFTR is 1108 s, 1435, and 1460 s, respectively. Network lifetime of LBFTR is highest compared to others. So, we can say that the proposed work (LBFTR) is outperforming w.r.t. network lifetime.

Figure 3 is representing the average behavior of the number of *dead nodes*. A comparison is done between traditional MHC and proposed work (i.e., MHC-FT and LBFTR). From the figure, it can be easily analyzed that MHC-FT and LBFTR outperform MHC. It can be verified from the figure that the number of dead nodes is increasing at a fast rate in traditional MHC in comparison with the proposed work.

Table 1 Simulation parameters

Simulation parameters	Value
Field size	1500×1500 (m)
Total number of nodes	135
Simulation period	1800 s
Sensing interval	30 s
Data packet size	50 B
Data aggregation energy	5 nJ/bit
Initial energy	0.5 J
Battery threshold	0.05 J
E_{elec}	50 nJbit
ε_{fs}	10 pJbitm^2
ε_{mp}	0.0013 pJbitm^4
$d_{threshold}$	$\sqrt{\varepsilon_{fs}/\varepsilon_{mp}}$

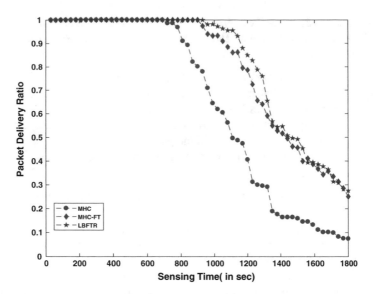

Fig. 2 Packet delivery ratio versus sensing time

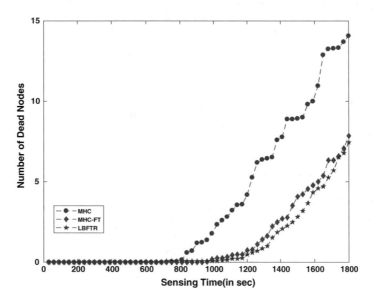

Fig. 3 Number of dead nodes versus sensing time

The reason of such a behavior is that in the case of traditional MHC, and next hop selection is not done in a better manner which results into some nodes are heavily loaded, and hence, they die earlier than other nodes in the network.

5 Conclusion and Future Scope

In this paper, we have proposed two techniques related to load balancing and fault tolerance for energy-efficient routing in wireless sensor networks. LPP-based formulation helps in better selection of next hops which results in enhanced network lifetime and reduction in the number of dead nodes. We have also presented a fault tolerance approach to ameliorate the packet loss. From simulation results, we have shown that our approach (i.e., LBFTR) is outperforming in comparison with simple hop-by-hop communication in terms of network lifetime and the number of dead nodes. However, we have considered only the routing approach in this paper. Clustering is a better approach to energy-efficient routing. In future, the clustering technique along with heterogeneous and mobile network can be applied for further improving the lifetime of the network.

Acknowledgements Priti Maratha acknowledges the support from the University Grant Commission, New Delhi, under the National Eligibility Test-Junior Research Fellowship scheme with Reference ID-3361/ (NET-JUNE 2015).

References

1. Rodríguez-Molina J, Martínez J-F, Castillejo P, López L (2013) Combining wireless sensor networks and semantic middleware for an internet of things-based sportsman/woman monitoring application. Sensors 13(2):1787–1835
2. Nayyar A, Puri V, Nguyen NG (2019) Biosenhealth 1.0: a novel internet of medical things (IoMT)-based patient health monitoring system. In: International Conference on Innovative Computing and Communications. Springer, Berlin, pp 155–164
3. Chang J-H, Tassiulas L (2004) Maximum lifetime routing in wireless sensor networks. IEEE/ACM Trans Netw 12(4):609–619
4. Inoue S, Kakuda Y, Kurokawa K, Dohi T (2010) A method to prolong the lifetime of sensor networks by adding new sensor nodes to energy-consumed areas. In: 2010 2nd international symposium on aware computing (ISAC). IEEE, pp 332–337
5. Bagci H, Yazici A (2013) An energy aware fuzzy approach to unequal clustering in wireless sensor networks. Appl Soft Comput 13(4):1741–1749
6. Al-Kiyumi RM, Foh CH, Vural S, Chatzimisios P, Tafazolli R (2018) Fuzzy logic-based routing algorithm for lifetime enhancement in heterogeneous wireless sensor networks. IEEE Trans Green Commun Netw 2(2):517–532
7. Ishmanov F, Malik AS, Kim SW (2011) Energy consumption balancing (ECB) issues and mechanisms in wireless sensor networks (WSNS): a comprehensive overview. Eur Trans Telecommun 22(4):151–167
8. Guleria K, Verma AK (2018) Comprehensive review for energy efficient hierarchical routing protocols on wireless sensor networks. Wirel Netw 1–25
9. Dhivya Devi C, Vidya K (2019) A survey on cross-layer design approach for secure wireless sensor networks. In: International conference on innovative computing and communications. Springer, pp 43–59
10. Kacimi R, Dhaou R, Beylot A-L (2013) Load balancing techniques for lifetime maximizing in wireless sensor networks. Ad hoc Netw 11(8):2172–2186
11. Mhatre V, Rosenberg C (2004) Homogeneous vs heterogeneous clustered sensor networks: a comparative study. In: ICC, pp 3646–3651

12. Heinzelman WR, Chandrakasan A, Balakrishnan H (2000) Energy-efficient communication protocol for wireless microsensor networks. In: Proceedings of the 33rd annual Hawaii international conference on System sciences, 2000. IEEE, p 10
13. Asorey-Cacheda R, Garcia-Sanchez A-J, García-Sánchez F, García-Haro J (2017) A survey on non-linear optimization problems in wireless sensor networks. J Netw Comput Appl 82:1–20
14. Abu-Baker A, Huang H, Johnson E, Misra S, Asorey-Cacheda R, Balakrishnan M (2010) Maximizing α-lifetime of wireless sensor networks with solar energy sources. In: Military communications conference, 2010-MILCOM 2010. IEEE, pp 125–129
15. Kulkarni RV, Venayagamoorthy GK (2011) Particle swarm optimization in wireless-sensor networks: a brief survey. IEEE Trans Syst Man Cybern Part C (Appl Rev) 41(2):262–267
16. Younis M, Akkaya K (2008) Strategies and techniques for node placement in wireless sensor networks: a survey. Ad Hoc Netw 6(4):621–655
17. Rodoplu V, Meng TH (1998) Minimum energy mobile wireless networks. In: 1998 IEEE International Conference on Communications, 1998. ICC 98. Conference Record, vol 3. IEEE, pp 1633–1639
18. Singh S, Woo M, Raghavendra CS (1998) Power-aware routing in mobile ad hoc networks. In: Proceedings of the 4th annual ACM/IEEE international conference on mobile computing and networking. ACM, pp 181–190
19. Kim D, Garcia-Luna-Aceves JJ, Obraczka K, Cano J-C, Manzoni P (2003) Routing mechanisms for mobile ad hoc networks based on the energy drain rate. IEEE Trans Mob Comput 2(2):161–173
20. Yen H-H (2009) Optimization-based channel constrained data aggregation routing algorithms in multi-radio wireless sensor networks. Sensors 9(6):4766–4788
21. Ok C-S, Lee S, Mitra P, Kumara S (2009) Distributed energy balanced routing for wireless sensor networks. Comput Ind Eng 57(1):125–135
22. Liu A, Ren J, Li X, Chen Z, Shen XS (2012) Design principles and improvement of cost function based energy aware routing algorithms for wireless sensor networks. Comput Netw 56(7):1951–1967
23. Habibi J, Aghdam AG, Ghrayeb A (2015) A framework for evaluating the best achievable performance by distributed lifetime-efficient routing schemes in wireless sensor networks. IEEE Trans Wirel Commun 14(6):3231–3246
24. Heinzelman WB, Chandrakasan AP, Balakrishnan H (2002) An application-specific protocol architecture for wireless microsensor networks. IEEE Trans Wirel Commun 1(4):660–670

A Novel Context Migration Model for Fog-Enabled Cross-Vertical IoT Applications

Ranjit Kumar Behera, K. Hemant Kumar Reddy and Diptendu Sinha Roy

Abstract With the maturity of Internet of things (IoT) paradigm, many innovative services are being conceived by integration of existing IoT services. Such services are termed as cross-vertical or cross-domain services. However, catering to real-time response requirement of such services is challenging and resource-constrained Fog nodes provide an alternative to cloud computing to realize end services. In this paper, we present a novel Fog resource aware and forecast based context migration model to address this delay requirement of such unified IoT applications by employing a Fog resource and forecast based mechanism among Fog nodes for minimizing system delay. Algorithms for context migration and required conditions for migrations been presented and simulation results carried out demonstrate the efficacy of the proposed methodology.

Keywords Internet of things · Cross-Vertical IoT applications · Fog computing · Context sharing · Service delay · Resource aware

1 Introduction

Internet of things (IoT) has established itself as one of the most promising information and communication technologies where billions of devices and things are connected seamlessly to deliver many intelligent applications across several domains namely, smart city, healthcare, transportation, and so on [1, 2]. Even some works have been done in the reliability aspects of IoT for service modeling [3]. But slowly, there has

R. K. Behera (✉) · K. H. K. Reddy
National Institute of Science and Technology, Berhampur, India
e-mail: ranjit.behera@gmail.com

K. H. K. Reddy
e-mail: khement.reddy@gmail.com

D. S. Roy
National Institute of Technology, Shillong, Meghalaya, India
e-mail: diptendu.sr@gmail.com

© Springer Nature Singapore Pte Ltd. 2020
A. Khanna et al. (eds.), *International Conference on Innovative Computing and Communications*, Advances in Intelligent Systems and Computing 1059,
https://doi.org/10.1007/978-981-15-0324-5_25

287

been a remarkable transformation from vertical applications to cross-vertical or cross-domain applications [4–6]. Even though cloud computing was initially used extensively for IoT applications [7], recently Fog computing emerged as viable alternate for delay sensitive real-time IoT applications, particularly to serve in cross-vertical domains [8–10].

Unlike Cloud data centers, most of the Fog nodes are heavily resource constrained due their inherent physical structure and can be deployed in a distributive manner across the edge [11, 12]. In our proposed model, the Fog nodes receive data from diverse applications and then processed it, and then sent to Cloud for further analysis. In this scenario, we have used the concept of context and context instances in the Fog node to handle massive data efficiently. Live Fog node migration balances the load by migrating applications from the over-loaded Fog node to under-utilized Fog node. However, unplanned migrations lead to resource contention, thereby degrading the performance of the application running in the Fog. Hence, in this paper we proposed an intelligent context migration technique that foresees the availability of resources in the destination Fog node prior to migration.

The remainder of this paper is organized as follows: Section. 2 presents the uses of Fog for real-time cross-vertical IoT application. Context management application migration problem in Fog node has been discussed in Sect. 3. Section 4 presents the forecast-based migration and resource aware and forecast based migration with corresponding algorithms. Simulation set up and result discussion are presented in Sect. 5, and the conclusion is given in Sect. 6.

2 Uses of Fog Layer for Real-Time Cross-Vertical IoT Application

In this paper, the concept of Fog has been used extensively for managing the latency sensitive real-time application particularly the cross-domain application. A general architecture for the same is shown in Fig. 1.

In the architecture shown in Fig. 1, it has been assumed that the IoT devices are only generating the data without processing power where as cloud is a standalone computational platform. In this scenario, Fog computing is used as an intermediate layer among these two. Within the Fog layer, Fog nodes are prearranged in a hierarchical order. These nodes are arranged such a way that the lower-level nodes are closer to IoT devices where as higher-level nodes are closer to cloud storage and at the same time, the capabilities of nodes like compute and storage are increased from lower level to higher level [13]. Each node in a particular level is directly associated with a node of immediate upper level.

Fig. 1 A overview of
Fog-layered architecture

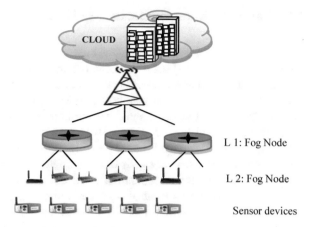

L 1: Fog Node

L 2: Fog Node

Sensor devices

2.1 Fog Node Architecture

In our paper, we presume, a Fog node is composed of three main modules: controller module, computing module and communication module as depicted in Fig. 2. Controller module is mainly responsible for monitoring and managing the operations of computing and communication module of a Fog node. It also maintains routing information and deployment information. Fog node can know which application modules are deployed within it from deployment list and can keep route-related information of other application modules in its routing information table, and these things are managed by a control unit. All the information is stored in Tcam. Computational module provides resources to execute application modules and each application modules is assigned to computing instances where resources like CPU, memory, and bandwidth are allocated according to the requirements. In a Fog node, when no MCIs are running, its computational component is turned off. In this case, the node only serves networking functionalities like routing, packet forwarding, etc., through its communication component. If the load of applications increases in the Fog layer, computational component of that node can be turned on again to handle the event.

Fig. 2 Component layout
diagram of Fog node

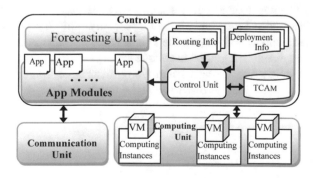

3   Context Management in Fog Node

As IoT devices generates humongous amounts of data, its management and storage in Fog layer needs a very careful consideration. To this end, in this paper, we are using context and context awareness. Context and context awareness were been extensively defined and used by so many authors previously [14, 15]. The learned information is used as a context instance and these are generated by processing the raw data. In our paper, we used these context instances to provide services across domains by sharing the related context instances. In this research paper, we use a context-sharing model proposed by Roy et al. [8] where a context instance can be transmitted from one Fog node to another depending upon the load and availability of the context instances as per the application's requirement. Although the authors [8] propose an efficient context-sharing model with minimum service delay, but they have not considered the context migration based on the availability of the resources. Hence, in this paper, we propose an intelligent resource aware context migration model for real-time cross-domain IoT applications.

3.1   Application Migration Problem

Let $F = \{f_1, f_2, \ldots, f_n\}$ represent the set of Fog nodes and $A(f_1) = \{A_1, A_2, \ldots A_k\}$ represent the set of applications hosted in a Fog node f_1. Let A^j represent the ith application hosted in the jth Fog node. For instance, when migration is initiated in f^j, the current resource requirement viz., computation and Tcam size of f^j are represented by Ω_{Comp} and Ω_{Tcam}, respectively. The migration controller considers the current resource requirement of application 1 in Fog node 1 represented as A_1^1 and it migrates into Fog node f^k. After migration, availability of computation and Tcam size of f^j represented as ω_{Comp} and ω_{Tcam}, respectively. It is evident that the load in Fog node is not constant rather it fluctuates in terms of computation and Tcam size. For instantaneous increase in load, resource requirements of f^k are increased by Δ_{Comp} and Δ_{Tcam} for the resources. Hence, the requirement of f^k is escalated to $\Omega_{\text{Comp}} + \Delta_{\text{Comp}}$ and $\Omega_{\text{Tcam}} + \Delta_{\text{Tcam}}$. Fog node f^k undergoes resource scarcity when $\Omega_{\text{Comp}} + \Delta_{\text{Comp}}$ and $\Omega_{\text{Tcam}} + \Delta_{\text{Tcam}}$ values exceeded ω_{Comp} and ω_{Tcam}, respectively. This is a symptom of faulty migration of the application among the Fog nodes. Hence, it is essential to design an efficient application migration technique among the Fog nodes (Table 1). Therefore, a novel-efficient forecast-based migration and resource-aware migration technique have been proposed and it is discussed in the following Sect. 4.

Table 1 Symbol notation

ω_{Tcam}	Available Tcam size of a Fog node
ω_{Comp}	Available computation of a fog node
Δ_{Comp}	Amount of computation for an application after sudden rise in demand
Δ_{Tcam}	Amount of Tcam size for an application after sudden rise in demand
Ω_{Tcam}	Required Tcam size for an application A for migration
Ω_{Comp}	Required computation for an application A for migration
$\text{Comp}_{\text{forcast}}$	Forecasted computation for migration
$\text{Tcam}_{\text{forcast}}$	Forecasted Tcam size for migration
R_{context}	Required context for an application
C_{Max}	Maximum context
$\text{Avl}_{\text{context}}$	Available context in a fog node for an application
LS_{App}	Latency sensitive application
A_{Comp}^{i}	Required computation for ith application
$\text{PMig}_{\text{time}}$	Predicted migration time
P_{time}	Processing time
SLA_{time}	Service level agreement time

4 Resource Aware and Forecast-Based Context Migration

To forecast the available resources in our paper, we have used four different forecasting techniques, namely weighted moving average, exponential smoothing average, Holt-Winter's technique, and autoregressive model. These techniques are selected because of different reasons. For example, weighted moving average technique is used to collect more recent resource requirements of each Fog node and then the collected information regarding resource requirements are weighted and averaged to predict the future resource requirements. Similarly, in Holt-Winter's technique, the information about the level, trend, and seasonal behavior of the past resource requirements are collected in the Fog node. Based on this information, the future resource requirements of the servers are predicted, and autoregressive model adaptively predicts the future resource requirement by correcting the prediction based on the past errors. Using the techniques mentioned above, the amount of required resources viz. computation and Tcam size are predicted. In the resource-aware migration module, the resources in a Fog node are classified into four categories based on the current and the combined forecasted outputs to identify the over-loaded Fog nodes. The four categories $\{C_0, C_1, C_2, C_3\}$ of the Fog nodes are defined as follows:

$$C_0 : L_{\text{Comp}} \text{ and } L_{\text{Tcam}}, \; C_1 : H_{\text{Comp}} \text{ and } L_{\text{Tcam}}$$
$$C_2 : L_{\text{Comp}} \text{ and } H_{\text{Tcam}}, \; C_3 : H_{\text{Comp}} \text{ and } H_{\text{Tcam}}$$

These four categories will help us to take the right decision when to migrate and if migrate then to which Fog node or at all we need to do migration or not. The details of Fog node migration have been explained in Algorithm 1. From Algorithm 1, it is clearly evident that we may not need migration always. For example, if required context is less than the available context and at the same time, if the same Fog node has some space to hold the missing context instances (i.e., if Flag==3) then we should simply do the context sharing. In our earlier research work, we presented a smart context-sharing model for the same. Similarly, when Flag==2, it can either employ a migration approach or simply context-sharing approach. But if Flag==1, then definitely we need to migrate to other Fog nodes. The detailed conditions for the above three flags has been explained in algorithm 2 and function CF_Migration() is presented in Algorithm 3.

```
Algorithm 1 : FoG _ Node _ Migration(Flag)
_ _ _ _ _ _ _ _ _ _ _ _ _ _ _ _ _ _ _ _ _ _ _
@ each Fog node f_j ∈ F
Flag = Conditions _ For _ Migration();
If (Flag == 1) then
            CF _ Migration();
            Else If (Flag == 2) then
                        If (PMig_time + P_time < SLA_time)
                                    CF _ Migration();
                        Else
                                    Smart _ Context _ Sharing();
                        Else If (Flag == 3) then
                                    Smart _ Context _ Sharing(); // refer SCS Algorithm[7]
                        Endif
            Endif
Endif
```

5   Simulation Setup and Result Discussion

The simulation setup for performance evaluation is done as the same way as our earlier paper [8]. We have compared the result of SCS model presented in [8] with our FRAFBCM model presented in this paper. The simulation results are shown in

Figs. 3 and 4. From the result shown in Fig. 4, it is clear that the number of migration in FRAFBCM model is less than the number of migration in SCS model with respect to the required maximum number of context instances.

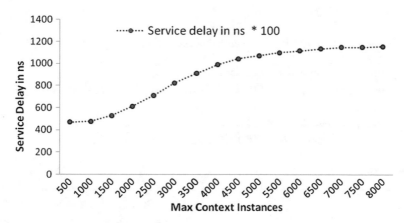

Fig. 3 Incurred service delay over max context instances

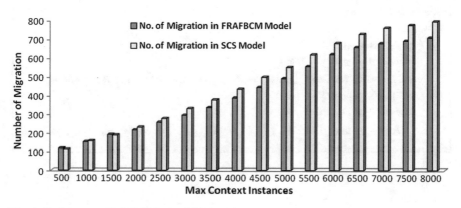

Fig. 4 Performance FRAFBCM over SCS model

Algorithm 2 : *Conditions _ For _ Migration*()

@ *each Fog node* $f_j \in F$

$$If \left(\begin{pmatrix} \omega_{Comp}(f_j) < \sum_{i=1}^{n} A_{Comp}^i \; \&\& \; LS_{App} \end{pmatrix} \| \\ \left(R_{context}(A_i) < Avl_{context}(A_i) \right) \&\& \left(C_{Max} - Avl_{context}(F) \right) < \left(R_{context}(A_i) - Avl_{context}(A_i) \right) \right)$$

$\qquad Flag = 1;$

$$Else\ If \left(\begin{pmatrix} cnt(C_3) > cnt(C_0) \| cnt(C_2) > cnt(C_1) \end{pmatrix} \&\& \\ \left(R_{context}(A_i) < Avl_{context}(A_i) \right) \&\& \left(C_{Max} - Avl_{context}(F) \right) < \left(R_{context}(A_i) - Avl_{context}(A_i) \right) \right)$$

$\qquad Flag = 2;$

$$Else\ If \left(\begin{pmatrix} R_{context}(A_i) < Avl_{context}(A_i) \end{pmatrix} \&\& \\ \left(C_{Max} - Avl_{context}(F) \right) > \left(R_{context}(A_i) - Avl_{context}(A_i) \right) \right)$$

$\qquad Flag = 3;$

Return *Flag*;

End.

Algorithm 3 : *CF _ Migration*()

@ *each Fog node* $f_i \in F$

For each App $A_i \in A$ *do*

If $Flag = 1$

$\quad$ *If* $(cnt(C_3) > cnt(C_0))$

$\quad$ *Migrate*(A_i) *to Fog node* f_j, *where*

$\quad$ $(cnt(C_3) < cnt(C_0)) \&\& \; Avl_{context}(A_i) >= R_{context}(A_i) \&\& \left(PMig_{time} + P_{time} \right) \le SLA_{time}$

$\quad$ *Else If* $(cnt(C_2) > cnt(C_1))$

$\quad\quad$ *Migrate*(A_i) *to Fog node* f_j, *where*

$\quad\quad$ $(cnt(C_2) < cnt(C_1)) \&\& \; Avl_{context}(A_i) >= R_{context}(A_i) \&\& \left(PMig_{time} + P_{time} \right) \le SLA_{time}$

End

6 Conclusion

In this research work, we have presented a novel context migration model to address the delay requirement of unified IoT applications by employing a resource aware and forecast based mechanism among Fog nodes. Algorithms are presented to minimize the service delay by minimizing the number of context migration and from the simulation results, it has been clear that it outperforms the SCS model.

References

1. Bisio I et al (2012) Smartphone-based user activity recognition method for health remote monitoring applications. PECCS
2. Bisio I et al (2017) Enabling IoT for in-home rehabilitation: accelerometer signals classification methods for activity and movement recognition. IEEE Internet Things J 4(1):135–146
3. Behera RK, Reddy KHK, Sinha Roy D (2018) Modeling and assessing reliability of service-oriented internet of things. Int J Comput Appl, 1–12
4. Intizar M et al (2017) Multi-layer cross domain reasoning over distributed autonomous IoT applications. Open J Internet Things
5. Soursos S et al (2016) Towards the cross-domain interoperability of IoT platforms. In: European conference on networks and communications (EuCNC). IEEE
6. Gyrard A et al (2015) Cross-domain internet of things application development: M3 framework and evaluation. In: 3rd International conference on future internet of things and cloud (FiCloud). IEEE
7. Behera RK, Gupta S, Gautam A (2015) Big-data empowered cloud centric internet of things. In: 2015 international conference on man and machine interfacing (MAMI). IEEE, pp 1–5
8. Roy DS et al (2018) A context-aware, fog enabled scheme for real-time, cross-vertical IoT applications. IEEE Internet of Things J
9. Dastjerdi AV et al (2016) Fog computing: principles, architectures, and applications. Internet of Things, 61–75
10. Renuka K, Das SN, Reddy KH (2018) An efficient context management approach for IoT. IUP J Inf Technol 14(2):24–35
11. Mahmud R, Kotagiri R, Buyya R (2018) Fog computing: a taxonomy, survey and future directions. In Internet Everything, Springer, Singapore, pp 103–130
12. Reddy KHK, Mudali G, Roy DS (2017) A novel coordinated resource provisioning approach for cooperative cloud market. J Cloud Comput 6(1)
13. Ashrafi TH et al (2018) IoT infrastructure: fog computing surpasses cloud computing. Intelligent communication and computational technologies. Springer, Singapore, pp 43–55
14. Gu T, Pung HK, Zhang DQ (2005) A service-oriented middleware for building context-aware services. J Netw Comput Appl 28(1):1–18
15. Baldauf M, Dustdar S, Rosenberg F (2007) A survey on context-aware systems. Int J Ad Hoc Ubiquitous Comput 2(4):263–277

Equity Data Distribution Algorithms on Identical Routers

Mahdi Jemmali and Hani Alquhayz

Abstract This paper focuses on the problem related to assigning several big data packages on different routers when seeking equity of sending time. It is challenging to find a good algorithm that can distribute big data on routers before sending them. We assume that all routers share the same technical characteristics. The problem is as follows. Given a set of big data, represented by its size in MB, the objective is to plan the assignment so that the minimum time sending gap exists between the routers. The objective function of the optimizing problem is the minimization of the size gap. This optimization problem is very *NP-hard*. We propose a new summarized network architecture, based on adding a new component: a scheduler. The scheduler applies several algorithms to search for a resolution to the studied problem. Four heuristics were developed, and experimental results are provided to allow a comparison between heuristics. Two classes of instances are provided. The results given by generated instances show that the performance of heuristics.

Keywords Big data · Routers · Equity distribution · Heuristics · Scheduling

1 Introduction

Nowadays, using networks have become more and more primordial in all domains. The sharing and communication necessities oblige users to use networks frequently [13]. Wireless networks must be protected to ensure that users can send data securely [8, 12]. The control of the data transmitted into the network is very important in order to guarantee a high level of security. Besides security, the transmission time for data is also very important when a lot of large-scale data need to be transmitted. A survey

M. Jemmali (✉) · H. Alquhayz
Department of Computer Science and Information, College of Science at Zulfi Majmaah
University, Al-Majmaah 11952, Saudi Arabia
e-mail: m.jemmali@mu.edu.sa

H. Alquhayz
e-mail: h.alquhayz@mu.edu.sa

© Springer Nature Singapore Pte Ltd. 2020 297
A. Khanna et al. (eds.), *International Conference on Innovative Computing
and Communications*, Advances in Intelligent Systems and Computing 1059,
https://doi.org/10.1007/978-981-15-0324-5_26

for networking for big data developed in [14]. The application of a new approach is developed in wireless networks as flexible real-time transmission scheduling in the presence of non-deterministic workloads [4]. The allocation of the limited wireless resource with dense ($RANs$) is studied in [1]. This research investigates a visualized RAN, where the CNC auctions channels at the beginning of scheduling slots to the mobile terminals (MTs).

Scheduling and wireless sensor networks were covered by many researchers, for example, [3] identify an appropriate metric based on sensor's data to measure the quality of WSN border crossing detection. While the authors in [11] use meta-heuristics to solve the resource-constrained project scheduling problem with optimal makespan.

The scheduling problem arises when networks attempt to give a suitable assignment. The well-known scheduling problem, $P||C_{max}$, is developed in [5, 6]. The authors developed a performed lower bounds and heuristics to solve the problem. An exact solution, based on branch and bound algorithms, was developed. Other application of scheduling problems is developed to solve a big data problem in smart city [9].

In this research, we introduce a network problem regarding the assignment of data on routers to ensure a fair distribution in terms of size and time. In a network where we want to transmit a data stream with different sizes, downloading different files or data received from different receivers take a long time. In fact, if the distribution is performed inefficiently, one receiver may quickly receive data while another receiver is left waiting. Our study proposes to include a component in the network architecture that plays the role of a Scheduler in distributing the data to be transmitted from the start to the different routers.

The paper is structured in four sections. Section 2 focuses on the problem description, detailing the objective of solving the proposed problem. Section 3 presents four heuristics as the algorithms for solving the proposed examples. The first algorithm is based on the longest size and the second on the smallest size. Section 4 shows the performance of the heuristics through undertaking an experimental study. Indeed, two classes of instances are generated and 540 instances were tested.

2　Problem Description

We present, in this section, a network problem based on scheduling algorithms. Suppose that we have several data packages that need to be transmitted via the network at the same time. The data are characterized by their large size. Several data packages will be transmitted by several users. The download time by the receiver can be very high due to the big data being transmitted. In this case, the good scheduling of the data to the routers facilitates the good reception at a better time. Indeed, the goal is not to make many others wait while the receiver downloads the data sent. Some

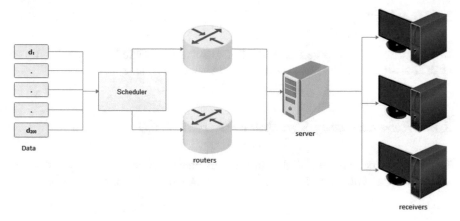

Fig. 1 Sending data in the presence of 2 routers and 3 receivers

notation is required as follows. Denoting by j the index of data d_j to be sent and by s_j the size of the corresponding data d_j with $j = \{1, \ldots, n_d\}$. The problem becomes how to send data having s_j through n_r routers. The number of data packages is n_d. Every data d_j can be assigned to only one router ro_i with $i = \{1, \ldots, n_r\}$. Denoting by t_j the cumulative sizes received by the router ro_i when data d_j are assigned. The total received sizes for each router ro_i after completion of all assignments is denoted by c_i. The maximum (minimum) given size after the completion of the scheduling on all routers is denoted by r_{max} (r_{min}). The total sizes received by each routers is indexed as follows: $r_1 \leq r_2 \leq \ldots \leq r_{n_r}$.

In Fig. 1, the new component, the scheduler, is responsible for distributing the data on the routers before the data start to be sent.

To evaluate the gap between the routers, we can set different indicators. We propose that, for each router, we need to reduce the received size by calculating $r_i - r_{min}$. Therefore, considering n_r routers, the total size gap is given in Eq. (1) below:

$$\text{Minimize} \sum_{i=1}^{n_r} [r_i - r_{min}]. \tag{1}$$

Equation (1) is the function that we have to obtain in the studied problem.

Denoting by $Sg_{max} = \sum_{i=1}^{n_r} [r_i - r_{min}]$, the total size gap between the routers, we denote by Sg_{max}^* the optimal solution for the studied problem. Using the standard three-field notation of [7], this problem can be denoted as $P||Sg_{max}$.

3 Heuristics

In this section, we propose four heuristics to solve the studied problem. We try to offer heuristics that give good results within a very acceptable time limit.

3.1 Non-decreasing Size Order Heuristic (NDS)

For this heuristic, we order all data in a non-decreasing order of size. After that, we assign the data with the smallest size to the router with the smallest total size.

3.2 Non-increasing Size Order Heuristic (NIS)

Contrary to NDS, we order all of the data in a non-increasing order of size. After that, we assign the data with the greatest size to the router with the smallest total size.

3.3 Sizing Multi-fit-Based Heuristic(SMF)

This SMF is based on the finding for the minimum capacity (minimum total size) such that all n_d data will be scheduled on the n_r routers.

This algorithm is based on the adaptation of the bin-packing algorithm (BPA). The router will be presented as a bin and the data as an item in the (BPA). For each fixed bin capacity, the first fit decreasing (FFD) method is used to fit the data to the bin.

At the beginning, we must order the data according to size, such that $s_1 \geq s_2 \geq \ldots \geq s_{d_n}$. The FFD algorithm schedules all of the data in succession to the lowest indexed router containing the data (regarding sizing capacity) within the capacity. Let $LB_{\max} = \max\left(s_1, s_{n_r} + s_{n_r+1}, \frac{\sum_{j=1}^{d_n} s_j}{n_r}\right)$ is the value given by applying the LPT rule for $P||C_{max}$. We run the FFD function more than once time. We denote by it the number of iteration of FFD. We set $it = 135$ and n_b is the number of bins returned by the FFD function. The following algorithm describes all of the steps taken to apply SMF.

Table 1 Instance of sizes for the heuristic SMF

d_j	1	2	3	4	5	6	7	8	9	10
s_j	55	59	94	30	51	95	43	44	100	70

Algorithm 1 Sizing multi-fit algorithm SMF

1: Set $I = 0$, $u = UB_{max}$ and $l = LB_{max}$.
2: Set $mid = \lfloor \frac{u+l}{2} \rfloor$, set $I = I + 1$.
3: Apply FFD with capacity C.
4: If we can assign all data d_n through the n_r routers, then set $u = C$ and go to 5; otherwise, set $l = C$ and go to 5.
5: If $I = it$ then STOP, otherwise go to 2.
6: If $n_n > r_n$, then the schedule given by NIS is taken; otherwise, the schedule given by FFD is taken. The taken schedule is noted by σ.
7: Determine the r_{min} of schedule σ.
8: $SMF = \sum_{i=1}^{n_r} [r_i - r_{min}]$.

Example 1 Let $d_n = 10$ and $n_r = 2$. The following table represents the size of each data package.

Considering the $P||C_{max}$ problem, the LPT rule gives the upper bound value of $UB_{max} = 323$ and the lower bound is $LB_{max} = 318$. Then, $mid = \lfloor \frac{323+318}{2} \rfloor = 322$. Applying the FFD function with a capacity of 322, the 10 items which are represented by the data in Table 1. The first router contains the following data $\{d_4, d_3, d_6, d_9\}$, and the second router contains the following data $\{d_1, d_2, d_5, d_7, d_8, d_{10}\}$. The first router has a total size 319; however, the second one has a total size of 322. Thus, $Sg_{max} = 641 - 2 \times 319 = 3$. On the other hand, the r_{min} obtained by LPT is 318, which means that the NIS heuristic gives $Sg_{max} = 5$, so, the result obtained by BMF is better than NIS.

3.4 Iteratively Solving Subset-Sum Problems Heuristic (ISS)

A greedy algorithm is utilized to develop this heuristic. This algorithm is based on the resolution of several subset-sum problems iteratively. We denote these problems by $(Ds)_k$ for $k = \{1, 2, \ldots, n_r - 1\}$ as follows:

$$(Ds)_k : \begin{cases} \min \sum_{p \in S_k} s_j y_j, \\ \text{subject to} \sum_{D_j \in S_k} s_j y_p \geq LB(S_k, n_r - k + 1), \end{cases} \quad (2)$$

with $y_j \in \{0, 1\}$ for all $D_j \in S_k$. where $S_1 = D$ and $S_{k+1} = S_k \setminus D_k$ where D_k is an optimal subset-sum for Ds_k and $k = 1, 2, \cdots, n_r - 1$. $LB(S, K)$ denote a valid lower bound on the makespan of the instance defined on $k \le n_r$ routers and a subset of data $S \subset D$.

Therefore, for the first router, we assign data until we reach LB on Ds_1. The remaining data with the remaining routers will constitute the second problem Ds_2 to solve the new SSP until LB is reached, and so on [5]. A pseudo-polynomial algorithm is utilized to find a solution to the subset-sum problem described above. This pseudo-polynomial algorithm is based on the dynamic programming algorithm and developed by [10]. After finishing the scheduling of all data, a r_{min} value will be returned and the Sg_{max} value will be deduced.

4 Experimental Results

In this section, we present the experimental results found after running all of all proposed heuristics. The assessment of the performance of the proposed algorithms is given after codding them in Microsoft Visual C++ (Version 2013). All of our experiments were obtained on an Intel(R) Xeon(R) CPU E5-2687W v4 @ 3.00 GHz and 64 GB RAM. The operating system used is Windows 10 with 64 bits.

The proposed procedures are tested on a set of test problems that are displayed as follows.

Two types of instances are generated in this research. We tested the results on a set of instances that was inspired as described in [2]. The data size s_j is generated according to the uniform distribution. Each one represents a class. These classes are:

- Class 1: s_j is generated from the discrete uniform distribution $U[30, 100]$.
- Class 2: s_j is generated from the discrete uniform distribution $U[50, 300]$.

The over all instances are based on the choice of d_n, n_r and $Class$. The choice of the pair (d_n, n_r) is shown in Table 2.

For each fixed triple $(d_n, n_r, class)$, we generate 10 instances of budget project. Based on the choice of (d_n, n_r) referred to Table 2, the total number of instances is 540.

We denote these by :

- UB the minimum value obtained after running all of the algorithms.
- U is the studied heuristic.

Table 2 Generation of (d_n, n_r)	d_n	n_r
	10	2, 3, 5
	20	2, 3, 5, 10
	50	2, 3, 5, 10, 25
	100	2, 3, 5, 10, 15, 25, 50
	300	2, 3, 5, 10, 15, 25, 50, 100

Table 3 Global comparison between heuristics

	NIS	NDS	SMF	ISS
Min	272	97	234	451
perc	50.4%	18.0%	43.3%	83.5%
AG_u	13.56	119.81	27.52	0.74
Time	-	-	0.017	0.001

Table 4 Behavior of AG_u and $Time$ according to n_d

n_d	NIS		NDS		SMF		ISS		Total	
	AG_u	Time	AG_u	Time	AG_u	Time	AG_u	Time	AG_u	Time
10	4.75	-	23.42	-	3.51	0.002	0.45	0.001	6.42	-
20	9.86	-	39.17	-	5.14	0.002	0.33	0.001	10.90	0.001
50	8.78	-	40.73	-	5.13	0.003	1.01	-	11.13	0.001
100	15.51	-	94.36	-	12.60	0.006	0.48	-	24.59	0.001
300	7.96	-	71.68	-	12.01	0.008	1.48	-	18.63	0.002

- Min the number of instances when $U = UB$.
- $G_u = \frac{U-UB}{U}$.
- AG_u is the average G_u for a fixed number of instances.
- $Time$ is the time spent executing the heuristics in corresponding instances. This time will be in seconds and we denote it by "-" if the time is less than 0.001 seconds.

We denote by $perc$ the percentage of all of the instances (540). In Table 3, we show the gap differences between each heuristic besides the average time consumed.

It clear from Table 3, that ISS is the best heuristic but not dominate the other heuristics. Indeed, in 83.5%, ISS is equal to the Min with $AG_u = 0.74$ in 0.001s. However, the NDS heuristic has a $perc$ equal to 18.0% and the gap is remarkably larger, at 119. The following table presents the variation of AG_u and $Time$ according to n_d. The maximum gap reach of 94.36 is obtained for $n_d = 100$ for the NDS heuristic. However, the minimum is around 0.33 and is obtained for $n_d = 20$ for the ISS heuristic. Remarkably, when n_d is larger, the time becomes less than 0.001s for the ISS heuristic. This is not observed for the heuristic SMF because the time increases when n_d increases (Table 4).

Figure 2 shows the behavior of the total AG_u according to n_d. This figure shows that the curve increases when $n_d = 100$ but, when n_d starts to increase after 100, the curve starts in decrease.

The variation of the gap according to the number of routers is shown in Table 5. Strangely, for ISS heuristic the maximum gap is obtained for $n_r = 3$. Usually, the gap increases when n_r increases for NIS and NDS.

From Table 6, we observe that the total gap is equal to 21.09 for $Class 1$ and equal to 59.73 for $Class 2$. This means that $Class 2$ is slightly more difficult than $class 1$.

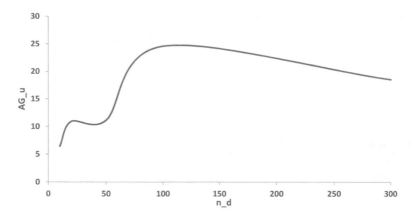

Fig. 2 Total AG_u for all instances and heuristics according to n_d

Table 5 Behavior of AG_u and $Time$ according to n_r

n_r	NIS		NDS		SMF		ISS		Total	
	AG_u	$Time$	AG_u	$Time$	AG_u	$Time$	AG_u	$Time$	AG_u	$Time$
2	8.15	-	49.94	-	6.89	0.004	0.97	-	16.49	0.001
3	9.83	-	45.45	-	7.02	0.003	1.02	-	15.83	0.001
5	9.61	-	45.31	-	6.42	0.004	0.97	0.001	15.58	0.001
10	8.53	-	45.75	-	6.25	0.004	0.92	0.001	15.36	0.001
15	10.66	-	55.98	-	8.51	0.004	0.61	-	18.94	0.001
25	12.91	-	69.48	-	10.48	0.005	0.51	0.001	23.35	0.001
50	12.35	-	78.31	-	10.58	0.005	0.67	-	25.48	0.001
100	11.04	-	76.38	-	10.14	0.006	0.81	-	24.59	0.002

Table 6 Behavior of AG_u and $Time$ according to $Class$

$Class$	NIS		NDS		SMF		ISS		Total	
	AG_u	$Time$	AG_u	$Time$	AG_u	$Time$	AG_u	$Time$	AG_u	$Time$
1	8.94	-	62.12	-	12.47	0.019	0.84	0.001	21.09	0.005
2	18.18	-	177.50	-	42.57	0.014	0.65	0.002	59.73	0.004

5 Conclusion

In this research, we proposed a new component that might be added to a wireless network to eliminate the wait time of the data receivers when a lot of data need to be sent at the same time through the same network by different users. We suppose that the data are large in size. The problem of the assignment of the data to the routers is NP-hard. To solve this problem, we proposed in this paper four heuristics. The two first ones are based on the dispatching rules, but the third heuristic is based on the multi-fit method. Finally, the last heuristic is based on the subset method, and the experiential results show its efficiency. The execution time for the best heuristic is highly acceptable.

Acknowledgements The authors would like to thank the Deanship of Scientific Research at Majmaah University for supporting this work.

References

1. Chen X, Han Z, Zhang H, Xue G, Xiao Y, Bennis M (2018) Wireless resource scheduling in virtualized radio access networks using stochastic learning. IEEE Trans Mobile Comput 1:1–1
2. Dell' Amico M, Martello S (1995) Optimal scheduling of tasks on identical parallel processors. ORSA J Comput 7(2):191–200
3. Devi CD, Vidya K (2019) A survey on cross-layer design approach for secure wireless sensor networks. In: International conference on innovative computing and communications. Springer, Heidelberg, pp 43–59
4. Gumusalan A, Simon R, Aydin H (2018) Flexible real-time transmission scheduling for wireless networks with non-deterministic workloads. Ad Hoc Netw 73:65–79
5. Haouari M, Gharbi A, Jemmali M (2006) Tight bounds for the identical parallel machine scheduling problem. Int Trans Oper Res 13(6):529–548
6. Haouari M, Jemmali M (2008) Tight bounds for the identical parallel machine-scheduling problem: Part ii. Int Trans Oper Res 15(1):19–34
7. Lawler EL, Lenstra JK, Kan AHR, Shmoys DB (1993) Sequencing and scheduling: algorithms and complexity. Handbooks Oper Res Manage Sci 4:445–522
8. Ma D, Tsudik G (2010) Security and privacy in emerging wireless networks. IEEE Wireless Commun 17(5)
9. Melhim LKB, Jemmali M, Alharbi M (2018) Intelligent real-time intervention system applied in smart city. In: 2018 21st Saudi Computer Society National Computer Conference (NCC), IEEE, pp 1–5
10. Pisinger D (2003) Dynamic programming on the word ram. Algorithmica 35(2):128–145
11. Roy B, Sen AK (2019) Meta-heuristic techniques to solve resource-constrained project scheduling problem. In: International conference on innovative computing and communications. Springer, Heidelberg, pp 93–99
12. Sarma HKD, Kar A (2006) Security threats in wireless sensor networks. In: Carnahan Conferences Security Technology, Proceedings 2006 40th Annual IEEE International, IEEE, pp 243–251
13. Sharma S, Mishra R, Singh K (2013) A review on wireless network security. In: International conference on heterogeneous networking for quality, reliability, security and robustness. Springer, Heidelberg, pp 668–681
14. Yu S, Liu M, Dou W, Liu X, Zhou S (2017) Networking for big data: a survey. IEEE Commun Surv Tutorials 19(1):531–549

Improved Leakage Current Performance in Domino Logic Using Negative Differential Resistance Keeper

Deepika Bansal, Bal Chand Nagar, Brahamdeo Prasad Singh
and Ajay Kumar

Abstract In this article, a new improved domino logic-based topology is proposed for achieving improved leakage current performance using negative differential resistance (NDR) keeper circuit. The NDR keeper is used to preserve the correct output level and reduced the power consumption with negative resistance. The proposed domino circuit is verified using Synopsys HSPICE simulator with 45 nm and 16 nm technology parameter provided by PTM model library. The simulation outcomes validate the improved performance of the proposed circuit in terms of leakage power consumption and power delay product. Simulation results show that the proposed NDR keeper circuit provides lower static and dynamic power consumption up to 26 and 30% respectively for 16nm technology, as compared to the domino circuits.

Keywords Domino logic · Negative differential resistance · Leakage current ·
Power consumption · MOSFET

1 Introduction

Speed, power, and area are the major concerned of VLSI industry in the nanometer technology regime. Reduction in leakage current is an important issue in VLSI. Domino logic is more suitable than the static CMOS logic, provides low delay and

D. Bansal (✉) · B. C. Nagar · B. P. Singh · A. Kumar
Manipal University Jaipur, Jaipur, Rajasthan 303007, India
e-mail: deepika.bansal386@gmail.com

B. C. Nagar
e-mail: balchandnagar@nitp.ac.in

B. P. Singh
e-mail: bpsinghgkp@gmail.com

A. Kumar
e-mail: ajay.kumar@jaipur.manipal.edu

B. C. Nagar
National Institute of Technology Patna, Patna, Bihar 800005, India

© Springer Nature Singapore Pte Ltd. 2020
A. Khanna et al. (eds.), *International Conference on Innovative Computing
and Communications*, Advances in Intelligent Systems and Computing 1059,
https://doi.org/10.1007/978-981-15-0324-5_27

small area. Domino logic circuits have wide usage in high-performance digital processing systems. They can be used in the area of high-speed computation, medical and communication systems where high density, energy-efficient, low power circuits are required [1]. The domino logic circuits are very fast and required less area than the static circuits. However, the main drawback of the domino circuits is sensitive to noise and consumes large power as compared to static circuits [2]. Various topologies are available to improve the noise and power performance such as keeper, high speed domino (HSD) [3], conditional keeper domino (CKD) [4], controlled keeper by current comparison domino (CKCCD) [5], leakage current replica keeper domino (LCRKD) [6], ultra-low power stacked (ULP-ST) domino logic [7], and efficient keeper for domino [8]. It is observed that these techniques do not resolve the leakage current problem very efficiently without compromising the delay and noise performance.

Circuits are designed with the help of MOSFETs that displayed the negative differential resistance (NDR) property in the literature [9]. Using these NDR properties, smart keepers has been realized. These keepers can be either a series-connected two-terminal device or three-terminal NDR device.

The domino-based proposed NDR keeper circuit configuration consumes low power than the conventional domino logic. The NDR keeper reduces leakage current along with correct output logic in the proposed circuit configuration. In this paper, a new negative differential resistor based smart keeper has been applied to overcome the leakage current problem in the domino logic circuits.

2   Proposed Circuit Configuration

The purpose of the proposed circuit is to reduce the leakage current of the domino logic using the negative differential resistance (NDR) circuit. Resistance can be elaborated in two categories in the nonlinear devices, first is Static resistance and second one is differential resistance. Static resistance (R) is the ratio of the voltage to current, and differential resistance (r) is the derivative of voltage with respect to resulting current. A positive resistance dissipates power when current pass and negative resistance cancel the power dissipation. Negative resistance is an unusual characteristic of some nonlinear electronic components.

The NDR is a property of some electrical circuits and devices where current decreases as voltage increases. NDR based keeper is applied on domino logic to overcome the charge leakage problem of the pseudo domino circuits [10]. The standard keeper circuit had been designed for the domino circuits. However, the standard keeper is bulky because of many transistors and provide short circuit path in domino circuit. Thus, the standard keeper is replaced with NDR circuit for better performance of the domino circuit. NDR circuit is characterized separately and found the negative resistance curve. NDR keeper is designed with two transistors: nMOS and pMOS as shown in Fig. 1.

The working principle of this circuit is basically charging the domino output node during the precharge phase only for high-frequency devices. This keeper works

Fig. 1 Proposed
NDR-based domino logic

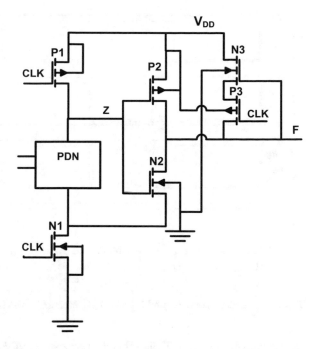

only one condition when output of domino circuit will be ON during the precharge phase. In the NDR keeper, N3 transistor is connected between the supply voltage and the source terminal of second keeper transistor N4 and operated by the node F. Second transistor N4 of keeper circuit is activated by CLK pulse to work in only precharge condition where this connection is also used to reduce the feedback loop. The simulation results show the improvement in speed and leakage current as compare to the previously designed circuits.

3 Results Analysis and Comparison

The proposed circuit is characterized with Synopsys HSPICE simulator. To verify the characteristics of the proposed NDR keeper circuit, the MOSFET parameters of 45 nm and 16 nm Predictive Technology Model (PTM) library are used [11]. The value of supply potential V_{DD} is 0.9 V and 0.7 V in the simulation for 45 nm and 16 nm technology, respectively. All the simulation is done for 8-bit domino OR gate logic as benchmark with 1fF load capacitance at normal room temperature (27 °C). The proposed circuit is operated at 100 MHz clock frequency. Figure 2 shows the timing analysis of 8-bit OR gate implemented using NDR keeper based domino logic.

The performance parameters of the proposed circuit are compared with domino, pseudo domino and standard keeper based topologies in tabulated form. The comparison is done with 45 nm and 16 nm technology parameter. Based on simulation

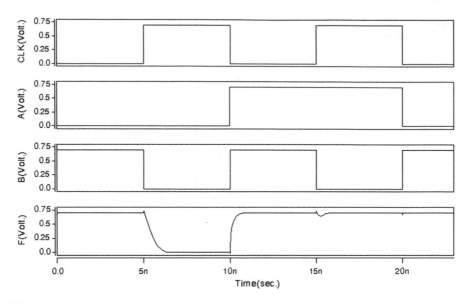

Fig. 2 Transient response of 8-bit OR gate using proposed NDR-based domino logic

results, we observe from Table 1 that the proposed NDR keeper topology has the low static and dynamic power consumption as compared to the domino, pseudo domino,

Table 1 Simulation results of the proposed NDR keeper based domino logic

CMOS technology	Design techniques	Performance parameters			
		Dynamic power dissipation (nW)	Static power dissipation (pW)	Delay (ns)	PDP (fJ)
45 nm	Domino circuit	184.89	9.92	98.10	18.49
	Pseudo domino circuit	162.43	9.22	99.89	16.23
	Standard keeper	213.00	9.34	NA	NA
	Proposed NDR keeper	144.58	8.91	107.2	15.50
16 nm	Domino circuit	59.39	4.32	7.10	42.47
	Pseudo domino circuit	43.09	4.10	11.40	49.12
	Standard keeper	73.23	4.75	NA	NA
	Proposed NDR keeper	41.89	3.18	10.84	45.40

and standard keeper topology. In comparison with published keeper based designs, the present NRD based keeper is found to offer significant improvement in terms of power and PDP but slightly increase the delay. The improvement in PDP confirms the energy efficiency of the proposed design.

4 Conclusion

In this article, a new keeper topology based on NDR circuit for domino circuit has been proposed, which can reduce leakage power. The 8-bit OR gate is validated with 45 nm and 16 nm PTM at 0.9 V and 0.7 V supply, respectively, as a benchmark circuit. A simulation result shows major improvement in the leakage current and acceptable speed for high-speed applications. The performance analysis of 8-input OR gate demonstrate that the proposed NDR circuit provides lower static and dynamic power consumption up to 11 and 22% respectively, and PDP improvement is 17% as compared to domino circuit.

References

1. Rabaey JM, Chandrakasan B, Nicolic B (2003) Digital integrated circuits: a design perspective, 2nd edn. Prentice-Hall, Upper Saddle River, NJ
2. Mohammad A (2018) A new leakage-tolerant domino circuit using voltage comparison for wide fan-in gates in deep submicron technology. Integr VLSI J 51:61–71
3. Anis MH, Allam MW, Elmasry MI (2002) Energy-efficient noise-tolerant dynamic styles for scale-down CMOS and MTCMOS technologies. IEEE Trans Very Large Scale (VLSI) Syst 10(2):71–78
4. Alvandpour A, Krishnamurthy R, Sourrty K, Borkar SY (2002) A sub 130 nm conditional keeper technique. IEEE J Solid-State Circ 37(5):633–638
5. Peiravi A, Asyaei M (2012) Robust low leakage controlled keeper by current-comparison for wide fan-in gates. IEEE Trans Very Large Scale (VLSI) Syst 45(12):22–32
6. Lih Y, Tzartzanis N, Walker KK (2007) A leakage current replica keeper for dynamic circuits. IEEE J Solid-State Circ 42(1):48–55
7. Dadoria AK, Khare K, Panwar U, Jain A (2018) Performance evaluation of domino logic circuits for wide fan-in gates with FinFET. Microsystem Technol. Available from: https://doi.org/10.1007/s00542-017-3691-3
8. Bansal D, Singh BP, Kumar A (2017) Efficient keeper for pseudo domino logic. Int J Pure Appl Math 117(16):605–612
9. Wu CY, Wu C-Y (1981) The new general realization theory of FET-like integrated voltage-controlled negative differential resistance devices. IEEE Trans Circuits Syst CAS-28(5):382–390
10. Fang T, Amine B, Zhouye G (2012) Low power dynamic logic circuit design using a pseudo dynamic buffer. Integr VLSI J 45:395–404
11. Predictive Technology Model. Accessed: 15 Sept 2017. [Online]. Available: http://ptm.asu.edu/

An Efficient Parking Solution for Shopping Malls Using Hybrid Fog Architecture

Bhawna Suri, Pijush Kanti Dutta Pramanik and Shweta Taneja

Abstract The abundant use of personal vehicles has raised the challenge of parking the vehicle in a crowded place such as shopping malls. This paper proposes an efficient parking system for shopping malls. To process the IoT generated parking data, a hybrid Fog architecture is adopted, to reduce the latency, where the Fog nodes are connected across the hierarchy. An algorithm is defined to support the proposed architecture and is simulated on a real-world use-case having requirements of identifying the nearest free car parking slot. The implementation is simulated for a shopping mall with a multilevel parking space. The simulation results have proved that our proposed architecture shows lower latency as compared to the traditional cloud architecture.

Keywords Cloud computing · Fog computing · Inter-Fog communication · IoT · Fog architecture · Latency · Smart building · Smart city · Mall parking · Smart parking

1 Introduction

The shopping malls typically have minimal parking spaces compared to the daily footfalls they receive. It is really challenging for the mall authorities to accommodate customers' cars efficiently at their inadequate parking slots. The traditional car parking systems usually required the car drivers to look for sufficient vacant space in the parking plot manually. The driver usually does not have any prior information and detail direction toward the available parking space. In most of the cases, there is no

B. Suri · S. Taneja
Bhagwan Parshuram Institute of Technology, New Delhi, India
e-mail: bhawnasuri12@gmail.com

S. Taneja
e-mail: shwetataneja@bpitindia.com

P. K. Dutta Pramanik (✉)
National Institute of Technology, Durgapur, India
e-mail: pijushjld@yahoo.co.in

© Springer Nature Singapore Pte Ltd. 2020
A. Khanna et al. (eds.), *International Conference on Innovative Computing and Communications*, Advances in Intelligent Systems and Computing 1059,
https://doi.org/10.1007/978-981-15-0324-5_28

313

manual or automated assistance to the driver in parking the car properly. This results in unnecessary wastage of considerable time and energy as well as congestion and chaos in the parking lot [1]. To overcome this problem, an efficient or smart parking system is required.

An Internet of Things (IoT) based smart parking approach can provide an efficient parking system. The 'things' in IoT are generally the sensors and actuators which continuously sense their environments and hence generate a huge amount of data [2]. Consumption of these data to generate meaningful information, near the source, would be the key enabler for viable real-time use cases such as smart parking [3].

Since the IoT devices lack sufficient computing resources, typically, these data are sent to a cloud data center for storing, processing, and analyzing. The latency involved in this arrangement affects the QoS for real-time applications. Using Fog computing for processing IoT data would allow real-time applications to have the data processed and analyzed close to the source which improves the response time significantly [4].

In the standard Fog architecture, the Fog nodes are connected in a hierarchical fashion. This paper presents the concept of a Hybrid Fog Architecture (HFA), to process IoT data of a shopping mall parking system, where the Fog nodes are connected across the hierarchy. This allows intercommunication between two Fog nodes at the same level, resulting in low latency. In this paper, we have implemented the solution to cater the parking problem where the latency time in searching and allocating parking space in an automated parking system should be minimized during peak hours, especially during the weekends, holidays, festive season, discounted sale offer, etc. An inefficient parking space allocation might create chaos and customer dissatisfaction which ultimately affects the business of the mall.

The simulation result shows that our proposed approach outperforms the traditional smart parking approach using cloud by far in terms of time efficiency while allocating parking slots to the cars.

Rest of the paper is organized as follows. Section 2 provides an extensive survey on different research works towards automated parking. Section 3 explains the proposed parking solution using hybrid fog architecture. The experiment and result are discussed in Sect. 4. Section 5 concludes the paper.

2   Related Works

In the domain of automated parking systems, different authors have proposed different methods.

A new method based on parking scene recognition was proposed in [5]. This method was based on three key steps: recognition of the parking scene, planning the parking path and controlling and tracking the parking path. It used intelligent techniques from machine vision and pattern recognition. An automated multilevel car parking system was designed and implemented in [6]. The authors have explained in brief about the proposed model and its software implementation. There is another work in which authors have proposed a solution for parallel parking [7]. Their method is based on the concept of reach control theory. They had designed a state space to solve this complex issue.

A new technology called compact automated parking system for congested areas was proposed [8]. The authors considered discrete and continuous lifts with two policies. For estimating the performance, they used travel time models and queuing networks. Another model of an automated car parking system was given in [9]. The proposed model was commanded by an android application. The basic idea to develop this model was to keep human interference at a minimum level.

In one of our previous works, we proposed an auxiliary connection that connects adjacent fog nodes at the same level in a hierarchical connection [10]. This was done to promote inter-fog communication. The proposed architecture achieved low latency as compared to traditional fog architecture in case of parking scenario.

In this paper, we have proposed a hybrid fog architecture in which the fog nodes are connected in a hierarchical manner. It is simulated on a real-world parking scenario, and the proposed architecture has shown a lower latency.

3 Proposed Hybrid Fog Architecture and Smart Parking

In traditional Fog computing, if the data is to be exchanged/transferred from one Fog node to another, it has to pass through a centralized coordinator, typically, the Cloud. This leads to an increase in the count of hops required to send the information in the form of a packet, to-and-fro between the Fog nodes, which is directly proportional to the overall latency in the application. Therefore, higher is the number of hops; the higher is the latency. Thus, it reduces the computation performance. Hence, we proposed an architecture HFA, where an auxiliary connection is built between the Fog nodes at the same level. The proposed architecture can be seen in Fig. 1.

In the malls, the parking is available at more than one floor which could be ground, basement 1, basement 2 and so on. This is a form of tree topology. Also, on a floor, there are many parking slots, and so the connections are spread across a floor. In HFA, both the topologies are merged to get an efficient parking solution. On each floor, there are sensor nodes or the fog nodes which are connected across and along with the level. Whenever a vehicle is about to reach the entrance gate of the parking area of the building or mall, the request for its allocation is then assigned to a Fog node at the entrance called the front node. The allocation calculated results are displayed on the screen connected to the front node.

In the malls, the multilevel parking is synonymous to a tree topology of fog nodes. Also, the fog nodes are installed on each floor. The fog nodes at each level in the tree are further connected via the auxiliary link. Thereby an HFA is proposed where both the topologies are merged to get an efficient parking solution. So, whenever a vehicle is about to reach the entrance gate of the parking area of the building or mall, the request for its allocation is then assigned to a Fog node at the entrance called the front node. The front node further propagates the allocation request at the same level

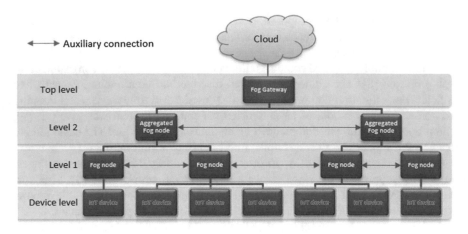

Fig. 1 Proposed hybrid fog architecture for multilevel parking in a shopping mall

else sequentially looks for the next available parking. The allocation results are then displayed on the screen which is connected to the front node. The space allocation algorithm that will be executed on the proposed skeleton is given in Fig. 2.

In this algorithm, the divide-and-conquer approach is applied to find a parking slot within an optimum time period, which is clearly depicted from the flow graph. In the proposed algorithm, there is an array of N fog nodes at each floor, on the ground floor at the entrance there is a special fog node called front node, i.e., Fog[0]. This Fog[0] is the root node of HFA. When the car is about to approach at the entrance gate the request for its allocation goes to the front node, i.e., Fog[0], and at this node, the vehicle number is saved. Since we have assumed that there is an array of fog nodes, hence the lb is 0, and ub is N. The request for finding the free parking lot is submitted to two logical daisy chains of Fog nodes from the front node. These two chains work parallel in opposite directions, i.e., from the lower bound lb to mid and from $N - 1$ to mid. Whomsoever finds the first free parking slot will return the slot directly to Fog[0] with the respective timestamp. As a result, there can arise four possibilities:

(i)   There is only one available parking on a floor, so only one daisy chain reverts with a positive outlook.
(ii)   The result comes from both the daisy chains but with different timestamps.
(iii) The result comes from both the daisy chains with the same timestamp; the jth index is allotted.
(iv)   When no available parking slot available at the Ground Floor, the request is then forwarded to the next floor.

The above process is repeated until the parking lot is found. Also, after every 5 min, the front node Fog[0] communicates with the aggregated Fog nodes which in turn updates the allocation table in the cloud.

```
1. Start
2. Input: front_node = Fog[0] = Vehicle_number requesting for service
          Intialise VisitedFogNodes[i] = FALSE, for all i ε {1,N}
            // N is total number of Nodes at a floor
          Initialise lb = 0, ub = N //lb = lower bound
                                    //ub = upper bound
          Assign VisitedFogNodes[0] = TRUE
3. Calculate mid = (lb+ub)/2
4. The request is then forwarded in parallel to 1 to mid and N to mid
```

```
for (j = 1 to mid)                    for (k = N-1 to mid)
  {                                     {
  if (VisitedFogNodes[j] == FALSE)      if (VisitedFogNodes[k] == FALSE)
       return (j, timestamp)                 return (k, timestamp)
  }                                     }
```

```
index = Min(timestamp for the jth index and the timestamp for the kth index
is retrieved at the front node)
5. if (j== index)
       { Fog[j] = Vehicle_number;  VisitedFogNodes[j] = TRUE; }
   else
       { Fog[k] = Vehicle_number; VisitedFogNodes[k] = TRUE; }
6. If result is not found
      the request is then forwarded to one floor up/down in the
      multilevel parking of the building and will start from Step 1
7. End
```

Fig. 2 Algorithm for the parking slot allocation in a shopping mall

4 Simulation Results and Discussion

The simulation is shown for parking scenario and the results for both the traditional method and our proposed architecture using python 3.7. To implement our algorithm, we have made the following assumptions.

The latency between the user interface node or the front node or Fog[0] and the first Fog node (the node on the ground floor) is, say, t ms.

We can calculate the time complexity of the above algorithm. As per the algorithm, the worst-case time complexity would be the traversal in one direction from the Fog[0] to $N/2$ nodes is:

$$O\left(t * \frac{N}{2}\right)$$

After serving for the allocation the worst-case time to revert back to the front node Fog[0] is:

$$O\left(t * \frac{N}{2} + t\right)$$

Hence the total time worst-case time for allocation is:

$$O\left(t * \frac{N}{2}\right) + t$$

Also, we can calculate the best-case time complexity as:

$$O(1) + O(1) = O(2) \text{ which is approximately } O(1)$$

Similarly, the average case time complexity can be calculated as:

$$1 + 2 + 3 + \cdots + \frac{N}{2} = O\left(t * (N + 2)\frac{N}{8} + t\right)$$

If the time taken for to-and-fro or the Round-Trip Time (RTT) of data between the Fog nodes is t ms, then the time taken from Fog[0] to the hybrid Cloud would obviously be greater than t ms.

Without loss of generality, we can assume that the Fog nodes connected to a LAN segment would be displaying a response time of 1 ms, i.e., $t = 1$ ms and a similar Fog node would show a response time of 50 ms while traversing the internet to a Cloud-hosted application. The data are showing the RTT after execution is shown in Table 1.

Table 1 is pictorially depicted in Fig. 3, where it is quite evident that the HFA-based smart parking system is better than a Cloud-based parking system.

Table 1 Comparison of round-trip time in Cloud-based and hybrid Fog based smart parking	Floor No.	Traditional cloud (ms)	Inter-Fog (ms)
	1	100	1
	2	100	2
	3	100	3
	4	100	4
	5	100	5
	6	100	6
	7	100	7
	8	100	8
	9	100	9
	10	100	10

Fig. 3 Comparison of the parking allotment efficiency using traditional cloud-based and hybrid Fog based systems

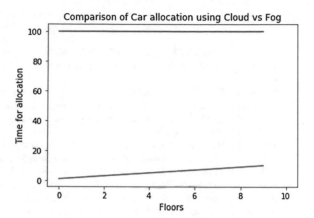

5 Conclusions

The increasing number of personal vehicles has caused major trouble in finding suitable parking space in shopping malls. IoT-based smart parking system has emerged as a viable solution. For effective use of IoT, the IoT data are needed to be processed. Due to the limited computing capability of the IoT devices, the data is sent to the Cloud for processing, storing, and analysis. This device-cloud communication involves high latency which can be overcome by adopting Fog computing where the IoT data are processed nearby to the source instead of sending to the Cloud. In Fog architecture, the devices and Fog nodes are generally arranged in a hierarchical fashion, but this incurs a delay in case of inter-Fog communication. To avoid this, in this paper, a hybrid Fog architecture (HFA) is proposed for communicating across Fog nodes where the Fog nodes are directly connected to other Fog nodes at same levels through the auxiliary connection. An algorithm is implemented for the proposed skeleton for the efficient parking solution problem. The simulation of the algorithm is also shown with best- and worst-case time complexities which offers better results as compared to the traditional cloud.

References

1. Khang SC, Hong TJ, Chin TS, Wang S (2010) Wireless mobile-based shopping mall car parking system (WMCPS). In: Asia-Pacific services computing conference (APSCC 2010), Hangzhou, China
2. Pramanik PKD, Pal S, Choudhury P (2018) Beyond automation: the cognitive IoT. artificial intelligence brings sense to the Internet of Things. In: Sangaiah AK, Thangavelu A, Sundaram VM (eds) Cognitive computing for big data systems over IoT: frameworks, tools and application. Springer, pp 1–37
3. Pramanik PKD, Choudhury P (2018) IoT data processing: the different archetypes and their security & privacy assessments. In: Shandilya SK, Chun SA, Shandilya S, Weippl (eds) Internet of Things (IoT) security: fundamentals, techniques and applications. River Publishers, pp 37–54
4. Pramanik PKD, Pal S, Brahmachari A, Choudhury P (2018) Processing IoT data: from cloud to fog. It's time to be down-to-earth. In: Karthikeyan P, Thangavel M (eds) Applications of security, mobile, analytic and cloud (SMAC) technologies for effective information processing and management. IGI Global, pp 124–148
5. Ma S, Jiang H, Han M, Xie J, Li C (2017) Research on automatic parking systems based on parking scene recognition. IEEE Access 5:21901–21917
6. Padiachy V, Kumar J, Chandra A, Prakash K, Prasad P, Prasad H, Mehta U, Mamun KA, Chand P (2015) Development of an automated multi-level car parking system. In: 2nd Asia-Pacific world congress on computer science and engineering (APWC on CSE), Nadi, Fiji
7. Ornik M, Moarref M, Broucke ME (2017) An automated parallel parking strategy using reach control theory. IFAC-PapersOnLine 50(1):9089–9094
8. Wu G, Xu X, De Koster R, Zou B (2018) Optimal design and planning for compact automated parking systems. Eur J Oper Res 273(3):948–967
9. Bonde DJ, Shende RS, Kedari AS, Gaikwad KS, Bhokre AU (2014) Automated car parking system commanded by Android application. In: International conference on computer communication and informatics (ICCCI), Coimbatore, India
10. Suri B, Taneja S, Bhardwaj H, Gupta P, Ahuja U (2018) Peering through the fog: an inter-fog communication approach for computing environment. In: International conference on innovative computing and communications, vol 56. Singapore, Springer

Sensor's Energy and Performance Enhancement Using LIBP in Contiki with Cooja

Shambhavi Mishra, Pawan Singh and Sudeep Tanwar

Abstract There are various sparing protocols that help to gather the information from various sensors and broadcast via network as represented in this paper. The protocol known as LIBP, it is a lightweight pathway helps to build a spanning routing tree with minimum distance. The distance between the routing tree to root node that based upon the scatter information via sporadic beaconing process. The information can be gathered through sensors via LIBP. The matching traffic record can flow from note to sink in a network. The simulation is done on the Contiki OS under a Cooja simulator. The LIBP outperforms the special version of RPL in the term of power consumption, scalability, and throughput in the CTP protocols.

Keywords IoT · LIBP · CTP · RPL · Efficiency · Security

1 Introduction

In the scenario of modern wireless communication, the rapidly growing technology is known as Internet of Things (IoT). The IoT is a heterogeneous computing element that can sense and identify the multi protocols platform. The platforms provide us to use or fetch the information from anywhere, anytime and anything [1]. Such as there are the various examples like pollution monitoring and traffic management (Fig. 1).

In the IoT setting the sensors reading are formulated as a problem for finding the paths for the traffic flows. The sensors reading is collected and flow from sink nodes to the gateways for further processing. The route from node to neighbor where

S. Mishra (✉) · P. Singh
Department of Computer Science and Engineering, Amity University, Noida, Uttar Pradesh, India
e-mail: shambhavimishra1000@gmail.com

P. Singh
e-mail: pawansingh51279@gmail.com

S. Tanwar
Department of Computer Engineering, Institute of Technology, Nirma University, Ahmedabad, Gujarat, India
e-mail: sudeep.tanwar@nirmauni.ac.in

© Springer Nature Singapore Pte Ltd. 2020
A. Khanna et al. (eds.), *International Conference on Innovative Computing and Communications*, Advances in Intelligent Systems and Computing 1059,
https://doi.org/10.1007/978-981-15-0324-5_29

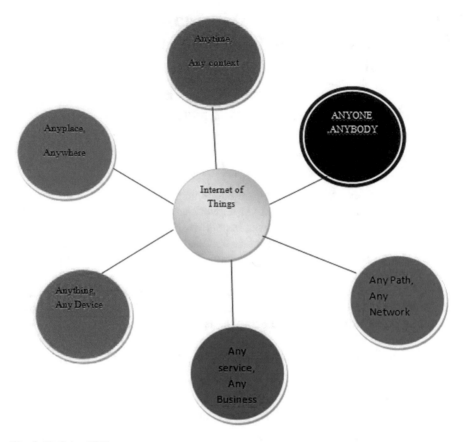

Fig. 1 Defining IOT

the packets carry the reading when we apply the mono-sink architecture along with unique paths and following the multi-hopping process.

This reduces the energy for each node if the data traffic flows directly from sink node. Depending on the IoT setting the spatio-temporal can constraints the sensors and applications. There is the various solution for the routing and maybe different as predictable to be organized and repairing with self-designing. The information is carried by the sensors and aggregated from node to exclusive sink that forms a roots node. While it relates to a gateway. This is typically held in the mono-sink IoT deployment where analyses are being done when the leaf node sends towards the root/sink node for the storing information. In the earlier the implementation of LIBP [2, 3] is done on the Tiny OS by using Tossim emulator [4]. In this paper, we have implemented the evaluation and performance of the energy efficiency in LIBP, CTP [5] and RPL [6] with help of the objective function and check the stinginess between them.

2   Least Path Interference Beaconing Protocol (LIBP)

2.1   Protocol Description

In this, the main aim of the LIBP is to stability the traffic flow. In such a manner it provides the best and energy-efficient nodes and carries less traffic over it. The simulation of LIBP [2, 3] was implemented by the help of the algorithm known as the LIBA where beaconing process was initiated through a sink node such as CTP. In this process, its occurrence on the sink node first to predictable within a one-hop distance and originated to neighbors within a repetitive process. The LIBP process includes the interference pattern where the nodes first select those parents that have smaller supporting children and have the least traffic flow of interferences. The researcher Bagula et al. [2] has projected the detailed information of network building process in his paper.

2.2   Protocol Description

LIBP: ALGORITHM OF SENSOR NODE
get(epoch); neighbor give the epoch id Ti = Clock (sync); synchronized time While (epoch! = 0) do
if (Ti mod Tf == 0) then Epoch + + ;
Select (parent(xy)); Compute (wi(xy)); Broadcast (wi(xy)); Else
Collect and forward sensor readings to parent(xy); if a faulty branch is announced by the gateway then; set epoch = 0;
endif
end while

An algorithmic solution for routing problem is LIBA verbalized where we use a time-bound by "epoch" BFS model for the traffic flow from nodes to sink will be carrying sensor node find out by routing path. The sensor node also represent the higher-level description of LIBP, where Tf is period of an epoch while "*mod*" is the modulo process. A compute cases are used at the beginning of the new epochs.

2.3   LIBP Implementation

LIBP is not implemented in Contiki OS [7], but RPL and CTP are previously applied. In the CTP, there is a high-level link-estimation module in Contiki network reference library that adapted to conform the LIBP ideas. The supporting children nodes conform to represent the interference that changed to relative by the ETX link metric.

The same communication stack and rime are underlying in the implementation of LIBP in Contiki OS [8].

3 Routing Protocol: CTP and RPL

3.1 Collection Tree Protocol (CTP)

The trickle algorithm [9] is the extended form of the CTP routing protocol [5]. The adaptive utilization of routing protocol in beaconing by the application of tickle algorithm to adapt and setup dynamically of network changes. The loop detection and protocol topology help to validate the CTP and make it reliable, robust, hardware independent. Data aggregation is one of the most important features in WSN. The estimated cost of the route to collection point can be maintained by implementing CTP and the metric known as ETX (expected transmission).

3.2 CTP Algorithm

Collection Tree Building algorithm
If ((REQ BUILD > 0) and (sta == SM)) then buil ← 1
else if (sta == HEAD) discard REQ BUILD End if
End if
If ((REQ BUILD > 0 and (Buil > 0)) or root short path, then
Insert Routing tuple into Routing Set A_next_add ← previous-hop A_dest_add ← <originator>
Jitt ← REQMAXJITT
Consider_Transmiting_REQ(loadngCTP_packet)
End if return true
End procedure

- REQ TRIGGER: This triggers the bidirectional routes in its locality.
- REQ BUILD: The route is built to the root after receiving permission from REQ.
- HELLO message: In LOADng-CTP protocol the HELLO message is used [10] that help to broadcast from route root and never promoted by the -hop neighbors. This is done for the identification of messages in the bidirectional routes.

4 Routing Protocol for LLNs (RPL)

4.1 Protocol Description

The routers are organized along with DODAG using the low power lossy network [3] routing protocol. The IETF results in a RPL [6] that allow to form an IPv6 standardization based on routing solution for LLNs. The problem known as ROLL, there is a specific formalization is made which is called IETF. Directed acyclic graph is a hands-on routing protocol which constructs its routes in a periodic interval. It can be run one or more time with each instance with the help of its own topology and unique apposite metrics.

4.2 RPL Algorithm

RPL algorithm "Pseudo code". n = 100; //network size
l = 1000; //network squared area side d = 100; //Locality radius
[g] = NL_T_LocalityConnex(n,L,dmax); //generation of a topology in respect with the Locality method i = NL_F_RandInt1n(length(g.node_x)); //selection of the source node
dw = 2; //display parameter ind = 1; //window index
g.node_diam(i) = 50; //node diameter g.node_border(i) = 10; //node border g.node_color(i) = 5; //node color [f] = NL_G_ShowGraphN(g,ind); //graph visualization
[dist,pred] = NL_R_Dijkstra(g,i); //application of NL_R_Dijkstra
ETX = 5;
[v] = NL_F_RandVector0nminus1 (length (g.head),ETX); //update of weight
v = v+1; g.edge_weight = g.edge_length; g.edge_length = v;
xc = 1/2; //area center yc = 1/2
[s] = NL_G_NodeClose2XY (g,xc,yc); //root node c = 5; //5 possible routes
[pred,dist,ra,DAG,DIO] = NL_R_RPL (g,s,c); //application of NL_R_RPL
[go] = NL_R_RPLPlot(g,pred); //highlight RPL tree ind = 1; //window index f = NL_G_ShowGraphN(go,ind); //graph visualization

NL_R_Dijkstra computes the shortest paths between all network nodes of the graph G towards the single source I in respect with the Dijkstra's algorithm. The displacement of the current node propagates minimum distances throughout the graph. N corresponds to the network size.

The vector D of size N provides the total distance between each network node and the source node I. The vector P of size N gives the predecessor node of each network vertex in order to reach the source node according to the shortest path.

NL_F_RandVector0nminus1: Generate a vector of integer values from the range [0, N − 1].

NL_G_NodeClose2XY: Find the closest node from a Geographic location.
NL_R_RPLPLOT HIGHLIGHTS THE DODAG TREE GENERATED BY THE
RPL ALGORITHM. THE PREDECESSOR VECTOR P IS GENERATED BY THE
RPL ALGORITHM STARTED FROM A DEFINED ROOT NODE S.

4.3 RPL Objective Functions

In the RPL, the objective function can measure the scalability, optimization, and constraints the link metric of a network. It operates the multiple OF for the same network
and same node. There is a set of rules that is allowed by the network administrator
which affect the network traffic flow. For example, in one section only single set of
rules is being executed that specify the paths with best ETX, non-encrypted, lowest
latency and used the evading battery nodes.

Objective Function ETX—The function (OF-ETX) [5] that broadly the field of
WSN that accept the link metric and incorporates the link latency and congestion
in the network. The expected number of transmission (ETX) defines the effective
transmit packets and acknowledge the packet via link in WSN. In hands-on relations
are

$$ETXroot = 0 \text{ means no predictable to send the packets}$$
$$ETXnode = ETXparent + ETXlink \text{ to parent}$$

To choose the route metric with squat ETX we use the OF-ETX and deliberate as
an additional component to the RPL.

Objective Function Zero—In the evaluation of ETX that is painstaking as an
established link of the metric in the field of WSNs whereas ETX and OF-0 is not an
extremely established. The OF-0 was projected by the IETF and main goal is that it
helps to get a good connectivity between an explicit set of nodes and better routing
communication selected by the parent nodes. The nodes are guided by the metric via
collecting parent nodes as a rank node. The position of constraints can be added to
the rank by OF-0 with a norm that can be seen in RFCs [11].

5 Performance Evaluations

Testing Environment—In the testing environment, we have used the Contiki OS
platform. In the simulation, UDGM environment is being used for the radio medium
choice. Tmote sky mote is been used for simulation on Cooja simulator.

Data Collection—Energy Estimation is worn for obtaining apiece component
power consumption. The collected metrices in this research can be done by executing
energy estimation [10] module in Contiki OS. The utilization count of energy for

the radio RX, LPM, TX (low power mode) and NPM (normal power mode) can be obtained by the metrices and known as awake node.

The calculation of power utilization is being done by the formula given below and used for the immediate functions:

$$f(x, y) = ((x \times 64) + (y \times 64)/1000) \tag{1}$$

To determine the power utilized given the energest RX TX LPM and APM values then calculate the power.

$$P = 3 \times \text{APM} \times f(1, 800) + \text{LPM} \times f(0, 545) + \text{TX} \times f(17, 700)$$
$$+ \text{RX} \times f(20, 0)/64 \times (\text{APM} + \text{LPM}) \div 1000 \tag{2}$$

In the Contiki OS under a Cooja framework, there is a data collection for online known as Shell collect view. It provides an exact status of nodes as a comprehensive fail, exact variables, and metadata. Various features such as power tracker, real-time radio duty cycle, etc.

Testing Variables—The experiments run on the objective function in RPL. RPL cannot be tested itself, as a result it requires objective functions such as OF-0 and OF-ETX. For the against of CTP and LIBP an objective function is needed to produce the result. The various implementation is already has been done on the Contiki OS such as OF-0 and OF-ETX but not a combination of RPL-0, RPL-ETX, LIBP, and CTP.

A. Methodology

Table 1 gives experimentation on runtime. In this experiment, to settle the network it usages a 2-min after it will run for 8-min mean total it will be 10-min run. In this, every node has a data packet that has a message or string known as HELLO that sends it to another node by containing this string. In this, at the period of 30 s every node sends the data to the 16 packets together from the sink for 8-min. The nodes density has been same in many of the technologies because routing protocol can be connected to the metrics and the topology size are increased. The scalability

Table 1 Simulation setup

Assessment attributes	Test values
Topology	175 * 175 m grid which placed 30 randomly nodes
Beacon interval	30 s (LIBP), and Adaptive (CTP, RPL)
Messaging interval	30 s
Message content	Hello message from node
Simulation runtime	10 min
LIBP	1
TX/INT range	50 m/100 m

test of the routing protocol where the 10 topologies are generated at random from a topology size of 10–100 with an increment of 10.

B. Result and Evaluation

The evaluations of CTP and RPL together with its objective function in contradiction of new implementation of LIBP on Contiki OS (Fig. 2).

Energy Profile—In Fig. 3, for average power consumption was taken of each node. When we compare the CTP and LIBP it seems that RPL is hungrier for the power consumption on an average of other both. The average radio duty cycle in Fig. 4 it represents in a particular stage the percentage of time is during the 10 min simulation runtime. The roughly power draw of TX and RX is same on many motes.

The sink nodes are classically powered as assumption is made by the RPL as shown in Fig. 5. As we compare the region of power consume of CTP and LIBP it consumes 4 mW whereas in RPL it would be 60 mW. Due to this, the RPL munch extra than one order of scale means extra energy consumption than CTP and LIBP (Fig. 6).

In Table 2, it represents the standard derivation of power consumption intended for routing protocol and describes the distribution of energy in the topology.

Routing Profile—The number of supporting children per node, dexterity of protocol and average path length is included by the routing metric for the individually routing protocol.

In average expanse for children that node could support as a path interference shown in Fig. 7. The expanse of time each node has referenced a parent by counting to obtain the value then averaging those values. To help in the energy distribution network is done by taking the smaller numeral of average child in a required metric

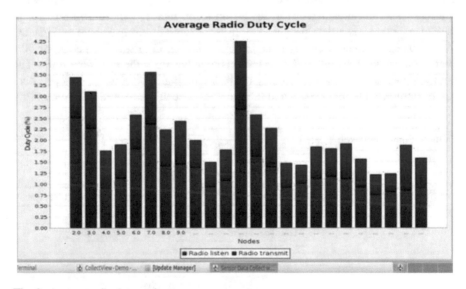

Fig. 2 Average radio duty cycle

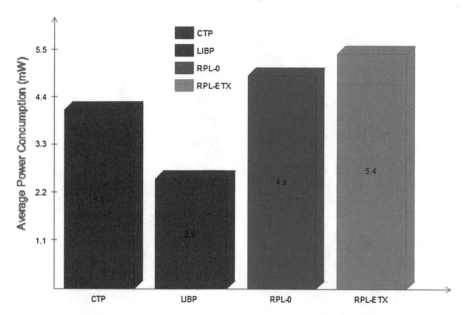

Fig. 3 Average power consumption

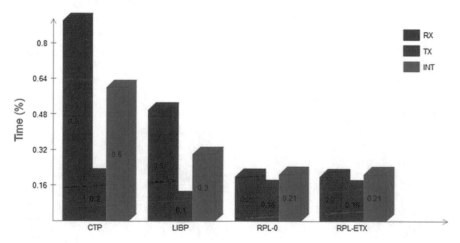

Fig. 4 Radio duty cycle

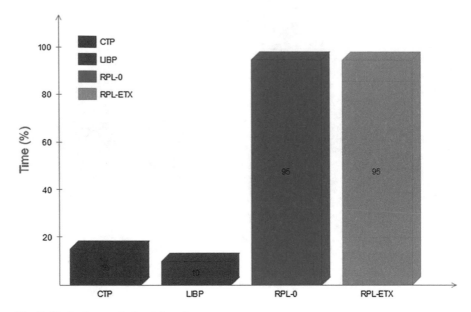

Fig. 5 Radio duty cycle for sink node

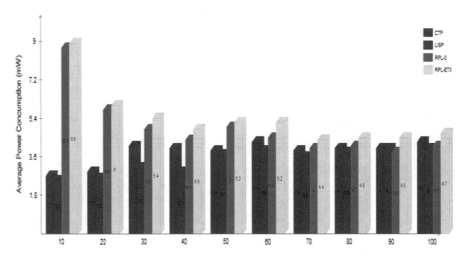

Fig. 6 Scalability for average power consumption

Table 2 Power distribution

Routing protocol	Mean (mW)	Standard derivation (mW)
CTP	4.16	0.454
LIBP	3.34	0.288
RPL-O	5.11	10.82
RPL-ETX	5.24	10.75

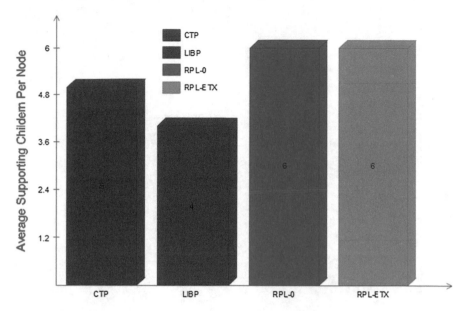

Fig. 7 Path interference

and help in leaving the nodes at battery life. If higher rate of loss packet is introduced in the network, due to high contention or interferences between networks.

The computing of TTL like attribute is obtained in the control plane packet by average path length. RPL and LIBP use TTL (time to live) while CTP usages TTL_MAX the average path length metric is given by the number of hops in each protocol respectively in Fig. 8. The lower path length results in lower latency between leaf and sink node while high average path for better energy distribution of energy (Fig. 9).

Recovery from Failure—The high contention fails node; how quickly protocol contingency came up routes shown in the Fig. 10. For the simulation, this experiment deleted 10 min in mark and chosen high degree of children. To find the alternative Parent/routes for the children some amount time required as represented in the graph and it deals with the catastrophic failure (Fig. 11).

Routing protocol in the ad hoc sense with agile represented in Fig. 12. There is a collection of 30 nodes excluding 1 node is out from the network to run the experiment. The seclude node is established after the run of the 10 min mark in the network. Time taken by the node to acknowledge a parent is shown by the graph in the network.

Traffic profile—In the control plane as different from data plane how much energy is used. As in the LIBP, the info is transmitted by piggybacks and count as the beacon sent.

The percentage of data packets is described in Table 3 collected by the sink node. The more than 99% transmission rate are positively composed by the CTP and LIBP while sending to sink node. 100% transmission from RPL, this is astonishing

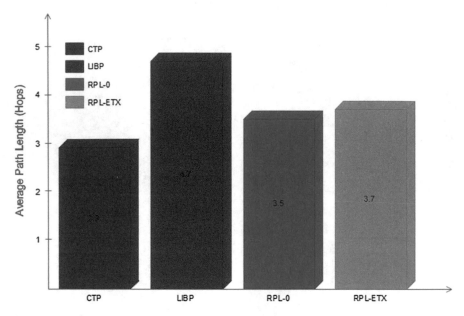

Fig. 8 Average path length

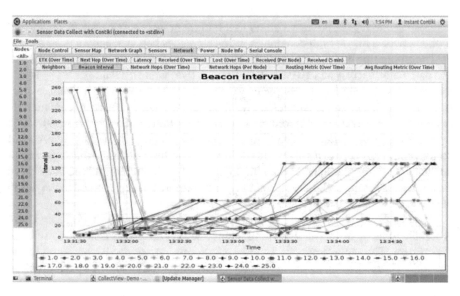

Fig. 9 Beacons intervals

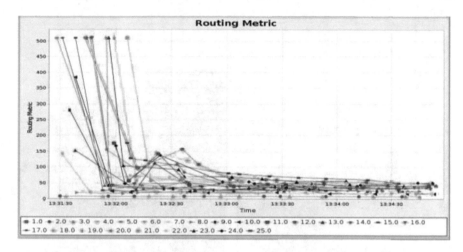

Fig. 10 Network recovery

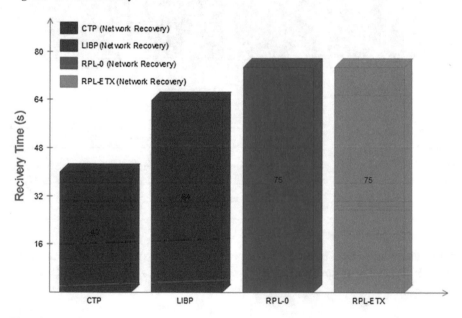

Fig. 11 Routing metric

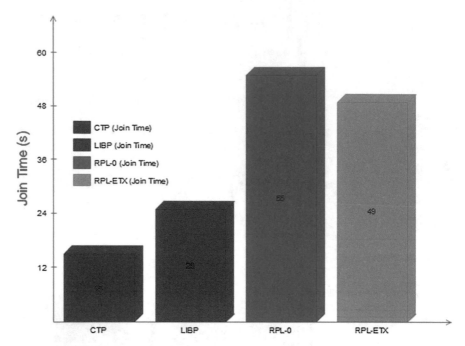

Fig. 12 Routing agility

Table 3 Successful transmission rate

Protocol	Success rate (%)
CTP	99.7
LIBP	99.7
RPL-O	100
RPL-ETX	100

transmission rate might be accredited to comprehensive the protocol communication that RPL is erected on the top of it [8]. On the rime, the top is LIBP and CTP (Fig. 13).

6 Results and Conclusion

This paper be presenting an evaluation between some routing protocols like RPL, CTP, and LIBP. In these three protocols are implemented their goalmouths but lack of insufficient landscapes of RPL such as mote to mote communication and in IPV6 RPL utilize security mechanisms with full-stack and also inherent heaviness can be credited to its fundamental protocol. Larger protocol with extra feature whereas, CTP

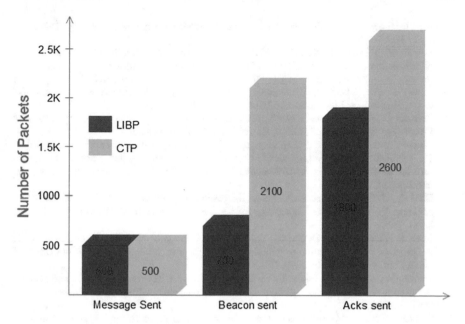

Fig. 13 Control packets sent

and LIBP are similar in the performance and both use the same essential communications stack. In this routing protocol obviously better modified for that type of scenario. In this routing protocol obviously better modified for that type of scenario. In this paper future work may be contain testing environment of topology lifecycle then see last topologies that how long it requires a re-charge or battery extra as well as exploratory the security mechanisms of RPL. From source to destination connection-oriented and IoT networks segment the equivalent routing paradigm of scenery traffic movements. Previously connection-oriented networks implemented the least path interference protocol [12–15], In this paper, LIBP protocol discloses the traffic engineering techniques which was formerly has been planned for outdated networks which contain to be refurbished to be useful and emerging IoT. The formerly methods that are envisioned aimed at gateway design [16], long-distance IoT deployment [17] and quality of service support [18] intended for the use in 6LoWPAN sceneries that permit flexibility and heterogeneity in the IoT sceneries.

References

1. Vasseur J, Dunkels A (2010) Interconnecting smart objects with IP, the next internet. Morgan Kaufmann. ISBN: 9780123751652
2. Bagula A, Djenouri D, Karbab EB (2013) Ubiquitous sensor network management: the least interference beaconing model. In: Proceedings of the IEEE 24th international symposium on

personal indoor and mobile radio communications (PIMRC), September 8–11, pp 2352–2356. ISSN 2166-9570

3. Bagula AB, Djenouri D, Karbab E (2013) On the relevance of using interference and service differentiation routing in the internet-of-things. In: Balandin S, Andreev S, Koucheryavy Y (eds) NEW2AN 2013 and ruSMART 2013. LNCS, vol 8121. Springer, Heidelberg, pp 25–35

4. Levis P, Lee N, Welsh M, Culler D (2003) TOSSIM: simulating large wireless sensor networks of tinyos motes. In: Proceedings of ACM, SenSys 2003, Los Angeles, CA, pp 126–137

5. Gnawali O, Fonseca R, Jamieson K, Moss D, Levis P (2009) Collection tree protocol. In: Proceedings of ACM, SenSys 2009, Berkeley, CA/USA, November 4–6

6. Winter T et al (2012) RPL: IPv6 routing protocol for low-power and lossy networks, RFC 6550

7. Dunkels A, Gronvall B, Voigt T (2007) Contiki—a lightweight and flexible operating system for tiny networked sensors. In: 29th annual IEEE international conference on local computer networks, pp 455–462

8. Dunkels A (2007) Poster abstract: Rime: a lightweight layered communication stack for sensor networks. In: European conference on wireless sensor networks (EWSN), Delft, The Netherlands

9. Levis P, Patel N, Culler D, Shenker S (2004) Trickle: a self regulating algorithm for code maintenance and propagation in wireless sensor networks. In: Proceedings of the USENIX NSDI Conference, San Francisco, CA

10. Dunkels A, Osterlind F, Tsiftes N, He Z (2007) Software-based online sensor node energy estimation. In: ACM Proceeding of the 4th workshop on embedded networked sensors (EmNets 2007), pp 28–32

11. Thubert P (2012) objective function zero for the routing protocol for low-power and lossy networks (RPL). Internet Engineering Task Force (IETF), Request for Comments 6552, 1–14 (2012)

12. Bagula A (2007) On achieving bandwidth-aware LSP/LambdaSP multiplexing/separation in multi-layer networks. IEEE J Sel Areas Commun (JSAC): Special issue on Traffic Engineering for Multi-Layer Networks 25(5)

13. Bagula A (2004) Hybrid traffic engineering: the least path interference algorithm. In: Proceedings of ACM annual research conference of the South African Institute of Computer Scientists and Information Technologists on IT research in developing countries, pp 89–96

14. Bagula AB (2006) Hybrid routing in next generation IP networks. Elsevier Comput Commun 29(7):879–892

15. Bagula A, Krzesinski AE (2001) Traffic engineering label switched paths in IP networks using a pre-planned flow optimization model. In: Proceedings of the ninth international symposium on modelling, analysis and simulation of computer and telecommunication systems (MASCOTS 2001), pp 70–77

16. Zennaro M, Bagula A (2010) Design of a flexible and robust gateway to collect sensor data in intermittent power environments. Int J Sensor Netw 8(3/4)

17. Bagula A, Zennaro M, Inggs G, Scott S, Gascon D (2012) Ubiquitous sensor networking for development (USN4D): an application to pollution monitoring. MDPI Sens 12(1):391–414

18. Bagula A (2010) Modelling and implementation of QoS in wireless sensor Networks: A multi-constrained traffic engineering model. EURASIP J Wireless Commun Netw, Article ID 468737, https://doi.org/10.1155/2010/468737

Analysis and Mitigation of DDoS Flooding Attacks in Software Defined Networks

Rajni Samta and Manu Sood

Abstract To analyze and evaluate the security of the latest network architectures like Software Defined Network (SDN) architectures is a significant step in protecting these against various security threats. The security of SDN assumes greater significance as this dynamic network paradigm, in addition to its great future potential, experiences various design complexities and common Open-flow shortcomings, such as the issues related to a centralized controller. There is no doubt that SDN has been perceived as a standout among the most common ideal models for the networks because of its property of isolation of control and information planes. However, various malicious activities have managed to affect the network performance. Distributed Denial of Service (DDoS) attack has been one of the most crucial issues as far as the dependability on the Internet is concerned. This attack makes the service of any host or hub connected to the network difficult due to a wide variety of its approaches by hampering the normal functioning of the network. The inherent simplicity of SDN makes it easily vulnerable to DDoS attacks. This paper presents the techniques to detect the presence of flooding DDoS attacks in SDN. Three types of techniques have been shown to be implemented for mitigation of these attacks in SDN. Besides, a comparison of the performance of traditional networks and SDN under this type of DDoS attack has been illustrated in terms of throughput and Round-Trip-Time. It has been shown through experimentation that performance of SDN's degrades drastically as compared to that of traditional networks under DDoS attacks.

Keywords Software Defined Network · Flooding DDoS attack · ICMP packets · RARP packets · Detection techniques · Botnet · Mitigation techniques · Throughput · Round-Trip-Time

R. Samta (✉) · M. Sood
Department of Computer Science, H.P. University, Shimla, India
e-mail: rajni95samta@gmail.com

M. Sood
e-mail: soodm_67@yahoo.com

© Springer Nature Singapore Pte Ltd. 2020
A. Khanna et al. (eds.), *International Conference on Innovative Computing
and Communications*, Advances in Intelligent Systems and Computing 1059,
https://doi.org/10.1007/978-981-15-0324-5_30

1 Introduction

Since the '90s, nothing significant had changed in traditional networks architectures but for adding just a few more protocols or new topologies to make networks faster and more reliable until the invention of Software Defined Networks (SDNs). Undoubtedly, traditional networks were at their best till the arrival of the era of clouds. The organizational needs changed with the passage of time and service providers needed to satisfy various network service requirements (like bandwidth, service quality, safety, security, reliability) for different sets of users. At the same time, these sets of users also needed the networks to be highly flexible, seamless and virtual. The traditional way of networking fails to accommodate new features like loose coupling among various software and hardware components and integration of generic network protocols into the manufacturer-proprietary devices. SDN networks support Open Flow and Programmability [1, 2]. Unlike traditional networks, in which adding a new feature into a network was very burdensome because devices were manufacturer-proprietary and were configured manually. Conversely, an SDN is enabled with capabilities to automatically handle these issues. It has become straightforward to add required features into the network not by the command-line interface but through an Application Programming Interface (API). What has changed in SDN is that the network forwarding devices are now merely dumb pieces of hardware which are controlled or instructed and are not doing anything on their own [2, 3]. These features have enabled the networks now with more flexibility and have made huge impact on the market with enhanced reliability. Also, SDN has given power to the organizations using these networks to customize the networks according to their needs.

1.1 API Controller

API controllers are the essential features of an SDN as these can be easily placed at different places (like physically apart systems or even mobile phones). What has gained more attention is the functional ability of a controller, i.e., how it can be made to perform operations with better efficiencies day-by-day.

1.2 Open Flow Support

Various network entities like Openflow compliant switches and controllers are placed physically apart from each other in SDNs and all of these need to communicate with one another [4]. It is interesting to know how the controller instructs these switches in various situations mainly using Openflow protocols. Some of the areas to explore in Openflow protocols are packet processing in these protocols, new versions of protocols, and design choices.

1.3 Security Aspects of SDN

As compared to the traditional network architectures, security threats specific to SDN also have a bearing on its security solutions [6], e.g., a compromised centralized controller can pose a grave threat to the whole network. The natural architectural features of an SDN, such as openness, flexibility, programmability, and dynamic nature make it vulnerable to some specific security breaches. On one hand, these features have made the network programmable and flexible for the service providers, but on the other, they have also made it easier for the attackers to tamper with the network policies. Some of the attack prone areas of SDN architectures that are vulnerable to security threats are: SDN controllers, SDN switches, links among the controllers and the switches, links among the controllers and application software [5] as shown in Fig. 1. Some of the attacks of significance that threaten the functioning of SDNs and are in a position to compromise various components of SDN are Denial of Service (DoS)/Distributed Denial of Service (DDoS) attack, man-in-middle attack, DNS spoofing, ARP poisoning, and DNS poisoning as shown in Table 1.

In this paper, we have analyzed the behavior of SDN facing the flooding based DDoS attacks. The flooding based DDoS attacks that can be used to tamper the SDN is Internet Control Message Protocol (ICMP) and Address Resolution Protocol

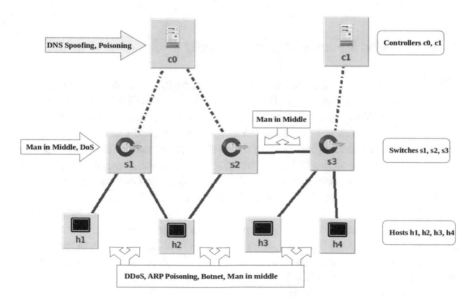

Fig. 1 Attack prone areas in the architecture of SDN [5]

Table 1 Attacks in the SDN architecture

Components of SDN architecture	Types of attack
Controller	DNS poisoning, DNS spoofing, Man-in-middle attack
Open flow switch	DDoS, DoS, man-in-middle attack
Flow tables	ARP poisoning, DoS, DNS spoofing attack
Nodes	DDoS attack, ARP poisoning, Botnet, man-in-middle attack

(ARP) packet floods. A traffic pattern inspection has been performed to differentiate between normal traffic and malicious traffic. It is possible to detect and mitigate the DDoS attacks in SDN using traffic pattern analysis, which has been shown in this paper, in addition, a comparative analysis of throughputs of SDN and traditional network has been presented to understand the performance of these networks while their security is being compromised. The comparison has been performed using the tools Mininet, Miniedit, Wireshark, and Xgraph. In the remainder of this paper, Sect. 2 presents various categories of DDoS attacks along with the classification of DDoS flooding attacks. In Sect. 3, detection and mitigation techniques for flooding based DDoS attacks in SDN have been discussed. In Sect. 4, an analysis of network traffic using ICMP, ARP, and Transmission Control Protocol (TCP) packets simulated in Mininet have been presented. Section 5 highlights the comparative analysis of network performance under various scenarios. This section also presents the performance of the network based on three mitigation techniques. Finally, Sect. 6 provides the conclusion and future work.

2   Distributed Denial of Service Attacks

DDoS is an extension to DoS attack, which has come up with few more features to tamper the network in many ways. The primary motive of the DDoS attack is to make the services of any node of a network unavailable much like the DoS attack. However, in a DDoS attack, multiple machines are used to pose such threats. By making use of various flooding techniques, DDoS attacks degrade the network services and can initiate one or more of the following attacks: Smurf attack, Fraggle attack, SYN flood attack, UDP fragmentation attack, DNS amplification flooding attack, or NTP amplification flooding attack [7]. Figure 2 shows the hierarchy of various categories of DDoS attacks [7] and Table 2 presents the classification of various DDoS flooding attacks along with their brief descriptions. The techniques that have been used in this work to analyze the behavior of SDN under DDoS flooding attack are ICMP FLOOD (ping request and reply) and ARP FLOOD. A ping service is used for a dedicated purpose in any network, which may sometimes play a role in effecting the DDoS attack; intentionally or through some other possibilities [7, 8].

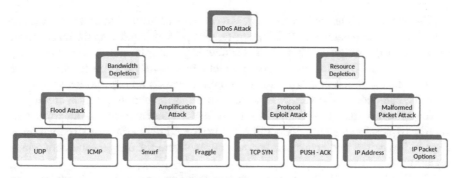

Fig. 2 Categorization of DDoS flooding attacks [7]

Table 2 Classification of DDoS Attacks

Smurf attack [9]	A substantial number of ICMP packets with source IP of a victim node are spread among the other nodes of the network, using an IP broadcast address. This procedure directs massive amount of reply packets causing the system blockage
Fraggle attack [9]	This attack is similar to the Smurf assault; however, it utilizes ill-conceived UDP movement rather than ICMP movement to accomplish a similar objective
SYN flooding attack [9]	An attacker transmits a packet with the SYN bit set for a three-way handshake in TCP SYN flood attack. In response, the victim sends the packet back at the sender's address with an SYN-ACK bit set. The attacker never reacts to the victim's packet; either intentionally or because the source address of the packet is forged. Therefore, the victim's TCP receiving queues would be filled up, denying new TCP connections to legitimate clients
UDP fragmentation attack [8, 10]	In a UDP fragmentation attack, attackers send massive numbers of UDP packets to occupy more bandwidth with large packets. The victim's resources are consumed in assembling these fragmented packets, and it renders the victim unavailable for most of the time
DNS amplification flooding attack [11]	In Domain Name System (DNS) amplification flooding attack, the zombies (heteronymous nodes) with forged source IP addresses generate the DNS queries that lead to a large volume of network traffic directed towards the targeted victim
NTP amplification flooding attack [12]	Network Time Protocol (NTP) amplification flooding attack is like a DNS amplification attack but uses NTP servers instead of DNS servers
Botnets [13]	Botnets can execute automated DDoS attacks. For this purpose, a remote malicious node deploys many honest/legitimate machines without their knowledge to clog any node of a network

The attacker might use this Network Layer Service/Protocol to swamp a targeted device with a large number of ICMP echo-request packets, rendering the inaccessible to the traditional traffic. If this process is adopted by multiple devices simultaneously, then it is called DDoS ICMP flooding attack. In this manner, the ICMP network layer protocol, which is utilized for the communication among the network components, can become the very reason for disrupting the communication among the server(s)/hosts in the network. When various devices generate large numbers of ping service requests to a victim, the victim will be forced to continuously reply to each request it receives and hence can become unreachable for the genuine networking node(s) [9].

The flooding based DDoS attack presented in this paper uses multiple autonomous nodes that are independent of any sort of control of the malicious node. Moreover, the DDoS attacks discussed here are not automated except for the Botnet based attacks. The automated DDoS flooding attacks can be implemented using botnets. Botnets act as the key players for many types of Internet attacks such as DDoS, Masquerade attacks, DNS Spoofing attacks, and phishing along with various other types of spam. The functionality of a botnet is that they use command and control (C&C) master-slave to spread the malicious activity from a remote place at any time. On the other hand, the legitimate users of those nodes would remain unaware of such activities. The capabilities of existing firewalls are inadequate at present to detect C&C based attacks, as the corresponding C&C techniques for botnets can be changed on purpose.

3   Detection and Mitigation Techniques for SDN Based DDoS Attacks

A number of DDoS detection techniques have been proposed in the literature. A brief overview of such detection techniques are given below.

3.1   Entropy Oriented Techniques

These techniques mainly depend on the distribution of some network features like flow tables to notice abnormal network activities. These take decisions based on the values calculates for the entropy. Various well-known parameters in such cases are destination/source IP addresses and port numbers to derive the values for entropy [14–16].

3.2 Machine Learning Techniques

Machine learning based detection techniques are widely used in traditional networking. Various techniques to detect the anomalies are Self Organized Maps, Artificial Neural Networks, and principles of fuzzy logic [17–20].

3.3 Network Traffic Pattern Analysis

This detection mechanism analyzes packet flow in the network for the type of pattern each packet has and the numbers of each type of packets present in the network link at any instance [21, 22].

3.4 Connection Rate

The connection rate based techniques keep track of the activities such as how many connections have been established in a window, and what is the rate of connection failure or success. The open-flow controllers such as POX, NOX, flood-light come with built-in algorithms to carry out such procedures [16, 18, 21–23].

3.5 Snort and Open Flow Integrated Techniques

It is an open-source mechanism for network intrusion detection and mitigation. Intrusion Detection System (IDS) as its integral part consists of hardware and software support to detect any malicious activity within a network [24, 25].

SDN based mitigation techniques have a variety of methods to overcome each of the DDoS attacks versions, and each technique has its own way of mitigating such attack. Some of the important mitigation techniques for SDN based DDoS flooding attacks have been summarized in Table 3.

4 Tools Used for Network Traffic Pattern Analysis

As shown in Fig. 3, in order to carry out the experimentation in this case, we have simulated the SDN environment in Ubuntu 18.04 along with it, Mininet tool, Open VSwitches, and POX controller have been used. In order to simulate a traditional network environment, we have used NS2. To capture the network traffic and to analyze it, Wireshark tool has been used. Various parameters that are analyzed by

Table 3 Mitigation techniques for SDN based DDoS flooding attacks

Control bandwidth [26]	Restricting traffic delivery rate to give moderate bandwidth to each link using the centralized controllers can help in the prevention of the DDoS flooding attacks
IP Address change [19, 25]	The simplest way to mitigate flood attacks is to change the IP address of the victim node to make the entire flooded traffic orphan. This mitigation technique works only when both, the occurrence of DDoS flooding attack and the victim, have been identified
Redirection [19, 21, 25]	Traffic in this type of mitigation technique is redirected to some unknown IP addresses to release the victim's dedicated link. It can be achieved using a honeypot security mechanism [27]
Drop packet [14, 20, 24]	After analyzing the pattern of malicious traffic and normal traffic, the packets having a pattern similar to the malicious packets can be dropped safely
Deep Packet Inspection (DPI) [19, 25]	DPI is carried out using controllers or gateways to analyze and manage the traffic in the network. Hence, this method of packet filtering keeps an eye on the type of service each packet carries
Block ports [19, 25]	Using a centralized controller, the network flow from attacker's port is obstructed to fail that service

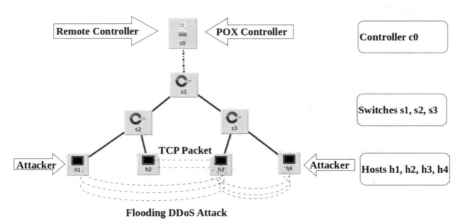

Fig. 3 Experimental setup for flooding DDoS attack

this tool are the network's Throughput, Round-Trip-Time (RTT), and Packet Flow. To plot the graph for SDN throughput, XGraph has been used. For generating flooding DDoS attack, Hping3 open source tool has been used. The types of packets that SDN networks mainly deal with are TCP, UDP, ICMP, ARP, and RARP packets. A flood created using a Ping Service from any host, or an open-flow switch would mainly contain an ICMP echo-request, reply packets and unnecessary ARP Packets. Otherwise, the normal traffic would consist of packets due to TCP/UDP depending on the type of service required by a network or the organization using that network.

4.1 Internet Control Message Protocol (ICMP) Packets

The traffic pattern has been analyzed using Wireshark while the host is under ping attack. Hping3 tool has been used for the purpose as shown in Fig. 4, allowing the attacker to attack from different IP addresses and generating a massive amount of ping requests and therefore making the victim unavailable completely, as shown in Fig. 5. The ICMP packets captured using Wireshark are shown in Fig. 6.

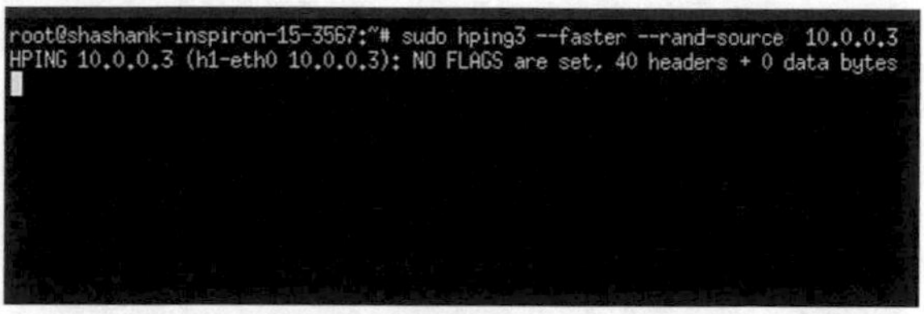

Fig. 4 Launching attack using HPING3

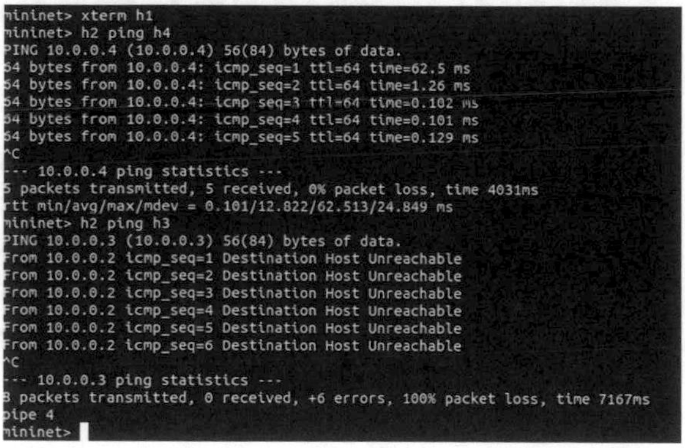

```
root@shashank-inspiron-15-3567:~# sudo hping3 --faster --rand-source  10.0.0.3
HPING 10.0.0.3 (h1-eth0 10.0.0.3): NO FLAGS are set, 40 headers + 0 data bytes
```

Fig. 5 Victim host unavailable

```
mininet> xterm h1
mininet> h2 ping h4
PING 10.0.0.4 (10.0.0.4) 56(84) bytes of data.
64 bytes from 10.0.0.4: icmp_seq=1 ttl=64 time=62.5 ms
64 bytes from 10.0.0.4: icmp_seq=2 ttl=64 time=1.26 ms
64 bytes from 10.0.0.4: icmp_seq=3 ttl=64 time=0.102 ms
64 bytes from 10.0.0.4: icmp_seq=4 ttl=64 time=0.101 ms
64 bytes from 10.0.0.4: icmp_seq=5 ttl=64 time=0.129 ms
^C
--- 10.0.0.4 ping statistics ---
5 packets transmitted, 5 received, 0% packet loss, time 4031ms
rtt min/avg/max/mdev = 0.101/12.822/62.513/24.849 ms
mininet> h2 ping h3
PING 10.0.0.3 (10.0.0.3) 56(84) bytes of data.
From 10.0.0.2 icmp_seq=1 Destination Host Unreachable
From 10.0.0.2 icmp_seq=2 Destination Host Unreachable
From 10.0.0.2 icmp_seq=3 Destination Host Unreachable
From 10.0.0.2 icmp_seq=4 Destination Host Unreachable
From 10.0.0.2 icmp_seq=5 Destination Host Unreachable
From 10.0.0.2 icmp_seq=6 Destination Host Unreachable
^C
--- 10.0.0.3 ping statistics ---
8 packets transmitted, 0 received, +6 errors, 100% packet loss, time 7167ms
pipe 4
mininet>
```

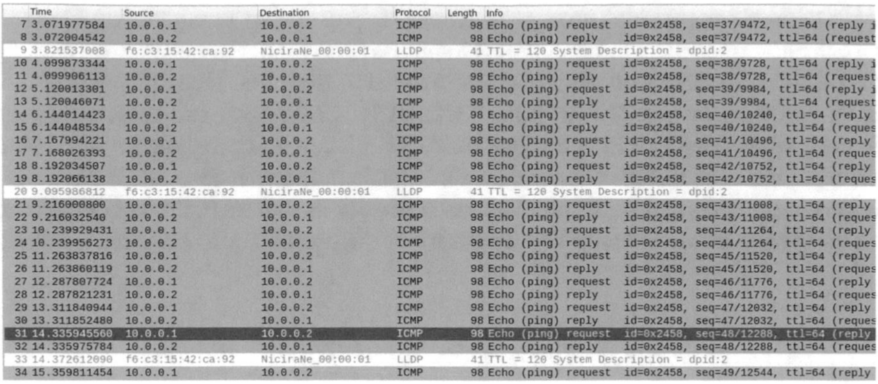

Time	Source	Destination	Protocol	Length	Info
7 3.071977584	10.0.0.1	10.0.0.2	ICMP	98	Echo (ping) request id=0x2458, seq=37/9472, ttl=64 (reply i
8 3.072004542	10.0.0.2	10.0.0.1	ICMP	98	Echo (ping) reply id=0x2458, seq=37/9472, ttl=64 (request
9 3.821537008	f6:c3:15:42:ca:92	NiciraNe_00:00:01	LLDP	41	TTL = 120 System Description = dpid:2
10 4.099873344	10.0.0.1	10.0.0.2	ICMP	98	Echo (ping) request id=0x2458, seq=38/9728, ttl=64 (reply i
11 4.099906113	10.0.0.2	10.0.0.1	ICMP	98	Echo (ping) reply id=0x2458, seq=38/9728, ttl=64 (request
12 5.120013301	10.0.0.1	10.0.0.2	ICMP	98	Echo (ping) request id=0x2458, seq=39/9984, ttl=64 (reply i
13 5.120046071	10.0.0.2	10.0.0.1	ICMP	98	Echo (ping) reply id=0x2458, seq=39/9984, ttl=64 (request
14 6.144014423	10.0.0.1	10.0.0.2	ICMP	98	Echo (ping) request id=0x2458, seq=40/10240, ttl=64 (reply
15 6.144048534	10.0.0.2	10.0.0.1	ICMP	98	Echo (ping) reply id=0x2458, seq=40/10240, ttl=64 (reques
16 7.167994221	10.0.0.1	10.0.0.2	ICMP	98	Echo (ping) request id=0x2458, seq=41/10496, ttl=64 (reply
17 7.168026393	10.0.0.2	10.0.0.1	ICMP	98	Echo (ping) reply id=0x2458, seq=41/10496, ttl=64 (reques
18 8.192034507	10.0.0.1	10.0.0.2	ICMP	98	Echo (ping) request id=0x2458, seq=42/10752, ttl=64 (reply
19 8.192066138	10.0.0.2	10.0.0.1	ICMP	98	Echo (ping) reply id=0x2458, seq=42/10752, ttl=64 (reques
20 9.095986812	f6:c3:15:42:ca:92	NiciraNe_00:00:01	LLDP	41	TTL = 120 System Description = dpid:2
21 9.216000800	10.0.0.1	10.0.0.2	ICMP	98	Echo (ping) request id=0x2458, seq=43/11008, ttl=64 (reply
22 9.216032540	10.0.0.2	10.0.0.1	ICMP	98	Echo (ping) reply id=0x2458, seq=43/11008, ttl=64 (reques
23 10.239929431	10.0.0.1	10.0.0.2	ICMP	98	Echo (ping) request id=0x2458, seq=44/11264, ttl=64 (reply
24 10.239956273	10.0.0.2	10.0.0.1	ICMP	98	Echo (ping) reply id=0x2458, seq=44/11264, ttl=64 (reques
25 11.263837816	10.0.0.1	10.0.0.2	ICMP	98	Echo (ping) request id=0x2458, seq=45/11520, ttl=64 (reply
26 11.263860119	10.0.0.2	10.0.0.1	ICMP	98	Echo (ping) reply id=0x2458, seq=45/11520, ttl=64 (reques
27 12.287807724	10.0.0.1	10.0.0.2	ICMP	98	Echo (ping) request id=0x2458, seq=46/11776, ttl=64 (reply
28 12.287821231	10.0.0.2	10.0.0.1	ICMP	98	Echo (ping) reply id=0x2458, seq=46/11776, ttl=64 (reques
29 13.311840944	10.0.0.1	10.0.0.2	ICMP	98	Echo (ping) request id=0x2458, seq=47/12032, ttl=64 (reply
30 13.311852480	10.0.0.2	10.0.0.1	ICMP	98	Echo (ping) reply id=0x2458, seq=47/12032, ttl=64 (reques
31 14.335945560	10.0.0.1	10.0.0.2	ICMP	98	Echo (ping) request id=0x2458, seq=48/12288, ttl=64 (reply
32 14.335975784	10.0.0.2	10.0.0.1	ICMP	98	Echo (ping) reply id=0x2458, seq=48/12288, ttl=64 (reques
33 14.372612090	f6:c3:15:42:ca:92	NiciraNe_00:00:01	LLDP	41	TTL = 120 System Description = dpid:2
34 15.359811454	10.0.0.1	10.0.0.2	ICMP	98	Echo (ping) request id=0x2458, seq=49/12544, ttl=64 (reply

Fig. 6 ICMP packets generated using ping service

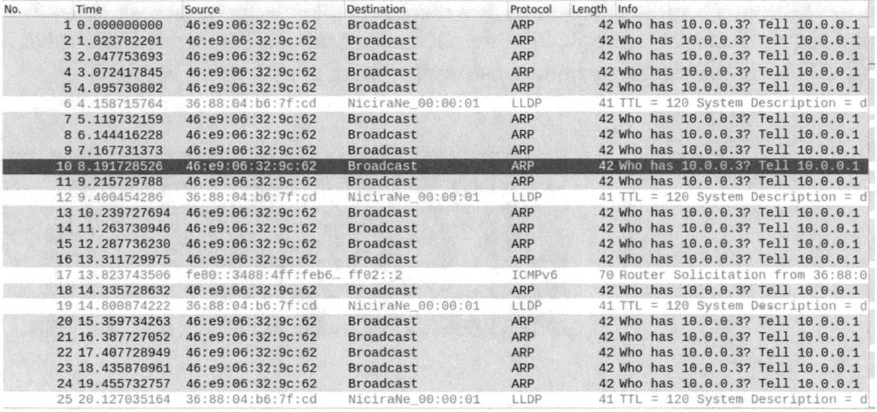

No.	Time	Source	Destination	Protocol	Length	Info
1	0.000000000	46:e9:06:32:9c:62	Broadcast	ARP	42	Who has 10.0.0.3? Tell 10.0.0.1
2	1.023782201	46:e9:06:32:9c:62	Broadcast	ARP	42	Who has 10.0.0.3? Tell 10.0.0.1
3	2.047753693	46:e9:06:32:9c:62	Broadcast	ARP	42	Who has 10.0.0.3? Tell 10.0.0.1
4	3.072417845	46:e9:06:32:9c:62	Broadcast	ARP	42	Who has 10.0.0.3? Tell 10.0.0.1
5	4.095730802	46:e9:06:32:9c:62	Broadcast	ARP	42	Who has 10.0.0.3? Tell 10.0.0.1
6	4.158715764	36:88:04:b6:7f:cd	NiciraNe_00:00:01	LLDP	41	TTL = 120 System Description = d
7	5.119732159	46:e9:06:32:9c:62	Broadcast	ARP	42	Who has 10.0.0.3? Tell 10.0.0.1
8	6.144416228	46:e9:06:32:9c:62	Broadcast	ARP	42	Who has 10.0.0.3? Tell 10.0.0.1
9	7.167731373	46:e9:06:32:9c:62	Broadcast	ARP	42	Who has 10.0.0.3? Tell 10.0.0.1
10	8.191728526	46:e9:06:32:9c:62	Broadcast	ARP	42	Who has 10.0.0.3? Tell 10.0.0.1
11	9.215729788	46:e9:06:32:9c:62	Broadcast	ARP	42	Who has 10.0.0.3? Tell 10.0.0.1
12	9.400454286	36:88:04:b6:7f:cd	NiciraNe_00:00:01	LLDP	41	TTL = 120 System Description = d
13	10.239727694	46:e9:06:32:9c:62	Broadcast	ARP	42	Who has 10.0.0.3? Tell 10.0.0.1
14	11.263730946	46:e9:06:32:9c:62	Broadcast	ARP	42	Who has 10.0.0.3? Tell 10.0.0.1
15	12.287736230	46:e9:06:32:9c:62	Broadcast	ARP	42	Who has 10.0.0.3? Tell 10.0.0.1
16	13.311729975	46:e9:06:32:9c:62	Broadcast	ARP	42	Who has 10.0.0.3? Tell 10.0.0.1
17	13.823743506	fe80::3488:4ff:feb6..	ff02::2	ICMPv6	70	Router Solicitation from 36:88:0
18	14.335728632	46:e9:06:32:9c:62	Broadcast	ARP	42	Who has 10.0.0.3? Tell 10.0.0.1
19	14.800874222	36:88:04:b6:7f:cd	NiciraNe_00:00:01	LLDP	41	TTL = 120 System Description = d
20	15.359734263	46:e9:06:32:9c:62	Broadcast	ARP	42	Who has 10.0.0.3? Tell 10.0.0.1
21	16.387727052	46:e9:06:32:9c:62	Broadcast	ARP	42	Who has 10.0.0.3? Tell 10.0.0.1
22	17.407728949	46:e9:06:32:9c:62	Broadcast	ARP	42	Who has 10.0.0.3? Tell 10.0.0.1
23	18.435870961	46:e9:06:32:9c:62	Broadcast	ARP	42	Who has 10.0.0.3? Tell 10.0.0.1
24	19.455732757	46:e9:06:32:9c:62	Broadcast	ARP	42	Who has 10.0.0.3? Tell 10.0.0.1
25	20.127035164	36:88:04:b6:7f:cd	NiciraNe_00:00:01	LLDP	41	TTL = 120 System Description = d

Fig. 7 ARP packets generated using HPING3

4.2 Address Resolution Protocol (ARP) Packets

The attacker using ARP Service would unnecessarily be generating a packet with the IP address of the victim, asking for the MAC address and pretending as if it wants to send some genuine information. This process would be carried continuously in bulk to clog the victim. Figure 7 depicts the ARP traffic pattern.

4.3 Transport Control Protocol (TCP) Packet

The TCP and UDP Packets are the forms of normal traffic in the network. These packets can also be converted into a flood attack if the Maximum Segment Size

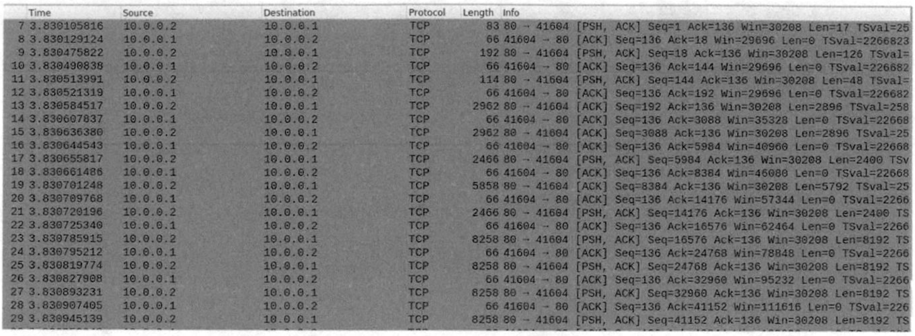

Time	Source	Destination	Protocol	Length	Info
7 3.830195816	10.0.0.2	10.0.0.1	TCP	83	80 → 41604 [PSH, ACK] Seq=1 Ack=136 Win=30208 Len=17 TSval=25
8 3.830129124	10.0.0.1	10.0.0.2	TCP	66	41604 → 80 [ACK] Seq=136 Ack=18 Win=29696 Len=0 TSval=2266823
9 3.830475822	10.0.0.2	10.0.0.1	TCP	192	80 → 41604 [PSH, ACK] Seq=18 Ack=136 Win=30208 Len=126 TSval=
10 3.830490838	10.0.0.1	10.0.0.2	TCP	66	41604 → 80 [ACK] Seq=136 Ack=144 Win=29696 Len=0 TSval=226682
11 3.830513991	10.0.0.2	10.0.0.1	TCP	114	80 → 41604 [PSH, ACK] Seq=144 Ack=136 Win=30208 Len=48 TSval=
12 3.830521319	10.0.0.1	10.0.0.2	TCP	66	41604 → 80 [ACK] Seq=136 Ack=192 Win=29696 Len=0 TSval=226682
13 3.830584517	10.0.0.2	10.0.0.1	TCP	2962	80 → 41604 [ACK] Seq=192 Ack=136 Win=30208 Len=2896 TSval=258
14 3.830607837	10.0.0.1	10.0.0.2	TCP	66	41604 → 80 [ACK] Seq=3088 Ack=136 Win=35328 Len=0 TSval=22668
15 3.830636380	10.0.0.2	10.0.0.1	TCP	2962	80 → 41604 [ACK] Seq=3088 Ack=136 Win=30208 Len=2896 TSval=25
16 3.830644543	10.0.0.1	10.0.0.2	TCP	66	41604 → 80 [ACK] Seq=136 Ack=5984 Win=40960 Len=0 TSval=22668
17 3.830655817	10.0.0.2	10.0.0.1	TCP	2466	80 → 41604 [PSH, ACK] Seq=5984 Ack=136 Win=30208 Len=2400 TSv
18 3.830661486	10.0.0.1	10.0.0.2	TCP	66	41604 → 80 [ACK] Seq=136 Ack=46080 Win=0 TSval=22668
19 3.830701248	10.0.0.2	10.0.0.1	TCP	5858	80 → 41604 [ACK] Seq=8384 Ack=136 Win=30208 Len=5792 TSval=25
20 3.830789768	10.0.0.1	10.0.0.2	TCP	66	41604 → 80 [ACK] Seq=136 Ack=14176 Win=57344 Len=0 TSval=2266
21 3.830720196	10.0.0.2	10.0.0.1	TCP	2466	80 → 41604 [PSH, ACK] Seq=14176 Ack=136 Win=30208 Len=2400 TS
22 3.830725340	10.0.0.1	10.0.0.2	TCP	66	41604 → 80 [ACK] Seq=136 Ack=16576 Win=62464 Len=0 TSval=2266
23 3.830785915	10.0.0.2	10.0.0.1	TCP	8258	80 → 41604 [PSH, ACK] Seq=16576 Ack=136 Win=30208 Len=8192 TS
24 3.830795212	10.0.0.1	10.0.0.2	TCP	66	41604 → 80 [ACK] Seq=136 Ack=24768 Win=78848 Len=0 TSval=2266
25 3.830819774	10.0.0.2	10.0.0.1	TCP	8258	80 → 41604 [PSH, ACK] Seq=24768 Ack=136 Win=30208 Len=8192 TS
26 3.830827908	10.0.0.1	10.0.0.2	TCP	66	41604 → 80 [ACK] Seq=136 Ack=32960 Win=95232 Len=0 TSval=2266
27 3.830893231	10.0.0.2	10.0.0.1	TCP	8258	80 → 41604 [PSH, ACK] Seq=32960 Ack=136 Win=30208 Len=8192 TS
28 3.830907405	10.0.0.1	10.0.0.2	TCP	66	41604 → 80 [ACK] Seq=136 Ack=41152 Win=111616 Len=0 TSval=226
29 3.830945139	10.0.0.2	10.0.0.1	TCP	8258	80 → 41604 [PSH, ACK] Seq=41152 Ack=136 Win=30208 Len=8192 TS

Fig. 8 Three-Way Handshake process (normal traffic)

(MSS) is intentionally kept very less (1 byte, say). It results into a massive overhead of headers and a large number of fragmentations of a single TCP segment. Figure 8 depicts the behavior of a normal traffic.

5 Analysis of SDN Under DDoS Flooding Attack and Mitigation Techniques

In this section, the evaluation, comparison, and analysis of the traditional and SDN networks under DDoS flooding attack have been presented.

5.1 Throughput

The network's throughput depends on how much bandwidth of the network is being used efficiently, i.e., effective bandwidth of the network. Here, after the DDoS flooding attack is affected in an SDN using the ARP and ICMP packets, the throughput (efficiency X bandwidth) of the SDN under attack has been shown to be dropped as expected in Fig. 9 and with normal traffic has been shown in Fig. 10.

5.2 Round-Trip-Time (RTT)

RTT represents the period in milliseconds which is equal to $(2 \times$ propagation time). In Fig. 11, the output of the Wireshark tool shows that the RTT is dense because the SDN network in under DDoS flooding attack (please refer to Fig. 3), whereas, when the network is not under attack, the RTT is normal as shown in Fig. 12.

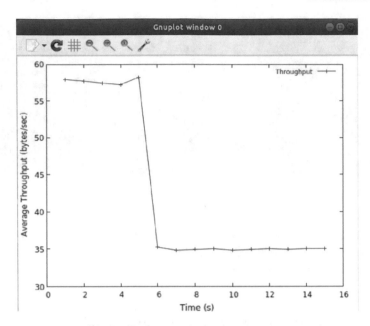

Fig. 9 Throughput under DDoS attack

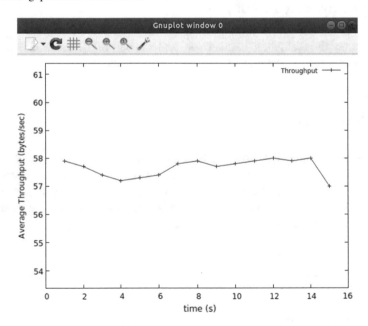

Fig. 10 Throughput under normal traffic

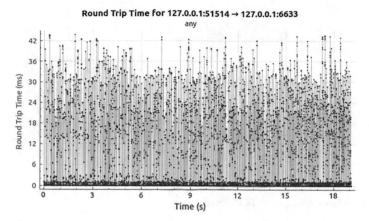

Fig. 11 RTT under DDoS attack

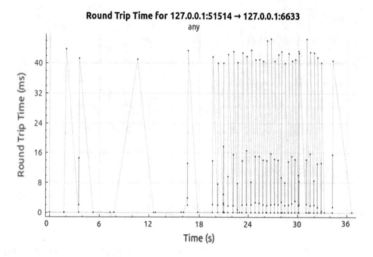

Fig. 12 RTT under normal traffic

5.3 *Flow Graph*

The tool Wireshark has reflected the overall I/O flow while the network is under attack. The TCP errors are shown using Blue Bar graph in Fig. 13 and on the other hand, the Normal traffic is shown using brown line graph in Fig. 14.

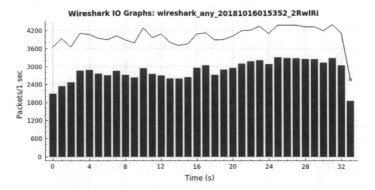

Fig. 13 I/O graph DDoS attack: TCP errors

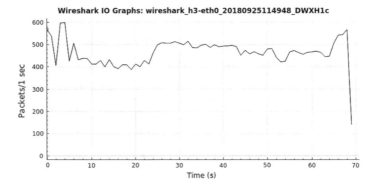

Fig. 14 I/O graph under normal traffic

5.4 Throughput (SDN Vs. Traditional Network)

It can be observed from Figs. 15 and 16 that the throughput of the SDN as compared to the traditional networks has degraded when the network underwent the flood attack.

The following mitigation techniques have been shown to be successfully simulated through our experiments.

5.5 Change in the IP Address of the Victim

As the attacker tends to clog any host's link by flooding packets, the simplest way is to change the IP address of the victim using the open flow centralized controller as shown in Fig. 17 and hence to make it unavailable to the attacker's access.

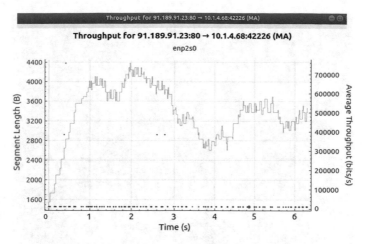

Fig. 15 Throughput—Traditional network

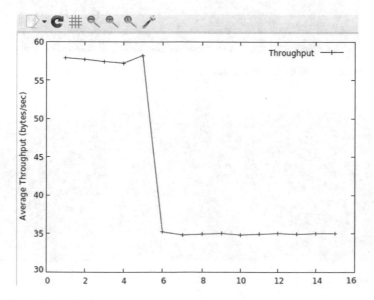

Fig. 16 Throughput—SDN

5.6 Block Port of Attacker

If an attacker has been identified by any means whatsoever then blocking the port of the attacker to fail the service can clog the link of any host. The services generated by the blocked port number are reported as reused, as shown in Fig. 18.

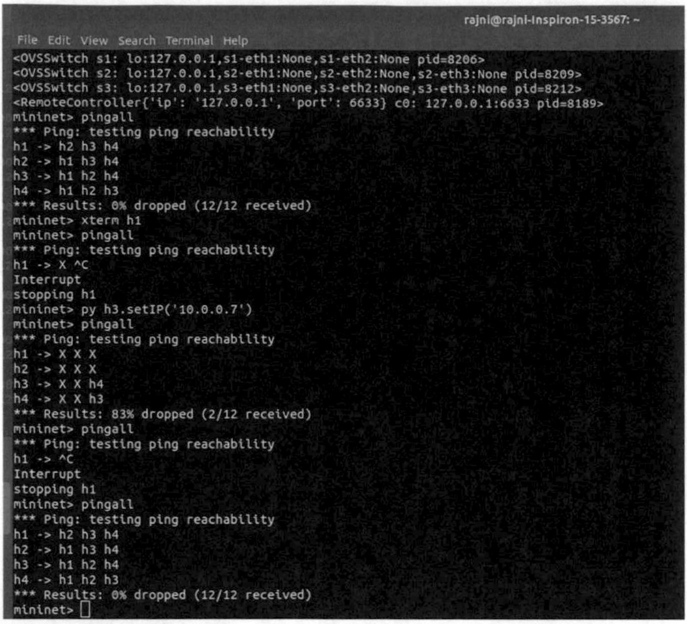

Fig. 17 Changing IP address of victim host

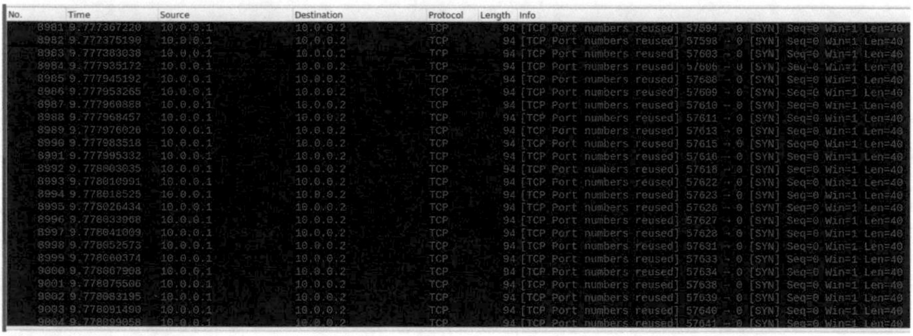

Fig. 18 TCP port reused (block port)

5.7 Redirection

Redirecting the flow of traffic created by the attacker to hamper the working of the network in an SDN to some alien address can also mitigate the effects of the attack. The redirection is done mainly using Openflow switches or centralized controllers. The controller is reporting the flood as an unknown element for the network after the redirection as shown in Fig. 19.

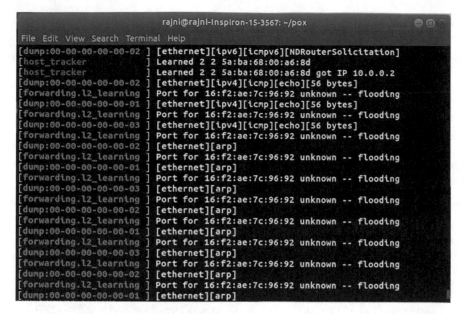

Fig. 19 Redirecting the attacker's traffic

6 Conclusion

SDN based controller is a crucial reason for the network's uniqueness, weaknesses, flexibility, programmability, and robustness. After analyzing the pattern of the packets in the network, the traffic can be managed appropriately. We have shown in this work the analysis of flooding DDoS attacks in both traditional and SDN scenarios. We have also presented a comparative evaluation of the two networks while under DDoS attack and parameters used for evaluation are throughput, RTT and traffic analysis. It has been revealed that the performance of SDN degrades as compared to that of traditional network when under DDoS attack unlike in normal scenarios, where SDN has performed better. Also, three mitigation techniques have been shown to successfully stop the DDoS attacker from flooding the victim. But in order to mitigate the DDoS attacks, changing the IP address of the victim host does not provide a permanent solution. It is because an attacker would somehow manage to get the IP address of the victim as long as it is part of the network (e.g. by sending RARP requests). However, blocking the port of the attacker can be a reliable but not an efficient technique. We are currently working on to find ways to detect and mitigate the botnet based automated flooding DDoS attacks using machine learning techniques. Various limitations of work are that the simulation is based on tree topology with four nodes. In this paper the flooding DDoS attack has been implemented only using ARP echo-request reply packets. Hence it would be interesting to perform the simulation on the real-life traffic for different topology and various number of hosts.

References

1. ONF, Software-defined networking: the new norm for networks, white paper. Available at https://www.opennetworking.org. Last accessed on 22 Dec 2018
2. Shenker S, Casado M, Koponen T (2011) The future of networking and the past of protocols. Open Networking Summit
3. Openflow Switch Specification v1.0–v1.4. Available at https://www.opennetworking.org/sdn-resources/onf-specifications. Last accessed on 22 Dec 2018
4. Open Networking Foundation, OpenFlow Switch Specification. Available at https://www.opennetworking.org/sdn-resources/onf-specifications/openflow. Last accessed on 22 Dec 2018
5. Gude N, Koponen T, Pettit J, Pfaff B, Casado M, McKeown N, Shenker S (2008) Threat analysis of software defined network, vol 38, no 3, pp 105–110
6. Sezer S, Scott-Hayward S, Chouhan P, Fraser B, Lake D, Finnegan J, Viljoen N, Miller M, Rao N (2013) Are we ready for SDN? Implementation challenges for software defined networks. Commun Mag IEEE 51(7):36–43
7. Mirkovic J, Reiher PL (2004) A taxonomy of DDoS Attack and DDoS defense mechanisms. Assoc Comput Mach 34(2):39–53
8. Kaufman C, Perlman R, Sommerfeld B (2003) DoS Protection for UDP-based protocols. In: Proceedings of the 10th association of computing machinery conference on computer and communication security—CCS'03, pp 2–7
9. Zargar ST, Joshi J, Tipper D, Member S (2013) A survey of defense mechanisms against distributed denial of service (DDoS). IEEE Commun Survey Tutorials 15(4):2046–2069
10. Shannon C, Moore D, Claffy KC (2002) Beyond folklore: observations on fragmented traffic. IEEE/ACM Trans Netw (TON) 10(6):709–720
11. Peng T, Leckie C, Ramamohanarao K (2007) Survey of network based defense mechanisms countering the DoS and DDoS problems. Assoc Comput Mach Comput Survey 39(1)
12. Czyz J, Kallitsis M, Papadopoulos C, Bailey M (2014) Taming the 800 pound gorilla: the rise and decline of NTP DDoS attacks. In: Proceedings of internet measurement conference, pp 435–448
13. Strayer WT, Lapsely D, Walsh R, Livadas C (2008) Botnet detection based on network behavior. In: Botnet detection, advances in information security, vol 36. Springer, pp 1–24
14. Giotis K, Argyropoulos C, Androulidakis G, Kalogeras D, Maglaris V (2014) Combining OpenFlow and sFlow for an effective and scalable anomaly detection and mitigation mechanism on SDN environments. Comput Netw 62:122–136
15. Wang R, Jia Z, Ju L (2015) An entropy-based distributed DDoS detection mechanism in software defined networking. In: Proceedings of IEEE Trustcom/BigDataSE/ISPA, pp 310–317
16. Mehdi S, Khalid J, Khayam S (2011) Revisiting traffic anomaly detection using software defined networking. In: Proceedings of 14th international conference on recent advances in intrusion detection, pp 161–180
17. Braga R, Mota E, Passito A (2010) Lightweight DDoS flooding attack detection using NO/Open Flow. In: Proceedings of the IEEE 35th conference on local computer networks. IEEE, Washington, pp 408–415
18. Dotcenko S, Vladyko A, Letenko I (2014) A fuzzy logic-based information security management for software defined networks. In: Proceedings of 16th international conference on advanced communication technology (ICACT). IEEE, pp 167–171
19. Chung CJ, Khatkar P, Xing T, Lee J, Huang D (2013) NICE: network intrusion detection and countermeasure. IEEE Trans Dependable Secure Comput 10(4):198–221
20. Dillon C, Berkelaar M (2014) OpenFlow (D) DoS mitigation. Technical Report. Available at http://www.delaat.net/rp/2013-2014/p42/report.pdf. Last accessed on 22 Dec 2018
21. Shin S, Porras P, Yegneswaran V, Fong M, Gu G, Tyson M, Texas A, Station C, Park M (2013) Fresco: modular composable security services for software defined networks. In: Proceedings of network and distributed System security symposium, pp 1–16

22. Jin R, Wang B (2013) Malware detection for mobile devices using software defined networking. In: Proceedings of GREE proceedings of second GENI research and educational experiment workshop. IEEE, Washington, pp 81–88
23. Schechter SE, Jung J, Berger AW (2004) Fast detection of scanning worm infections. In: Proceedings of international workshop on recent advances in intrusion detection. Springer, Berlin, Heidelberg
24. Chin T, Mountrouidou X, Li X, Xiong K (2015) Selective packet inspection to detect DoS flooding using software defined networking (SDN). In: Proceedings of IEEE 35th international conference on distributed computing systems workshops (ICDCSW). IEEE, pp 95–99
25. Xing T, Huang D, Xu L, Chung CJ, Khatkar P (2013) Snort-Flow: a OpenFlow-based intrusion prevention system in cloud environment. In: Proceedings of 2nd GENI research and educational experiment workshop, GREE 2013, pp 89–92
26. Piedrahita AFM, Rueda S, Mattos DMF, Duarte OCMB (2015) FlowFence: a denial of service defense system for software defined networking. In: Proceedings of global information infrastructure and networking symposium (GIIS), Guadalajara, pp 1–6
27. Spitzner L (2002) Honeypots, tracking hackers, 1st edn. Addison Wesley, Boston, MA, USA
28. Grizzard JB, Sharma V, Nunnery C, Kang BB, Dagon D (2007) Peer-to-peer Botnets: overview and case study. In: Proceedings of USENIX HotBots '07, pp 04–03

Analysis of Impact of Network Topologies on Network Performance in SDN

Dharmender Kumar and Manu Sood

Abstract As compared to traditional networks, software-defined networks (SDNs) have made the communication process more flexible, dynamic, and agile employing its unique features such as centralized control, direct programmability, and physical separation of the network control plane from the forwarding plane or data plane. As the control plane has control over several devices, the process of separation and controlling different devices has made SDN different from the traditional networks. Communication is a vital part of any network. To obtain the best communication results in an SDN, it is essential to analyze and evaluate the performance of different topologies being used. It would be interesting to find out which of these topologies can be used in SDN environment to establish the best communication and to obtain better results if not the best. In this paper, we propose to find out the best topology among four possible topologies in SDN through simulation in Mininet. This selection of best topology is based upon the evaluation and analysis of various network parameters such as throughput, round-trip time, end-to-end delay, bandwidth, and packet loss with/without link down. Based on the values of these parameters through our limited experiments for this paper, we identify the topologies that provide the best and the worst communication results in SDN. Four different types of topologies have been shown to be simulated through Mininet and Wireshark for SDN for the purpose of comparison of this performance analysis.

Keywords Software-defined network · OpenFlow · Mininet · Wireshark · Control plane · Throughput · Round-trip time

D. Kumar (✉) · M. Sood
Department of Computer Science, H.P. University, Shimla, India
e-mail: thakur.dhermender44@gmail.com

M. Sood
e-mail: soodm_67@yahoo.com

© Springer Nature Singapore Pte Ltd. 2020
A. Khanna et al. (eds.), *International Conference on Innovative Computing and Communications*, Advances in Intelligent Systems and Computing 1059,
https://doi.org/10.1007/978-981-15-0324-5_31

357

1 Introduction

The traditional way of network communication, familiar now since long, is straight-forward. The conventional approach of networking is characterized by its network functionality feature that is mainly implemented in a dedicated device/appliance. The dedicated device/appliance in traditional networking means one or multiple switches, routers, and/or application delivery controllers [1–5]. Most of the functionality within these devices is implemented through dedicated hardware. As there is a strong coupling of control and data plane in traditional networks, so data controlling and forwarding policies are provided in the same layer in this approach. This coupling of control and data plane in a single device has led to a certain set of complexities, e.g., once a data forwarding policy has been defined initially in the traditional network and is required to be changed later on, the only way to make changes in that policy is by altering the device configuration disrupting the communication temporarily. A few other significant limitations of traditional networking are (a) time-consuming setup that is error-prone, (b) requirement of high level of expertize in handling the multi-vendor environments, (c) lesser reliability and security of networks, etc. These factors have a vital contribution to make in the evolution of SDN as a new emerging networking paradigm.

1.1 Software-Defined Networking

SDN has emerged as a new buzzword for networking over a past few years and has started to play an essential role in overcoming the shortcomings of traditional networks. Unlike the conventional approach of networking, SDN has been founded on the principle of decoupling of data flow from its control in terms of two separate planes—control plane and data planes. These planes are not tied to a particular hardware or networking device/appliance as in traditional network [1–5]. The data forwarding policy in SDN is quite different from that of conventional networks. Here, the controller plays a vital role in controlling and forwarding the policies related to data in the network since a controller can dynamically be configured through programming in order to make changes in the data forwarding policies [6]. Hence, a new forwarding policy can be enforced on the fly in SDN to override the already existing forwarding policy. The control plane specifies paths for the traffic along with other control parameters, and the data plane implements these decisions so as to forward the traffic toward the earmarked destinations as per this forwarding policy.

Some of the features of significance in the SDN are: (a) the arrangement of the control plane that makes is completely separate from the data plane, (b) a centralized authority in the form of a controller and the corresponding architecture of the network, (c) open interfaces among the devices/appliances in the control as well as data planes, (d) programmability features of the external applications, etc.

Table 1 Comparison of a traditional network with SDN [7]

S. No.	Characteristics	Traditional network	SDN
1	Control and data plane	Coupled, both reside on the same network element	Both are decoupled, not residing on the same element
2	Protocol	Variety of protocol used	Mostly OpenFlow protocol is used
3	Reliability	Less than SDN	Higher reliability
4	Security	Lesser secure	High security
5	Energy consumption	High energy consumption, because the network is rigid	Less energy consumption due to the inherent flexibility

The architecture of SDN has been explained in detail in [7]. This architecture encompasses the whole networking platform. There exist three layers in the architecture of SDN which are application, control, and data layer/infrastructure layer. Application layer and control layer communicate through northbound API, whereas the control layer and data layer communicate through southbound API. Application layer provides the abstract view of applications, and the control layer plays a vital role in SDN. The control layer consists of an OpenFlow controller network operating system. This layer provides an interface to both the upper and lower layers. As it allows for OpenFlow facility, this feature makes SDN different from traditional networks. We can programmatically configure the SDN controller to change the configuration at any time. This functionality is not available with conventional networks. Data is forwarded through the data/infrastructure layer.

1.2 Comparison of SDN and Traditional Networks

The key differences between both the networking paradigms are illustrated in Table 1.

1.3 Significance of Network Topologies

The topology of the networks plays an important role in managing the networks as these affect many characteristics of the networks, e.g., overall performance, level of complexity of communication policies, reliability, the efficiency of data transmission, etc. [8, 9]. There are many different types of network topologies [10], all having their own merits and demerits. As such, there exists no single topology that is ideal for all types of networking requirements. Therefore, selection of the best-fit topology needs some serious evaluations. Bandwidth, throughput, end-to-end delays, packet loss, round-trip time (RTT), etc., are some of the parameters that affect the performance of the topology of a network [11]. After the careful examination of the literature related to the SDNs available till date, it has been observed that the performance of these networks has not been compared for different possible topologies.

Mininet is one of the tools available for the simulation of various SDN topologies. In this paper, we explore the possibilities of comparing the performance of various topologies that can be implemented through Mininet. In order to find out the best and the worst performing topologies, we simulate four of the possible topologies out of the six available in Mininet. The results of the simulation carried out on Mininet have been extended with the help of another tool for drawing the graphs, and the name of the tool is Wireshark. The main contributions of this paper are:

- Worst topology for SDN, as far as the RTT was concerned, is the tree topology because it showcased maximum round-trip time 65.72 ms.
- Also, the worst topology for SDN as far as the average throughput was concerned is the single topology because it gave the minimum throughput of 20,300 bps.
- The best topology for SDN as far as the RTT was concerned is the reversed topology because it showcased minimum round-trip time of 13.55 ms. Similarly,
- the best topology for SDN as far as the average throughput was concerned is linear topology because it gives the maximum throughput of 31,000 bps.

2 Proposed Work

Through simulated experiments, we evaluate the performance of various SDN topologies based on network parameters such as bandwidth, throughput, round-trip time, end-to-end delay, and packet loss. Following are the various topologies that are possible to be created in Mininet SDN [12, 13] along with the commands for their creation in Mininet:

(a) Minimal: Minimal topology is the basic topology in SDN having one switch and two hosts by default. Command to simulate minimal topology in Mininet is **sudo mn - -topo = minimal**.

(b) Single Topology: It is the simplest topology that has one switch and "n" hosts. Command to simulate single topology in Mininet is **sudo mn - -topo = single, 3** with one switch, one controller, and three hosts as shown in Table 2. One of

Table 2 Network elements used for simulation of different topologies

S. No.	Topology	Number of servers	Number of switches	Number of hosts	Numbers of controllers
1	Single	1	1	2	1
2	Linear	1	3	2	1
3	Tree	1	7	7	1
4	Reversed	1	1	2	1

these three hosts can be made to act as a server using a Mininet command.

(c) Linear Topology: Linear topology is another kind of topology in SDN having "*n*" switches and "*n*" hosts. Command to simulate linear topology in Mininet is **sudo mn - -topo = linear, 3** with three switches, one controller, and three hosts as shown in Table 2. One of the three hosts can be made to act as a server using a Mininet command.

(d) Tree Topology: It is a multi-level topology having "n" levels with every switch has exactly two hosts. Command to simulate tree topology in Mininet is **sudo mn - -topo = tree, 3** with seven switches at three levels, one controller, and eight hosts as shown in Table 2. One of these eight hosts has been configured to act as a server using a Mininet command.

(e) Reversed Topology: This topology is quite like the single connection topology, with the reversed order of connection between hosts and the switch. Command to simulate reverse topology in Mininet is **sudo mn - -topo = reversed, 3** with one switch, and three hosts. One of the hosts has been made to act as a server using a Mininet command.

(f) Torus Topology: Torus topology is like mesh topology SDN. Mainly used in a parallel computer system. Command to simulate the torus topology in Mininet is **sudo mn - -topo = torus, 3, 3**.

The numerical values that follow all the topology commands specify the number of switches, numbers of hosts, and the value of level in case of tree topology. These values are not fixed and vary according to the number of switches/hosts/levels as per the requirements of the network.

3 Simulation Environment and Results

This implementation has been carried out by the system where the operating system used was Ubuntu 18.04 LTS. The minimum configuration required for the laptop/computer for this implementation was 2 GB RAM with a minimum of 6–8 GB of free space on the hard disk. The SDN controller used was Mininet controller POX, and the language used is Python. The version of the Mininet used is 2.3 and that of the Wireshark is 2.6.3. For the purpose of experiments, we chose four topologies, namely single, linear, tree, and reversed. The minimal topology is very simple and does have any bearing on our results. Reason for not selecting the torus topology was that it was a non-switched topology. Also, we did not use ring topology since the switches are irrelevant in these types of topologies. Moreover, in a present-day networking scenario, a ring topology is very rarely used. The topologies chosen for our experiments have been created in Mininet, and the numbers of switches, as well as hosts used for each of these topologies in our experiments, are listed in Table 2.

In all the four topologies under consideration, we created one of the hosts as a server (i.e., "h1"), and one of the client hosts (e.g., "h2" or "h3," etc.) was made to a request the server download a file. We used the same bandwidth for all topologies.

Command to set the bandwidth in Mininet is --**link tc, bw = value**. Thus, using the commands mentioned, we created all the four topologies—single, linear, tree, and reversed in Mininet environment and performed the operation of downloading the same file from a local-host server to local client using Mininet to get the results of simulation for each of the topologies.

Figure 1 depicts the linear topology creation in Mininet. Here, "h1" to "h3" are the hosts in this topology, "s1" to "s3" are the switches, and "c0" is the controller. Figure 2 exhibits the execution of commands to make a local-host "h1" as a server using the command **h1 python –m SimpleHTTPServer 80 &** [12, 13]. In similar ways, we performed the file downloading function in all the four topologies to have the values of various parameters shown in Tables 2, 3, and 4 as the outcomes of these simulations.

The implementation and results details of different topologies for comparing all topologies on the basis of different network parameters have been obtained with the help of Mininet and Wireshark [13–17]. Mininet has been to get the results after the simulation experiments have been performed, and Wireshark has been used to obtain the tables and graphs from these results. These results are specific to downloading a particular file from the server, which is the local host nodes and made to act as an HTTP server in our experiments. One specific file has been set to be downloaded from that server by another specific client node of that network for all the topologies under consideration. In all four cases, we obtain the values of six network parameters as listed in Table 3. The purpose of the experiments was to find out a single best topology among four that could provide the best result when all five network parameters have been taken into account. Figure 3 is the screenshot of a table captured through Wireshark and displays the details of the packets transmitted in linear topology during the downloading of the file. It can be seen that the file was downloaded by dividing

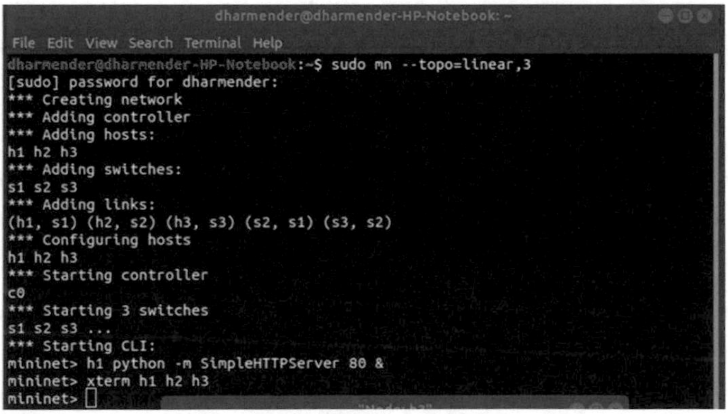

Fig. 1 Linear topology creation in Mininet SDN environment

Fig. 2 Linear topology view after downloading the file from local host

Table 3 Simulation results obtained for various topologies

Parameters / Topology	Bandwidth (Gb/sec)	End-to-End Delay (ms)	Throughput (b/sec)	Round-trip time (ms)	Packet loss (without link down)	Packet loss (after link down)
Single	26	7.2799	20,300	15.032	0	66
Linear	26	23.41	31,000	46.83	0	66
Tree	26	32.36	24,100	65.72	0	25
Reversed	26	6.26	24,000	13.550	0	Network unreachable

Table 4 Parameter values in experiments for simulation of different topologies

Topology / Parameters	Single	Linear	Tree	Reversed
Segment range (bytes)	0–1500	01–1500	0–120	0–1500
Number of segments	3	7	12	3
RTT (ms)	15.032	46.83	65.72	13.55
Average throughput (bps)	20,300	31,000	24,100	24,000

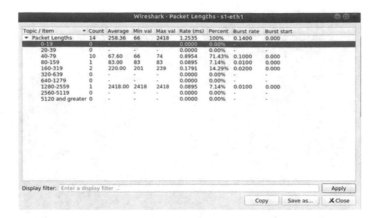

Fig. 3 Details of TCP packets for linear topology as captured through Wireshark

it into 14 packets of various sizes. In a similar manner, this file was also downloaded in other three topologies by dividing it into a variable number of packets of different sizes.

3.1 Throughput and Round-Trip Time Graph

Figures 4, 5, 6, 7, 8, 9, 10, and 11 as presented below have been produced with the help of Wireshark. Figure 4 depicts the throughput of our experiment for the linear topology and plots the throughput of local download of the file from server "h1" against time as well as "segment length." The throughput rises abruptly after a lapse of a time period of about 17.5 ms and for segment lengths ranging from 0 to 1500

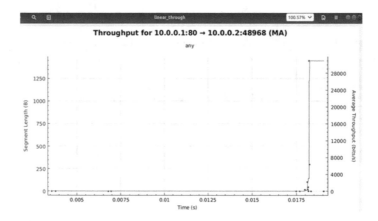

Fig. 4 Throughput graph for linear topology

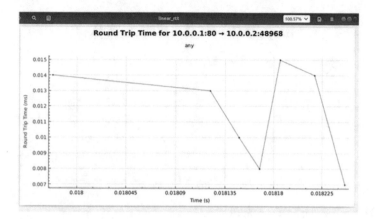

Fig. 5   Round-trip time graph for linear topology

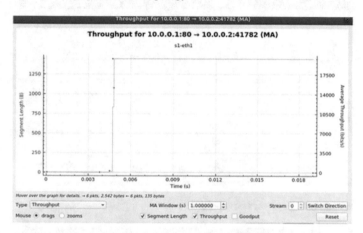

Fig. 6   Throughput graph for a single topology

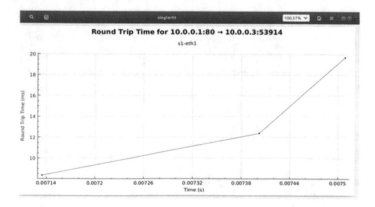

Fig. 7   Round-trip time graph for single topology

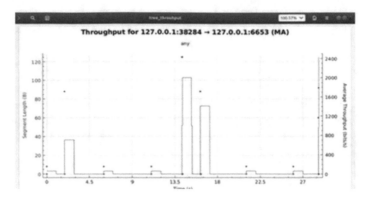

Fig. 8 Throughput graph for tree topology

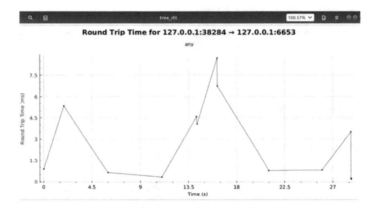

Fig. 9 Round-trip time graph for tree topology

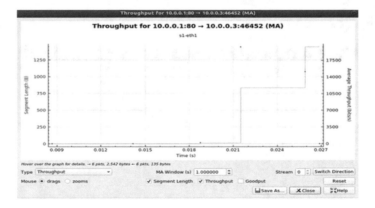

Fig. 10 Throughput graph for reversed topology

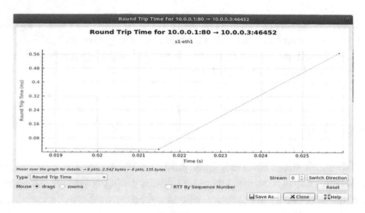

Fig. 11 Round-trip time graph for reversed topology

bytes. Average throughput of 31,000 bps (as shown in Table 4) has been found for segment lengths of 1450 bytes in this experiment. Figure 5 shows the round-trip time graph for this topology. This graph plots the RTT for various segments of the file being downloaded with respect to time. This particular graph shows the RTT for seven segments of this file so as to show the RTT for the download of the complete file which is 46.83 ms as shown in Tables 3 and 4. The throughput and RTT graphs for rest of the three topologies obtained through our experiments can also be described in a similar manner with the help of Figs. 6–11 to present the graphs of throughput and round-trip time for all the four different topologies using Wireshark [16, 17].

As discussed, Figs. 4 and 5 represent the throughput and round-trip time graph for linear topology, respectively.

Figures 6 and 7 represent the throughput and round-trip time graph for single topology, respectively.

Figures 8 and 9 represent the throughput and round-trip time graph for tree topology, respectively.

Figures 10 and 11 represent the throughput and round-trip time graph for reversed topology, respectively.

From the analysis of results of our experiments, in the setup under consideration, we observed that the worst topology for SDN, as far as the RTT was concerned, was the tree topology because it showcased maximum round-trip time of 65.72 ms. Also, the worst topology for SDN as far as the average throughput was concerned was the single topology because it gave the minimum throughput of 20,300 bps. The best topology for SDN, as far as the RTT was concerned, was the reversed topology because it showcased minimum round-trip time of 13.55 ms. Similarly, the best topology for SDN, as far as the average throughput was concerned, was the linear topology because it gives the maximum throughput of 31,000 bps.

Based on these results, we can say that for SDN, (a) the linear topology is the best among all four topologies because of the maximum average throughput and moderate RTT, and (b) the tree topology is the worst with maximum RTT and low average throughput. Moreover, we cannot find out the best and worst topologies on the bases of parameters of packet loss when the link is down (Table 3). In such a case, we have to rely only on throughput and round-trip time for proposing the best and the worst topologies among all four.

4   Conclusion

Critical SDN applications involving fine-grained traffic engineering and fast failovers demand fast interaction among the switches, controller(s), and hosts. The generations of control messages and the execution of control operations can be variable in SDN under different topologies. Therefore, our aim in this paper was to analyze the possible SDN topologies to find out the best-controlled communication topology under this environment. We found that there was no single topology in SDN environment which exhibited the best outcome for all the network parameters under consideration. To be precise, on the basis of the values of throughput, best topology found was the linear topology but it had a moderate RTT. As far as the worst performing topology was concerned, on the basis of our results, we concluded that tree topology was the worst with the highest RTT and low throughput. Due to the time constraints, our experiments were limited and as such, many different sets of experiments have been left for future as the generalization of our findings requires deeper experimentations and analysis of the related results. In the future work, instead of downloading only a single file that too of the same size for all the topologies, more files of the same size and/or files of different sizes can be downloaded to expand the scope of investigations. It will also be interesting to investigate the effects of fixing the size of the packet for all the four topologies while downloading a single/multiple files.

References

1. Kreutz D, Ramos FMV (2015) Software-defined networking: a comprehensive survey. IEEE/ACM Trans Audio Speech Lang Process 103(1):1–76
2. Farhady H, Lee HY (2015) Software-defined networking: a survey. Comput Netw 81:1–95
3. Badotra S, Singh J (2017) A review paper on software defined networking. Int J Adv Res Comput Sci 8(3)
4. Sezer S, Scott-Hayward S, Are we ready for SDN? Implementation challenges for software-defined networks. IEEE Commun Mag 51(7):36–43. Available https://doi.org/10.1109/MCOM.2013.6553676
5. Astuto B, Nunes A (2014) A survey of software-defined networking: past, present, and future of programmable networks. IEEE Commun Surv Tutor 16(3):1617–1634
6. Yeganeh SH, Tootoonchian A (2013) On scalability of software-defined networking. IEEE Commun Mag 51(2):136–141
7. Sood M, Nishtha (2014) Traditional verses software defined networks: a review paper. Int J Comput Eng Appl 7(1)

8. Perumbuduru S, Dhar J (2010) Performance evaluation of different network topologies based on ant colony optimization. Int J Wirel Mob Netw (IJWMN) 2(4). http://airccse.org/journal/jwmn/1110ijwmn12.pdf. Last Accessed on 31 Dec 2018

9. Lee DS, Kal JL (2008) Network topology analysis. Sandia report, SAND2008-0069, Sandia National Laboratories, California. Available https://prod-ng.sandia.gov/techlib-noauth/access-control.cgi/2008/080069.pdf. Last accessed on 31 Dec 2018

10. Meador B, A survey of computer network topology and analysis examples. Available https://www.cse.wustl.edu/~jain/cse567-08/ftp/topology.pdf. Last accessed on 31 Dec 2018

11. Gallagher M, Effect of topology on network bandwidth. Masters Thesis, University of Wollongong Thesis Collection, 1954–2016, University of Wollongong, Australia. Available https://ro.uow.edu.au/cgi/viewcontent.cgi?referer, https://www.google.com/, https://redir=1&article=3539&context=theses. Last accessed on 31 Dec 2018

12. Kumar D, Sood M (2016) Software defined networks (SDN): experimentation with Mininet topologies. Indian J Sci Technol 9(32). https://doi.org/10.17485/ijst/2016/v9i32/100195

13. Mininet walkthrough. Available http://mininet.org/walkthrough/. Last Accessed on 31 Dec 2018

14. Barrett R, Facey A (2017) Dynamic traffic diversion in SDN: test bed vs Mininet. In: International conference on computing, networking and communications (ICNC): network algorithms and performance evaluation. https://doi.org/10.1109/iccnc.2017.7876121

15. Guruprasad E, Sindhu G, Using custom Mininet topology configuring L2-switch in openday-light. Int J Recent Innov Trends Comput Commun 5(5):45–48. ISSN: 2321-8169

16. Biswas J, Ashutosh (2014) An insight into network traffic analysis using packet sniffer. Int J Comput Appl 94(11):39–44

17. Wireshark complete tutorial. Available https://www.wireshark.org/docs/wsug_html/. Last Accessed on 31 Dec 2018

18. Hegde R (2013) The impact of network topologies on the performance of the In-Vehicle network. Int J Comput Theory Eng 5(3). Available http://ijcte.org/papers/719-A30609.pdf. Last accessed on 31 Dec 2018

Traffic Congestion Visualization by Traffic Parameters in India

Tsutomu Tsuboi

Abstract This study focuses on traffic flow analysis in developing country (India) based on a month period. The data is collected by traffic monitoring cameras in a city and summarized as major traffic flow parameters such as traffic density, traffic volume, vehicle velocity, occupancy, and headway. In general, traffic negative impact becomes big issues in developing countries. The main purpose of the study is to analyse traffic congestion condition in detail by visualization analysis for traffic flow characteristics based on real measured data. We took one month real traffic data in the city and analyze traffic congestion level with using occupancy parameter and traffic density relationship based on time zone. These analyses become important to understand real traffic condition, especially in developing countries. The reason is that traffic congestion becomes big issues in terms of economical loss, environment destruction, and traffic fatality growth these days. This study helps to analyze traffic condition and find out solution for Indian traffic congestion.

Keywords Traffic flow · Traffic congestion · Basic traffic diagram · Developing country traffic

1 Introduction

The transportation becomes more key topics these days as mobility serves. There are two side movements in the advanced countries and the developing countries. In major developing countries such as India and China, traffic congestion becomes urgent issue in terms of environment destruction by CO_2 emission, unnecessary fuel consumption by traffic jam, and social loss by increasing traffic accidents. The economic growth in these developing countries is growing rapidly compared with improvement of transportation infrastructure. It is well known that the reason of traffic congestion in the developing countries is too much vehicles in their cities.

T. Tsuboi (✉)
Global Business Development Office, Nagoya Electric Works Co. Ltd., 29-1, Mentoku Shinoda, Ama-Shi 490-1294, Aichi, Japan
e-mail: t_tsuboi@nagoya-denki.co.jp

© Springer Nature Singapore Pte Ltd. 2020
A. Khanna et al. (eds.), *International Conference on Innovative Computing and Communications*, Advances in Intelligent Systems and Computing 1059,
https://doi.org/10.1007/978-981-15-0324-5_32

371

But, it is not so much analysis about these traffic conditions. The author had a chance to have transportation management project in Ahmedabad city of Gujarat state in India since October 2014, which is founded by Japan International Cooperation Agency or JICA. The ITS system function is quite simple. One is traffic monitoring camera which is installed in the city and collected real-time traffic flow data by video recognition. Two is cloud networking which receives traffic data from the camera by internet access and stores the data in the cloud server. Three is traffic information sign board called various message sign(VMS) which displays traffic condition at its location by analysis result of collected traffic data [1, 2].

1.1 ITS System in Ahmedabad

The total system configuration is shown in Fig. 1. There are fourteen traffic monitoring cameras and four VMSs in Ahmedabad city of Gujarat state where it is located in the most west part of India continent. The population of Ahmedabad city is around 8 million in 2015, and automotive registration becomes over 3 million. There is big issue about traffic jam because of rapid economic growth. The west side of the city is so called new city or new developing city, and it becomes more buildings and peoples. Therefore, ITS system has been installed at this west side of the city. In Fig. 1, the traffic monitoring camera and VMS location show Cam# number and VMS# number. The VMS also has traffic monitoring camera, so total number of camera is ten plus VMS camera 4.

1.2 Traffic Flow Data

From the ITS system, we have several kinds of data such as traffic density (K), traffic volume (Q), average vehicle velocity (V), occupancy (OC), and headway (HW). The major parameter for defining traffic congestion condition is OC, which is defined by the percentage ratio between total measurement time (t) of the vehicles to a certain block of road section under certain moment. The formula is shown in (1).

$$OC = \frac{1}{T} \sum_i t_i \times 100(\%) \tag{1}$$

where T is time of measurement and t_i is detected time of vehicle i [3].

When number of existing vehicle of a certain section is N, average length of vehicle is $\bar{l}$, and formula (2) is given.

$$OC = 100 \frac{Q}{V} \bar{l} = 100 K \bar{l} \tag{2}$$

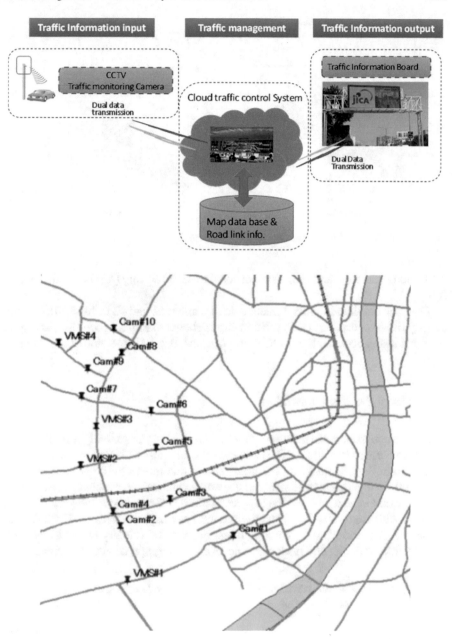

Fig. 1 Ahmedabad ITS system and system installation location in the city

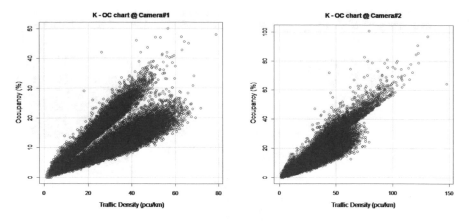

Fig. 2 Example of *K*–OC chart at Camra#1 and Camera#2

Therefore, occupancy (OC) is proportional to traffic density (*K*) and traffic volume (*Q*).

From the one-month measurement data of Camera#1 and #2 in June 2015, traffic density (*K*) to occupancy (OC) relationship is shown in Fig. 2. According to Fig. 2, the relationship between *K* and OC is proportional, but the dispersion of data is seen.

1.3   Daily Traffic Condition

In order to understand real traffic condition, it is necessary to see daily traffic changes of traffic volume and velocity. The two cases at Camera#1 and #2 of traffic condition are shown in Figs. 3 and 4 [4]. The graph shows one-month traffic volume changes from 7:00 am to 6:00 am in next day, and average vehicle velocity changes from 7:00 am to 6:00 am in next day. We see two peek volume of each case. One peek happens between 7:00 am to 10:00 am, and the second peek happens between18:00–20:00. In case of Camera#1, there is not so much changes of velocity but at Camera#2, there is drop at the second peek of traffic. And, each graph shows classification by

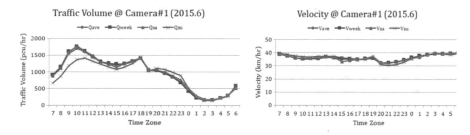

Fig. 3 Daily traffic changes at Camera#1 (June 2015)

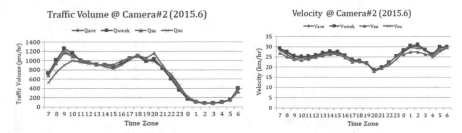

Fig. 4 Daily traffic changes at Camera#2 (June 2015)

four cases—total one month average, weekday, Saturday, and Sunday. According to Figs. 3 and 4, we see traffic jam in Camera#2 from 18:00 to 20:00 because the average vehicle velocity goes down to less than 20 km/h.

2 Traffic Jam Analysis

The current VMS in Ahmedabad city displays each traffic condition by categorized three phases such as smooth, light congestion, and heavy congestion based on the occupancy level. The occupancy of smooth condition is up to 20% and 20–30% is light congestion. The VMS shows heavy congestion over 30% occupancy level. From Figs. 2 to 4, it is difficult to judge about traffic congestion. Therefore, in this chapter, we analyze the traffic jam by using other traffic parameters.

2.1 Time Zone Analysis

As for traffic jam analysis, we use K–OC chart based on time zone basis because traffic jam condition is changed by time to time. Here is example of classification of time zone in Table 1. There are six time zones such as $T1$ is defined time frame from 7:00 am to 10:59 am which is most congested time frame as we see in Fig. 4. And, $T4$ is defined time frame from 19:00 to 22:59 which is the second congested

Table 1 Time zone classification

Zone name	Time zone
$T1$	7:00–10:59
$T2$	11:00–14:59
$T3$	15:00–18:59
$T4$	19:00–22:59
$T5$	23:00–2:59
$T6$	3:00–6:59

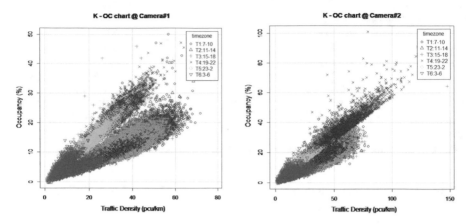

Fig. 5 Time zone-based K–OC chart at Camera#1 and #2

time frame. In case of using this, time zone classification for K–OC chart, time zone-based K–OC chart is shown in Fig. 5. In this paper, we focus on the characteristics at Camera#1 and #2. From Fig. 5, we see traffic jam which happens most likely at time zone $T4$ above 20% occupancy level. Therefore, it is valid to use occupancy level as for traffic congestion judgement at this time.

2.2 Time Zone and Q–OC Characteristics & V–OC Characteristics

When we look at formula (2), it is necessary to other parameters (Q, V) effect to occupancy (OC). The traffic volume is proportion to occupancy, and velocity is inversely proportion to occupancy. The time zone-based Q–OC and V–OC chart at Camera#1 are shown in Fig. 6 and those at Camera#2 are shown in Fig. 7.

According to those two graphs, Figs. 6 and 7, the traffic congestion occurs above 20% occupancy. Therefore, it is necessary to change the example of traffic congestion threshold level like Table 2. It is necessary to check between real traffic condition compared with occupancy level in order to adjust between the real traffic condition situation and value of the occupancy level. It is important to show the traffic condition to VMS for the drivers to feel display contents confidence level based on the current traffic condition.

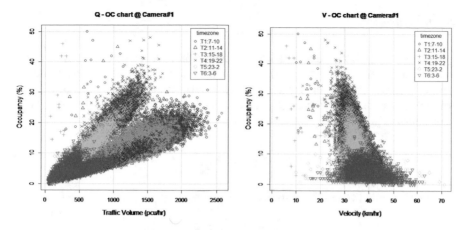

Fig. 6 Time zone-based Q–OC and V–OC chart at Camera#1

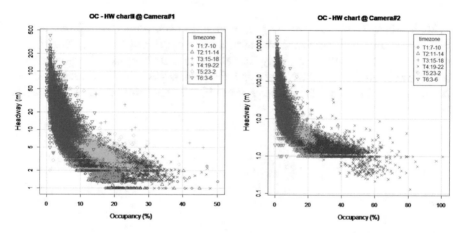

Fig. 7 Time zone-based Q–OC and V–OC chart at Camera#2

Table 2 Traffic congestion classification example

Congestion level	Current occupancy (%)	Proposed occupancy (%)
Smooth	>20	>10
Slight congestion	20–30	10–20
Heavy congestion	30<	20<

3 Discussion

In the previous section, we only check the relationship between traffic density, traffic volume, and velocity to occupancy. There is also another traffic parameter such as headway (HW) which is the length between the fronts of vehicle to that of the next following vehicle. When the average HW value defines $\bar{h}$, the following formula (3) is established [5].

$$\bar{h} = \frac{1}{K} = \frac{V}{Q} \tag{3}$$

From formula (2) and (3), formula (4) is given.

$$\bar{h} = \frac{1}{K} = \frac{100\bar{l}}{OC} \tag{4}$$

So, headway (HW) is inversely proportional to occupancy (OC). Based on this relationship, Fig. 8 shows OC–HW chart for Camera#1 and #2. From Fig. 8, it is not clear to define traffic occupancy for each time zone because of large variation of headway. In order to make clear image of occupancy for time zone, the headway axis is converted to logarithm in Fig. 9. From Fig. 9, it becomes clear that Table 2 is valid.

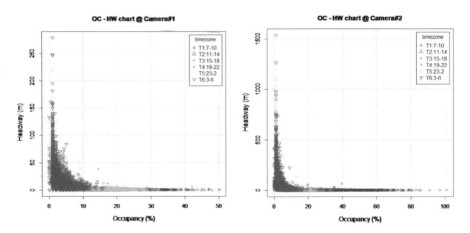

Fig. 8 Time zone-based OC–HW chart at Camera#1 and #2

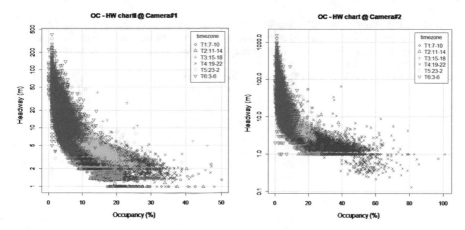

Fig. 9 Time zone-based OC–HW logarithm chart at Camera#1 and #2

4 Conclusion

In this study, author collects one-month traffic data at the actual city in India. And, based on major traffic parameters relationship among traffic density, traffic volume, average velocity, occupancy, and headway, it makes validation by using traffic occupancy level for traffic congestion. As for Ahmedabad traffic condition, it is useful to use occupancy at three levels—smooth, light congested, and heavy congested condition by up to 10%, 10–20%, over 20% of occupancy value. It is also important to check actual traffic congestion between the result of occupancy level from the analysis and drivers' confidence at each location.

In this study, we analyze 14 location traffic conditions in Ahmedabad, but it is necessary to collect more other location data and different timing, hopefully a year.

Acknowledgments This study also underwent the ID16667556 of the International Science and Technology Cooperation Program (SATREPS) challenges for global challenges in 2016.
Special appreciation to Mr. C. Kikuchi and Mr. B. Mallesh of Zero-Sum ITS India for providing traffic data in Ahmedabad.

References

1. Tsuboi T, Oguri K (2016) Traffic flow analysis in emerging country. Inf Process Soc Jpn J 57(4):1284–1289
2. Tsuboi T, Oguri K (2016) Analysis of traffic flow and traffic congestion in emerging country. Inf Process Soc Jpn J 57(12):2819–2826
3. Kawakami S, Matsui H (2007) Transport engineering. Morikita Publishing Co., pp 102–103

4. Tsuboi T, Yoshikawa N (2017) Traffic flow analysis in emerging country (India), CODATU VXII. http://www.codatu.org/bibliotheque/doc/codatu-xvii-presentation-traffic-flow-analysis-in-emerging-country-india-tsutomu-tsuboi-nagoya-electric-works-co-ltd-noriaki-yoshikawa-cyber-creative-institute-co-ltd/
5. Kawakami S, Matsui H (2007) Transport engineering. Morikita Publishing Co., pp 108–109

Modified Genetic Algorithm with Sorting Process for Wireless Sensor Network

Václav Snášel and Ling-Ping Kong

Abstract The genetic algorithm is widely used in optimization problems, in which, a population of candidate solutions is mutated and altered toward better solutions. Usually, genetic algorithm works in optimization problem with a fitness function which is used to evaluate the feasibility and quality of a solution. However, sometimes, it is hard to define the fitness function when there are several optimization objectives, especially only one solution can be selected from a population. In this paper, we modified genetic algorithms with a novel-sorting process to solve the above problem. Two algorithms, the classic genetic algorithm and newly proposed recently M-Genetic algorithm, are simulated and altered by embedding the novel-sorting process. Besides, both the algorithms and their alteration versions are applied into wireless sensor network for locating Relay nodes. The sensor node loss and package loss number are reduced in genetic algorithms with our sorting process compared to the original ones.

1 Introduction

Wireless sensor network (WSN) is a system of sensors for monitoring and recording the environment and organizing data [1]. Those sensor nodes measure area conditions like temperature, sound, pollution levels, and so on, but have limited low power and constrained resources [2]. Therefore efficient power consumption is a very important issue in wireless sensor networks [3]. Relay node is responsible for data packet fusion and fault tolerance incorporation; its usage is unavoidable [4]. The Relay node connects the sensor nodes to the Sink node (data processing center) as a bridge via multi-hop routing [5]. They also reduce the burden of sensing nodes

V. Snášel (✉) · L.-P. Kong
Faculty of Electrical Engineering and Computer Science,
VSB-Technical University of Ostrava, Ostrava, Czech Republic
e-mail: Vaclav.Snasel@vsb.edu

L.-P. Kong
e-mail: konglingping2007@163.com

© Springer Nature Singapore Pte Ltd. 2020   381
A. Khanna et al. (eds.), *International Conference on Innovative Computing and Communications*, Advances in Intelligent Systems and Computing 1059,
https://doi.org/10.1007/978-981-15-0324-5_33

and provide energy-efficient data transferring paths and extend the network lifetime for the wireless sensor network. Hence, the position of the Relay node and communication topology are needed carefully designed [6]. Swarm optimization algorithm, collective behavior of a self-organized system, is wildly used for optimizing problems [7]. This kind of system consists of a population of simple artificial individuals communicating frequently with one another and updating based on the good ones. Examples of swarm optimization algorithms in used include ant colony algorithm (ACO) [8], particle swarm optimization (PSO) [9], cat swarm optimization (CSO) [10], bacterial foraging optimization algorithm (BFOA) [11], shuffled frog leading algorithm (SFLA) [12], and so on. Genetic algorithm [13] is widely used to produce reasonable and feasible solutions to optimization and search problems. For the main running part, it is required to select a section of individuals from a population as the parents for producing the offsprings with crossover and mutation operators [14]. PageRank [15] is designed to measure the importance of website pages, which also is a way of ordering in other fields. The main contribution of this paper includes proposing a new optimization algorithm based on genetic algorithm, which introduces a way of improving selection operator and reduces the difficulty for designing fitness function.

The rest of paper is organized as follows: Sect. 2 introduces the new sorting process and presents the combination procedure of sorting process with GA and MGA two algorithm. Section 3 gives the results of comparison experiments tested on two algorithms and their sorting version algorithms. Section 4 concludes the paper.

2   Modified Genetic Algorithm

Genetic algorithm is a metaheuristic, which is commonly used to produce solutions by mutation and crossover operations. In many genetic algorithm relevant applications, a fitness function is designed to measure the feasibility and quality of a solution. While, due to more than one optimization objective is required to optimized, which makes it hard to design a well-performed fitness function that can completely address the problem. More difficultly, sometimes, only one solution can be selected from a population as the final answer. A novel-sorting method for a population is presented in this section. This sorting method initially creates a $N \times N$ matrix and computes their pagerank ratings, where N is the population size, and those solutions with higher ratings are considered as the better solutions. Then those better solutions are selected as the crossover parents, subsequently, parents generate offsprings.

In our simulation network, 100×100 units square area with 100 randomly deployed sensor nodes are created, and 20 Relay nodes are prepared to arrange with calculated positions. A relay node, as a special type of communication node, is used in multi-hop routing network. It is more powerful than a common sensor node in transmission distance and energy source. So, the cost of relay node is much higher than a common node. Since the relay node often works as a connection station between the Sink node and sensing nodes, the locations of relay nodes are very important, which directly affects the transmission consumption and delivering time. To evaluate the position of Relay nodes, two parameters are computed, the average distance between Relay nodes to Sink node, noted as D_{RS} and the average distance between common nodes to Relay nodes, noted as D_{RN}. The first parameter is simple, the total distance dividing the Relay number is the D_{RS}. Each Relay node is responsible for transmitting task of some common sensor node, the distance from a Relay node to all its responsible nodes divides the responsible number is one RN distance, then calculates all the rest of Relay nodes, their summation is the D_{RN}.

Create sorting matrix

Suppose there is a population with N individuals, and each individual represents a Relay node's deployment solution. Meanwhile, the $D_{RS}[i]$ and $D_{RN}[i]$ are calculated for each solution, where i is in the range of [0-(N-1)]. The matrix, noted as A, will be $N \times N$ size, and $A_{i,j}$ is the number of better distances. For example, two individuals i and j, if $D_{RS}[i]$ is better than $D_{RS}[j]$, and $D_{RN}[i]$ is better than $D_{RN}[j]$, then $A_{i,j} = 2$, $A_{j,i} = 0$. If $D_{RS}[i]$ is better than $D_{RS}[j]$, and $D_{RN}[j]$ is better than $D_{RN}[i]$, then $A_{i,j} = 1$, $A_{j,i} = 1$. After creating the matrix A, the ratings of N solutions can be obtained by power method of Pagerank [16], where the $\alpha = 0.85$ is used in our simulation.

The main loop

1. In our simulation, two genetic algorithms are simulated, the classic one GA and MGA, MGA algorithm is called MGA-RNP in [17], while in the paper, we solve a different problem, so the initialization part is modified, but MGA adopts the three-parent crossover operation and Elite individuals preservation strategies from MGA-RNP. The basic process of GA starts with a randomly initialized population, then using the fitness function to evaluate the population and get their fitness value. Next, to select candidate parents from the population and generate the offsprings under crossover and mutation operation. Subsequently, to check the feasibility of solutions, if not, repair the unreasonable solutions. Finally, re-evaluate the population and check the terminal condition, if it is satisfied, then stop, if not, continue the crossover, mutation and afterward process. The flowchart of a basic GA will be the Fig. 1 without the PageRank sorting process. MGA is a modified version of GA. In MGA algorithm, it uses three-parent crossover instead of two-parent crossover along with the heterogeneity operator to make the search efficient. For both modified genetic algorithm, named as sortGA and sortMGA, they share all the same initialization process and other updating processes with GA and MGA except for adding the PageRank sorting part.
2. Initialization, we use two different initialization ways to produce the MGA and GA population. For GA, the individual is a two-dimensional array of Relay node number size, where the network is two-dimensional. And, an individual represents all the Relay nodes' positions. For MGA, a string of N size chromosome with values varied in $(1 - N)$, each bit represents a sensor node, and its value noted its cluster belonged. And, the center of each cluster is one location of a Relay node.

Fig. 1 Flowchart of genetic algorithm with the sorting process. The PageRank sorting creates a unique rating list of a population. In GA, the half population are selected as the parents, and those offsprings will replace the other half population. In MGA, all the population could be parents, and offsprings replace the population

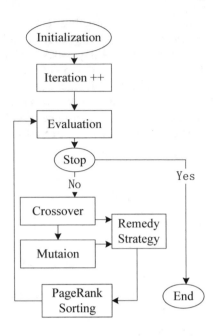

3. Evaluation, two parameters are considered to evaluate the quality of an individual, they are D_{RS} and D_{RN}. According to the equation about the relation between distance and transmission cost. Both parameters are to be minimized. To keep the best one individual, the one with the minimum summation of D_{RS} and D_{RN} is stored.

4. Crossover and Mutation, In the crossover process GA, select the half population with the lowest summation value of D_{RS} and D_{RN} as the parents to produce the offsprings, and the middle (single) point crossover operator is adopted in GA. On the contrary, the sortGA and sortMGA will choose the half population with the highest ratings based on PageRank sorting process as the parents to producing offsprings. And, all the offsprings will replace the other half of the population. Hence, the population size will be a constant number. In MGA, the three consecutive individuals are chosen as parents and generate three offsprings, each offspring will be replaced by one of solution from Elite set in 40% probability, and the offsprings will replace the population.

5. Remedy, to an individual, the Relay node's positions could not cover 80 percent of sensor nodes, the individual need to be repaired. For both methods, calculate the number of unworking Relay node that does not cover any sensor nodes. Then, calculate the uncovered sensor nodes number. The center of some uncovered sensor nodes (uncovered nodes number/unworking Relay node number) will replace one of the unuseful Relay node.

6. Terminal, to check the terminal condition. If the final stop criterion is satisfied, then stop. Otherwise, go to the Evaluation step and continue.

3   Experiment Results

The simulation area is 100×100 units size, and 100 sensor nodes are randomly deployed in it. Sink node, the data processing center, is located at the center of the area. The Relay nodes number is defined as 20. The communication radius of sensor nodes is 30 units. The Relay node could transmit the packet to Sink node directly. The population size is 32, run 100 iteration times. The mutation rate is 0.02 in all the genetic algorithms. The Elite set is a half size of the population. The power for a Relay node is 14 times to a common node. The power consumption is calculated by Eq. 1 [18], where $\alpha = 0.45$ and $\beta = 0.0001$ and r is the Euclidean distance. A total of 5000 packets are transmitted. Once there is an event, sensor node after sensing the event will transmit the packet to its closest Relay node. If the Relay node dies due to out of energy, the node will choose the other close Relay node to retransmit the packet or choose the closest neighbor to do the delivery.

$$E = \alpha + \beta \times r^2 \qquad (1)$$

Figure 2 shows the network structure of four algorithm results; they are GA_100, which is the result of the classic genetic algorithm applied in 100 sensor node network,

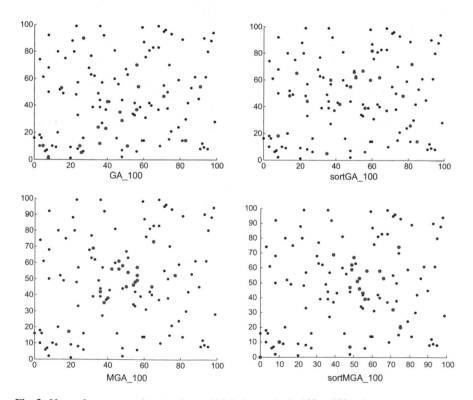

Fig. 2 Network structure of 100 nodes and 20 Relay nodes in 100×100 units area

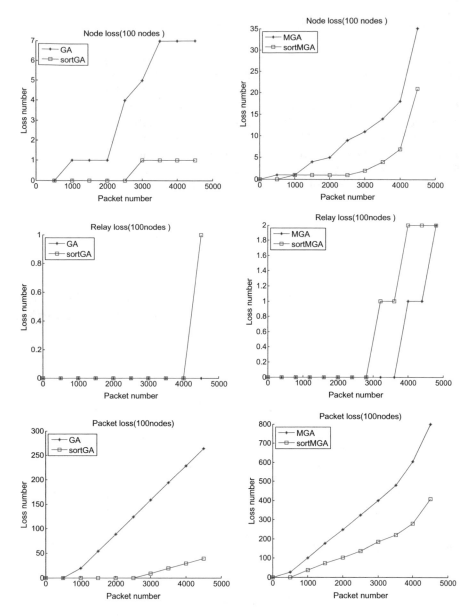

Fig. 3 Data transmission diagram. X-coordinate is the packet transmitted number, Y-coordinate records the loss number (Sensor node loss, Relay node loss, Packet loss)

sortGA_100 is the genetic algorithm with the PageRank sorting process. MGA_100 is the result of MGA, the PageRank sorting version of algorithm is denoted as sortMGA_100. In the figure, the black node is the sensor node, and the Relay nodes are shown in red. The processing center sink node is in the center, does not show in the figure. The final fitness values for their results are: GA_100 = 168.5559, sortGA_100 =172.9464, MGA_100 = 156.1127 and sortMGA_100 = 153.5046. The sorting process has a bigger fitness value than the algorithm without sorting process for GA, as shown in the flowchart of sorting algorithm version, a sorting process always keeps those high rating individuals instead of the fitness value best individuals. But, as mentioned above, the fitness value, sometimes, could not express the problem well and completely, the trends to minimize the fitness value is correct, but it does not mean a little higher value individual is the worst result. We could see the sensor node loss, Relay node loss, and packet loss number of the transmission results from Fig. 3.

Figure 3 is the packet transmission results. It lists the comparison result of sensor node loss, Relay node loss and packet loss number in GA VS sortGA and MGA VS sortMGA. X-coordinate records the packet transmitted number, and Y-coordinate expresses the loss number. From the figure, the red curve presented MGA goes up higher than sortMGA, which could be translated as the loss numbers of sensor node and packet are bigger in MGA, while the Relay node loss numbers are equal. From the other pair GA and sortGA two comparison algorithms, it shows there is one Relay node loss for sortGA, and there is no Relay node loss for GA. However, the sensor node loss number is higher in GA than sortGA, and the lost packet number is much higher in GA. Based on the transmission data, the fitness function helps the population convergent to the better solutions, but sometimes for some applications, it can not distinguish the quality of solutions with a small gap in fitness values.

4 Conclusion

In this paper, we propose a new form of an optimization algorithm, combining the sorting method based on PageRank with genetic algorithm. The parents for producing the offsprings are selected from the population sorting list with higher ratings instead of the fitness values. The experiment simulates the GA and MGA two algorithms and applied the sorting method into these two algorithms. The results show, even the final solution does not own the best fitness value, but the solutions obtained by sorting version are more usefulness. As explained above, sometimes, it is difficult to design a well-performed fitness function with several objectives, and only one solution could be chosen. The fitness function orients the convergence of solutions, and the sorting method could adjust the feasibility of solutions.

Acknowledgements This work was supported by the ESF in "Science without borders" project, *reg.nr.C Z*.02.2.69/0.0/0.0/16_027/0008463 within the Operational Programme Research, Development and Education.

References

1. Moghadam RA, Keshmirpour M (2011) Hybrid ARIMA and neural network model for measurement estimation in energy-efficient wireless sensor networks. In: International conference on informatics engineering and information science. Springer, Berlin, Heidelberg, pp 35–48
2. Almajidi AM, Pawar VP, Alammari A (2019) K-means-based method for clustering and validating wireless sensor network. In: International conference on innovative computing and communications. Springer, Singapore, pp 251–258
3. Leu JS, Chiang TH, Yu MC et al (2015) Energy efficient clustering scheme for prolonging the lifetime of wireless sensor network with isolated nodes. IEEE Commun Lett 19(2):259–262
4. Vallimayil A, Raghunath KMK, Dhulipala VRS et al (2011) Role of relay node in wireless sensor network: a survey. In: 2011 3rd international conference on electronics computer technology (ICECT). IEEE, vol 5, pp 160–167
5. Gupta SK, Kuila P, Jana PK (2016) Genetic algorithm for k-connected relay node placement in wireless sensor networks. In: Proceedings of the second international conference on computer and communication technologies. Springer, New Delhi, pp 721–729
6. Azharuddin M, Jana P K (2015) A GA-based approach for fault tolerant relay node placement in wireless sensor networks. In: 2015 third international conference on computer, communication, control and information technology (C3IT). IEEE, pp 1–6
7. Esmin AAA, Coelho RA, Matwin S (2015) A review on particle swarm optimization algorithm and its variants to clustering high-dimensional data. Artifi Intell Rev 44(1):23–45
8. Sun Y, Dong W, Chen Y (2017) An improved routing algorithm based on ant colony optimization in wireless sensor networks. IEEE Commun Lett 21(6):1317–1320
9. Du KL, Swamy MNS (2016) Particle swarm optimization. In: Search and optimization by metaheuristics. Birkhäuser, Cham, pp 153–173
10. Chandirasekaran D, Jayabarathi T (2017) Cat swarm algorithm in wireless sensor networks for optimized cluster head selection: a real time approach. Cluster Comput 1–11
11. Kumar S, Nayyar A, Kumari R (2019) Arrhenius artificial Bee Colony Algorithm. In: International conference on innovative computing and communications. Springer, Singapore, pp 187–195
12. Luo J, Li X, Chen MR et al (2015) A novel hybrid shuffled frog leaping algorithm for vehicle routing problem with time windows. Inf Sci 316:266–292
13. Elhoseny M, Yuan X, Yu Z et al (2015) Balancing energy consumption in heterogeneous wireless sensor networks using genetic algorithm. IEEE Commun Lett 19(12):2194–2197
14. Gupta SK, Kuila P, Jana PK (2016) Genetic algorithm approach for k-coverage and m-connected node placement in target based wireless sensor networks. Comput Electr Eng 56:544–556
15. Arasu A, Novak J, Tomkins A et al (2002) PageRank computation and the structure of the web: experiments and algorithms. In: Proceedings of the eleventh international World Wide Web conference. Poster Track, pp 107–117
16. Langville AN, Meyer CD (2011) Google's PageRank and beyond: the science of search engine rankings. Princeton University Press
17. George J, Sharma RM (2016) Relay node placement in wireless sensor networks using modified genetic algorithm. In: 2016 2nd international conference on applied and theoretical computing and communication technology (iCATccT). IEEE, pp 551–556
18. Marta M, Cardei M (2009) Improved sensor network lifetime with multiple mobile sinks. Pervasive Mob Comput 5(5):542–555

Secure MODIFIED AES Algorithm for Static and Mobile Networks

Yogesh Khatri, Rachit Chhabra, Naman Gupta, Ashish Khanna
and Deepak Gupta

Abstract Security has been a primary concern for each type of network to provide secure communication among the users of different types of network. Presently, many techniques are present for providing security in networks but Advanced Encryption Standard (AES) has been proved to be the most prominent, keeping the data security as the major factor in data transmission. In the presented exposition, a modified variant of AES algorithm named as Modified Advanced Encryption Standard (MAES) has been introduced for secure data transmission in wired and various wireless networks namely, MANET, VANET, and FANET. The proposed technique focuses on both prevention and detection of the security attacks on the network. Theoretically, it is difficult to break the security or crack the key in the assumed network. The proposed protocol takes 2^{256} computations as compared to the basic AES standard which take 2^{32} in case of differential fault analysis. The simulation results show that the proposed MAES outperformed AES in terms of security against attacks such as side channel attacks.

Keywords AES · MANET · FANET · VANET · Security

Y. Khatri (✉) · R. Chhabra · N. Gupta
Maharaja Agrasen Institute of Technology, Guru Gobind Singh Indraprastha University, Delhi, India
e-mail: yogeshkmait@gmail.com

R. Chhabra
e-mail: Chhabrarachit@gmail.com

N. Gupta
e-mail: namangupta0227@gmail.com

A. Khanna · D. Gupta
Computer Science and Engineering, Maharaja Agrasen Institute of Technology, Delhi, India
e-mail: ashishkhanna@mait.ac.in

D. Gupta
e-mail: deepakgupta@mait.ac.in

© Springer Nature Singapore Pte Ltd. 2020
A. Khanna et al. (eds.), *International Conference on Innovative Computing and Communications*, Advances in Intelligent Systems and Computing 1059,
https://doi.org/10.1007/978-981-15-0324-5_34

389

1 Introduction

The secure data transmission is the need of the hour for various types of networks. There are different types of wireless networks that are being used at present for different purposes. Some of them are MANET, VANET, FANET, etc. MANET is a widely popular network based on mobile devices and it has some of the suggested methods for security [1] and resource allocation [2]. VANET is a vehicular based ad hoc network and its working can be understood by [3]. FANET is a relatively a recent developed network and it has huge applications in many diverse fields [4]. As considering the present scenario where there are a large number of powerful resources available to everyone at a very low cost, the security predefined methods of transmission cannot be fully trusted. Hence, we need to either create a new security technique or improve the predefined ones such that even if a collaboration of large number of resources is used then also there is no threat to our data. First, we have to choose algorithm on basis of energy consumption, memory usage, degree latency for different bandwidths [5]. So here, we have chosen to improve the AES technique in which the side attacks and the implementation defects are creating a potential threat in the security of the standard which must be nullified in order to remain as the most secured symmetric transmission standard.

Here the key size is chosen as 256 bit and number of rounds in the encryption and decryption will be fourteen and the key which will be used in the implementation will be different from what user has provided so as to avoid any possibility of weak key leading to a potential threat to the security of the algorithm.

The organization of this paper is as follows: Sect. 2 discusses the related work from the literature survey. Section 3 presents background of Advanced Encryption Standard (AES) algorithm. The proposed Modified AES algorithm is described in Sect. 4. Section 5 presents the static performance analysis of MAES. Simulation setup and Experimental results with discussions are presented in Sect. 6. Finally, Sect. 7 concludes the paper along with the potential future work.

2 Related Work

The modifications in AES that have been suggested before are studied here in order to understand the issue which leads to a potential threat to the standard. In 2010 the first modification was proposed in AES by Jie and Shurui [6] which was to optimize the algorithm to make it suitable for the smartphone as it only focuses on the optimization hence no change for the security of algorithm were suggested. It discourages the uses of for loops and the macrostructure considering the storage and the speed of the smartphone. Similar modifications were proposed to improve the performance of AES but a breach in AES in future was not considered as a possibility by them. Hence, we skip such papers and move to the first paper which considered the scope in improvement of security of AES algorithm. Another improvement proposed by Rana and Kumar [7] was that in order to increase the security of AES, the key size is increased to 320 bits and the number of rounds for it to be 16. It can be considered as a significant improvement in AES but it improves the security for only against the brute-force attacks which should not be considered as a threat as the existing AES cannot be broken by brute-force in 10^{53} years, even by the supercomputers. So, if there is no dramatic improvement in the performance of currently available resources, we do not need 320 bits key and 16 rounds. In 2015, one more paper was proposed by Sridevi Sathya Priya and Karthigai Kumar [8] which was based on improving AES performance by introducing parallelism in the mix columns operation. Hence, it improves the performance of AES which is a crucial aspect if we consider AES to be secure.

In 2017, Balasubramanya Raju [9] introduced a concept of dynamic shuffling in which instead of initialization vector, it adds one byte to the encrypted text and shuffles the encrypted byte as post-processing of the encryption operation and the re-ordering of the block as pre-processing for generating the encrypted ciphertext. It introduces additional byte in substitution of the initialization vector which produces better performance. Here the concept of additional byte which increases the security can only be considered as a substitute of initialization vector and no further security improvement was introduced by this.

The most recent improvement in AES is based on making it more adaptable for modern appliances on the basis of IoT concept. In 2018, Chowdhury et al. [10] proposed to modify the S-Boxes from matrix to a single dimensional array in order to reduce the energy consumption. As it is similar to the above papers based on its performance hence it also neglected the possibility of a breach in AES in future.

In 2010 [11], a new differential fault analysis was proposed by Chong Hee Kim on AES which determines that if there is a fault introduced in AES and the attacker know the position of the fault in the algorithm then the number of searches required to break AES reduces from 2^{256} to 2^{32} this is a very serious concern as any defect in the implementation of AES by the user can introduce a weakness in the algorithm and also as proposed by Hercigonja [12] there is a flexibility in AES and potential of side-channel attacks in AES by reducing the complexity of deducing the key by a great extent.

3   AES Algorithm

In each encryption process, there are fourteen rounds of encryption. In every round, a different 128-bit round key is used which is evaluated from the original key. Each round comprises of four sub-processes namely, Sub Bytes, Shift Rows, Mix Columns, and Add Round Key. The first round is depicted in Fig. 1.

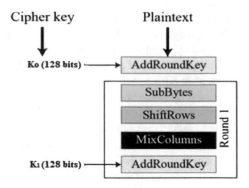

Fig. 1 AES encryption process Round 1

A. Byte Substitution

The substitution of the 16 input bytes is done by values from the fixed table (S-box) which is predefined. The result is in a matrix containing four rows and four columns.

B. Shift Rows

Every row of the matrix is shifted towards the left. Any elements that are left out are inserted on the right side of row. Shift operation is performed as follows:

- The first row remains unchanged.
- The elements of the second row are shifted one position towards the left.
- The elements of the third row are shifted two positions towards the left.
- The elements of the fourth row are shifted three positions towards the left.
- The output of the shift operations is a new matrix with rows re-arranged.

C. Mix Columns

Each existing column of the matrix is now transformed using a mathematical function. The input of the function is the four bytes of the column and it outputs new four bytes which are used to replace the previous column.

D. Add Round Key

The resultant sixteen bytes of the matrix are now treated as 128 bits and are used in XOR operation with the round key of 128 bits. In the last round, it produces the ciphertext. In the other rounds, the resultant 128 bits are used as sixteen bytes for the next round.

The decryption process is very similar to encryption except that the individual operations are performed in reverse as explained in [13].

4 The Proposed Method

A. The Algorithmic Concept

The proposed method overcomes the weakness due to side attacks and the weak implementation of AES by the user. In order to avoid the threat to reuse of caching of initialization vector, the concept of dynamic initialization vector generation is included in the algorithm. A random one-dimensional 16-byte array is generated to be used as the initialization vector for each different encryption process.

Side attacks and differential fault analysis clearly depicts the weakness in AES due to the static nature of the key but as AES is a symmetric key encryption technique so the key has to be same at both ends.

To overcome this difficulty, we use salt and the key provided by user as input to generate the key for each encryption process. As salt is generated for each encryption process hence the key for each encryption is different so no side attacks can be helpful to deduce the key as no similar pattern is generated.

The process of decryption works as the initialization vector and the salt are transferred with the message so as to generate the key used in encryption at the decryption node as the process of using initialization vector is non-reversible so even if the unauthorized node intercepts the initialization vector then also the message previous to the initialization vector usage cannot be generated by the unauthorized node.

Pseudo Code of the proposed Modified AES algorithm is presented in Algorithm 1.

Algorithm 1: Modified Advanced Encryption Standard (MAES) Algorithm

Input: Password, W(plain text)
Output: Encrypted Cipher text in States array

1. Salt←random(array[])
2. InitVector←random(array[])
3. Key←key-generation(Password, Salt)
4. States[][] = W
5. AddRoundKey(States,W[0])
6. for j = 1 to 13
   ByteSubstitution(States)
   ShiftRows(States)
   MixColumns(States)
   AddRoundKey(States, &w[j*4])
   end for
7. ByteSubstitution(States)/*last round of encryption
8. ShiftRows(States) without the mix column
9. AddRoundKey(States,&w[56]) operation*/

B. Working of MAES

Algorithm 1 shows working for a single encryption of one plain text. Here password is the key given by user. The modifications in the algorithm can be realized from the step before starting of encryption in the above algorithm. The first step is the creation of dynamic salt which is an important input in creation of the dynamic key to be used in encryption. The second step is to prevent attacks due to the reuse of initialization vector which is assigned a 16-byte random generated array. In the next step, we create the key to be used for encryption by the combination of salt and the static key which is the represented here as password. As the key needs to be secured hence this static key or password will be only provided to the authorized nodes of network. After key generation, the rounds are similar to that of AES the addition of round key, substitution of S-Boxes, shifting of rows, mixing of columns and after completion of end round the last XOR operation with the round key.

5 Performance Analysis

In this section, static performance analysis of MAES in comparison with AES is presented. Comparison of MAES with AES using differential fault analysis is shown in Fig. 2 and is discussed below.

The grey portion represents faulty bytes. Here S-S^{12} state, S'-S^{13} state, S''-S^{14} state, S'''- is the output after the encryption.

Figure 2 shows the process of Differential fault analysis for AES and the fault analysis of MAES will be similar to it except for a few points as discussed below. Here using the two pairs of faulty and correct ciphertexts generated by the fault in between 11th and 12th Mix Columns K^{14} can be found and by following a similar procedure by introducing a fault in between 10th and 11th Mix Column a pair of faulty and correct ciphertext can be generated which will help in generating 2^{32} candidates for K^{13}. This procedure will break the key in 2^{32} searches [11] if there was no modification in AES but as in our algorithm the first defense is created using a dynamic initialization vector which makes it very difficult to get the correct pair of ciphertext for the hacker and even if the hacker gets the two pair of correct and faulty ciphertext for K^{14} and one pair for K^{13} all the three pairs will be encrypted by three different dynamic keys so the above combination will no longer work for reducing the complexity to break the key. Comparison between AES and Modified AES is shown in Table 1.

Here the difference can be understood by studying the resultant number of possible subkeys and the original key in AES and MAES after applying the differential fault analysis on it. The difference arises due to the dynamic nature of initialization vector and the dynamic key generation by using salt which leads to different value of key for different rounds of differential value analysis.

Fig. 2 Comparison using differential fault analysis

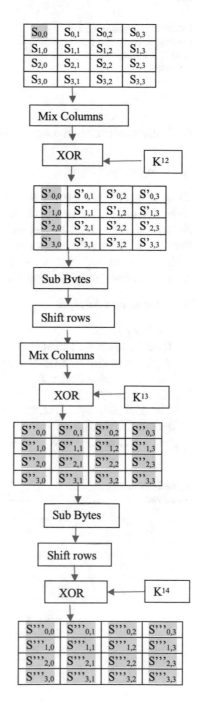

Table 1 Difference in no. of possible keys of AES and MAES in differential fault analysis

	AES	Modified AES
K^{14} key	1 candidate key	2^{256} candidate key
K^{13} key	2^{32} candidate keys	2^{32} candidate keys
Original key	2^{32} candidate keys	2^{256} candidate key

6 Simulation Setup and Result Analysis

A. Simulation Setup

To dynamically analyze the MAES algorithm, "*THE ONE*" simulator is used to perform simulation experiment. The major classes involved in simulation includes Node.java, World.java, Router.java, Map.java, MessageGenerator.java, DirectDelivery.java. Three parameters are used namely, delivery rate, encryption time and avalanche effect, to compare the performance between MAES and AES. The simulation parameters used are shown in Table 2.

B. Results and Analysis

1. **Delivery Rate**
   It can be observed from Fig. 3 that the delivery rate of the AES and the modified AES is very similar besides some minor difference for some time periods hence the modification does not bring any change in the delivery rate of the algorithm.
2. **Encryption Time**
   The time taken by the modified AES was observed to be near 4.4 s for packet size around 5 kB and on increasing more than 5 kB the time taken is linearly dependent on the size of the packet and increases as the size of the packet increases (Fig. 4).

Table 2 Simulation parameters

Parameter	Value
Area	$4500 \times 3400 \text{ m}^2$
Communication range of drone	700 m
No. of nodes	40–100
Routing protocol	Epidemic router
Interface	Radio
Recharge energy	3000 μJ/s
Scan energy	0.92 μJ/s
Transmit energy	0.02 μJ/s
Receive energy	0.02 μJ/s
Movement model	Shortest path map-based movement

Fig. 3 Delivery rate versus time for MAES and AES

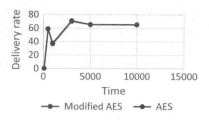

Fig. 4 Delivery rate versus time for MAES and AES

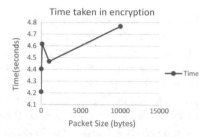

3. **Avalanche Effect**

   Avalanche effect is a property which means that a slight change in input should result into a large change in the output which is cipher code. If a cipher has a low avalanche effect then the analysts can simply guess the plain code by analyzing the ciphertext.

   So here we generate the avalanche effect for various inputs and check whether avalanche is sufficient for use or not. Here we keep the key as constant but we change the plaintext is changed by 1 byte. For example, plain text is *ABCDE-FGHIJKLMNOP* and different cipher text generated by changing one character were:

- *y9eXnY3ZCacf/iLyLVV6xA*
- *5uERVI3a+pAuIDYzll6otg*
- *bW6wcciVnHOdzBjkxGzQfg*
- *AY2xNBQX3shTuZgbKoYP4w.*

   The avalanche value is calculated as:

$$= \frac{48.3 + 52.33 + 56.7 + 47.2}{4}$$

$$= 51.325$$

As we can see that the avalanche effect is high enough to be used. This test is performed for different values of key parameters and the result exist in the range of 41–59.

Hence, mere analysis of ciphertext cannot result in cracking of the key for our modified AES algorithm. Here the avalanche effect is compared to the results of existent AES as given in [14]. As here dynamic key is used so even if the most recent supercomputer is used for cracking the AES key it is useless as the key changes within a very short period of time so crack a key which exists for only about 4–5 ms the time taken will be of order 1.0844×10^{53} years [15].

7 Conclusion

We have studied the existent AES algorithms and the past proposed modifications in AES. The AES needs to be updated for the new attacks that did not exist at the time of creation of AES and there is a need to protect it against side attacks and the threats due to static nature of the key as well as due to weakness created by implementation error by the users. So, we suggested to make the key of AES dynamic and the key creation will be implemented by the algorithm so as to avoid any error from user input. As well as there is a need for using dynamic initialization vector to prevent caching attacks. The proposed system has been analyzed on basis of avalanche effect, delivery rate, time taken in encryption as well as against the differential fault analysis and the results indicate that the above-mentioned changes protect MAES against the new attacks which pose a threat to the existent AES algorithm and preserves its complexity.

References

1. Doss S, Nayyar A, Suseendran G, Tanwar S, Khanna A, Son LH, Thong PH (2017) APD-JAFD: accurate prevention and detection of jelly fish attacks in MANET. IEEE
2. Khanna A, Singh AK, Swaroop A (2016) A token based solution to group local mutual extension problem in mobile ad hoc networks. Springer
3. Kaur EG, Singh S (2016) Technique to control data dissemination and to support data accessibility in meagerly connected vehicles in vehicular ad-hoc networks (VANETS). Int J Adv Res Comput Sci
4. Bekmezci I, Sahingoz OK, Temel S (2012) Flying ad-hoc networks (FANETs): a survey. Elsevier
5. Azam S, Shanmugam B, Mota AV (2017) Comparative analysis of different techniques of encryption for secured data transmission. IEEE
6. Jie L, Shurui L (2010) A modified AES algorithm for the platform of smartphone. In: International conference on computational aspects of social networks
7. Rana SB, Kumar P (2015) Development of modified AES algorithm for data security. Elsevier
8. Sridevi Sathya Priya S, Karthigai Kumar P (2015) FPGA implementation of efficient AES encryption. IEEE
9. Balasubramanya Raju BK, Krishna A, Mishra G (2017) Implementation of an efficient dynamic AES algorithm using ARM based SoC. IEEE
10. Chowdhury AR, Mahmud J, Kamal ARM (2018) MAES: modified advanced encryption standard for resource constraint environments. IEEE
11. Kim CH (2010) Differential fault analysis against AES-192 and AES-256 with minimal faults. IEEE

12. Hercigonja Z (2016) Comparative analysis of cryptographic algorithms. Int J Digit Technol Econ 1:127–134
13. Mewada S, Sharma P, Gautam SS (2016) Exploration of efficient symmetric AES algorithm. IEEE
14. Nejad FH, Sabah S, Jam AJ (2014) Analysis of avalanche effect on advance encryption standard by using dynamic S-Box depends on rounds keys. In: International conference on computational science and technology
15. Al-Mamun A, Rahman SSM, Shaon TA, Hossain MA (2017) Security analysis of AES and enhancing its security by modifying S-Box with an additional byte. Int J Comput Netw Commun (IJCNC) 9:20

A New Efficient and Secure Architecture Model for Internet of Things

Ahmed A. Elngar, Eman K. Elsayed and Asmaa A. Ibrahim

Abstract Internet of Things (IoT) is considered as one of the latest intelligent communication technologies in the world. IoT has been growing in different architectural designs which contributes to the connection between heterogeneous IoT devices. Although, there are many different IoT architectures have been proposed, but they are still suffering from many challenges such as standardization, security, and privacy. This paper will focus on reviewing some existing IoT architectures used to provide security and privacy for IoT networks. Moreover, this paper will introduce a novel IoT security architecture model named IoT-EAA. IoT-EAA tends to satisfy different requirements of IoT security. Hence, the novelty of the proposed IoT-EAA architecture model comes from an important layer called the security layer. This layer employs an external and internal security service against different attacks for each layer within the IoT-EAA architecture model. The proposed architecture model is compared to existing IoT architectures based on several criteria such as security, power and time consumption. Therefore, the proposed IoT-EAA increases security and decrease the power and time consumption, which consider an important contribution to improving the performance of IoT architecture.

Keywords Internet of Things (IoT) · IoT architecture · IoT security · Privacy

A. A. Elngar (✉)
Faculty of Computers and Artificial Intelligence, Beni-Suef University, Salah Salem Str., Beni
Suef City 62511, Egypt
e-mail: elngar_7@yahoo.co.uk

E. K. Elsayed · A. A. Ibrahim
Faculty of Science, Al-Azhar University, Nasr City, Cairo, Egypt
e-mail: emankaran10@azhar.edu.eg

A. A. Ibrahim
e-mail: asmaa_ah@yahoo.com

© Springer Nature Singapore Pte Ltd. 2020   401
A. Khanna et al. (eds.), *International Conference on Innovative Computing
and Communications*, Advances in Intelligent Systems and Computing 1059,
https://doi.org/10.1007/978-981-15-0324-5_35

1   Introduction

These days, the Internet of Things (IoT) is getting one of the newest research topics. IoT enables different devices which used to interact with each other via the Internet. Therefore, it ensures the device to be smart and send information to a centralized system, which will check and take some actions according to the task given to it [1]. IoT technology means the inter-connection of vast heterogeneous frameworks and systems networks. IoT can improve the traditional systems, which converted to intelligent systems by intelligent controls. Thus, it provides exact data more than offline and online analysis. The main infrastructure components of IoT architectures are sensors, actuators, computer servers, and communication network forms. Since IoT devices have different designs, deployments, and maintenance, so it suffers from various problems and many weaknesses in its software and hardware [2]. The most important requirements of IoT architectures could be classified into two categories: which are called IoT specific requirements and IoT security requirements. The IoT specific requirements which consider an important part for the implementation of efficient IoT architectures. Whereas, IoT security requirements which consider the main requirements for achieving the IoT security architectures, as shown in Fig. 1. Requirements that take into account an important part of the implementation of

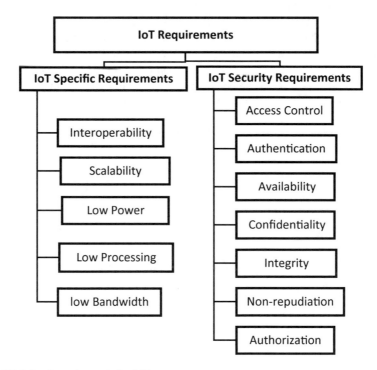

Fig. 1 High-level requirements for IoT

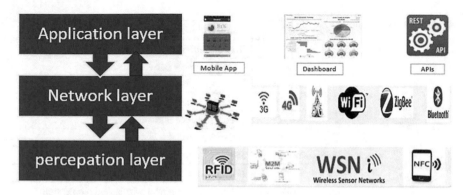

Fig. 2 Layers of IoT

efficient IoT architectures. During the IoT, the security requirements that address key requirements for the realization of IoT security architectures, as shown in Fig. 1.

Since there is no established a well-defined for IoT architecture. But, generally, IoT architecture divided into three layers: Perception layer, Network layer, in addition to the Application layer [3], as shown in Fig. 2.

Perception layer: the principal goal of the perception layer is to gathering information from things through perceiving the physical properties utilizing numerous sensing technologies (for examples RFID, WSN, GPS, and NFC, etc.). In addition, the perception layer is in charge of transforming the information into digital signals, which are more appropriate for network transmission.

Network layer: this layer in charge of managing the received data from the perception layer. In addition, the application layer responsible for data transmission to the upper layer uses various network technologies, such as wireless/wired networks and local area networks (LANs). The main data transmission media are 3G/4G/5G, LTE, Zigbee, Bluetooth, WIFI, UMB, etc. A large amount of data is transported by the network. It is therefore important to provide a healthy middleware to store and process this huge amount of data. To achieve this goal, the main technology "cloud computing" is used in this layer. This technology provides a reliable and dynamic interface in which data can be processed after storage.

Application layer: this layer always at the front of the IoT architecture in which through IoT potential will be utilized. The application layer utilizes the handle data by the network layer. In addition, this layer offers developers with the necessary tools, such as (actuators), to realize the vision of IoT. Therefore, IoT has an impressive vision in various applications. For example, Smart Transportation, Waste Management, Data Management, Intelligent Parking, Logistics Management, Identity Verification, Location-based Services, Intelligent Cities, etc.

Security in IoT networks is much more problematic than with traditional networks. Since then, there has been a wide range of communication protocols, standards and features for various devices, all of which have been identified as critical. Without guaranteeing the security and confidentiality of the Internet of Things, it is therefore

Table 1 Nomenclature

IoT	Internet of things
WI-FI	Wireless fidelity
WSN	Wireless sensor network
RFID	Radio frequency identification
HTTP	Hypertext transfer protocol
MQTT	Message queue telemetry transport
IP	Internet protocol
IPSec	Internet protocol security
IPV6	IP Version 6
6LWPAN	IPV6 with low-power wireless personal area networks
M2M	Machine to machine
QoS	Quality of service
DoS	Denial-of-service

considered likely that it will be accepted by specialists. When security is considered a major issue, they worry about more devices and their communication. This document has proposed a new IoT security architecture model called IoT-EAA. IoT-AEE which composed of five layers which address the physical of terminal equipment in the Hardware Layer, as well as the sensor networks for wireless transmission, computer networks, and mobile communication transmission is introduced in the network layer. Also, the Service Application Layer, Connectivity Management Layer, and Security Layer have presented in IoT-EAA architecture. Moreover, IoT-EAA architecture has facilities to secure the IoT networks through the security layer, which tends to secure each layer and decrease the power and time consumption. Also, it will help academic researchers and industrial designers to improve the performance of IoT environments.

This paper will be organized as follows. Section 2 presents the literature review. Section 3, introduce the proposed IoT-EAA architecture model. The performance evaluation for IoT-EAA architecture is explained in Sect. 4. Finally, we supply conclusions in Sect. 5 (Table 1).

2 Literature Review

In recent years, there are many approaches introduced to the security of IoT [4]. In [5] a new suggested layer of security was added to the Communication Reference Model which increase the performance of different layers. The security layer can also include the End-to-End encryption; because they depend on the security layer to validate the data and confirmation of its sources.

In [6] a single IoT architecture design based on the Mobility-First networks has been introduced, which resolves security issues and builds confidence in the secure operation of the Internet of Things. The author introduces a new layer in the architecture, called IoT middleware, which includes heterogeneous hardware in local IoT systems to the global Mobility-First networks.

In [7], authors proposed a model of security management for the IoT system to select suitable protocols and security algorithms. The proposed model aims for data protection, transmission mediums, protocols, and applications to deny different threats and attacks. Furthermore, providing the requirements for security and securing various applications.

As well as the European Union introduced the IoT-A project conducted in the period between 2010 and 2013 [5]. This project introduced into two IoT models which are architecture reference model (IoT-ARM) and IoT-A Communication Reference Model.

A. **Architecture Reference Model (IoT-ARM)**: The FP-7 European project developed IoT-ARM, which was established in a reference model to serve as a basis for IoT architectures [5]. Unfortunately, this model has some disadvantages. As shown in Fig. 3, this model cannot address security device issues or manage communications between devices.

B. **IoT-A Communication Reference Model**: it works on the internet model [5], and it has four layers, the team of IoT-A developed the internet model to be suitable for the IoT environment. But this model could not address the problems of interoperability between heterogeneous devices, such as security and quality

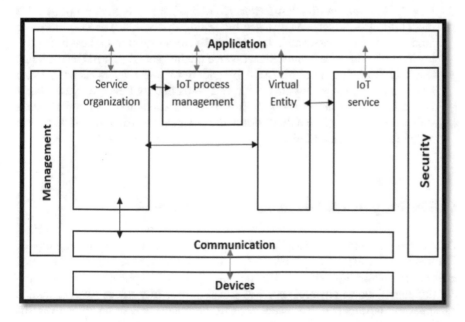

Fig. 3 IoT functional model

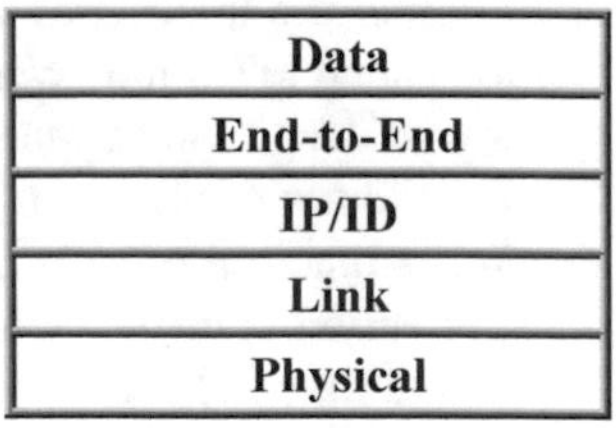

Fig. 4 IoT-A
communication reference
model [5]

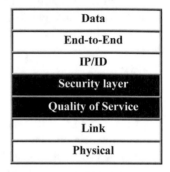

Fig. 5 New IoT-A
communication reference
model [8]

of service, etc. IoT-A Communication Reference Model divided into five layers are presented in Fig. 4.

C. **New IoT-A Communication Reference Model**: was presented in [8]. In this model, the IoT-A has been changed by adding two levels. The new reference model for IoT communication has been built from lowest to highest in the following order: physical, connection, service quality, security, IP/ID, end-to-end, and data. Modification of the IoT-A by adding two new layers called: Service Level Layer, which is classified as Layer 2, and Security Level, which is classified as Layer 3, as shown in Fig. 5. This model focused on IoT Communication Reference Model only, but don't secure the other section from the IoT-A project called IoT Functional Model.

3　The Proposed IoT-EAA Architecture Model

This section contains the description of the proposed IoT-EAA architecture model. The main aim of the proposed IoT-EAA is to solve different security issues which exist at the bottom layer through the top one of the IoT applications. These security problems involve physical security issues, the security issues for information transmission, information processing security problems, and so on. Moreover, the proposed aims to detect various attacks in each IoT architecture layers, which consider a good contribution to secure the IoT networks.

Figure 6 shows the IoT-EAA security architecture model, which divided into five

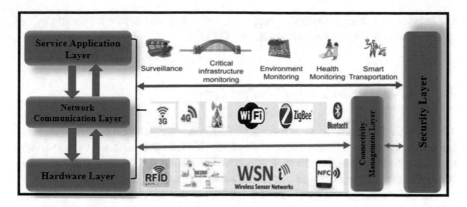

Fig. 6 IoT-EAA architecture

layers, three main layers (Hardware Layer, Network Communication Layer, and the Service Application Layer) that used to deliver different services between the IoT smart objects. As well as, two crosscutting/vertical layers (Connectivity Management Layer and Security Layer). In the following subsections, each layer in the proposed architecture is described in details.

3.1 IoT-EAA Hardware Layer

This layer is responsible for gathering all kinds of information through physical equipment and identifies the physical world. This layer includes different devices, RFID tags, and suitable readers/inscriptions for the data. Such as cameras, GPS, sensors, terminals, and wireless sensor networks (WSN), etc.

3.2 IoT-EAA Network Communication Layer

It is the most developed layer of the proposed **IoT-EAA** architecture model. In this **IoT-EAA** network communication layer has a responsibility to send and receive data between two devices through various communication networks such as the mobile network, internet or another type of networks. This layer is a convergence of the internet and communication-based networks. As well as it is responsible for confirming unique addressing and abilities for routing the unified integration of heterogeneous devices in a single supportive network. **IoT-EAA** network communication layer comprises network interfaces, network management, communication channels, information, and conservation. Also, the key function of **IoT-EAA** Network communication layer is transferring the acquired information from the hardware layer to the Service Application layer this happened through transmission technologies like as Bluetooth, WIFI, Zigbee, WiMAX, 3G, GSM, etc.

3.3 IoT-EAA Service Application Layer

This layer is considered as a top layer for **IoT-EAA** architecture model and responsible for the interface with the end-users and provides the personalized based services according to the user-related needs. Moreover, the main responsibility for this layer is the link for the major gap between users and applications.

The IoT-EAA Service Application Layer has expanded across several fields such as smart cities, smart energy, intelligent buildings, smart transport, smart industry, and smart health. This service allowed by diverse technologies such as sensing, radio frequency identification (RFID), wireless sensor network (WSN), storage, localization, and cloud.

3.4 IoT-EAA Connectivity Management Layer

This layer is a cross-cutting layer which considers an important layer of the proposed **IoT-EAA** architecture. It takes responsibility for managing the connection between the devices by defining the mechanisms for message exchanges between the communicating endpoints. Also, this connectivity management layer handles the Network manager which able to assist an HTTP server and/or an MQTT broker to speak to all devices. As well as, the capacity to aggregate and combine communications from several devices. And, route communications between particular devices using different protocols like IPv6, IPv4, DDS, MQTT, etc. [9]. Moreover, it supports the adaptation of tradition protocols such as 6LWPAN protocol [10]. This layer provides complex data processing, uncertain information (such as restructuring, cleansing, and combining), Quality of Service (QoS), etc. [11].

3.5 IoT-EAA Security Layer

This layer considers the most important layer in the IoT-EAA architecture model. The purpose of this layer is to trust all IoT-EAA architecture layers by providing information security protection for tag privacy, sensor data security, and data transmission. In this section, we introduce two subsections. The first subsection tends to explain some attacks, whereas the second subsection can present different security method in all IoT-EAA architecture layers.

3.5.1 IoT-EAA Attacks

This section introduces various attacks that facing IoT-EAA architecture model. IoT-EAA architecture maybe attacked physically in the hardware layer or attacked from within its network in the network communication layer, or attacks during the communication between devices in the connectivity management layer, and finally maybe attacked from applications on the system in the service application layer. The following subsections will refer in brief the most important attacks:

The Attacks Which are Facing the Hardware Layer

The attacks which are facing the hardware layer are in the following:

Tag Cloning: as the tags are implemented in several objects, that is an unmistakable and the data could be read and modified with several strategies of hacking, in this manner it can be simply taken through any cybercriminal that they can imitate the label and, Therefore, cooperate so that the reader can not differentiate between the original and the compromised tags [12].

Spoofing (fake node): Spoofing occurs when an attacker broadcasts fake information to the RFID system. When this happens, its originality is assumed to be false, this will make it appear from the original source [13]. By using this technique, the attacker gains full access to my system, which makes him vulnerable.

Jamming: RFID tags can likewise be compromised through the type of DOS attack in which communication through RF signals is disturbed by overloaded noise [14].

Attacks in the Network Communication Layer

It comprises of wireless sensor network (WSN) that reliably transfers sensor data to its destination. Security concerns related with this layer are discussed below:

Sybil Attack: This attacker causes the node to have multiple existing identities for a single node. Therefore, much of the system can be affected by the generation of erroneous information for redundancy [15].

Sinkhole Attack: This is a sort of attack in which the adversary makes the compromised node look tempting to the neighboring nodes, because all the data of a specific node is diverted to the compromised node which causes the packets drop, i.e., all traffics are silenced, while the system is believed that the data was received on the alternative side. In addition, this attack has additional consequences energy consumption, which canister a sensible reason a DoS attack [16].

Sleep Deprivation Attack: The sensor nodes of the wireless sensor network are powered with an extra battery for a long period of time. Nodes are therefore limited to monitoring sleep routines to extend their life. This type of attack keeps the nodes awake, minimizing burnout and battery life, and closing nodes [17].

Denial of Service (DoS) Attack: The sort of attack in which the network is flooded by a large amount of useless traffic through an attacker, causing the depletion of system resources used. Because of that, the network converts to unfeasible to the users [18].

Malicious Code Injection: This attack type considers a serious attack during which the attacker threatens a node to introduce malicious code into the system, which may even lead to a complete shutdown of the network or, in the worst case, the attacker be able to get total control of the network [19].

The Attacks of the Connectivity Management Layer are Discuses Blow

Unauthorized Access: Middleware Layer gives diverse interfaces for the applications and facilities for data storage. Undoubtedly, the attacker can damage the system by denying access to the associated IoT services or by deleting the existing data. Subsequently, unauthorized access could be fatal to the system [20].

DoS Attack: It's like the DoS attack described in the Network communication layer, that is, it closes down the system which reasons for the inaccessibility of the services [20].

Malicious Insider: This kind of attack happens when a person inside manipulates the data to gain benefits from private users or the benefits of an outsider. Data could be extracted simply and then that has changed on purpose from the inside [20].

Man-in-the-Middle Attack: This is a form of interception that attacks channel communications, allowing an unauthorized person to monstrously monitor or control all personal communications between the two parties. The unauthorized party can even simulate the character of the victim and communicate normally to receive additional information [21].

The Various Attacks of the Service Application Layer Are Described Below

Malicious Code Injection: An attacker can use the attack on an end user's system with different hacking strategies that allow him to inject malicious code of any type into the system to steal the user's personal information.

Spear-Phishing Attack: This attack is an e-mail attack that asks the victim, a high-ranking person, to open the e-mail used by the enemy, and then use the credentials of that victim on false pretense of recovering confidential information can access.

Sniffing Attack: In this kind of attacks, a sniffer application into the system could be introduced by the attacker to force an attack on the system, which could pick up network information causing system corruption [22].

3.5.2 Security for IoT-EAA

The simplest definition of the security target of IoT is to protect the data collected since the collected data from physical devices might likewise contain sensitive user information. So, at any IoT system, the security should be adaptable to data-associated attacks and give confidence and data security and privacy.

There are the variety of attacks that facing each IoT architecture layer, so we introduce a new security layer that describes the used security protocols and security mechanisms suitable for the attacks facing each layer of the IoT-EAA architecture. The following subsections we describe the used protocols and security mechanisms of the IoT each security layer.

Security Method for Hardware Layer

There are many different protocols for securing the hardware layer such as IEEE 802.11 AH which is suitable for wireless communication. As well as, it is a lighter protocol and satisfy for consuming low power and time. IEEE 802.11 AH uses the encryption algorithm to provide confidentiality and privacy. The appropriate security mechanisms of the composition of this layer are Key Management (PKI) for (WSN) and secure key algorithms which using symmetric key algorithms which provide low power consumption as well as, IPsec for (RFID). In addition to, Anonymity and risk assessment which suitable for all types of sensors used to protect the private information of users and detect network errors.

Security Method for Network Communication Layer

The security mechanisms used in this layer are concerned with the security for data transmission using a Cryptography system like symmetric key cryptography algorithm which consumes time and low power. The most suitable protocol in the network communication layer is 6LoWPAN which is used to encapsulate IPv6. 6LoWPAN can reduce the transmission overhead problem using a set of features like low-bandwidth, low power consumption, low-cost, mobility, as well as, scalable networks and long sleep time.

Security Method for Connectivity Management Layer

The security mechanisms used in this layer concerned with the Secure MQTT (SMQTT) protocol with encryption based on lightweight encryption algorithms to achieve low power and time. The backbone of the IoT connectivity management security is the MQTT server over SSL (Secure Sockets Layer), which protects the sensitive information over the IoT network. In addition to SSL protects the IoT connectivity management using encryption algorithms to secure sensitive information, authentication, critical security and data integrity for the interface of application and personal information about users. The most appropriate protocol of the connectivity management layer is the Message Queue Telemetry Transport (MQTT) was progressed by IBM and aims to lightweight M2M communications. That use publishes/subscribe protocol which keeps running over the TCP stack and meets better the IoT requirements than request/response in COAP. Just as the cross-network authentication mechanism that used to secure the transmission protocols which decrease the complexity of network management.

Security Method in Service Application Layer

The diversity of application layer of IoT, similar to the integration of multiple applications, various systems, data multiform, diversiform sources, and different requirements for the similar data, has the capability to cause privacy leakage of information. Biometrics considers the most suitable security measure for the service application layer [7] because it can prevent internal and external attacks and protect all data between the service application layer and users.

4 Performance Evaluation and Comparison for IoT-EAA

In IOT-EAA architecture model, the IoT network protects using different protective measures at each layer. Figure 7 explains how the security layer in IoT-EAA helps

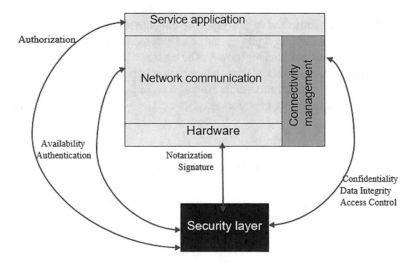

Fig. 7 Security steps at each IoT-EAA architecture layer

for securing the IoT architecture.

The performance of the proposed IoT-EAA architecture model is measured according to many factors such as QoS support for applications, security, privacy, mobility, and manageability.

Some of the solutions for IoT security and privacy are achieved by the proposed (IoT-EAA) architecture model, such as End to End protection of data lifecycle which makes sure the data of IoT environment, end-to-end data protection is provided in a complete network.

The security layer effects for each layer in IoT-EAA architecture as in the following points:

- **Service application layer**: the security layer in IoT-EAA effects in this layer by secure the execution environment. It refers to secure managed-code, runtime environment designed for protection against oblique applications.
- **Network communication layer**: the affection of the security layer of IoT-EAA secures the data communication. It includes authenticating communicating peers and the availability. Security for network communication layer refers to ensuring that unauthorized persons or systems can't deny access or use it to authorize users.
- **Connectivity management layer**: IoT-EAA security layer affected in this layer by ensuring confidentiality and integrity of communicating data (for network communication layer). It prevents repudiation of a communication transaction and protects the identity of communicating entities (Access control).
- **Hardware layer**: the security layer of IoT-EAA helps for security in hardware that includes security computation, secure communication, and service multi-party computation.

Table 2 introduces the comparison between various IoT architecture according to some criteria such as security, Interoperability, power and time consumption. In addition to IoT-EAA can manage large volume of data than other architecture.

Table 2 Comparison of different IoT architecture

IoT compared criteria	IoT architectures			
	IoT-ARM [5]	IoT-A [5]	New IoT-A [8]	Proposed IoT-EAA
Security	Partially	Partially	Partially	Totally
Power and time consumption	High	High	High	Low
Interoperability	✗	✗	✔	✔
Management of large volumes of data	✗	✗	✗	✔
Device discovery and management	✗	✗	✗	✔
Scalability	✔	✔	✔	✔
Redundancy and disaster recovery	✔	✔	✔	✔

From the previous table **partially** refer to all security concerns that are not considered in each IoT architectures layer as in references [5, 8]. But, **Totally** refer to all security concerns take into consideration in each IoT-EAA architecture layer, **high** refers to the IoT architectures in reference [5, 8] have high power and time consumption, but IoT-EAA architecture have **low** power and time consumption, the (✗) symbol refers to the IoT criteria not satisfy, and (✔) symbol refer to the IoT criteria satisfy.

Finally, Table 3 summarizes different attacks, problems, and security requirements which required for each IoT-EAA layer. Moreover, different security methods for solving the security attacks in IoT architecture.

Table 3 Specific attacks for each layer and their security methods

IoT layer	Attacks/problems	Security requirements	Security methods/mechanisms
Hardware	Node capture, Fake node (spoofing.), Replay attack, node jamming, routing threats, RFID tag cloning, others	Authentication Notarization/signature	Data encryption IP-Sec. Mechanism Cryptography technology
Network communication	Sybil attack Sinkhole attack Sleep deprivation Denial of service (DoS) Malicious code injection	Availability/authentication	End to end Authentication Network Virtualization technology Flooding Detection
Connectivity management	Man-in-the-middle Malicious insider Unauthorized access DoS attack, other	Data integrity/confidentiality/access control	Key management Secure routing protocol Asymmetric and symmetric cryptosystem
Service application	Malicious code injection Spear-phishing attack Sniffing attack	Authorization	Biometrics and access control lists (ACLs) using IPS, antivirus and anti-spam and firewall

5 Conclusion

Internet of things (IoT) considers the most significant innovation in the world and is a promising innovation that improves our life. IoT facing multiple challenges for security and privacy related to issues facing all layers in IoT architecture. Therefore, it is vital to establish a sound security architecture and mechanism of the IoT which affects the potential development of the IoT industry security. In this paper, a way of designing a trustable IoT security architecture model is proposed called IoT-EAA, which is depends on the exceptional security requirements of the IoT, wishing to provide theory reference to build a trustable information security system of the IoT. IoT-EAA architecture model includes an important layer named as security layer which utilizes several security services against attacks that face each layer within the IoT-EAA architecture model. The proposed model helps to decrease the power and time consumption, which consider a good contribution to enhancing the performance of the IoT network.

References

1. Farooq MU, Waseem M, Khairi A, Mazhar S (2015) A critical analysis on the security concerns of internet of things (IoT). Int J Comput Appl 111(7)
2. Soumyalatha SGH (2016, May) Study of IoT: understanding IoT architecture, applications, issues and challenges. In: 1st International conference on innovations in computing & networking (ICICN16), CSE, RRCE. Int J Adv Netw Appl
3. Rani P, Sri Lakshmi G (2017) IoT vulnerabilities and security. Int J 2:6
4. Aldosari HM (2015) A proposed security layer for the Internet of things communication reference model. Procedia Comput Sci 65:95–98
5. Bassi A, Bauer M, Fiedler M, Kranenburg RV (2013) Enabling things to talk. Springer-Verlag GmbH
6. Liu X, Zhao M, Li S, Zhang F, Trappe W (2017) A security framework for the internet of things in the future internet architecture. Future Internet 9(3):27
7. Kamel SOM, Hegazi NH (2018) A proposed model of IoT security management system based on a study of internet of things (IoT) security. Int J Sci Eng Res 9(9):1227 ISSN 2229-5518
8. Alhamedi AH, Snasel V, Aldosari HM, Abraham A (2014, July) Internet of things communication reference model. In: 2014 6th International conference on computational aspects of social networks (CASoN). IEEE, pp 61–66
9. Zhang Y (2011, September) Technology framework of the internet of things and its application. In 2011 International conference on electrical and control engineering. IEEE, pp 4109–4112
10. Fremantle P (2014) A reference architecture for the internet of things. WSO2 White paper
11. Kim JT (2015) Requirement of security for IoT application based on gateway system. Communications 9(10):201–208
12. Burmester M, De Medeiros B (2007, April) RFID security: attacks, countermeasures and challenges. In: The 5th RFID academic convocation, the RFID journal conference
13. Mitrokotsa A, Rieback MR, Tanenbaum AS (2010) Classification of RFID attacks. Gen 15693:14443
14. Li L (2012, May) Study on security architecture in the Internet of Things. In: Proceedings of 2012 international conference on measurement, information and control, vol 1. IEEE, pp 374–377

15. Douceur JR (2002, March) The sybil attack. In: International workshop on peer-to-peer systems. Springer, Berlin, Heidelberg, pp 251–260
16. Ahmed N, Kanhere SS, Jha S (2005) The holes problem in wireless sensor networks: a survey. ACM SIGMOBILE. Mobile Computing and Communications Review 9(2): 4–18
17. Bhattasali T, Chaki R, Sanyal S (2012) Sleep deprivation attack detection in wireless sensor network. arXiv preprint arXiv:1203.0231
18. Padmavathi DG, Shanmugapriya M (2009) A survey of attacks, security mechanisms and challenges in wireless sensor networks. arXiv preprint arXiv:0909.0576
19. Fulare PS, Chavhan N (2011) False data detection in wireless sensor network with secure communication. Int J Smart Sens Ad Hoc Netw (IJSSAN) 1(1):66–71
20. Farooq MU, Waseem M, Khairi A, Mazhar S (2015) A critical analysis on the security concerns of internet of things (IoT). Int J Comput Appl 111(7)
21. Padhy RP, Patra MR, Satapathy SC (2011) Cloud computing: security issues and research challenges. Int J Comput Sci Inf Technol Secur (IJCSITS) 1.2:136–146
22. Thakur BS, Chaudhary S (2013) Content sniffing attack detection in client and server side: a survey. Int J Adv Comput Res 3(2):7

Desmogging of Smog Affected Images Using Illumination Channel Prior

Jeevan Bala and Kamlesh Lakhwani

Abstract Visibility restoration of smoggy images plays a significant role in various computer vision applications. However, designing an efficient desmogging technique is still a challenging issue. The majority of existing researchers have designed restoration model for rainy, dusty, foggy, hazy, etc., images only. Therefore, these approaches perform poorly for smoggy images. In this paper, a novel illumination channel prior is proposed to restore smoggy images in a significant way. A gradient magnitude-based filter is also utilized to refine the transmission map. The proposed desmogging approach is compared with the existing visibility restoration approaches over ten real-time smoggy images. The subjective and quantitative analysis reveals that the proposed desmogging approach outperforms others.

1 Introduction

Smog contains a combination of fog and haze present in the atmosphere [1]. Smog generally occurs in winter season when warm water cools quickly due to low temperature and also at the same time pollution is present in the environment. Designing a novel desmogging approach is an ill-posed problem. Therefore, not much work is found in the literature to remove smog from images [2]. Even though the existing dehazing and desmogging approach can be applied to remove smog from images, these restoration approaches are not so effective for smoggy images [3–6].

A novel gain intervention-based filter has been designed and implemented in [7]. It has an ability to restore the images in an efficient manner. Fourth-order partial differential equations-based anisotropic diffusion model is used in [8]. This model can be utilized during the desmogging process. An integrated dark and bright channel prior-based model can restore smoggy images in an efficient manner [9]. An

---

J. Bala · K. Lakhwani (✉)
Computer Science and Engineering, Lovely Professional University, Jalandhar, India
e-mail: kamlesh.lakhwani@gmail.com

J. Bala
e-mail: jeevan94.47@gmail.com

© Springer Nature Singapore Pte Ltd. 2020
A. Khanna et al. (eds.), *International Conference on Innovative Computing and Communications*, Advances in Intelligent Systems and Computing 1059,
https://doi.org/10.1007/978-981-15-0324-5_36

image enhancement technique based on dark channel prior and gamma correction is implemented in [10]. An approximation radiance darkness prior is designed and implemented in [11]. It has been found from [12] that the models discussed in [7–11] can be used to restore the smoggy images. However, these models are effective only for smoggy images with low degree of smog.

This paper makes the following contributions:

- An illumination channel prior is proposed to restore smoggy images. This is achieved by replacing the dark channel prior with illumination channel. Therefore, it allows the proposed approach to evaluate the transmission map and atmospheric light in an efficient manner. It has also an ability to handle sky region and gradient reversal artifact issues with existing restoration approaches.
- An edge-preserving filter is proposed for accurately refining the transmission map. Further, it is improved via a newly proposed edge-preserving loss function.
- As existing restoration models designed for dehazing and defogging are not so effective for smoggy images, therefore in this work, a modification of restoration model is also proposed.
- Extensive experiments are conducted on real-world smoggy images. In addition, comparisons are performed against several recent restoration approaches.

The remaining structure of the paper is as: The review of existing restoration approaches is discussed in Sect. 1. Section 2 presents the proposed desmogging model. Section 3 presents experimental set-up, result, and discussions. The conclusion is demonstrated in Sect. 4.

2   Proposed Desmogging Approach

This section discusses the designed desmogging model. Figure 1 demonstrates the overall flow of the designed desmogging model.

2.1   Depth Map Estimation

Initially, an illumination channel is designed to estimate depth information from smoggy image (I_m) as

$$I^d(p, d) = \delta_{y \in \Psi(p,q)} \cdot \delta_{c \in (r,g,b)} \cdot I_m^c(l)^{\Sigma \Sigma} \tag{1}$$

Here, I_m^c is the available color channel of I_m. δ represents the illumination channel prior. $\Psi(p, q)$ shows the local window.

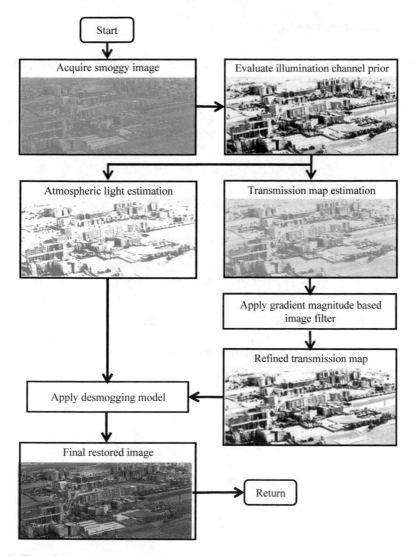

Fig. 1 Flow of the proposed approach

2.2 Atmospheric Light

Atmospheric light (A_1) plays an important role to restore the smoggy image, it can be calculated as [13]:

$$A_1(p, q) = I_m^- \max_c \left(I_m^c \right)^{\Sigma}.$$ (2)

2.3   Transmission Map

Transmission map ($\tilde{t}$) is another building block of desmogging model and it is achieved by:

$$\tilde{t}(pq) = 1 - \min_{y \in \Psi(p,q)} \cdot \min_c \frac{I_m^c(y)^\Sigma}{A_l^c} \tag{3}$$

2.4   Coarse Atmospheric Light Estimation

The coarse atmospheric light ($A_{\text{viel}}\ (p, q)$) estimation is achieved by [13]:

$$A_{\text{viel}}(p, q) = \beta \min_{y \in \Psi(p,q)} \cdot \min_c \frac{I_m^c(y)^\Sigma}{A_l^c} \tag{4}$$

In this paper, gradient magnitude-based filter is utilized to refine t as:

$$\tilde{t}(p, q) = \sigma(p, q) - J^t O^f \cdot |t - \sigma(p, q)|^\Sigma \tag{5}$$

Here, $\sigma\ (p, q)$ is standard deviation.

2.5   Restoration Model

Finally, the restored image (A_r) is obtained by using the restoration model as:

$$A_r(p, q) = \frac{I_m(p, q) - A_l}{\max\left(\tilde{t}(p, q), l\right)} + A_l \tag{6}$$

3   Results and Discussions

To evaluate the effectiveness of the proposed desmogging model seven existing restoration models are considered. These approaches are DCP [13], CAP [14], CoD [15], WGIF [16], LTQ [17], L_1 norm [18], and FVID [19] on dataset obtained from [20].

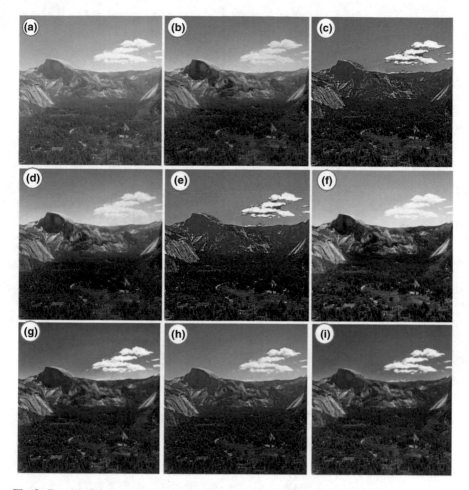

Fig. 2 Results of desmogging approaches for Im2 **a** Input image, **b** DCP [13], **c** CAP [14], **d** CoD [15], **e** WGIF [16], **f** LTQ [17], **g** L_1 norm [18], **h** FVID [19], and **i** proposed approach

Figure 2 shows that the proposed model provides better visual quality with minimum artifacts associated with the existing approaches.

Table 1 shows the analysis of contrast gain (CG) that shows that the average enhancement in CG over the existing approaches is 0.06324.

Table 2 shows the comparison of proposed approach based on percentage of saturated pixels (ρ). It indicates that the average reduction of the proposed approach over the existing approaches is 0.06793.

Table 1 Comparative analysis of proposed approach using contrast gain (CG) images DCP [13], CAP [14], CoD [15], WGIF [16], LTQ [17], L_1 [18], FVID [19], and proposed improvement (in %)

Im1	0.0216	0.1089	0.1471	0.1643	0.1591	0.1486	0.1719	0.1862	1.2479
Im2	0.0132	0.1368	0.1426	0.1359	0.1652	0.1549	0.1718	0.1864	1.3197
Im3	0.1253	0.1350	0.1691	0.1554	0.1996	0.1885	0.2180	0.2360	1.4107
Im4	0.0195	0.0931	0.1211	0.1466	0.1539	0.1523	0.1622	0.2019	1.2694
Im5	0.0279	0.0456	0.1104	0.1650	0.1794	0.1199	0.1801	0.1926	2.0138
Im6	0.1178	0.1612	0.2266	0.1323	0.1009	0.1834	0.2198	0.2301	1.5861
Im8	0.0520	0.0465	0.1674	0.1569	0.1780	0.1461	0.1659	0.1922	1.7641
Im7	0.0273	0.0899	0.1049	0.1729	0.1633	0.1798	0.2156	0.2200	1.1972
Im9	0.0689	0.0989	0.1407	0.1911	0.1602	0.1355	0.2010	0.2199	1.2397
Im10	0.0303	0.0711	0.1376	0.1242	0.1502	0.1533	0.1300	0.1698	1.4620

Table 2 Comparative analysis of proposed approach based on percentage of saturated pixels (ρ) image DCP [13], CAP [14], CoD [15], WGIF [16], LTQ [17], L_1 [18], FVID [19], and proposed reduction (in %)

	DCP [13]	CAP [14]	CoD [15]	WGIF [16]	LTQ [17]	L_1 [18]	FVID [19]	Proposed	Reduction
Im1	0.2317	0.1283	0.1341	0.1216	0.0932	0.1246	0.0872	0.0643	1.0375
Im2	0.1648	0.1241	0.1037	0.0874	0.0764	0.0548	0.0479	0.0321	1.0572
Im3	0.1246	0.1219	0.1146	0.1091	0.0986	0.0846	0.0767	0.0613	1.1603
Im4	0.1327	0.1448	0.1215	0.1091	0.0942	0.0982	0.0864	0.0687	1.1421
Im5	0.1641	0.1456	0.1164	0.0947	0.0749	0.0847	0.0742	0.0643	1.2158
Im6	0.1479	0.1694	0.1497	0.1254	0.1169	0.0948	0.0847	1.0935	1.1554
Im7	0.1467	0.1142	0.0986	0.0912	0.0872	0.0846	0.0846	0.0449	1.0891
Im8	0.1348	0.1497	0.1356	0.1081	0.0946	0.0925	0.0879	0.0740	1.0844
Im9	0.1672	0.1587	0.1461	0.1315	0.1267	0.1257	0.1069	0.0962	1.1348
Im10	0.1649	0.1597	0.1347	0.1169	0.1026	0.0964	0.0864	0.0492	1.0299

4 Conclusion

An efficient desmogging approach has been proposed in this paper. The proposed approach uses two new concepts namely illumination channel prior and refined trilateral filter. The dynamic threshold has been used to reduce the color distortion rate. The experimental results show that the proposed approach is able to mention the colors of smog-free image. On the basis of results obtained, we can conclude that the proposed approach is applicable for real-time applications.

This work opens several research directions for future studies. Firstly, supervised learning-based approaches can be applied to estimate the transmission map and atmospheric veil. Secondly, machine learning models can be integrated with desmogging approach to improve their performance. It is also worth to investigate the applicability of proposed approach for other computer vision problems such as remote sensing imaging, outdoor video surveillance, underwater image analysis, etc.

References

1. Fan X, Wang Y, Tang X, Gao R, Luo Z (2016) Two-layer gaussian process regression with example selection for image dehazing. IEEE Trans Circ Syst Video Technol 99:1
2. Draa A, Bouaziz A (2014) An artificial bee colony algorithm for image contrast enhancement. Swarm Evol Comput 16:69–84
3. Guo F, Peng H, Tang J (2016) Genetic algorithm-based parameter selection approach to single image defogging. Inf Process Lett 116(10):595–602
4. Kim J-H, Jang W-D, Sim J-Y, Kim C-S (2013) Optimized contrast enhancement for real-time image and video dehazing. J Vis Commun Image Represent 24(3):410–425
5. Li Z, Zheng J (2015) Edge-preserving decomposition-based single image haze removal. IEEE Trans Image Process 24(12):5432–5441
6. Fattal R (2014) Dehazing using color-lines. ACM Trans Graph (TOG) 34(1):13
7. Pathak Y, Arya K, Tiwari S (2018) Low-dose ct image reconstruction using gain intervention-based dictionary learning. Mod Phys Lett B 32(14):1850148
8. Pathak Y, Arya K, Tiwari S (2018) Fourth-order partial differential equations based anisotropic diffusion model for low-dose ct images. Mod Phys Lett B 32(25):1850300
9. Singh D, Kumar V (2018) Single image haze removal using integrated dark and bright channel prior. Mod Phys Lett B 32(04):1850051
10. Zheng L, Shi H, Gu M (2017) Infrared traffic image enhancement algorithm based on dark channel prior and gamma correction. Mod Phys Lett B 31(19–21):1740044
11. Xiang R, Wu F (2018) Single image haze removal with approximate radiance darkness prior. Mod Phys Lett B:1840086
12. Bala J, Lakhwani K (2019) Performance evaluation of various desmogging techniques for single smoggy images. Mod Phys Lett B:1950056
13. He K, Sun J, Tang X (2011) Single image haze removal using dark channel prior. IEEE Trans Pattern Anal Mach Intell 33(12):2341–2353
14. Zhu Q, Mai J, Shao L (2015) A fast single image haze removal algorithm using color attenuation prior. IEEE Trans Image Process 24(11):3522–3533
15. Li J, Zhang H, Yuan D, Sun M (2015) Single image dehazing using the change of detail prior. Neurocomputing 156:1–11
16. Li Z, Zheng J, Zhu Z, Yao W, Wu S (2015) Weighted guided image filtering. IEEE Trans Image Process 24(1):120–129

17. Wang W, Yuan X, Wu X, Liu Y (2017) Fast image dehazing method based on linear transformation. IEEE Trans Multimedia 19(6):1142–1155
18. Cui T, Tian J, Wang E, Tang Y (2017) Single image dehazing by latent region-segmentation based transmission estimation and weighted l1-norm regularisation. IET Image Proc 11(2):145–154
19. Galdran A, Vazquez-Corral J, Pardo D, Bertalmío M (2017) Fusion-based variational image dehazing. IEEE Signal Process Lett 24(2):151–155
20. Tarel JP (2011) Single image visibility restoration comaprison. http://perso.lcpc.fr/tarel.jean-philippe/visibility/index.html. Dec 2011

Construing Attacks of Internet of Things (IoT) and A Prehensile Intrusion Detection System for Anomaly Detection Using Deep Learning Approach

Marjia Akter, Gowrab Das Dip, Moumita Sharmin Mira, Md. Abdul Hamid and M. F. Mridha

Abstract An abundance of physical instruments around a group of countries which are now associated to the hyperspace, collecting or sharing data known as the internet of things (IOT). As the statistic of IoT devices increases, new security and privacy dare will be confronted for both home and office devices. An intrusion detection system (IDS) helps to detect the malicious system to get notified when any malicious flurry or anomaly occurred in the system. In this paper, we dispute four types of attacks of IoT ambiance. We have proposed such a model that recuperates from attacks like DoS (Denial of Services), DDoS (Distributed Denial of Services), R2L (Remote 2 Local), U2R (User to Root), and probe attack. Our model mainly focused on the security of home-based appliances like air-condition, fan, light, television, oven, refrigerator, printer, heater, washing machine, geysers, electric stove, and others electronic devices. We have developed an algorithm by using deep learning approach to dispute attacks and give security to the user. Deep learning is divergent from regular machine learning approach which has self-taught techniques (STL) that represents data such as images, video or text, without using human domain knowledge. They have more ductile architectures that comprehend from raw data and can increase their accuracy level when acquires more data. Our model analyses six features a server to identify whether it is malicious or not. Self-taught technique of deep learning has been approached in our paper. We have used NSL-KDD dataset for training and testing.

M. Akter · G. D. Dip · M. S. Mira · Md. Abdul Hamid · M. F. Mridha (✉)
Department of CSE, University of Asia Pacific, Dhaka, Bangladesh
e-mail: firoz@uap-bd.edu

M. Akter
e-mail: marjiah0016@gmail.com

G. D. Dip
e-mail: dipdasnew@gmail.com

M. S. Mira
e-mail: mowmitasharmin1997@gmail.com

Md. Abdul Hamid
e-mail: ahamid@uap-bd.edu

© Springer Nature Singapore Pte Ltd. 2020
A. Khanna et al. (eds.), *International Conference on Innovative Computing and Communications*, Advances in Intelligent Systems and Computing 1059,
https://doi.org/10.1007/978-981-15-0324-5_37

427

Keyword Internet of Things · Intrusion detection · Attacks · Deep learning

1 Introduction

An apparatus of interrelated computing devices, mechanical machines that can distinguish the difficulty and can transfer details through a network within in seconds alludes as the internet of things (IOT) [1]. It is expected that 45 billion devices will be interconnected by the end of 2020. With the progress of the Internet, people change their lifestyle, work, study, and effort [2]. As technology changes, our standard of maintenance has changed. To make this change a step further, we are trying to apparatus the IoT appliance on a smart home. On this network, all the electronic devices of a home are connected through a server. As a result, users can control their domicile appliances from anywhere on the planet. That's why no one else can control devices without the owner's desire. As a result, someone cannot misuse the device. That's why Smart Home is a Better alternative from Home. As IoT is giving us the most desirable and preferable life, it also has some drawbacks. We may have faced some attacks while working with the IoT devices.

Attacks may have come in form of packets, requests, emails, etc. These attacks can hamper our ambiance of smart home. Some of attacks are given below which we are going to solve by our algorithm:

DoS (Denial of service): it is one kind of attack which sends so many requests through botnet to a server. All servers can afford only a limited number of requests at a certain time. But if the server gets too many requests during a certain time, the server will lose its regular working ability. So, attackers use this attack to destroy a server. Mail bomb, UDP storm are the different kind of DoS attacks.

Remote to Local (R2L): Remote to Local (R2L) is one genre of network or server attacks by the attackers. Attackers do not use this type of attacks to take control of a network or computer or device they use it to find the weaker point of a network or computer so that, they can easily take control of the flowing network or computer or device. Guess password, send mails are the different kind of R2L attack.

User to Root (U2R): U2L is a kind of attack, which is used by the outsiders or hackers or attackers to get access to the root of a server or network. Buffer overflow, SQL attack are the examples of this type attack.

Probing: Probing is one kind of attack which is used by the hackers to determine the weakness of a network. The hackers run some scan in a system to determine the weakness. Quick scan, instant scan, mscan are some types of probing.

In past years, we have seen cyberattacks on our banking, health, educational, home and government-related systems. So, the cybersecurity for systems is a must now. In this paper, we are going to develop a security system for home appliances. There are many researches on this topic [3] but most of them used machine learning approach hence provide us some algorithms [4, 1]. Deep learning is a part of machine learning but provides us more accurate result than machine learning. We have used deep learning algorithm for more efficient outcomes.

As known, Anomaly is something that differs from behaving normal or expected. Anomaly detection is the detection of abnormal behavior of networks and other devices. Machine learning is used to develop anomaly detection models and there are two algorithms which are shallow learning and deep learning [5]. In our case, deep learning shows eligibility to give better performance from raw data. We are using neural network which is composed of multiple nodes. Node takes input and performs operation which gives better output for anomaly detection. It also increases the performance of the system by taking more data.

In other paper [18], they have proposed a deep learning-based method which is related to smart city (social internet of things). They have used fog networks and their network is divided by Stochastic gradient descent (SGD). They did work on some attacks. But they did not work on distributed denial of services. They only analyzed the traffic to detect attacks. As traffic can be manipulated so it will not be efficient enough.

In our paper, we have focus on the distributed denial of services attack. There are other attacks which we have construed. We have built a model using self-taught technique (STL) that can successfully detect the attacks and anomaly behavior. We have built two algorithms which are rule-based one for detection of anomaly or attack and the other is for notifying the user about the anomaly behavior. Our results are more accurate than [18].

We have structured the paper as follows: Sect. 1 describes the introduction and different attacks of IoT. Section 2 describes literature review and Sect. 3 explained our proposed approach and the algorithm, respectively. Section 4 presents the methodology and Sect. 5 discusses the result analysis. Section 6 describes comparison with other models and related graphs. Finally, Sect. 7 concludes our work.

2 Literature Review

In this section, we have reviewed some other paper related to IoT, IDS, and their attacks. In [6], authors have discussed about various NIDS and their utilization. Mainly they have focused on signature-based and anomaly-based NIDS. They also talked about advantage, disadvantages of these two NIDS and attacks and threats it can tackle. In [4], authors have proposed a mechanism that detects malicious nodes in MANET. They have also prevented the black hole, flooding, and selective packet drop from attacker nodes and also improve the performance of the network. Authors in [7] have elaborately talked about IDS. Their contents are detection methodology, technology, virtual machines and short and clamAV. They also discussed about the requirements and designs. In [8], issues related to internet of things have been discussed. They have discussed about an IDS which is developed by using machine learning. This model is applied in IoT smart home with IoT devices. It can detect Denial of services attack and also network scanning problems. In [5], authors describe an evolution of deep learning approaches to IDS. They evaluate three models and show their outcomes and accuracy. In [3], authors proposed a model based on a deep

learning approach for intrusion detection which uses recurrent neural networks. They
have shown RNN-IDS model can improve the performance of IDS and identify the
intrusion type. In [1], the IoT networks, their attacks, and on the basis of application
survey is described. Cryptographic techniques are used to secure the data of IoT.
They have also described about lightweight algorithms. In [9], authors have shown a
deep learning-based approach to develop a NIDS. They have used Self-taught Learning (STL) which is a deep learning-based technique. They compared metrics include
the accuracy, precision, recall, and f-measure values of other related works. Also
in [10], the model of an intrusion detection system is implemented based on deep
learning, and Long Short-Term Memory (LSTM) technique which is applied to a
Recurrent Neural Network (RNN). They confirmed that it is effective for NIDS. In
[11], authors focused on intrusion detection systems (IDS) that support host-based
data sources that detect attacks on network. Discussed IDS types include various
locations and types of IDSs. In [12], authors have developed a model using RNN
intrusion detection that can detect seen and unseen threats of a network with low false
alarms. So far it has superiority over others developed system. In [13], for novelty
detection, authors created a new model which depends on sensor signals based on
Levene's test which probes the homogeneity of variances of samples which is taken
from the exact population and mixed with other autocorrelation features. They have
used fog computing platform. In [14], authors developed a technique for wireless
networks. They describe a novel construction methodology from two-dimensional
geometry by showing regular Hexagon's property. In [15], a deep learning approach
to cybersecurity to detect the attacks in social IoT is introduced. They have also compared their model with other works. In [2], authors proposed a cloud-centric vision
of Internet of Things for worldwide implementation. This enables technologies and
application domains that are likely to affect IoT research which is discussed in the
paper. They used "Aneka" for their implementation. In [16], authors have constructed
an IDS model using deep learning approach. They also used long short-term memory
architecture to RNN. In [17], authors explained the threats of the IoT and used an
Artificial Neural Network (ANN) for removing these threats. This paper focuses on
the categorization of normal and threat types on an IoT Network. In [18], authors
have proposed a deep learning-based model and showed an algorithm. They also
have compared it with deep model and shallow model.

3   Proposed Model

In the data pre-processing part, we change the data format from nominal to binary
which is referred to as one hot encoding (Fig. 1).

Then we normalize the data which organizes data in such a way that reduces
dependency and duplicity of data. In our case, normalization is scaling the range of
attributes to scale the range in [0, 1]. The formula is given below:

Fig. 1 Flow chart of our proposed model

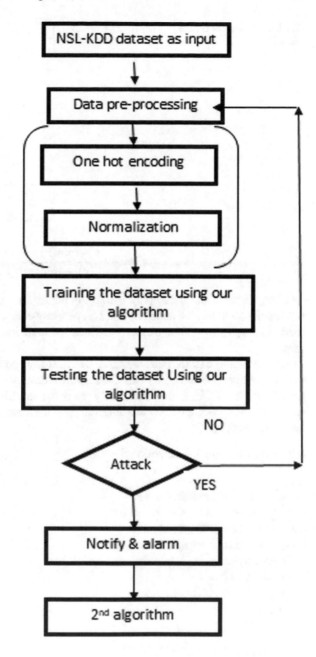

Fig. 2 Flow chart of testing
dataset

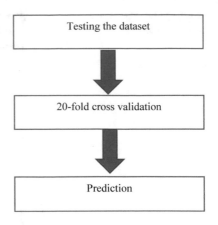

$$x' = \frac{x - \min(x)}{\max(x) - \min(x)} \qquad (1)$$

Then we train the dataset by using our algorithm and finally test our dataset whether it can detect the attack or not. If it gets any attack our second algorithm will work. In second algorithm, user can unblock the malicious IP address or just see the report.

In our testing dataset part, we have used 20-fold cross-validation to make our model more precise and better classification. We have done our whole deep learning model in Weka. Weka previously only used for machine learning models. After adding the new package deep learning for java in Weka, we get to know with some new features (Fig. 2).

Algorithm 1. Intrusion detection algorithm

Input:
C1=Destination_host_serror_rate
C2=Destination_host_srv_rerror_rate
C3=Destination_host_rerror_rate
C4=Destination_host_srv_serror_rate
S=Source bytes
D=destination bytes
Output:
O=output
Algorithm:

1. Input data (C1,C2,C3,C4,S,D)
2. If ((C1 or C2 or C3 or C4)=1)
3.   O=Block ip address &notify_user
4. Else
5. (s=0 && d=0)
6. O=Block ip address & notify_user

In our first algorithm, we have 6 inputs. If c1, c2, c3 or c4 has 1 value it will block ip_address. Otherwise, it will go to else check the source and destination bytes. If both of them are zero. It will block ip-address. Otherwise, the server is safe.

Algorithm 2. Notify algorithm

Input:
U=user input
Output:
p=output
Algorithm:

1. if (u=0)
2. O=Retrieve data & unblock IP address
3. else
4. O=seen_report

After notifying the user, our second algorithm will start its work. The user will see the report about the attack and also, he can unblock the malicious IP.

4 Methodology

In cross-validation data divided into k validation each part of the data is equal to each other. The Mean Squared Error (MSE) is calculated for each fold. The same procedure will be done for k validations. When the process completes, we get the estimates named k for the test error as MSE1, MSE2...MSEk. The cross-validation of k-fold is calculated by averaging these test errors as

$$cv = \left(\sum \text{MSE} \right) \tag{2}$$

Typically, k assumed as 10 in k-fold cross-validation, but we have used 20-fold cross validation.

Then, we calculate the performance of the algorithm in terms of precision, recall, and F-score. Precision of a model is calculated with the equation

$$\text{precision} = \frac{\text{true positive}}{\text{true positive} + \text{false positive}} \tag{3}$$

Recall of a model is measured with the equation

$$\text{Recal} = \frac{\text{true positive}}{\text{true positive} + \text{false positive}} \tag{4}$$

F-score takes into account both precision and recall of a model and is the harmonic mean of precision and recall. It is measured using equation:

Table 1 Accuracy (%) of our model

Class	Precision (%)	Recall (%)	F-measure (%)
Normal	99.9	99.9	99.9
Probe	99.04	99.9	99.23
DoS	99.01	99.82	99.67
DDoS	87.21	88.12	87.21
U2R	80.01	92.87	82.1
R2L	76.70	93.67	69.02

$$\text{precision} = \frac{\text{precision} * \text{recall}}{\text{precision} + \text{recall}} \qquad (4)$$

5 Result Analysis

Table 1 shows the outcome of our model. It shows that our proposed model was successful in detect anomaly and different type of attack flood (i.e. DoS, DDoS, R2L, U2L, probe).

6 Comparison

In this section of our paper, we are discussing our model's compatibility. We will be comparing our model with a deep learning model proposed by Diro and Chilamkurti [18]. As we can see in Table 1, our model has higher rate of recall, precision, and f-measure of normal class and attacks than the compared one (Table 2).

Table 2 Accuracy (%) of deep and shallow model

Models	Class	Precision (%)	Recall (%)	F1 measure (%)
Deep model	Normal	99.52	97.43	98.47
	DoS	97	99.5	98.22
	Probe	98.56	99	98.78
	R2L.U2R	71	91	80
Shallow model	Normal	99.35	95	97
	DoS	96.55	99	97.77
	Probe	87.44	99.48	93
	R2L.U2R	42	82.49	55.55

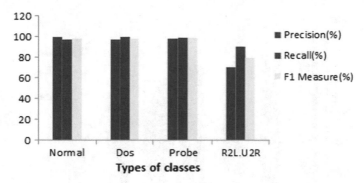

Fig. 3 Deep model

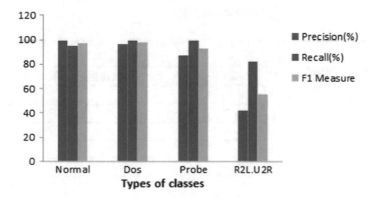

Fig. 4 Shallow model

The graphs shown below demonstrates the differences between our model and their model. As they did not do work for DDoS attack, we are assuming it as zero (Figs. 3, 4 and 5; Table 3).

This table has the values of classes which are shown in the charts of Figs. 6, 7 and 8, respectively.

7 Conclusion

In our paper, we have designed a system of intrusion detection system using deep learning method in which malicious attacks like DoS, DDoS, U2L, R2L, and probe have been detected and demolished. Our proposed model gives positive results against the mentioned attacks. Our model is able to update data and add new dataset. In today's world security towards our home appliance has become a challenge. We

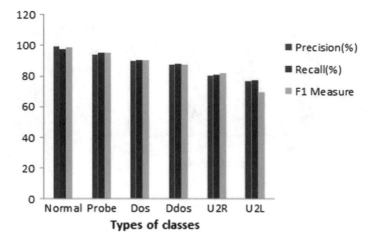

Fig. 5 Proposed model

Table 3 Class numbers in
y axis

Classes	Number in *y* axis
DoS	1
Probe	2
U2R	3
R2L	4
Normal	5

Fig. 6 Precision comparison

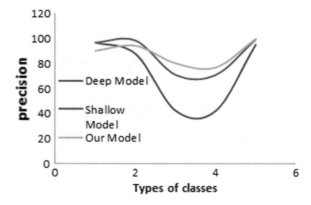

Fig. 7 Recall (%)
comparison

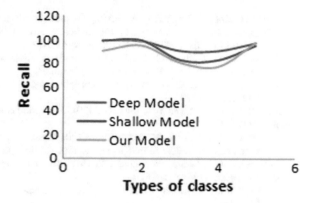

Fig. 8 F1 measure (%)
comparison

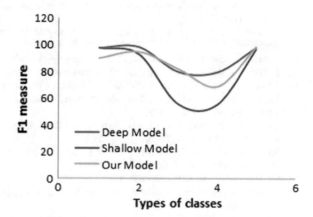

hope that our proposed model can successfully solve attack issues related to home devices. Since it can also add data to the system, it will be able to solve new attacks after training.

Acknowledgements This paper is supported by The Institute for Energy, Environment, Research and Development (IEERD), University of Asia Pacific (UAP).

References

1. Sharma A, Singh A (2017) Review on Internet of Things attacks and their countermeasure using lightweight cipher algorithms. Int J Adv Res Comput Sci 8(5), May–June 2017
2. Gubbi J, Buyya R, Marusic S, Palaniswami M (2013) Internet of Things (IoT): a vision, architectural elements, and future directions (30-01-2013)
3. Yin C, Zhu Y, Fei J, He X (2017) A deep learning approach for intrusion detection using recurrent neural networks (07-11-2017)

4. Singh O, Singh J, Singh R () An intelligent intrusion detection and prevention system for safeguard mobile adhoc networks against malicious nodes. Indian J Sci Technol 10(14). https://doi.org/10.17485/ijst/2017/v10i14/110833, April 2017
5. Lee B, Amaresh S, Green C, Engels D Comparative study of deep learning models for network intrusion detection. SMU Data Sci Rev 1(1), Article 8
6. Punia A, Vatsa VR (2017) Current trends and approaches of network intrusion detection system. JCSMC 6(6):266–270
7. Liao H-J, Lin C-HR, Lin Y-C, Tung K-Y (2012) Intrusion detection system: a comprehensive review (27-08-2012)
8. Mridha MF, Hamid MA, Asaduzzaman M (2017) Issues of Internet of Things (IoT) and an intrusion detection system for IoT using machine learning paradigm
9. Niyaz Q, Sun W, Javaid AY, Alam M (2016) A deep learning approach for network intrusion detection system (24-05-2016)
10. Ponkarthika M, Saraswathy VR (2018) Network intrusion detection using deep neural networks (28-04-2018)
11. Glass-Vanderlan TR, Iannacone MD (2018) A survey of intrusion detection systems leveraging host data (18-05-2018)
12. Elsherif A (2018) Automatic intrusion detection system using deep recurrent neural network paradigm
13. Raafat HM, Shamim Hossain M, Essa E, Elmougy S, Tolba AS, Muhammad G, Ghoneim A (2017) Fog intelligence for real-time IoT sensor data analytics (25-09-2017)
14. Abdul Hamid M, Abdullah-Al-Wadud M, Hassan MM, Almogren A, Alamri A, Kamal ARM, Mamun-Or-Rashid M (2017) A key distribution scheme for secure communication in acoustic sensor networks (10-06-2017)
15. Diro AA, Chilamkurti N (2017) Distributed attack detection scheme using deep learning approach for Internet of Things (23-07-2017)
16. Kim J, Kim J, Thu HLT, Kim H (2016) Long short term memory recurrent neural network classifier for intrusion detection (17-02-2016)
17. Hodo E, Bellekens X, Hamilton A, Dubouilh P-L, Iorkyase E, Tachtatzis C, Atkinson R (2016) Threat analysis of IoT networks using artificial neural network intrusion detection system (13-05-2016)
18. Diro AA, Chilamkurti N (2017) Distributed attack detection scheme using deep learning approach for Internet of Things (12-06-2017)

Study of Application Layer Protocol for Real-Time Monitoring and Maneuvering

Devesh Mishra, Ram Suchit Yadav, Krishna Kant Agrawal and Ali Abbas

Abstract The motes in the sensor network with having low processing capability and limited storage space combined to form a tiny smart sensor with the restricted power source cannot be directly used as things in the present network of an Internet of things. Millions of different devices placed anywhere at any time performing any job can be connected, and transfer of data can be established in the Internet of things. In a network of billions of devices connected together, security of data transmitted can only be provided if the protocol at the application layer is lightweight to be accessible by tiny motes. Application layer comes at the top of the TCP/IP heap. In this chapter, two major protocols of application layer HTTP and MQTT have been discussed and tested with a real-time operated hardware module with the capability to Internet connectivity. The arena for testing both the protocols consists of the transmission and reception of data in the network of tiny motes connected with the network of the Internet.

Keywords HTTP · MQTT · IoT · TCP/IP · ESP

1 Introduction

The nodes in the IoT speak with each other based on their IP address within the network. The present depositions of things in IoT are a part of megastructures within the volume of a network. It is expected to have around 1.6 ZB of data is expected till 2020 which leads technology towards big data analysis. The bulk ingredients of IoT consist of billions of minute, low-cost nodes as things which possess on board stored power supply, microcontroller unit for processing and transmission with a sensor as analog or digital input. Definite protocols are used for transmission of data between

D. Mishra (✉) · R. S. Yadav · A. Abbas
Department of Electronics & Communication, University of Allahabad, Prayagraj, India
e-mail: deveshbbs@gmail.com

K. K. Agrawal
Department of Computer Science & Engineering, ABES Institute of Technology (ABESIT), Campus 2, Ghaziabad, India

© Springer Nature Singapore Pte Ltd. 2020
A. Khanna et al. (eds.), *International Conference on Innovative Computing and Communications*, Advances in Intelligent Systems and Computing 1059, https://doi.org/10.1007/978-981-15-0324-5_38

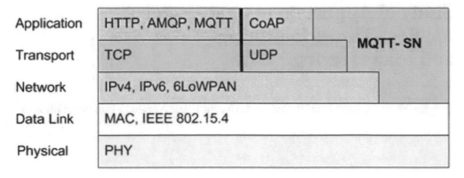

Application	HTTP, AMQP, MQTT	CoAP	MQTT- SN
Transport	TCP	UDP	
Network	IPv4, IPv6, 6LoWPAN		
Data Link	MAC, IEEE 802.15.4		
Physical	PHY		

Fig. 1 Protocols on IoT stack

heterogeneous nodes which varies from physical level to application level [1]. While talking about the transfer of data from one node to another node on the Internet of thing with a breakdown of coverage area, TCP/IP is used for the congregation. The facilitation for formatting of data is provided at the application layer with its challenges in the existing network architecture. Mainly there are three protocols at the application layer to control the transmission of data followed by Hyper Text Transfer Protocol (HTTP), Message Queue Telemetry Transmission (MQTT) and Constrained Application Protocol (CoAP) as shown in Fig. 1. This new-fangled group of interconnected things cannot be diluted in present-day Internet infrastructure. The CoAP provides an insubstantial substitute to HTTP which is the most commonly used application layer protocol available in the present scales of the Internet. The paradigm of this most common protocol is based on a request–response analogy. This is a lightweight protocol useful in current scenario for the available interconnected devices with a low number of devices in comparison with the devices estimated for machine-to-machine communication, or in IoT, it will flood the network and generate congestion without having capability to handle the burst. The HTTP is developed to the relocation of data, imageries, and diverse types of data formats. But in IoT, it only deals with the transmission of sensor data. For which another lightweight application layer protocol is being used named as MQTT, it is designed by IBM which does not include large overhead due to the application of publish–subscribe instead of request–response strategy. It is a more secure protocol because it uses broker mechanism for transmission of data in between client's authentication. MQTT also provides different levels of QoS which helps in handling the untrustworthy network. MQTT is very resourceful at multicasting [2].

2 MQTT

MQTT stands for message queue telemetry transport in which telemetry refers to tell in addition with metering or can be understood as a remote measurement. MQTT is a lightweight machine-to-machine communication protocol. It gets initiated from message queueing architecture of IBM. The IBM Company is inventor of MQTT and in fact is does not contain any queueing in it. It uses the publish/subscribe mechanism to transfer data from the device to the client or broker. MQTT is utilized by Facebook messenger to minimize battery usage [3].

2.1 Definition

MQTT is a lightweight protocol for transmission and reception of data from smart sensor motes in IoT network. In MQTT, all the concerned devices are connected to a single server which is known as a broker in MQTT context and forms its topology. The transmission of data is always initiated by the client to the broker. For data transmission, the client side uses a simple Internet connection, while the broker acts as a bridge between clients which is publically accessible on the Internet. It buffers the data collected from clients and forwards the message to one or multiple clients [4].

The message transmitted from a client needs to have a unique subject, and this allocation of the subject is defined as a topic. Queues are formulated based on such topics in the broker. Every client not only publishes a message to the topic's queues but also subscribes the message of a particular topic. MQTT does not standardize anything regarding the content of the message it solely depends upon the client and the broker to approve the format of message content [5].

2.2 Overview

It has been described above that MQTT is a lightweight publish/subscribe protocol between client and broker. The clients in the network can subscribe or can publish the content. Brokers as servers allocate messages between different clients to publish or subscribe as per requirement. In MQTT protocol, both clients and brokers can transmit and can receive which makes this asymmetric protocol. Due to short & simple header structure and ease of implementation makes this protocol very much suitable of IOT application which as limited energy constrained crafting this as a lightweight protocol [6] (Fig. 2).

During transfer of data between client & Broker, the first quality of Service is only responsible for transmitting the message to the from sender. The first one is there should be the transmission of data at least once. If it is not delivered re-transmission occurs at application layer. The second quality of service promises message delivery. It depends upon the acknowledgment of the message it is not received; then, it will retransmit the message. The last quality of service ensures that each message should be transmitted and received at once to avoid loss and repetitive messages [8].

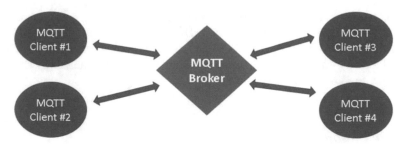

Fig. 2 MQTT operation [7]

3 HTTP

HTTP stands for Hyper Text Transfer protocol which is an application layer protocol which is used to deliver data to WWW. HTTP provides a standardized way of transmission and reception of data among computers and other devices connected in a network through the Internet. HTTP operates on a request–response pattern. When the client sends a request which is further processed and responded back by the server, both the server and the client can communicate any kind of text, document, or multimedia data with each other. The client and the server making communication with can recognize each other only at request–response duration; because of this, neither the client nor the browser can recollect information. The client sends the request to the server, while the server responds to the client [9].

3.1 HTTP Experimentation

Proper estimation of HTTP application layer protocol in the IoT environment has been performed in this experiment of controlling LED light through HTTP with the application of ESP 8266 NodeMCU as a hardware module and Arduino IDE version 1.8.5. And a micro-USB to USB cable for programming and power supply. NodeMCU has a system on chip I.C. i.e Tensilica L106 32-bit processor along with a Wi-Fi transceiver and 11 GPIO (General Purpose Input Output Pins) in GPIO pins 1–16 there are few missing pins also 6, 7, 8, 11. The NodeMCU has its own API and SDK. Our experiment has been conducted over a built-in LED on NodeMCU board connected at GPIO 13 or at D7. The onboard LED will get switch OFF at logic high (1) and get switch ON at logic low (0) [10] (Figs. 3 and 4).

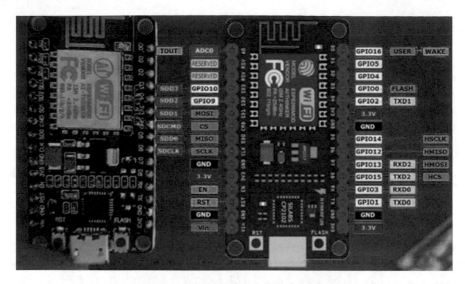

Fig. 3 Pin description of NodeMCU

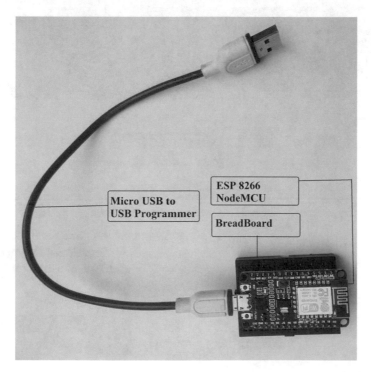

Fig. 4 NodeMCU module with programmer

4   Organization and Operation

NodeMCU is placed over a breadboard coupled with USB to micro-USB programmer connected with a laptop with Arduino IDE installed in it. The IoT module NodeMCU 12E version of ESP8266 gets connected with the Internet (access point) with the unique IP address provided to the module by the network displayed over the serial monitors of the Arduino IDE. Upon using HTTP, the URL is passed over the browser which displays the texts and documentation provided to the NodeMCU in HTML. The SSID and password of the current network need to be entered and server address is filled up into the browser when the enter key is pressed HTML content is displayed in response from the server [11] (Fig. 5).

4.1   Program Code Formulation

Start Server and Print IP Address over Serial monitor

```
server.begin();
  Serial.println("Server started");
  Serial.print("Use this URL to connect: ");
  Serial.print("http://");
  Serial.print(WiFi.localIP());
```

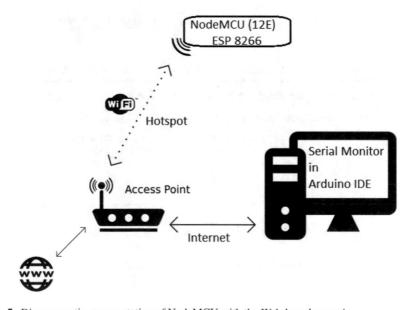

Fig. 5 Diagrammatic representation of NodeMCU with the Web-based operation

```
Serial.println("/");
IP = WiFi.localIP();
Serial.println(IP); }
```

HTML Response

```
client.println("HTTP/1.1 200 OK");
client.println("Content-Type: text/html");
client.println(""); // do not forget this one
client.println(" < !DOCTYPE HTML > ");
client.println(" < html > ");
client.println(" < h2 >  ip  =  ");
client.println(WiFi.localIP());
client.println(" </h2 > ");
client.print("Led pin is now: ");
if(value == HIGH) {
client.print("On"); } else {
client.print("Off");}
client.println(" < br > <br > ");
client.println(" < a href =\"http://");
client.println(WiFi.localIP());
client.println("/LED = ON\"\" > <button > Turn On  </button >
</a > ");
client.println("  < a href =\"/LED = OFF\"\" > <button >
Turn Off  </button > </a > <br /> ");  client.println(" </html
> ");
delay(1);
Serial.println("Client disonnected");
Serial.println("");}
```

4.2 Output Obtained

The above code has been written over Arduino IDE and got successfully compiled and uploaded upon the NodeMCU which is a regular process of transferring data over Wi-fi module. The IP address of the NodeMCU is obtained from the serial monitor, a real time output window within Arduino IDE with URL that has been provided in real serial monitor as shown in Fig. 6. The obtained URL has been put over the browser and a HTML page is obtained with switches to turn ON/OFF in built LED and clearly stating the state of switches. It has also been mentioned earlier that the built-in LED of NodeMCU operates on active low due to its architecture; therefore, pushing on the button over HTML page will switch OFF the led and vice versa.

It is shown in Fig. 7 the HTML page with URL obtained over serial monitor has the IP address of the NodeMCU. As per the functionality of the module, it has

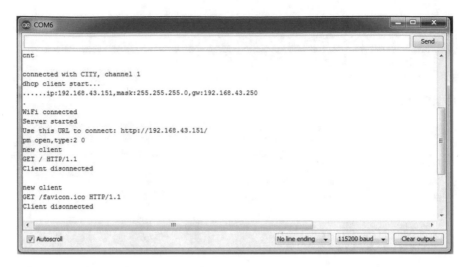

Fig. 6 IP and URL obtained on the serial monitor

Fig. 7 HTML page showing switch OFF status

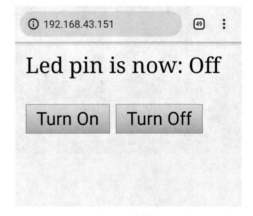

active low phenomenon; therefore, when the switch is turned OFF, LED will glow as shown in Fig. 8. Similarly in Figs. 9 and 10 which portraits the same HTML page with switch ON status and the inbuilt LED is put out. Figure 11 shows a graph which demonstrates the working of this switching action of the LED. It has instant peaks when the switch is pressed momentarily and comes down as voltage drops.

Fig. 8 Glowing inbuilt LED over NodeMCU

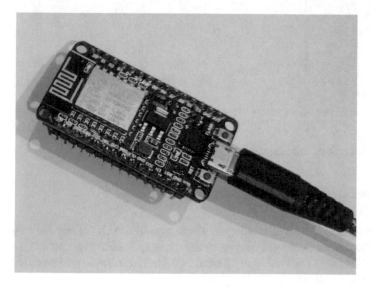

Fig. 9 Inbuilt LED is OFF

Fig. 10 HTML page
showing ON status

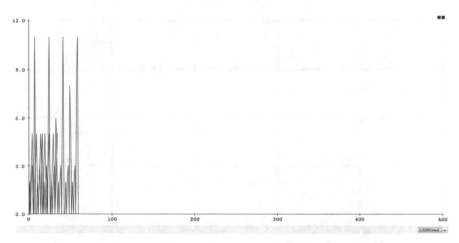

Fig. 11 Serial plotter showing turn ON/OFF status

5 Concluding Observation

HTTP is used in IoT for elementary applications only. With the increasing number of
IoT devices, it became necessary to switch the application layer protocol to MQTT
from HTTP. MQTT is used in PUSH and PULL requests leveraged by cloud industry
champions like IBM. MQTT is a data-centric protocol which follows publish–subscribe mechanism and provides small message time with 3 QoS, while HTTP is
data-centric with request/response system and large message size which makes this
system more complex (Table 1).

Table 1 Comparison sheet [12]

Parameters	MQTT	HTTP
Design	Data-centric	Document-centric
Pattern	Publish/subscribe	Request/response
Complexity	Simple	Complex
Message	Small message	Large message
Service level	3 quality messages	All messages are same
Libraries	30–100 KB	Depend upon application
Distribution	1 to 0, 1 to 1, 1 to n	1 to 1 only
Design	Data centric	Document centric
Pattern	Publish/subscribe	Request/response

References

1. Gopi Krishna1 P, Sreenivasa Ravi K (2017) Implementation of MQTT protocol on low resourced embedded network. Int J Pure Appl Math 116(6), 161–166. ISSN: 13118080
2. Gündoğan C, Kietzmann P, Lenders M, Petersen H, Schmidt TC, Wählisch M (2018) NDN, CoAP, and MQTT: a comparative measurement study in the IoT. In: Proceedings of ACM ICN. ACM
3. Diogo P, Lopes NV, Rei LP (2017) An ideal IoT solution for real-time web monitoring. © Springer Science+Business Media, New York
4. Papageorgas P, Piromalis D. Wireless sensor networking architecture of polytropon: an open source scalable platform for the smart grid. ISSN 1876-6102. ©2014 Elsevier Ltd. This is an open-access article under the CC BY-NC-ND license
5. Grokhotkov I (2017) ESP8266 Arduino core documentation. Release 2.4.0, 14 May 2017
6. Shaout A, Crispin B (2018) Using the MQTT protocol in real time for synchronizing IoT device state. Int Arab J Inf Technol 15(3A), Special Issue
7. https://websupport.ewon.biz/sites/default/files/aug-073-0-en-getting_started_with_mqtt_on_ewon_flexy.pdf
8. Mishra D, Agrawal KK, Yadav RS, Srivstava R (2018) Design of out pipe crawler for oil refinery based on analysis & classification of locomotion and adhesion techniques. In: IEEE UPCON, India, 2018, pp. 1–7. https://doi.org/10.1109/upcon.2018.8596988
9. Minerva R (2016) Towards a definition of Internet of Things.: IoT IEEE initiative chair, 8–10 June 2016
10. Kesarwani S, Mishra D, Srivastava A, Agrawal KK (2019) Design and development of automatic soil moisture monitoring with irrigation system for sustainable growth in agriculture (March 20, 2019). Available at SSRN: https://ssrn.com/abstract=3356380 or http://dx.doi.org/10.2139/ssrn.3356380
11. Mishra D, Agrawal KK, Yadav RS (2018) Contemporary status of machine prognostics in process industries based on oil and gas: a critical analysis. In: International conference on research in intelligent and computing in engineering (RICE) 2018. https://doi.org/10.1109/rice.2018.8627898
12. Manohar HL, Reuban Gnana Asi T (2018) Data consumption pattern of MQTT protocol for IoT applications. ©Springer Nature Singapore Pte Ltd.

Review of WSN and Its Quality of Service Parameters Using Nature-Inspired Algorithm

Cosmena Mahapatra, Ashish Payal and Meenu Chopra

Abstract Wireless sensor networks have become the focus of many recent researches focusing on topics like energy optimization, compression schemes, self-organizing network algorithms, routing protocols, quality of service management, security, energy harvesting and many more. The three major concern revolves around efficient energy usage, service quality and security management. To achieve success in these domains, it is imperative to have WSN optimization. Also, in applications like vehicular ad hoc networks and body area sensor networks, there can be conflict between these concerns and hence requires some trade-off. Because of these heavy energy expenditure and data processing needs, there is a requirement to review which WSN-based research has been done for optimizing the same through the use of bio-mimetic strategy-based optimization techniques which encompass numerous optimization algorithms. Thus, this paper reviews the various researches done to optimize quality of service parameters of wireless sensor networks and hence also aims to classify the challenges which are faced by these nature-inspired algorithms in WSN environment and thus identify future scope to consider a more comprehensive approach toward the subject matter.

Keywords Wireless sensor networks · Nature-inspired algorithms · Challenges · QoS · Optimization

C. Mahapatra (✉) · A. Payal
University School of Information, Communication & Technology, Guru Gobind Singh
Indraprastha University, New Delhi, India
e-mail: cosmenamahapatra1@gmail.com

A. Payal
e-mail: ashish@ipu.ac.in

C. Mahapatra · M. Chopra
Vivekananda Institute of Professional Studies, Guru Gobind Singh Indraprastha University,
New Delhi, India
e-mail: meenu.mehta.20@gmail.com

© Springer Nature Singapore Pte Ltd. 2020　　　　　　　　　　　　　　　　451
A. Khanna et al. (eds.), *International Conference on Innovative Computing
and Communications*, Advances in Intelligent Systems and Computing 1059,
https://doi.org/10.1007/978-981-15-0324-5_39

1   Introduction

Sensor networks are made up of miniature sensor nodes having components responsible for sensing, processing of data and communication. Numerous sensor nodes when deployed either inside or close the phenomenon create a sensor network. In order to allocate random deployment in areas which are either inaccessible or under emergencies, it becomes futile to predetermine or fix the positions of sensor nodes. All the same, it indicates the necessity to have self-organizing sensor network protocols and algorithms. Another notable feature is the collaborative property of sensor nodes. There is an onboard processor fitted on them that generates the capability to perform local computations in order to convert the raw data into partially processed data [1, 2].

Sensor network demonstrates the following characteristics—

1. Constrained resources of energy, bandwidth, memory size, power, etc.
2. Limited control after deployment due to no supervision, antagonistic environment, self-configurability, etc.
3. Massive distribution owing to numerous nodes which are rapidly deployed.
4. Specific needs like pattern of communication (data centric versus node centric), cost-efficient, robust design, effortless, etc.
5. Varied quality features like reliability, information quality and sensor fidelity.

In contrast to conventional networks, sensor networks show quite different features and quality measures. Due to excessive assimilation of sensor nodes and their precise application objectives, there is no strictly defined solution for all situations [3, 4].

Quality of service is a repetitive word in literature having many connotations and meanings given by different authors. Generally, QoS in WSN can be explained from two angles—application-specific and network specific. Application-specific takes into consideration such QoS factors that are particular to an application, like measuring sensor node, deploying it, area of coverage and number of active sensor nodes. On the other hand, network viewpoint implies efficient use of network resources like power usage and bandwidth to meet application needs of the supporting communication network.

Wireless sensor networks are being increasingly employed in many applications with advanced developments in the field of wireless networks and multifunctional sensors. A consistent or exact monitoring service is available for applications of all classes by the WSN. Hence, QoS can be a significant means to ensure that the specific requirements in regard to different classes of applications are fulfilled. Conventional QoS procedures employed in wired networks are not enough for WSNs due to restrictions like limited resources and active topology. A significant issue to deal with WSN is provision of QoS factor guarantee in applications running on real time. Hence, new mechanisms are needed in middleware which can not only sustain QoS over a long period but also adapt whenever QoS and the state of application alters. A trade-off between performance metrics like capacity of network or throughput, energy usage and delay in delivery of data while designing middleware could help offer QoS in WSN [5].

Quality of service has been defined as a group of required services that the network has to fulfill during transportation of a flow. 'Flow' in turn is the packet stream traveling from a source to a unicast or multicast destination, with a related QoS. Putting in other words, QoS can be described as a quantifiable service, typified by packet loss profitability, accessible bandwidth, end-to-end delay, etc. Through Service Level Agreement (SLA) with the network users, service providers are able to deliver such QoS. For instance, users can demand a minimum 2M bandwidth from the network in case of certain traffic flows [6].

This paper in Sect. 2 throws light on the various quality of service challenges that wireless sensor network designers face in order for their effective implementation. Section 3 gives a brief overview of various classes of nature-inspired algorithms focuses and Sect. 4 focuses on the literature survey which is comprised of only those research articles which have dealt with WSN QoS challenges effectively. These papers have been identified by the effectiveness of their algorithms, ease of implementation and verifiability. Challenges faced by programmers while implementing various nature-inspired algorithms and its future scope has been covered in Sect. 5 of this paper. The paper concludes in Sect. 6.

2 QoS Challenges in Wireless Sensor Networks

To ensure implementation of QoS, several quantifiable metrics, like bandwidth, loss probability, cost, delay, etc., have to be put into place to describe the service needs. As sensor network, unlike IP network, effortlessly supports many service types, different QoS can be available. Service types can vary from Constant Bit Rate (CBR), having the promise of bandwidth, delay and delay jitter to Unspecified Bit Rate (UBR) providing nearly no promises. Although the issues surrounding QoS in sensor networks are much similar to general wireless networks, they are still distinctive in characteristics [7]

A summary of factors to be considered for managing QoS traffic in WSN is as follows:

Restricted Bandwidth—In contrast to general wireless networks, it assumes more significance. A blend of real-time and non-real-time traffic often bursts through the sensor networks. Allocating the present bandwidth only to QoS traffic is not feasible. A compromise in quality of image or video would be required to adjust non-real-time traffic. Furthermore, to meet the QoS needs, many concurrent independent routes will be required to divide the traffic. Establishing unconnected routes for a particular flow could not only be intricate but tricky too in sensor networks because of varying factors. The governing factors could be restricted sources of energy and computation, as well as probable amplification of collisions among the sensors' communication.

Eliminating Redundancy—The data produced in the sensor networks create redundancy which, for unrestricted traffic, can be removed easily through the simple aggregation functions. Conversely, the process could become complex when data aggregation is performed for a QoS traffic. Excessive energy could be expended to evaluate images and video streams. Grouping of regulations that function at system and sensor level would be needed to allow the possibility of computationally aggregate QoS data. For instance, selective aggregation of imaging data can be done for traffic produced by unidirectional sensors due to relative similarity in the images. Besides, the magnitude of QoS traffic for a given time also affects the issue. To avoid overhead becoming dominant, in low traffic, doing away with data aggregation is more resourceful. Even though it becomes complex while performing data aggregation of images and video data, it can still be worthwhile from the vantage point of network performance, for a given data size and transmission frequency [8].

Compromise between energy and delay—Taking into account the direct mathematical relation between radio's transmission power and square of distance, multi-hop routing is generally utilized in a noisy or non-flat environment. Though there is much reduction in energy consumption during collection of data through multi-hop approach, magnification of the accumulated packet delay becomes an issue. Queuing of packet causes a delayed transmission; therefore, when the number of hops gets increased, there is not only a slowed packet delivery but also makes the analysis and managing of delayed traffic worse. Hence, there has to be a trade-off between energy efficiency and delivery requirements in the QoS routing of sensor data. Due to redundancy in data routing, which is inevitable while coping with generally elevated inaccuracy rate in wireless transmission, the compromise between energy expenditure and interrupted packet delivery can become more complex to handle [9].

Restriction in size of buffer—The processing and storage properties of sensor nodes are restricted. Intermediary relay nodes in the multi-hop routing play an important role in accumulating the inward bound packets before passing them on to the next hop. This buffering of multiple packets is more advantageous in WSN because of multiple reasons. As much energy is expended during the process of radio transition (between transmission and reception), it is more efficient to store up multiple packets before dispatching them further. Additionally, while aggregating and merging data, multiple packets are formed. Routing of QoS data through multiple hops generally needs lengthy sessions and thus buffering of much more data, especially in cases of delay. Constrained buffer size will proportionately increase the variation in delay that packets endure during transmission on either same or different routes. This will in turn make the medium access scheduling complicated and further QoS requirements difficult to achieve.

Sustaining numerous types of traffic—Many technical issues crop up when diverse set of sensors are included in routing. To illustrate, there are different sensors for measuring environmental metrics like temperature, humidity and pressure. Then, there are sensors to detect movement through acoustic signatures to capture the imagery or video of moving entity. Either the present sensors can be added with these functionalities, when needed, or the new sensors can be independently deployed in addition to the current ones. The reading produced from these assorted sensors is at varying rates, in accordance with the imposed constraints and several models of data delivery. Thus, having a heterogeneous setting creates difficulty in routing of data [10, 11].

Pattern of Traffic in WSN—In contrast to conventional networks, the WSNs show exclusive asymmetric pattern of traffic, most commonly because of the WSN function of collecting data. The sensor nodes constantly relay their information to the base station whereas there is only occasional transmission of control messages by the base station to the sensor nodes. Also, widespread difference in traffic pattern can be produced by different applications.

3   Brief Review of Natured-inspired Algorithms

Nature-inspired algorithms, as the name suggests, have their roots in the Mother Nature. Every natural process is optimized with an efficient strategy, taking into account the intricate inter-relations among organisms of all kinds and species. Be it rain, forestation, diversity, adaptations or ecological balance, all phenomenon are carried out with utmost optimization. The approach behind these phenomenons is simple yet delivers amazing results. Considering nature as the instructor, researchers are attempting to imitate nature in their technology. These algorithms can illustrate and solve complicated relationships through easy preconditions without intensive knowledge of the search space. Nature has evolved through many problems since years and therefore can guide the computer science field with ways to handle its own problems. Hence, problem-solving is possible through simple mapping between nature and technology. Thus, these bio-inspired computing has found variety of applications in the zones of networking, security, robotics, data mining, production engineering and so on [12, 13].

The basic biological evolutionary mechanism and its processes became the foundation of evolutionary algorithms. These are the computational techniques inspired by Darwin's theory of evolution characterized by natural selection (modifications in each subsequent descent) in species giving them suitable features and adaptability in accordance with environment. Evolutionary process involves modifications and proliferation of genetic material that is responsible for certain suitable traits. This very process is simplified and put into investigative computational systems to achieve an effective adaptive system. Fields of application for Evolutionary Computation are Population Ecology and Genetics, Developmental Biology and Co-evolutionary Biology [14].

Physical processes or systems like music, metallurgy, cultural evolution, avalanches, etc. have inspired to form physical algorithms. They are usually related to the domains of Metaheuristics and Computational Intelligence and do not categorically belong to any of the actual bio-inspired techniques like swarm, neural or evolution. With a combination of local (neighborhood) and global search methods, these are stochastic optimization techniques. external optimization, simulated annealing, harmony search, cultural algorithm, etc. are some of the examples [15].

Algorithms that represent a problem or investigate a problem space through a probabilistic style of feasible solutions are called Probabilistic Algorithms. Certain algorithms belonging to Metaheuristics and Computational Intelligence could come under probabilistic category with their implicit style of using the tools of probability in solving a problem [16]. Probability algorithm is an expansion of the Evolutionary Computation domain which considers forming model out of all possible solutions. A number of iterations are run which alternatively create possible solution in the problem space from the probabilistic model while decreasing the set of probable solutions in the probabilistic model. Examples include Bayesian optimization algorithm, population-based incremental learning, etc.

Another notable nature-inspired algorithm is based on collective intelligence found in the collaborative activities of a number of homogeneous organisms in the ecology. It is called swarm intelligence and can be observed in the flock of birds, ant colonies, fish schools, etc. It is a dispersed phenomenon in nature characterized by decentralization and self-organization, used to solve daily problems of hunting food, evading the predator or relocating/migrating. All the participating organisms have the necessary information or the information itself might be available in the environment through chemicals (pheromones for ants) or behavior (dancing of bees, closeness in birds/fishes). The concept can be divided into two broad categories of ant colony optimization and particle swarm optimization. Former explores the occurrence of stigmergy and searching activities of ants, while latter considers the behaviors of herding, schooling and flocking in nature. These algorithms, much like evolutionary, are adaptive in nature and are used for searching and optimization [17, 18].

Taking cue from the natural biological immune system that works primarily to protect the host organism from many threats of pathogens (viruses/bacteria/parasite/pollen) and toxic substances, Immune Algorithms have been devised for computational purposes. The external threats (pathogens) are first detected by the immune system so that they can be expelled by the body through elimination. It requires the basic process of identifying host cells (belonging to own body) from the external ones (harmful to body). The pathogens have to go through a multilayered architecture comprising of sub-systems of coordinated activities of many organs and systems in body [19]. Body's innate and acquired immunity come into play to protect the cells. Examples of such algorithms are Artificial Immune Recognition System and Clonal Selection Algorithm, etc.

4 Literature Review

Saleem et al. [20] proposed a model and conclusions of an ant-based autonomous routing algorithm, using NS2 simulator, for the sensor network. The algorithm is a self-organized routing mechanism based on delay, energy and velocity. The considered parameters helped in enhancing the complete data output. It also demonstrated efficient energy consumption in real-time traffic. In addition to it, there was no deadlock problem owing to algorithm's capacity to avert permanent loops in the running networks. Overall, the bio-inspired algorithm improved sensor network needs like energy expenditure, degree of success and time.

The authors in this study [21] conducted a large-scale survey on the routing protocols for WSN in accordance with the swarm intelligence. Authors extensively discussed the principles of swarm intelligence and how it is applied for routing. The surveyed protocols are classified according to a novel taxonomy that has been proposed in the paper. There is critical scrutiny regarding the field status and questions raised on methodology of presenting and empirically evaluating the algorithms. Future scope pointed toward two directions—one to complement simulation-based studies with mathematical models and other to carry forward the simulation toward real hardware implementation.

Authors of this study [22] performed a comparative analysis between bio-inspired and geometric field system through two algorithms. One of the algorithms is already established called as compass routing protocol or DIR, while the other one, based on bees' communications, is proposed by the authors and named as Bee routing protocol or Bee RP.

The results of the comparison showed that the two algorithms, despite having different tools, were complementary in action and used same directional principle toward destination in routing. Future study proposed studying Bee RP in the direction of asymmetric links and addition to heterogeneous networks.

In this paper, authors [23] presented a new mathematical model for quality of service route determination to let a sensor establish an optimal path for minimum expenditure of resources within the mentioned QoS restrictions. The mathematical model employed Lagrangian relaxation mixed-integer programming technique for outlining important parameters and suitable objective functions to control the adaptive QoS restricted route discovery process. LINGO mathematical programming language was used to simulate the performance through a trade-off between energy consumption and QoS needs. The suggested approach demonstrated reduced energy use and end-to-end delays through optimized sharing of resources in the sensor network in relation to existing routing algorithms.

In another study [24], the authors reviewed and brought to notice many area detection and coverage crisis in sensor network. Authors sorted out various scenarios to apply sensor node movement in enhancing the network coverage. Bio-inspired evolutionary algorithms like PSO and genetic are included for the study. Issues and agenda of controlling sensor node coverage are discussed too. A number of coverage control algorithms have been presented based on coverage and sensing models. More than one sensor node is used in the new sensing model along with new node mobility control. There are added features which increase network lifetime and reliability of data. Future study is directed toward reducing energy consumption in coverage with holes and in mobile nodes through development of sensor node movement strategies.

This paper [25] presented a routing protocol for WSN called RED-ACO (reliability, energy and delay—ant colony optimization) to create a route supporting QoS with multiple constraints, improve reliability of the network and decrease energy expenditure as well as delay through ant colony algorithm. Both user and network perspectives with defined limitations are taken into consideration. The search range for the ant to choose its subsequent hop-sensor is restricted. Also the rules for select-ing the next hop-sensor are defined. Authors proved that in comparison with the current routing protocols, the presented algorithm performed better in many scenar-ios on end-to-end delay and packet delivery ratio. Algorithm's performance improved over the period of time in terms of end-to-end delay.

This paper [26] evaluated performance analysis for a number of optimization techniques viz. Genetic, Particle Swarm, Bacterial Foraging and Hybrid approach of GA-PSO. To optimize the QoS of WSN, authors used many algorithms and protocols. Initially, the authors adopted GA-PSO and BFO separately on WSN set up followed by hybrid approach of GA-PSO. Comparisons are used through routing. There was better performance reported for GA-PSO, namely on factors like network lifetime, end-to-end delay, bit error rate, throughput and routing overhead. The focus was efficient packet transferring with as little error rate as possible so that there are reduced chances of node failure and increased network lifetime. Future scope is suggested through using any other optimization algorithm for deploying the same issue of load balancing. Implementing hybridization of other optimization techniques has been suggested to improve WSNs.

Authors [27] proposed a hybrid bio-inspired algorithm for congestion control in large-scale wireless sensor networks. Initially, a scheme based on C-LV model is used to decrease the congestion in WSN and for maintaining fairness among sensor nodes. Further, PSO algorithm has been utilized too for improving the C-LV scheme through optimizing the parameters for reducing end-to-end delay and helping the system adapt in accordance with change in number of sensor nodes. Results verified the positive effects of using this scheme in terms of reduced congestion and maintained fairness for each sensor node. Moreover, with any number of sensor nodes, the transmission by all sensor nodes was at an optimal rate. There was rapid adaptation by the PSO algorithm positively affecting the system adaptability too. Authors proposed future work of evaluating the method on actual industrial set up and also investigating congestion control with other bio-inspired algorithms (Table 1).

Table 1 Summarized table of nature-inspired algorithms and their effect on WSN QoS

S. No	Nature-inspired algorithms	Parameters
1	Ant-based autonomous routing algorithm	Efficient routing
2	Bee routing protocol	Efficient routing
3	Lagrangian relaxation mixed-integer programming technique	Energy QoS and route Determination
4	Particle swarm optimization and genetic algorithm	Increase in node coverage
5	Reliability, energy and delay—ant colony optimization	Improve reliability and decrease energy and delay
6	Hybrid GA-PSO	Increased network lifetime, minimum error rate, maximum throughput

5   Challenges Faced during Implementation of Nature-Inspired Algorithms in WSN and Its Future Scope

Although through extensive literature survey, it can be safely summarized that nature-inspired algorithms have been able to effectively deal with various QoS related optimization of WSN parameters yet the journey has not been without its challenges and limitations [28, 29]. This section tries to put together the said points and thus form a basis for its further study [30–34].

No one algorithm fits all: It is unfortunate that no one algorithm has proven to solve all the QoS related optimization problems. For example, while one algorithm might be successful in tackling network coverage, but it may not effectively reduce the network error rate. Much research is left to come upon this one single powerful algorithm.

Limited processing speed and space: Wireless sensor networks have limited processing power which may prove to be a hindrance for those nature-inspired algorithms which have major processing and memory requirement.

Major Success based on Simulations: Majorly a nature-inspired algorithm has been proven to work successfully in WSN through simulations and statistical analysis. Real-world implementation of the same is still not seen in more then 90% of the cases.

Competition from Geometric and Mathematical Algorithms: Nature-inspired algorithms are facing major competition from geometrical and other mathematical-based algorithms which are also in the race to reach optimization in majorly all fields.

Randomness in execution: Nature-inspired algorithms are stochastic in nature, which means the solution given by them is guaranteed only to be nearer to optimal solutions. This is too a major challenge faced by nature-inspired algorithms.

The above-given points actually form the basis of future scope of the study for this paper. All points covered under challenges faced by nature-inspired algorithms have valid basis and may be taken up for further study.

6   Conclusions

In current times, WSNs have assumed a significant role of efficient data selection and their delivery. QoS factors like efficient energy usage are critical for WSN which are already restricted by small powered batteries. The intricacy and dependence of corporate operations on WSNs need utilization of effective routing methods and protocols to deliver the promise of network connectivity and energy efficiency because researchers are increasingly turning to nature-inspired algorithms to optimize the

WSN QoS operations. However, as seen in the study done by us, implementation of nature-inspired algorithm in WSN environment does not come without its own quota of road blocks and brickbats. Much research is left to be done in order to find one algorithm which can have efficient results in optimizing the various QoS parameters of WSN so that the same can be put into use in real-world applications.

Acknowledgements Foremost, we would like to express our sincere gratitude to the Doctoral Research Committee of Guru Gobind Singh Indraprastha University (GGSIPU), New Delhi: Prof. Dr. Pravin Chandra, Prof. Dr. C. S. Rai, Prof. Dr. Amrinder Kaur, Prof. Dr. B. V. R. Reddy, Prof. Dr. Amit Prakash, Dr. Anurag Jain and Dr. Rahul Johari, for their encouragement, insightful comments, and hard questions. Our sincere thanks also goes to Vivekananda Institute of Professional Studies (VIPS) Respected Chairman sir Shri. Dr. S. C. Vats, Vice Chairman Shri. Suneet Vats, Shri. Vineet Vats and rest of VIPS management. We also wish to thank Dr. Ravish Saggar, Dr. Shubra Saggar, Dr. Ashish Khanna, Dr. Deepak Gupta for their constant motivation and moral support. This would be a right place to thank Dr. Tania Mahapatra for giving her invaluable time to help make this manuscript comprehendible. Last but not the least; we would like to thank our parents and God for supporting us throughout our life.

References

1. Gante D, Aslan M (2014) Smart wireless sensor network management based on software-defined networking. In: 27th biennial symposium on Communications (QBSC), pp 71–75
2. Moon Y, Lee J, Park S (2008) Node management and implementation. In: 10th international conference on advanced communication technology, IEEE, pp 1738–9445
3. Mahmood M, Seah W, Welch I (2015) Reliability in wireless sensor networks: a survey and challenges ahead. Comput Netw 79:166–187
4. Choi Y, Hong YG (2016) Study on coupling of software-defined networking and wireless sensor networks. In: 8th international conference on ubiquitous and future networks (ICUFN), pp 900–902
5. Yamsanwar Y, Sutar S (2017) Performance analysis of wireless sensor networks for QoS. In: 2017 international conference on science, pp 120–123
6. Ezdiani S, Acharyya IS, Sivakumar S, Al-Anbuky A (2017) Wireless sensor network softwarization: towards WSN adaptive QoS. IEEE Internet Things J
7. Wang J (2014) Trust-based QoS routing algorithm for wireless sensor net-works. In: Wang H (ed) 26th Chinese control and decision conference (CCDC)
8. Akkaya K, Younis M (2005) Energy-aware and QoS routing in wireless sensor networks. Springer Cluster Comput J 8:179–188
9. Torregozal JP (2006) Quality of service aware route discovery for wireless sensor networks. In: ICE-ICASE, Busan, pp 2153–2157
10. Ning GZ, Song Q, Zhang L (2016) A qos-oriented high-efficiency resource allocation scheme in wireless multimedia sensor networks. IEEE Sens J
11. Ehsan S, Hamdaoui B (2012) A survey on energy-efficient routing techniques with QoS assurances for wireless multimedia sensor networks. IEEE Commun Surv Tutorials 14(2):265–278
12. Kapur R (2015) Review on nature inspired algorithms in cloud computing. In: Proc. of IEEE international conference on computing communication and automation (ICCCA-2015) School of Computer Science and Engineering Galgotias University Uttar Pradesh India, pp 15–16
13. Yang XS (2014) Nature-inspired optimization algorithms. Elsevier, Amsterdam
14. Vikhar PA (2016) Evolutionary algorithms: a critical review and its future prospects. In: 2016 proc. of international conference on global trends in signal processing, information computing and communication, Dec, pp 22–24

15. Paul A, Paul AM, Ghosh K (2016) Communication con-verging towards adaptive intelligence: a survey in 2nd international conference on computational intelligence and networks (CINE), pp 3–12
16. Fei Z, Li B Shaoshi Yang, Chengwen Xing, Hongbin Chen, Lajos Hanzo (2017) A survey of multi-objective optimization in wireless sensor networks: metrics algorithms and open problems. Commun Surv Tu-torials IEEE 19(1):550–586
17. Birattari DM (2010) Ant colony optimization. In: encyclopedia of machine learning, Springer
18. Zhang WGW (2010) A comprehensive routing protocol in wireless sensor net-work based on ant colony algorithm. In: 2010 second international conference on networks security wireless communications and trusted computing (NSWCTC), vol 1, pp 41–44
19. Yang F (2010) An improved artificial immune algorithm. In: 6th international conference on natural computation, pp 2837–2841
20. Saleem K, Fisal N, Hafizah S, Kamilah S, Rashid R (2009) Ant based self-organized routing protocol for wireless sensor networks. Int J Commun Networks Inf Secur 1(2):42–46
21. Saleem, Caro, Farooq (2011) Swarm intelligence based routing protocol for wireless sensor networks: survey and future directions,. Inf Sci 181(20):4597–4624
22. Aksa Benmohammed (2012) A comparison between geometric and bio-inspired algorithms for solving routing problem in wireless sensor net- work. Int J Networks Commun 2(3):27–32
23. Hasan MZ, Wan TC (2013) Optimized quality of service for real-time wireless sensor networks using a partitioning multipath routing approach. J Comput Networks Commun 2013:1–18
24. Abbasi M, Latiff Chizari H (2014) Bioinspired evolutionary algorithm based for improving network coverage in wireless sensor networks. Sci World J 2014:1–8
25. Deepa Visalakshi K (2016) A self-organized QoS-aware RED-ACO routing protocol for wireless sensor networks. Middle-East J Sci Res 24:224–230
26. Kaur M, Sohi (2018) Comparative analysis of bio inspired optimization techniques. In wireless sensor networks with GA-PSO Approach,. Indian J Sci Technol 11(4):1–10
27. Royyan, Ramli, Lee, Kim (2018) Bio-inspired scheme for congestion control in wireless sensor networks. In: 2018 14th IEEE international workshop on factory communication systems (WFCS)
28. Saunhita S and Mini M (2018), Optimized relay nodes positioning to achieve full connectivity in wireless sensor networks, Springer Science, Wirel Pers Commun, Springer, pp-1521–2540
29. Yadav, Saneh and Phogat, Manu. (2017). Study of nature inspired algorithms. Int J Comput Trends Technol. 49. 100–105. https://doi.org/10.14445/22312803/ijctt-v49p115
30. DengYi Zhang and WenHai Li (2011). Research on quality of service in wireless sensor networks. In Software Engineering and Service Science
31. Oreku GS (2013) Reliability in WSN for security: mathematical approach. In: 2013 international conference on computer applications technology (ICCAT), pp 1–6
32. Polastre J, Hill J, Culler D (2004) Versatile low power media access for wireless sensor networks. In: Proc. ACM SenSys' 04, pp 95–107
33. Ruan, Zhu, Chew (2017) Energy-aware approaches for energy harvesting pow-ered wireless sensor nodes. IEEE Sens J 17(7):2165–2173
34. Zhi-jie Han (2014) A novel wireless sensor networks structure based on the SDN. Int J Distrib Sens Networks 2014(7):1–7. Article ID 874047
35. Distefano S (2012) Evaluating reliability of WSN with sleep/wake-up interfering nodes. Int J Syst Sci 44:10–1793

Adjacency Cloud-Oriented Storage Overlay Topology Using Self-organizing M-Way Tree

Ajay Kumar and Seema Bawa

Abstract This paper proposes a self-organizing and scalable storage approach named adjacency COS overlay topology (ACOT). This topology is based on balanced multi-way tree organization. ACOT removes the traditional static and centralized storage management and adopts the dynamic and highly scalable data accessibility for the cloud-oriented storage system. A Hadoop test bed has been created for experimental review and result in analysis on various aspects like dynamicity, scalability, effectiveness, and bandwidth evaluation of proposed topology. Our experiments have proven that the ACOT topology offers the dynamic and highly scalable for cloud-oriented storage systems.

Keywords Cloud computing · Self-organizing · Cloud storage · M-way tree

1 Introduction

The grid and cloud systems are concerning the design of self-organizing storage center so that the underlying resources can respond dynamically to improve demand levels in a more flexible manner [8]. It may also be used to provide data protection and security when workload moves during migrations, provisioning or enhancing performance. Many enterprises are switching their business to cloud infrastructures can improve service-level agreement (SLA), quality-of-service, reduce costs and make more efficient consumption of energy [10]. Most of the dynamic infrastructure is service-oriented and market-oriented and supports the end users in the highly responsive way [9]. The dynamic infrastructure also provides high availability requirements and long-term business continuity to facilitate cloud computing. The service-oriented architecture (SOA) [5] is one of the most popular concepts that represents discrete

A. Kumar (✉) · S. Bawa
Thapar Institute of Engineering and Technology, Patiala, Punjab, India
e-mail: akumarphd@thapar.edu

S. Bawa
e-mail: seema@thapar.edu

© Springer Nature Singapore Pte Ltd. 2020
A. Khanna et al. (eds.), *International Conference on Innovative Computing and Communications*, Advances in Intelligent Systems and Computing 1059,
https://doi.org/10.1007/978-981-15-0324-5_40

463

functionality, adheres to a published contract, and is cohesive, independent, and standard-based. Most of the services are published in the public registries and discover these services to the web services using standardized protocols. The composition in SOA refers to invoking services in a particular sequence to represent complex business process flow. Cloud computing intends to provide everything as a service, and thus the notion of duty in cloud computing is much broader than that of SOA. SOA aims to make the cloud computing platform flexible, extensible, and reusable. The Hadoop is one of the popular platforms to achieve self-organizing capability for large-scale storage systems [14]. The rest of this research paper is organized as follows: Sect. 2 explores the brief literature review on self-organizing storage systems in various computing eras. The proposed ACOT topology has been explained in Sect. 3. An experiment details and results analysis are presented in Sect. 5. Section 6 followed by conclusion and future scopes.

2 Related Work

This section discusses significant research work done so far regarding self-organizing storage model in the various computing eras. Comparatively, analysis has also been highlighted in this section. Self-organizing is an autonomous concept designed to make more efficient resource invocation and runtime estimation in network environments. It describes an ability of the storage system to organize their components into a working framework without need of extra controlling and external help. Generally, the self-organizing system has the ability to reconstruction or add and remove parts of the system without need for human intervention. The recent related works have been reviewed in this research paper which is mentioned in Table 1. The proposed model, objectives, approaches, and their computational complexity have been systematically exhibited in Table 1.

3 The Proposed Adjacency COS Overlay Topology (ACOT)

We have proposed and designed a flexible and self-organizing overlay topology named adjacency COS overlay topology (ACOT) for large-scale storage systems. ACOT is a novel approach in cloud environments, introducing a balanced multi-way COS overlay topology. It removes centralized approaches for traditional static storage management and adopts the dynamic and scalable data accessibility in the cloud environments. It is very challenging to meet the requirements requested by different applications among thousands of COS nodes. Traditionally, all the cloud based storage systems are divided into different domains each domain has one COS representative (CR) node that cooperates with other domains. Here, each COS

Table 1 Summary of self-organizing storage model

Prevailing model	Objectives	Approaches	Environment	Time complexity	References
S-TREE	Illustrate the data clustering and compression	Supervised learning	Cluster computing	$O(\log N)$	[3]
Adaptive resource allocation model	Adaptive resource allocation in dynamic environment	Ant-based control algorithm	Distributed computing	$O(\log N^2)$	[2]
Safari	Scalable ad hoc network routing and integration services	Hybrid routing algorithm	Ad hoc network	$O(\log N)$	[6]
SSA_G	Resource optimization in dynamic environment and improve the system performance	Election algorithm	Cloud computing	$O(N^{1/2})$	[12]
SOS cloud	Reduce the service-level agreement (SLA) violation cost	Bio-inspired algorithm	Cloud computing	–	[4]
SOPSys	Reduce the network join and discovery cost	M-ways balanced tree	Decentralized peer-to-peer network	$O(\log N)$	[7]
CCPS	PC cluster based storage system	M/M/1 queuing network model	Cloud computing	–	[17]
ecoCloud	Virtual machine (VM) consolidation assignment and migration of VMs	Bernoulli trials	Cloud computing	–	[13]
AETOS	Improve the performance and enhance runtime estimation strategies	Tree overlay topologies	Distributed computing	$O(\log N)$	[15]
Cloud cell	Improve the storage capacity and mobility performance	Fuzzy-logic-based decision framework	Cloud computing	$O(2^{N+1})$	[1]
Proactive dynamic secure data scheme (P2DS)	Constraints data access and deal with dynamic threats	Attribute-based semantic access control algorithm	Cloud computing	–	[16]
Named data networking (NDN)	To store and merge data interests for vehicles routing networks. Reduce the high communication cost for packet forwarding	Multi-way tree	Ad-hoc network	–	[11]

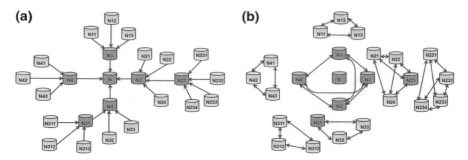

Fig. 1 Two tiers representation of ACOT. **a** Shows the first tier using balanced multi-way tree structure. Each node has unique id. **b** Represents the different storage domain formation. Each domain has one representative node that coordinates to respective leaf nodes

node has their identical functionalities and capabilities that can join or leave the domain dynamically. The ACOT is a new and more flexible topological designed for self-organizing cloud-oriented storage system. It is based on balanced multi-ways (m-way) tree organization. In general, m-way tree has been conveniently defined with small value of m. But in our case, m is usually very large. Each node corresponds to a physical COS device and m represents the maximum number of COS devices that can be linked with a cloud domain.

The value of m could be maximized as per the demand of storage service in the particular cloud domain. An ACOT is divided into two layers—(1) interior and (2) exterior. The interior layer focuses on dynamic, scalable, and self-organized virtual COS nodes and also solves the routing problems. The exterior layer forms with various physical cloud domains and each domain has one COS representative (CR) node that cooperates with other domains with the help of domain. The architectural view of ACOT is shown in Fig. 1. The first level of ACOT is balanced m-way tree organization that manipulates COS nodes very efficiently. The root COS node will be most responsible because it makes the list of most trusted COS node that are publically available through cloud domain. Whenever a new COS is interested to join the ACOT first time, it has to make a request to root node in order to receive the list of active domains and establish the connection with nearest domain and find the appropriate location id with the help of Algorithm 1. Algorithm 2 focuses on joining the COS node and maintains the balancing factor of the storage topology. ACOT is based on trusted and stable COS nodes. Any COS node can leave the domain without any restriction. The technical specification of joining and leaving process is defined in Algorithms 2 and 3, respectively. The list of notations are shown in Table 2.

Algorithm 1 : Routing and naming mechanism

Input: S_{id}, D_{id}, msg
Output: $depth, NC_{id}$

1: Procedure $Path(S_{id}, D_{id}, msg)$
2: **if** $D_{id} = NC_{id}$ **then**
3:    $Set(msg)$
4: **end if**
5: $depth = LCP(NC_{id}, D_{id})$ ▷ find the depth with the help of node prefix id
6: **if** $depth = NC_{level}$ **then**
7:    $N_{id} = D_{id}[NC_{level}]$
8:    **if** $N_{id} = 0$ **then**
9:       $STA(NC_{id}, D_{id}, S_{id}, msg)$ ▷ destination node is an ancestor node
10:    **else**
11:       $STN(NC_{id}, N_{id}, S_{id}, msg)$
12:    **end if**
13: **else if** $depth > NC_{level}$ **then**
14:    $STA(NC_{id}, D_{id}, S_{id}, msg)$ ▷ destination node is neither children nor adjacent node
15: **else**
16:    $Cid = D_{id}[NC_{level} + 1]$
17:    $STD(NC_{id}, D_{id}, S_{id}, msg)$ ▷ destination node is either descendant or child node
18: **end if**
19: End Procedure

Algorithm 2 : Joining new COS node

Input: NC_{id}, msg
Output: NC_{ip}

1: Procedure $Join(NC_{ip}, msg)$
2: **if** $C_{accept} = 1$ **then**
3:    **if** $C_{remain} > 0$ **then**
4:       $NC_{id} = C_{max} - C_{remain}$
5:       $STD(NC_{id}, D_{id}, S_{id}, msg)$ ▷ move on leaf node
6:    **else if then**$C_{remain} = 1$
7:       $RJN(NC_{ip}, N_{ip})$ ▷ join neighbor as ancestor
8:    **else**
9:       $RJA(NC_{ip}, A_{ip})$ ▷ join as ancestor node
10:    **end if**
11: **else**
12:    **if** $L_{level} = 1$ **then**
13:       $RJA(NC_{ip}, A_{ip})$ ▷ join as ancestor
14:    **else if** $L_{level} > NC_{level}$ **then**
15:       $RJD(NC_{ip}, C_{ip})$ ▷ join as child node
16:    **else**
17:       $RJN(NC_{ip}, N_{ip})$ ▷ join neighbor as child
18:    **end if**
19: **end if**
20: End Procedure

Algorithm 3 : Leaving COS node

Input: S_{id}, D_{id}, msg
Output: D_{id}, CN_{id}

1: Procedure $Leave(S_{id}, D_{id}, msg)$
2: $depth = LCP(S_{id}, D_{id})$
3: **if** $depth! = 0$ **then**   ▷ not a leaf node
4:    $Delete(D_{id})$
5:    $NC_{ID} = C_{MAX} - depth$ ▷ create new COS id
6:    $STN(NC_{id}, D_{id}, S_{id}, msg)$
7:    $Update()$ ▷ update tree after deletion
8: **else**
9:    $Delete(S_{id})$
10:    $STA(NC_{id}, D_{id}, S_{id}, msg)$
11:    $Update()$
12: **end if**
13: End Procedure

4 Complexity Analysis

In this section, we have been analyzed the complexity of proposed algorithms afore-mentioned in Sect. 3. There are some important properties of ACOT that has been defined as follows:

Definition 1 We assume ACOT has order of m and height h. Let N number of COS nodes are connected with ACOT. Then the following properties have been satisfied:

1. $2(\frac{m}{2})^{h-1} - 1 <= N <= m^h - 1$
2. $log_m(N + 1) <= h <= log_{\frac{m}{2}}(\frac{n+1}{2}) + 1$.

Definition 2 The root has at most m sub-domain, and the structure of each sub-domain has also at most m sub-domain. Each COS node has upto $m - 1$ pairs and m children.

Definition 3 For the height h, the maximum number of COS node is $m^h - 1$ where m denotes the branching factors of the tree.

Definition 4 To handle overflow at physical domain D_i, perform middle entry $m = \lceil \frac{N+1}{2} \rceil$ in D_i directly goes to the ancestor in right place.

Definition 5 Each node has a degree a or b. The number of nodes is between $a^h - 1$ and $b^h - 1$ where h is the height. Hence, the height between $\lceil log_b(N + 1) \rceil$ and $\lceil log_a(N + 1) \rceil$ for n nodes.

ACOT takes constant time, i.e., $O(1)$ on each level to perform various operations. There are $O(logN)$ different levels. Therefore, insertion takes $O(log_k N)$ time, where k indicates the branching factors and N specifies the all active nodes in a complete domain. In the worst case, one leaf node is placed to another sub-domain

Table 2 List of notations used

Notation	Description
COS	Cloud-Oriented Storage
msg	Message
$depth$	Depth of the particular COS node
id	It is unique and self-explainer id of COS node such as N232
S_{id}	Source COS node ID
D_{id}	Destination COS node ID
N_{id}	Neighbor COS node ID
C_{id}	Child COS node ID
NC_{id}	It is a new COS node ID that want to join the network
NC_{level}	Current level of new COS node
L_{level}	Last level (leaf COS node level)
C_{max}	Maximum number of children COS node
C_{remain}	Shows the status of remaining child COS slots
NC_{ip}	IP address of new COS node
N_{ip}	IP address of neighbor COS node
A_{ip}	IP address of ancestor COS node
C_{ip}	IP address of child COS node
C_{accept}	Shows the status of either child node is accepted or not
$LCP(NC_{id}, D_{id})$	It is a method to find the Longest Common Prefix path between new COS node to destination COS node
$STA(NC_{id}, D_{id}, S_{id}, msg)$	Send to ancestor COS node
$STN(NC_{id}, D_{id}, S_{id}, msg)$	Send to neighbor COS node
$STD(NC_{id}, D_{id}, S_{id}, msg)$	Send to descendant/child COS node
$RJA(NC_{ip}, A_{ip})$	Redirect to join ancestor COS node
$RJD(NC_{ip}, C_{ip})$	Redirect to join descendant COS node
$RJN(NC_{ip}, N_{ip})$	Redirect to join neighbor COS node

and $2\log_k N - 1$ hopes will traverse. Figure 2 shows the number of nodes that are actively involved in routing process. Basically, it shows the capacity of proposed ACOT in order to their height and branching factors.

5 Implementation Details and Results Analysis

This section deploys the test benchmark to demonstrate the complexity and workload design efficiency of the proposed ACOT topology. The Hadoop framework is the most popular open-source developed by Apache and Yahoo! foundation. In this paper, we use Hadoop 2.7 running on JDK 1.8.0. We have changed some parameter's

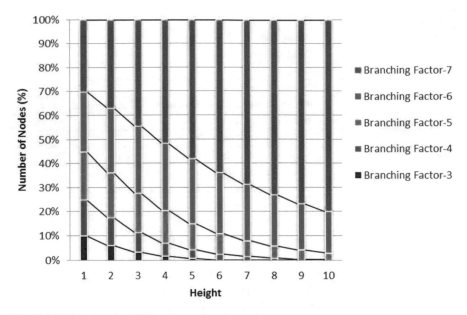

Fig. 2 Number of active COS nodes involves in routing process

Table 3 Configuration parameters setting

Parameter	Value	Remarks
Data block size	256 MB	
Maximum heap size	512 MB	Individual task executor JVM
Maximum heap size	1.5 GB	DataNode JVMs
Enable "Rack awareness"	True	It increases the fault tolerance
Enable block access token	True	For kind secure operation
Enable cross-origin support (CORS)	True	It supports all kind of web services
Methods allowed	GET and POST	
Enable ACL	True	Support for access control list
Enable log aggregation	True	
Buffer size per node	512 MB	
Cluster size	1000 nodes	

value in Hadoop system which is shown in Table 3 and rests of are default configuration settings. Besides, we configured the system to run two Map instances and a single Reduce instance simultaneously on each COS node. All input and output data of each benchmarks have been stored in the Hadoop Distributed File System (HDFC). It uses a central job tracker and master HDFS demon to coordinate node activities. The central job tracker and master HDFS daemon of Hadoop system coordinate the COS node activities and to ensure that these daemons do not affect the performance

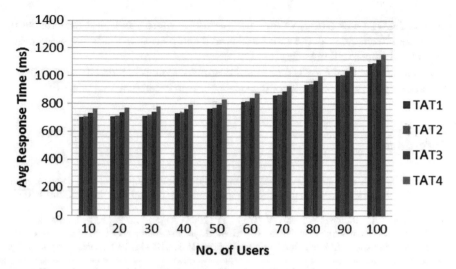

Fig. 3 Average response time (in four turnaround time TAT = 50 ms, TAT = 65 ms, TAT = 80 ms, and TAT = 100 Ams)

of other active COS node. We found the 512 MB buffer size per COS node is more effective for our experiment. Testing benchmarks have been deployed in Hadoop on the cluster of 1000 COS nodes. We have used three different networks: (1) Research Lab-II, (2) software engineering laboratory, and (3) creative computing laboratory of Thapar Institute. All the computers are connected through wireless access points. We have executed three times of each experiment, and results are reported the average of all three trials. First, we executed on single node to each experiment and then separately performed at different cluster sizes to measure the amount of data processed and seems also how much time taken to perform the experiment. Experimental results have been stored in the distributed file system. The time of reply is the critical metrics to analyze the performance of storage management in cloud environments. It specifies the time between user request submission and application delivered. Figure 3 illustrates the average response time evaluate the performance in four turnaround times (TAT = 50 ms, TAT = 65ms, TAT = 80 ms, and TAT = 100 ms). We have also measured bandwidth in two cases load vs. throughput and delay vs. throughput. In both cases, we got better results as compared to static allocation.

6 Conclusion and Future Work

In this paper, a more flexible, self-organizing, and highly scalable topology named ACOT has been designed and implemented in the Hadoop environment. This topology is based on balanced multi-way tree organization. ACOT removes the centralized and static approaches for cloud-oriented storage systems in cloud environments. In general, ACOT takes $O(\log_k N)$ time for insertion, where k indicates the branching factors and N specifies all the active nodes in a complete domain. In the worst case, $2\log_k N - 1$ hopes will traverse.

Future scopes of the ACOT in cloud computing include dynamic and scalable data access policy, storage integration services from heterogeneous platforms.

References

1. Alsedairy T, Qi Y, Imran A, Imran MA, Evans B (2015) Self organising cloud cells: a resource efficient network densification strategy. Trans Emerging Telecommun Technol 26(8):1096–1107
2. Andrzejak A, Graupner S, Kotov V, Trinks H (2002) Algorithms for self-organization and adaptive service placement in dynamic distributed systems. Internet Systems and Storage Laboratory Publisher: HP Laboratories. https://doi.org/10.1.1.10.431. Web Access: http://hpl.hp.com
3. Campos MM, Carpenter GA (2001) S-tree: self-organizing trees for data clustering and online vector quantization. Neural Netw 14(4–5):505–525
4. Caprarescu BA, Calcavecchia NM, Di Nitto E, Dubois DJ (2010) Sos cloud: self-organizing services in the cloud. In: International conference on bio-inspired models of network, information, and computing systems. Springer, pp 48–55
5. Deka GC (2014) A survey of cloud database systems. IT Prof 16(2):50–57
6. Du S, Khan A, PalChaudhuri S, Post A, Saha AK, Druschel P, Johnson DB, Riedi R (2008) Safari: a self-organizing, hierarchical architecture for scalable ad hoc networking. Ad Hoc Netw 6(4):485–507
7. Istin MD, Visan A, Pop F, Cristea V (2010) Sopsys: self-organizing decentralized peer-to-peer system based on well balanced multi-way trees. In: 2010 International conference on P2P, parallel, grid, cloud and internet computing (3PGCIC). IEEE, pp 369–374
8. Kumar A, Bawa S (2012) Distributed and big data storage management in grid computing. arXiv preprint arXiv:1207.2867 (2012)
9. Kumar A, Bawa S (2012) Virtualization of large-scale data storage system to achieve dynamicity and scalability in grid computing. In: Advances in computer science, engineering & applications. Springer, pp 323–331
10. Kumar A, Bawa S Generalized ant colony optimizer: swarm-based meta-heuristic algorithm for cloud services execution. Computing pp. 1–24 (2018)
11. Li X, Wang S, Wu W, Chen X, Xiao B (2018) Interest tree based information dissemination via vehicular named data networking. In: 2018 27th International conference on computer communication and networks (ICCCN). IEEE, pp 1–9
12. Lin W, Qi D (2010) Research on resource self-organizing model for cloud computing. In: 2010 International conference on internet technology and applications. IEEE, pp 1–5
13. Mastroianni C, Meo M, Papuzzo G (2013) Probabilistic consolidation of virtual machines in self-organizing cloud data centers. IEEE Trans Cloud Comput 1(2):215–228
14. Pavlo A, Paulson E, Rasin A, Abadi DJ, DeWitt DJ, Madden S, Stonebraker M (2009) A comparison of approaches to large-scale data analysis. In: Proceedings of the 2009 ACM SIGMOD international conference on management of data. ACM, pp 165–178
15. Pournaras E, Warnier M, Brazier FM (2014) Adaptive self-organization in distributed tree topologies. Int J Distrib Syst Technolog (IJDST) 5(3):24–57
16. Qiu M, Gai K, Thuraisingham B, Tao L, Zhao H (2018) Proactive user-centric secure data scheme using attribute-based semantic access controls for mobile clouds in financial industry. Future Gener Comput Syst 80:421–429
17. Yee TT, Naing TT (2011) Pc-cluster based storage system architecture for cloud storage. Int J Cloud Comput: Serv Archit (IJCCSA) 1(3):117–128

A Capacitated Facility Allocation Approach Based on Residue for Constrained Regions

Monika Mangla and Deepak Garg

Abstract Allocation of services has observed widespread applications in real life. Therefore, it has gained comprehensive interest of researchers in location modeling. In this paper, the authors aim to allocate p capacitated facilities to n demand nodes in constrained demand plane. The allocation is aimed to minimize the total transportation cost. The authors consider continuous demand plane, which is constrained by the presence of barriers. Here, the authors present a residue-based capacitated facilities allocation (RBCFA) approach for allocation of capacitated facilities. Finally, an illustration of RBCFA is presented in order to demonstrate its execution. Authors also perform tests to validate the solution, and the tests yield that suggested approach outperforms traditional approach of allocation. It is observed that although the achievement by RBCFA is not significant for few resources, achievement is significant as the number of resources rises.

Keywords Facility location · Barriers · Visibility graph · Constrained demand plane · Convex hull

1 Introduction

Location modeling has been an interesting area of research since its inception due to its widespread application in various domains [1, 2]. Moreover, location and allocation problems have been studied inseparably in the literature [3, 4]. Location problem optimally locates p facilities in the demand plane consisting of m existing facilities $EX = \{Ex_1, Ex_2 \ldots Ex_m\}$ such that the weighted distance from p facilities to m

M. Mangla (✉)
CSED, Thapar University, Patiala, India
e-mail: manglamona@gmail.com

Faculty of CSED, LTCoE, Navi Mumbai, India

D. Garg
CSED, Bennett University, Greater Noida, India

© Springer Nature Singapore Pte Ltd. 2020
A. Khanna et al. (eds.), *International Conference on Innovative Computing and Communications*, Advances in Intelligent Systems and Computing 1059,
https://doi.org/10.1007/978-981-15-0324-5_41

473

existing facilities is minimized [5]. This location problem can be generalized by Fermat–Weber problem (FWP) [6]. Moreover, few exclusive variants of the FWP are available in the literature to simulate different realistic models [7, 8]. Among these variants, capacitated Weber problem (CFWP) is a popular variant of FWP in which each facility has the capacity constraint [8, 9]. CFWP considers the allocation where involved resources have capacity constraint and thus limits the number of demand nodes it handles. For uncapacitated facilities, a demand node is always allocated to its nearest facility. However, capacity constraint of resource restricts the allocation of every demand nodes to its nearest resource.

In the literature, CFWP has been further categorized as single-source capacitated FWP (SSCFWP) and multiple-source capacitated FWP (MSCFWP) [10, 11]. Similarly, there exists a variant of FWP where demand plane is constrained known as constrained FWP. In constrained FWP, location of facilities is restricted to permissible area barring constraints. Various categories of constraints are forbidden region, congested region, or barriers [12]. In forbidden region, location of resources is prohibited but passing through this region is permitted. On the contrary, congested region represents the area where location of facility is not permitted but it is permitted to pass through at an additional cost. Barriers represent the region which restricts location of facility as well as travel through it. The presence of a hill or river in the demand plane may be considered as an example of a barrier for locating a company warehouse. Constraints in CFWP further intricate the allocation problem and thus require a specialized approach. In this paper, the authors consider SSCFWP with barriers in the demand plane.

The paper has been organized as follows: In Sect. 2, the authors present the problem formulation for SSCFWP in constrained region. Related work in the related field is discussed in Sect. 3. The authors proposed a novel residue-based approach in Sect. 4. Section 5 focuses on the computational simulation and results. Finally, Sect. 6 concludes the paper and provides some future research avenues.

2 Problem Formulation

In order to mathematically formalize the SSCFFWP, the authors define following parameters:

n	number of demand point s
m	number of facilities
req_i	requirement of demand node i
$EX = \{Ex_1, Ex_2 \ldots Ex_m\}$	existing resources in demand plane
C_j	capacity of facility j
$dloc_i$	location of demand point i in 2d plane
$floc_j$	location of facility j in 2d plane
$z_{ij} = 1$	if demand point i is connected to facility j, 0 otherwise

The objective function for SSCFWP is defined as follows:

$$\min g = \sum_{i=1}^{n} \sum_{j=1}^{m} d\left(\text{dloc}_i, \text{floc}_j\right) z_{ij} \tag{2.1}$$

$$\sum_{j=1}^{m} z_{ij} = 1 \text{ for } i \in \{1, 2, \ldots n\} \tag{2.2}$$

$$\sum_{i=1}^{n} \text{req}_i z_{ij} \leq C_j \text{ for } j \in \{1, 2, \ldots m\} \tag{2.3}$$

$$z_{ij} \in \{0, 1\} \text{ for } i \in \{1 \ldots n\}, j \in \{1 \ldots m\} \tag{2.4}$$

Equation (2.1) minimizes the total weighted distance from demand points to allocated resources. Equation (2.2) stipulates that each demand point obtains its required demand. Constraint (2.3) represents that demand served by a facility to all connected demand nodes never exceeds its capacity. Subsequently, constraint (2.4) maintains that each demand node gets its entire demand from one facility only, an absolutely necessary condition for SSCFWP.

In order to represent constrained SSCFWP, the objective function in Eq. (2.1) is rewritten as follows:

$$\min g = \sum_{i=1}^{n} \sum_{j=1}^{m} d_\beta\left(\text{dloc}_i, \text{floc}_j\right) z_{ij} \tag{2.5}$$

where d_β represents the barrier distance which does not passes through a barrier.

3 Related Work

Inclusion of barrier in the location models was incepted by Katz and Cooper [13] by considering one circular barrier. The authors in [14] aimed to locate facilities in constrained plane so as to minimize average Manhattan travel distance. This work is further extended in [15] by considering p-median model in the presence of barriers and convex forbidden regions. For this problem, the authors in [15] establish that optimal solution for this problem is restricted to finite discrete locations.

The line of research continues in [16, 17] for p-center location model with a fixed line barrier in the demand plane. This line barrier is used to simulate the presence of rivers, borders, etc., in real life. Solution of line barrier problem is based on the number of passages (bridges, flyovers, etc.) through that barrier. The authors in [16] presented a mixed integer nonlinear programming model (MINLP) for this constrained allocation problem having line barrier with 1 connection. Further, the authors in [18] considered line barrier with k connections and presented a MINLP to address this problem. The proposed approach is validated by a numerical example [18]. Thereafter, the authors in [19] extended the line by considering probabilistic line barrier and presented a MINLP model. In probabilistic model, starting point of the line barrier is uniformly distributed.

Thereafter, the authors in [6] devised a procedure for locating a facility in the presence of convex polygonal forbidden regions. The suggested procedure iteratively solves a number of unconstrained problems and finally terminates at local optimum [6]. Later, it is extended to include convex polygonal barriers [20, 21]. However, barriers can be convex or non-convex, but non-convex polygons are excluded based on the findings in [22, 23]. In [23], the author proved that non-convex polygon can be easily replaced by its convex hull (CH) without influencing its solution. Consequently, in this paper, the authors consider convex barriers to address allocation of resources to demand nodes in the constrained demand plane. The suggested approach can also be implemented for the non-convex barriers as a result of work in [22, 23].

4 Proposed Approach for Constrained SSCFWP

This section begins by presenting the basic terms and definitions. Thereafter, the authors introduce some basic properties upon which the proposed approach, i.e., RBCFA, is based.

Let β_i represent a convex polygonal barrier. Here, $\beta = \bigcup_{i=1}^{n} \beta_i$ represents the entire barrier region in the demand plane. Now, location of resource is limited to a feasible region $F = R^2 \setminus \text{int}(\beta)$. **Barrier distance** $d_\beta(X, Y)$ is defined as the minimum distance between X and Y so that it does not pass through interior of barrier β as

$$d_\beta(X, Y) = \min\{\text{path}(X, Y) : \quad \text{path}(X, Y) \in F\} \tag{4.1}$$

Here, Eq. (4.1) represents that barrier distance passes through only F (feasible region) without entering the barrier region β. From the definition of $d_\beta(X, Y)$, it is clear that

$$d_\beta(X, Y) \geq d(X, Y) \tag{4.2}$$

where $d(X, Y)$ represents the Euclidean distance, the smallest distance between X and Y. From Eq. (4.2), it is evident that barrier distance is lower bounded by Euclidean distance. The barrier distance has been illustrated in Fig. 1.

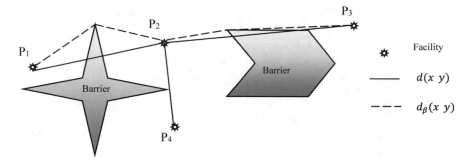

Fig. 1 Illustration of barrier distance and barrier touching property

In Fig. 1, polygons and stars represent barriers and existing facilities, respectively. Dashed line and solid line represent the barrier distance d_β and Euclidean distance, respectively. In Fig. 1, p_2 and p_4 are directly visible. It implies $d_\beta(p_2, p_4) = d(p_2, p_4)$. However, p_1 and p_2 are not directly visible and therefore $d(p_1, p_2)$ passes through β. In such cases, barrier distance $d_\beta(p_1, p_2) > d(p_1, p_2)$.

Moreover, it is evident from Fig. 1 that if p_i and p_j are not visible, $d_\beta(p_i, p_j)$ has breaking points only at the vertices of barrier polygons. This property is called **Barrier Touching Property (BTP)** [24]. BTP can be noticed in Fig. 1 for pair (p_1, p_2) as it has turning point at vertex of the barrier. Thus, according to BTP, $d_\beta(x, y)|x, y \in F$ consists of line segments with turning points only at the vertices of barrier polygons if x is not visible to y. This BTP leads to generation of visibility graphs which are utilized in the suggested approach RBCFA. The following subsection explains the visibility graph and its employment in RBCFA.

4.1 Visibility Graph

Visibility graph (VG) can be defined as undirected graph $VG = \langle V, E \rangle$ with $V = EX \cup B$ where EX is set of existing facilities and B is set of vertices of $CH(\beta)$. Here, $CH(\beta)$ represents convex hull (CH) of barrier region β. Here, Fig. 2 represents the VG for facilities and barriers of Fig. 1. CH of the barrier region is shown by red solid line in Fig. 2. Here, $arc(x, y) \in E|x, y \in V$ if x is directly visible to y, i.e., Euclidean path between x and y doesn't pass through $int(\beta)$. This arc, known as visibility arc, is represented using dashed line in Fig. 2.

However, if x is not directly visible to y, $arc(x, y) \notin E(VG)$. In such case, $d_\beta(x, y)$ starts from x, is followed by some vertices of B, and finally terminates at y. Thus, the path $path_\beta(x, y)$ can be shown as follows:

$$path_\beta(x, y) = x \rightarrow \text{some vertices of } B \rightarrow y \tag{4.3}$$

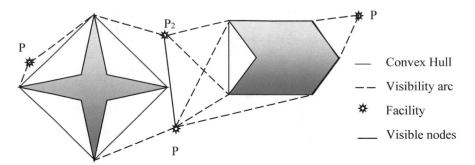

Fig. 2 Visibility graph

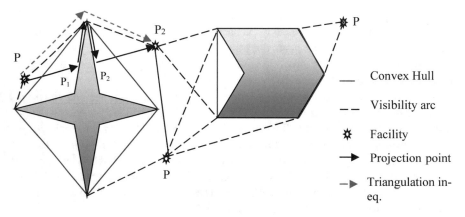

Fig. 3 Illustration of projection point

Now, in $\text{path}_\beta(x, y)$, the first vertex of B which appears after x is called the projection point of x toward y, i.e., p_{xy} [25]. Based on this definition of projection point, $d_\beta(x, y)$ is reformulated as:

$$d_\beta(x, y) = d\left(x, p_{xy}\right) + d_{\partial \text{CH}(B)}\left(p_{xy}, p_{yx}\right) + d\left(p_{yx}, y\right) \qquad (4.4)$$

Equation (4.4) holds if $\text{arc}(x, y) \notin E(\text{VG})$. However, if $\text{arc}(x, y) \in E(\text{VG})$ holds true, $d_\beta(x, y)$ is expressed as follows:

$$d_\beta(x, y) = d(x, y) \qquad (4.5)$$

Equation (4.5) represents the scenario of visible nodes. However, Eq. (4.4) represents the scenario where x and y are not visible. In such case, $d_\beta(x, y)$ comprises $d\left(x, p_{xy}\right)$, shortest distance from p_{xy} to p_{yx} along boundary of $\text{CH}(B)$, and finally $d\left(p_{yx}, y\right)$. This scenario is illustrated for p_1 and p_2 in Fig. 3 using black solid arrows

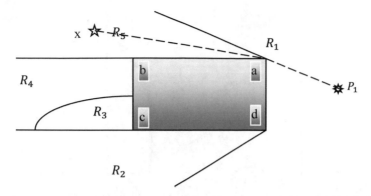

Fig. 4 Partitioning of demand plane with respect to p_1

as p_1 and p_2 are not directly visible. This is further optimized with help of triangular inequalities so as to generate the path represented using green dashed arrows in Fig. 3. Thus, VG can be employed to determine barrier distance $d_\beta(x, y)$ whether x and y are visible or not visible.

Thereafter, the demand plane is partitioned with respect to projection point [26]. For instance, Fig. 4 represents five partitions of demand plane with respect to p_1 in the presence of rectangular barrier.

In such partitions, every point from a partition has same projection point for p_1. Thus, based on these partitions, $d_\beta(x, p_1)|\forall x \in R_2$ always passes through projection point of x for p_1, i.e., a. Consequently, Eq. (4.4) can be rewritten as:

$$d_\beta(x, p_1) = d(x, a) + d_{\partial CH(B)}(a, a) + d(a, p_1) \forall x \in R_2 \qquad (4.6)$$

Similar portioning is obtained for all points. All these partitions with respect to different points are superimposed to obtain the overall partitions of the demand plane. Figure 5 represents the overall partitions for Fig. 4. In Fig. 5, each partition is represented by different colors. From Fig. 5, it is clear that there exist various types of relations among these partitions as: (i) sharing a single vertex, (ii) sharing an edge, and (iii) partitions connected along ∂B(barrier). Whenever two regions share a vertex represented by cbv, Eq. (4.7) holds true

$$\text{cbv}(r_i, r_j) \in E(G) \text{ if } \exists x, \ ys \cdot tr_{ix} = r_{jy} \qquad (4.7)$$

Similarly if two regions share an edge represented by cbe, Eq. (4.8) holds true

$$\text{cbe}(r_i, r_j) \in E(G) \text{ if } \exists x, \ y \, s \cdot t \, (r_{ix}, r_{ix+1}) = (r_{jy}, r_{jy+1}) \qquad (4.8)$$

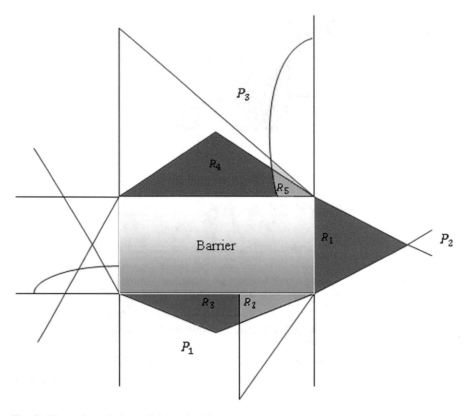

Fig. 5 Illustration of planar divisions in [1]

Any pair of regions for which (4.7) and (4.8) do not hold true, these regions are considered to be connected along boundary of barrier. The term for representing this type of relation is $cb\partial B$.

$$cb\partial B(r_i, \ r_j) \in E(G) \text{ if } \exists \ x, \ ys \cdot tr_{ix} \xrightarrow{\partial B} r_{jy} \tag{4.9}$$

Here in Eq. (4.9), $r_{ix} \xrightarrow{\partial B} r_{jy}$ represents that there exists a path along boundary of barrier ∂B from r_{ix} to r_{jy}. This planar partitioning and their association are utilized to optimize the distance functions in RBCFA presented in the following subsection.

4.2 RBCFA for Constrained Region

In CFWP, each resource j has some maximum capacity C_j. In addition, each resource j also has some balance (residue) capacity represented by res_j. Thus, it is evident that

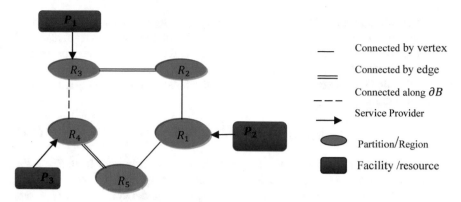

Fig. 6 Graphical representation for planar division

res_j is upper bounded by C_j, i.e., $\text{res}_j \leq C_j$. Initially, when no demand point is allocated to resource j, we have $\text{res}_j = C_j$. Thereafter, whenever a demand node is allocated to resource j, residue capacity res_j decreases. Demand nodes can be allocated to resource j as long as $\text{res}_j > 0$. In RBCFA, authors propose a novel metric *residue ratio*, ratio of residue capacity to its total capacity. This *residue ratio* is expressed as:

$$R_j = \text{res}_j / C_j | j \in \{1, 2 \ldots m\}, R_j \in [0, 1] \qquad (4.10)$$

In RBCFA, the authors integrate residue ratio and VG during allocation. The motive behind including R_j is to ensure that future demand also gets fair probability to be served by its nearest resource, thus optimizing the total transportation cost. In the beginning, usage of residue ratio may not result in significant optimization. But, it becomes noticeable as $\sum_{j=1}^{m} C_j$ approaches $\sum_{i=1}^{n} \text{req}_i$.

The fundamental step of RBCFA is transformation of planar partitions (Fig. 5) into an undirected graph $G = V|E|V = P \cup R$. This graph G considers various kinds of relations among partitions as discussed above in Eqs. (4.7), (4.7), and (4.9). These relations among the partitions are represented using distinct edge notations. The planar division for Fig. 5 is graphically transformed to equivalent graph in Fig. 6 as follows:

Figure 6 denotes partitions and facilities by blue ovals and red rectangles, respectively. In Fig. 6, regions r_i and r_j are connected using solid line if $\text{cbv}(r_i, r_j)$. Similarly $\text{cbe}(r_i, r_j)$ and $\text{cb}\partial B(r_i, r_j)$ are represented using double solid line and dashed line, respectively.

Furthermore, an edge $(p_j, r_k) \in E(G)$ (a edge among a facility and region) indicates that all demand nodes in r_k have p_j as their nearest facility. Consequently, each demand node in r_k is allocated to p_j unless its demand exceeds residue capacity res_j. We use the term service provider (SP) to represent this scenario. In such a case facility p_j coincides with a vertex of r_k. Now, based on above description, the following equation is derived:

$$(p_j, r_k) \in E(G) \text{ if } \exists x \ s \cdot t \ r_{kx} = p_j \tag{4.11}$$

This condition implies $SP(r_k) = p_j$ and $p_j \in r_k$. Thus, it is clear from Fig. 6 that p_1, p_2, and p_3 are service provider for r_3, r_1, and r_4, respectively. Therefore, each demand point from r_1 is preferably allocated to p_2, while demand $\leq res_2$. However, as per the proposed approach, the residue ratio $R_j \in [0, 1]$ is also considered during allocation. Accordingly, whenever a demand node $i \in r_k$ arises it is allocated to a nearest resource p_j iff $req_i \leq res_j$ and $R_j \geq \theta$ (threshold). Otherwise, it needs to allocate demand node to some other resources. This other resource is determined based on the following Eq. (4.12).

$$\min_i f(R_i, d_\beta(x, p_i)) = \alpha \ d_\beta(x, p_i) + \beta / R_i | x \in r_y, p_i \in r_z \tag{4.12}$$

Here, α and β are constants which may be fixed in accordance to the service. Now, the objective function in Eq. (2.5) is again modified as follows:

$$\min f(x) = \sum_{i=1}^{m} w_i d_\beta(X, Ex_i)$$

where $d_\beta(x, y) | x \in r_i, y \in r_j$ is determined using the following Eq. (4.13).

$$d_\beta(x, y) = \begin{cases} d(x, \vartheta) + d(\vartheta, y) & |cbv(r_i, r_j) \in E(G) \\ d(x, y) & |cbe(r_i, r_j) \in E(G) \\ d(x, u) + d_{\partial B}(u, v) + d(v, y) | u \in r_i, v \in r_j, cb\partial B(r_i, r_j) \in E(G) \end{cases} \tag{4.13}$$

Here, $d_{\partial B}(x, y)$ represents distance between x and y along boundary of barrier region ∂B. The algorithm for proposed approach is given subsequently:

> **RBCFA: Algorithm for allocation a demand node x to an existing facility**

1. Find containing Region $r_c \mid x \in r_c$ for demand node x.
2. Set $p_{sp} = SP(r_c)$
3. If $p_{sp} \neq Null$

        If $res(p_{sp}) \geq req_x$ and $R(p_{sp}) \geq \theta$

                Allocate x to p_{sp}

                Update balance and residue ratio of p_{sp}

        else

                $S = \{y \mid y$ is Delaunay neighbor of $p_{sp}\}$

                $S' = \{z \mid z \in S \text{ and } res_z \geq req_x\}$

                $X = \min_{j \in S'} f(R_j, d_\beta(SP(r_j), x)) \mid p_j \in r_t$

                Allocate x to p_X

                Update balance and residue ratio of p_X

      Otherwise go to step 4

4. $S =$

      $\{y \mid y \in P,$

      y is reachable from x through partitions $x \xrightarrow{R} y\}$

             $S' = \{z \mid z \in S \text{ and } res_z \geq req_x\}$

             $X = \min_{j \in S'} f(R_j, d_\beta(SP(r_j), x)) \mid p_j \in r_t$

             Allocate x to p_X

             Update balance and residue ratio of p_X

5. If $\frac{1}{m}\sum_{i=1}^{m} R_i < \delta$

             Add a new resource to the plane

In RBCFA, step 1 evaluates the containing partition of the demand node. Step 2 and step 3 find service provider (SP) for the selected partition. Thereafter, if the necessary condition satisfies, then the demand node is allocated to the SP. Otherwise, Eq. (4.12) is evaluated for Delaunay neighbors of the SP and the resource is selected accordingly. It is followed by updating the balance capacity and residue ratio of the selected resource.

Moreover, if containing region does not have a SP, all reachable resources are considered into S. Furthermore, $res_z \geq req_x \mid z \in S$ is evaluated and taken into S'. The resource is chosen among S' using Eq. (4.12). Barrier distance d_β among demand node and corresponding region evaluated using Eq. (4.13). Finally, Step 5 ensures that average residue capacity does not go beneath δ (threshold). Thus, RBCFA considers residue ratio during allocation and optimizes the objective function. The simulation of RBCFA has been presented in the subsequent section.

5 Simulation and Results

In order to simulate RBCFA, the authors consider a popular problem given by Butt and Cavalier [6]. The planar decomposition and corresponding graphical representation have already been demonstrated in Figs. 5 and 6, respectively. The authors choose this instance in order to have clear understanding of the readers. The coordinates of the facilities and barrier are considered as follows:

Barrier B_1	(114,164) (272,164) (272,224) (114,224) (114,164)
Facility P_1	(191,107)
Facility P_2	(291,187)
Facility P_3	(155,263)

Furthermore, the coordinates of the partitions (Fig. 5) are as follows:

Partition R_1	(272,164)(291,187)(272,224)(272,164)
Partition R_2	(231,136)(272,164)(229,164)(231,136)
Partition R_3	(114,164)(191,107)(231,136)(229,164)(114,164)
Partition R_4	(114,224)(194,224)(191,250)(155,263)(114,224)
Partition R_5	(191,250)(194,224)(272,224)(191,250)

The capacity of each resource, i.e., C_j is shown in Table 1.

Similarly, each partition has a quantum of demand that specifies the demand from corresponding partition. This quantum of demand is shown in following Table 2.

Now as already discussed, this planar decomposition of demand plane is based on projection points. Now, from Fig. 5, it is clear that path from demand point $x|x \in r_i$ to resource p_j passes through projection point ρ_{ij}. For instance, projection point from r_5 for p_2 denoted by $\rho_{5,2}$ is the top right corner of rectangular barrier. This projection point aids in further optimization of the barrier distance given in Eq. (4.13). Now, whenever a node $x|x \in r_i$ generates request, ρ_{ij} is used to evaluate $d_\beta(x, p_j)$ as

Table 1 Resources and capacity limit

Resource	1	2	3
Capacity	35	30	35

Table 2 Partitions and their quantum of demand

Partition	1	2	3	4	5
Quantum of demand	16	–	–	6	9

Table 3 Values of dβ(ρij,pj)

	p_1	p_2	p_3
r_1	80.10(272,164)	0	116.5(272,224)
r_2	0	19.17(272,164)	118.55(272,164)
			71.20(114,164)
r_3	0	19.17(272,164)	118.55(272,164)
			71.20(114,164)
r_4	82.09(272,224)	19.43(272,224)	0
	110.99(114,224)		
r_5	82.09(272,224)	19.43(272,224)	0
	110.99(114,224)		

follows:

$$d_\beta\left(x, p_j\right) = d\left(x, \rho_{ij}\right) + d_\beta\left(\rho_{ij}, p_j\right)|x \in r_i \tag{5.1}$$

In Eq. (5.1), $d_\beta\left(\rho_{ij}, p_j\right)$ is static. As a result, $d_\beta\left(\rho_{ij}, p_j\right)$ can be evaluated once and used repetitively. The value of d_β for this instance is presented in Table 3. In Table 3, two entries indicate that ρ_{ij} may be selected in accordance with location of x. Thus, it helps to evade execution of shortest path routine every time. This helps in reduction of execution time for RBCFA.

The authors implemented the proposed algorithm RBCFA in MATLAB for number of demand points ranging from 5 to 100. The proposed algorithm is compared with $M/M/1$ queue model in classical queuing system [3]. The demand nodes are uniformly distributed in the demand plane. This uniform distribution causes variations in objective function value during various executions. Thus, the authors executed the algorithm 5 times for same parameters and considered the average results. Table 4 represents the average results. The authors consider the average distance and CPU time for validating RBCFA. The major advantage of the proposed algorithm is the transformation into graph problem as there are several well-established algorithms

Table 4 Values of Favg and CPU time

n	F_{min}	F_{max}	F_{avg} (Existing)	F_{avg} (RBCFA)	CPU time (existing)	CPU time (RBCFA)
5	15	65	44.8	44.8	1.693	1.774
10	15	65	35.8	35.8	11.173	11.7
20	15	65	30.421	27.737	16.492	15.005
30	13	65	22.733	15.767	17.028	16.878
40	13	65	18.653	15.575	23.967	23.721
50	13	65	13.120	11.140	30.878	30.137
100	13	65	11.249	10.218	42.43	39.896

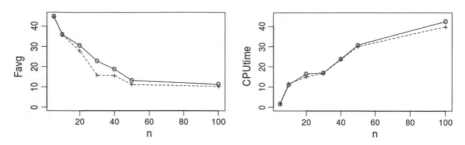

Fig. 7 f_{avg} and CPU time for traditional approach and RBCFA

for graphs in existence. Additionally, in order to find the barrier distance, barrier distance for each resource to each possible projection point is evaluated only once and may be repeatedly used which further optimizes the execution cost. This reduction in execution cost is achieved due to additional memory of $O(m.n)$ which is definitely advisable. The achievement of the proposed algorithm in comparison with traditional method in terms of F_{avg} and CPU time is presented in Table 4 and Fig. 7.

From Table 4 and Fig. 7, it is evident that improvement of RBCFA in terms of F_{avg} is not obvious initially but it becomes significant with increase in number of demand nodes. It is also noticed that CPU time of proposed algorithm is comparable with classical queuing model. The performance of RBCFA with existing approach in terms of F_{avg} and CPU time is demonstrated in Fig. 7. In Fig. 7, red line and blue line represent RBCFA and existing approach, respectively. Figure 7 clearly demonstrates that RBCFA outperforms the traditional allocation approach.

6   Conclusion and Future Scope

Here, in this paper, the authors consider SSCFWP for constrained plane. The authors utilize BTP to find projection point in the demand plane. Based on projection point, the demand plane is partitioned into multiple regions. The proposed approach suggests using the residue ratio of capacitated resource during allocation. The authors simulated the proposed approach and compared with classical $M/M/1$ queuing model. Consequently, it is validated that RBCFA yields optimized results by including residue ratio.

The proposed RBCFA works for SSCFWP, but it can be extended for MSCFWP also. For MSCFWP, greedy approach may be employed for allocation. In this paper, the authors considered the demand plane consisting of static barriers. The research work can be further extended in the direction of considering dynamic barriers.

References

1. Rohaninejad M, Navidi H, Nouri BV, Kamranrad R (2017) A new approach to cooperative competition in facility location problems: mathematical formulations and an approximation algorithm. Comput Oper Res 83:45–53
2. Toro EM, Franco JF, Echeverri MG, Guimarães FG (2017) A multi-objective model for the green capacitated location-routing problem considering environmental impact. Comput Ind Eng 110:114–125
3. Hajipour V, Fattahi P, Tavana M, Di Caprio D (2016) Multi-objective multi-layer congested facility location-allocation problem optimization with Pareto-based meta-heuristics. Appl Math Model 40(7–8):4948–4969
4. Davoodi M, Mohades A, Rezaei J (2011) Solving the constrained p-center problem using heuristic algorithms. Appl Soft Comput 11(4):3321–3328
5. Aneja YP, Parlar M (1994) Algorithms for Weber facility location in the presence of forbidden regions and/or barriers to travel. Transp. Sci. 28(1):70–76
6. Butt SE, Cavalier TM (1996) An efficient algorithm for facility location in the presence of forbidden regions. Eur J Oper Res 90(1):56–70
7. Pfeiffer B, Klamroth K (2008) A unified model for Weber problems with continuous and network distances. Comput Oper Res 35(2):312–326
8. Oncan T (2013) Heuristics for the single source capacitated multi-facility Weber problem. Comput Ind Eng 64(4):959–971
9. Quevedo-Orozco, DR, Ríos-Mercado RZ (2015) Improving the quality of heuristic solutions for the capacitated vertex p-center problem through iterated greedy local search with variable neighborhood descent. Comput Oper Res 62:133–144
10. Canbolat MS, Wesolowsky GO (2012) A planar single facility location and border crossing problem. Comput Oper Res 39(12):3156–3165
11. Sarkar A, Batta R, Nagi R (2007) Placing a finite size facility with a center objective on a rectangular plane with barriers. Eur J Oper Res 179(3):1160–1176
12. McGarvey RG, Cavalier TM (2003) A global optimal approach to facility location in the presence of forbidden regions. Comput Ind Eng 45(1):1–15
13. Katz IN, Cooper L (1981) Facility location in the presence of forbidden regions, I: Formulation and the case of Euclidean distance with one forbidden circle. Eur J Oper Res 6(2):166–173
14. Larson RC, Sadiq G (1983) Facility locations with the Manhattan metric in the presence of barriers to travel. Oper Res 31(4):652–669
15. Batta R, Ghose A, Palekar US (1989) Locating facilities on the Manhattan metric with arbitrarily shaped barriers and convex forbidden regions. Transp. Sci. 23(1):26–36
16. Mirzapour SA, Wong KY, Govindan K (2013) A capacitated location-allocation model for flood disaster service operations with border crossing passages and probabilistic demand locations. Math Probl Eng 2013
17. Klamroth K (2001) Planar Weber location problems with line barriers. Optimization 49(5–6):517–527
18. Shiripour S, Amiri-Aref M, Mahdavi I (2011) The capacitated location-allocation problem in the presence of k connections. Appl. Math. 2(8):947
19. Shiripour S, Mahdavi I, Amiri-Aref M, Mohammadnia-Otaghsara M, Mahdavi-Amiri N (2012) Multi-facility location problems in the presence of a probabilistic line barrier: a mixed integer quadratic programming model. Int J Prod Res 50(15):3988–4008
20. Akyüz MH (2017) The capacitated multi-facility weber problem with polyhedral barriers: efficient heuristic methods. Comput Ind Eng 113:221–240
21. Canbolat MS, Wesolowsky GO (2012) On the use of the Varignon frame for single facility Weber problems in the presence of convex barriers. Eur J Oper Res 217(2):241–247
22. Hamacher HW, Klamroth K (2000) Planar Weber location problems with barriers and block norms. Ann Oper Res 96:191–208
23. Butt SE (1995) Facility location in the presence of forbidden regions and congested regions

24. Klamroth K (2002) Planar Weber location problems with line barriers. Optimization 49(5–6):517–527
25. Hamacher HW Klamroth K (1999) Planar location problems with barriers under polyhedral gauges
26. Ghosh S, Mount D (1991) An output-sensitive algorithm for computing visibility graphs. SIAM J Comput 20(5):888–910

Routing and Security Issues in Cognitive Radio Ad Hoc Networks (CRAHNs)—a Comprehensive Survey

Anshu Dhawan and C. K. Jha

Abstract Cognitive Radio has been considered as a potential contender to interpret the problems of limited availability of spectrum and inefficiency of spectrum usage. Cognitive radio ad hoc networks (CRAHNs) are integrated with both cognitive and spectrum properties therefore routing is the most important parameter that needs to be addressed. Due to higher mobility of nodes, cooperative communication among nodes and multiplicity in the available channels makes the routing protocols for CRAHNs more vulnerable to security attacks. Hence, routing and security are important considerations that need to be addressed individually. It is essential to design a secured spectrum-aware routing protocols to offer healthier stable routing performance and enhanced security. In this paper, the challenges and solutions for routing and security of CRAHNS are outlined. Fundamentally, CRAHNs are more susceptible to security threats because of its intrinsic nature. Hence in order to validate it, an algorithm has been proposed in the paper to check the vulnerability of multi-channel CRAHNs.

Keywords Cognitive radio · Cognitive radio ad hoc network · Routing protocols · Spectrum awareness · Security

1 Introduction

In order to meet continuously growing requirement of mobile users in wireless world, cognitive radio (CR) technology have been proposed [1, 2, 3]. CR is a transceiver that intelligently changes its transmission parameters so that wireless communications get the capability to select available wireless channels opportunistically [1, 4]. The federal communications commission (FCC) [5] has defined a CR as, "a radio that can change its transmitter parameters based on interaction with the environment in

A. Dhawan (✉) · C. K. Jha
Department of Computer Science, Banasthali Vidyapith University, Rajasthan, India
e-mail: anshu.dhawan27@gmail.com

C. K. Jha
e-mail: jchandra@banasthali.ac.in

© Springer Nature Singapore Pte Ltd. 2020
A. Khanna et al. (eds.), *International Conference on Innovative Computing and Communications*, Advances in Intelligent Systems and Computing 1059,
https://doi.org/10.1007/978-981-15-0324-5_42

which it operates". In order to handle the growing need of mobile traffic data FCC has approved unlicensed devices to use licensed bands [6]. In this scenario, both licensed and unlicensed users exploit spectrum dynamically to solve the problem of spectrum scarcity and inefficiency.

Ad hoc networks and their amalgamation with CR devices ultimately lead to cognitive radio ad hoc networks (CRAHNs) [6, 7]. In CRAHNs, every participating node is equipped with a CR technology allowing them to communicate in an intelligent manner. There are two types of participating nodes in such networks. One group of nodes is of primary users (PUs) that form primary network and other group of nodes is of secondary users (SUs) that form secondary network or CR network. Primary network can be wired or wireless network in which nodes or PUs operate in a licensed spectrum band. PUs has priority over SUs in accessing the spectrum portions. To resourcefully exploit the spectrum holes, SUs and PUs communicate with each other through certain communication protocols. Since SU cannot predict the influence of PU's activities on the entire network with its local observation, it necessitates the requirement of schemes so that the observed information can be exchanged among nodes to broaden the knowledge on the network [6]. To enhance the spectrum efficiency in such networks, spectrum management functionalities [8] viz. spectrum sensing, spectrum decision, spectrum sharing, and spectrum mobility needs to be incorporated into the classical layering protocols and implemented through the cross-layer paradigm.

Paradoxically, CRAHNs is accompanied by certain issues and challenges. Routing is one of the important issues which need to be addressed because of the inclusion of the spectrum properties. The routing module necessitated to have good coupling with spectrum management functions so that more precise decisions can be made based on the dynamically changing cognitive environment [3]. Also, the security in CRAHNs is the important concern to be discussed as there is cooperative communication between two types of users which makes them even more vulnerable to security threats.

Rest of the paper is organized as follows. Section 2 talks about issues and challenges involved in CRAHNs. Section 3 comprises of survey on solutions to the routing issues in CRAHNs due to inclusion of cognitive, and spectrum properties in CRAHNs. Further, Sect. 4 incorporates survey on security aspects in CRAHNs. Finally, Sect. 5 wrap up the paper by concluding it.

2 Problem Statement

Data routing is a challenging task due to varying link-quality, frequent topology changes and intermittent connectivity caused by the movement of PUs in the network [9]. Additional features in CRAHNs calls for unique issues and challenges. In this section, specifically, two issues concerning the CRAHNs, i.e., routing and security are talked about. The following Table 1 shows various characteristics of CRAHNs along with the challenges involved.

Table 1 Characteristics and challenges involved in CRAHNS

Characteristics of CRAHNs	Challenges involved in CRAHNs
High probability of incomplete network topology	More collisions among nodes
No pre decided channel paths	Different spectrum availability as per Pus
Multi-hop/Multi-spectrum Architecture	Nodes use different channels and bands to form end-to-end route; collaboration needed between routing module and spectrum allocation
Node mobility	Disconnection which further increases due to PU presence

2.1 Routing Challenges in CRAHNs

The characteristics of CRAHNs such as incomplete topology, multi-hop, multi-spectrum environment, no pre-decided channel paths and node mobility in presence of PUs calls for certain issues which needs to be dealt while designing a routing protocol for CRAHNs. The topology of such networks is highly influenced by the behaviour of the PUs as shown in Fig. 1; thus, the entrances of PUs or change in PU's behaviour determine the topology pattern of the network. They have node mobility issue as shown in Fig. 2a, b. They have mobility issue because of two obvious scenarios. If node goes out of range or there is a sudden appearance of PU, disconnection happens which leads to mobility. So, node mobility should be handled considering both the situations.

Conventional routing metrics like hop count, throughout, delay, time a packet delivery ratio, and congestion are not sufficient for taking routing decisions and comparing the outputs to find a stable and an optimal path. Hence, it is another challenge to define and compute routing metrics for the CRAHNs. Therefore, there

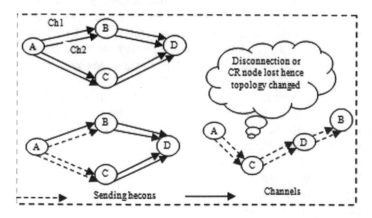

Fig. 1 Network topology

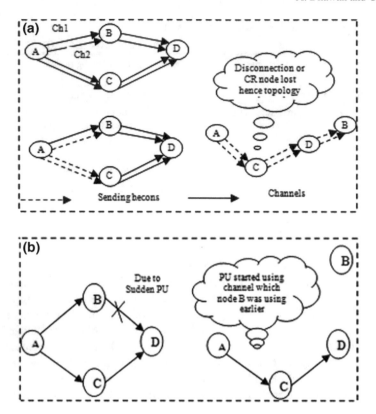

Fig. 2 Node mobility

is necessity to incorporate the new metrics like path/route stability, spectrum availability, PU presence, etc. along with the conventional measures in order to attain better routes. Spectrum-awareness should be included in the routing module of the CR protocols in order to incorporate these new measures. Also, the conked-out paths may render the channel unstable due to sudden appearance of a PU. This leads to unpredictable route failures that can be handled through rerouting procedures. Thus, an effective and efficient solution is required to reinstate "conked-out" paths without degrading the route quality.

Challenges as stated in Table 1 need to be addressed in order to attain better performance of the network and better route stability in the dynamic spectrum. Therefore, the routing module should be designed with the inclusion of spectrum-awareness properties for CRAHNs. The routing module should be aware of the external environment and should be tightly coupled with cognitive cycle of spectrum management [10]. Authors [11] have talked about various routing challenges in CRAHNs and has classified the routing protocols into various groups based on delay, link stability, throughput, location, energy-aware protocols, etc.

2.2 Security Challenges in CRAHNs

As enumerated earlier, CRAHNs inherit features of both ad hoc networks and CR networks. Inclusion of cognitive properties and spectrum sharing among PUs and SUs emerges more security issues. Due to intrinsic nature of CRAHNs, the prospects for attackers to attack are more as compared to usual wireless networks. Therefore, routing protocols designed for CRAHNs are vulnerable to multiple threats. This is the prime motivation for picking up the security parameter for this survey. Many authors have proposed numerous security attacks for ad hoc networks and cognitive radio networks (CRNs) on different layers of the protocol stack [12]. The solutions to the routing and security issues have been explored in the subsequent sections.

3 Survey on Solutions to the Routing Issues in CRAHNs

The routing solutions for CRAHNs may perhaps be categorized as complete spectrum knowledge and limited spectrum knowledge. In case of complete spectrum knowledge, availability of spectrum among any two nodes is informed to all other nodes of the network. This can be accomplished by maintaining central database that specifies channel availability over time and space. The solutions that are put together for complete spectrum knowledge case are commonly based on a graph abstraction of the CRN. However, generally, they are not realizable for implementation but have been used to derive routing performance benchmarks. In such scenarios, the routing module is not tightly coupled with the spectrum management functionalities. While in case of limited spectrum knowledge, information about spectrum availability together with conventional network information such as the routing metrics, node mobility, etc. is dealt among the network nodes. Grounded on this, the limited spectrum knowledge-based routing protocols have been further categorized into protocols that intend to decrease end-to-end delay, exploit throughput and path stability. Furthermore, there are numerous probabilistic methods in which CR users opportunistically transmit over the entire band available during the small unused gaps [13].

Further, down in the section, we have presented routing protocols proposed for CRAHNs, which are ranging from improvement of existing ad hoc on-demand distance vector (AODV) based routing protocol to entire newly designed protocols based on characteristics of CRAHNs.

Zeeshan et al. [14] have assumed each node to be equipped with single transceiver. It has talked about problem of CCC, deafness problem and routing for CRAHNs. Backup channel and cooperative channel-switching on-demand routing protocol for multi-hop cognitive radio ad hoc networks (BCCCS) proposed that nodes should not use more than one channel for the same flow to handle deafness problem. Authors further proposed to choose one channel as CCC among all available channels. Each node stays in CCC interface while in non-transmission state. However, in transmission state node switches periodically to CCC channel to check for control message. They have introduced concept of backup channels to handle sudden appearance of PU on an established path. Backup channel information is shared and stored in neighbour nodes. Each neighbour node also switches to same channel in cooperative manner in case of PU appearance.

In novel routing metric for multi-hop cognitive wireless networks (WHAT) Chen et al. [15] have introduced a new routing metric. In order to extend basic AODV, new metric has included parameters like spectrum-awareness, path stability, and weighted hop count. Channel usage for spectrum-awareness, channel-switching frequency for stability and path length for hop count are used while selecting path between any source and destination nodes. Destination node receives multiple paths and evaluates all received paths in terms of suggested routing metric. Routing metric is shown to establish an optimal, loop-free and consistent path.

Lin et al. [16] have proposed a routing protocol which is aware of the spectrum properties. It creates the map of the spectrum as per the usage of PUs. This protocol uses link metric and path metric for launching the path. Authors have claimed that spectrum map information improves the path stability as per CRAHNs scenario.

Xu et al. [17] have presented a protocol which facilitates node traffic information prior to advance forwarding RREQ. Node exchanges control packet called cognitive packet with its neighbours to get local traffic information. Node uses present local traffic information to estimate the future traffic and delays route request packet to schedule as per traffic information so that it should traverse through best optimal path. Destination send route reply packet on the path from which it receives the first route request. Moreover, Traffic-Aware Routing Protocol for Cognitive Network (TARP) also uses learning technique to adjust the route lifetime and thereby meets the quality of service demand.

Ramanathan et al. in [18] have proposed a multi-path routing approach to generate shortest path between source and destination using modified Dijikstra's algorithm. To discover feasible multi-paths, the metric used in algorithm is interference among the channels. Algorithm requires knowledge of topology to generate graph with multiple edges labeled with frequency of each channel (to determine minimum interference) as input of Dijikstra's algorithm.

Kim et al. [19] have projected a key to improve performance of the metric. It creates a guild of node's geographical location information, spectrum-awareness as well as availability of other resources like transmission power, delay, etc. at each link. A source node or intermediate node as per protocol generates a route request and looks for its neighbours to reply depending on whether it is nearest to destination and satisfies required resources availability in route request or not. A weight metric is defined for each link that will evaluate maximum if neighbouring node is nearest to destination. Authors claimed to improve delivery ratio and throughput but they compared their performance with location-based routing protocols of ad hoc networks rather than other spectrum-aware protocols of CRAHNs.

In [20] authors have proposed a spectrum-aware routing protocol based on AODV. They have proposed to use TVWS (television white spaces) to improve the performance of routing in the presence of PUs. They emphasized SUs located in remote locations; using TVWS can communicate by including SUs of intermediate geographic regions. Intermediate SU uses a geo-location database to find the availability of TVWS of next neighbouring secondary node towards the destination. SUs' usage of TVWS may cause interference to PUs so the scheme protects PUs using anti-interference geo-location database. There is need to see the delay acquisition of this scheme in comparison to other spectrum-aware routing schemes as this scheme uses the database to assist SUs for operating in TVWS.

In [21], authors introduced a spectrum-aware routing protocol (SARP) based on AODV. SARP proposed that each node in network computes a metric called path availability time using spectrum information. Path availability time for the path is period before any link on path is failed. Authors claim that using this metric, established path is more stable and improves the network performance. They proposed a new route maintenance procedure in case of appearance of PU on established path. They assumed a channel usage model and presented the results of rerouting but they did not present the results for metric like packet delivery ratio to study the performance of protocol.

In [22], authors have proposed a cooperative routing protocol that considers the concern of end-to-end throughput in multi-hop scenario. According to the authors, relay nodes aids in cooperative transmission if their inclusion in path leads to high throughput and low delay. Authors have taken into account channel utilization, and bandwidth of channels with the help of these parameters, cost of direct transmission or cooperative transmission at a specific channel with relay nodes is calculated. Upper limit of capacity with cooperative benefit between two adjoining nodes is used to achieve low cost in terms of throughput and delay in entire route discovery. Authors have assumed that nodes know about channel utilization.

Cacciapuoti et al. [23] have presented cognitive on-demand distance vector (CAODV) to improve performance of AODV by exploiting spectrum diversity so that a path may be established through available different channels if same channel throughout the path is not available. They have talked about intra- and inter-route spectrum diversity. Former employs diverse channels on one path while latter employs same channel across different paths.

Rahman et al. [24] have extended CAODV and presented dual diversity cognitive routing protocol (D2CARP). The work proposed by them not only exploits spectrum diversity but also path diversity. Path diversity means to keep path through multiple channels while maintaining multiple disjoint paths. D2CARP broadcasts route reply on all available channels for a path through multiple channels and thus incurs communication overhead in comparison to CAODV. D2CARP improves in delivery ratio and end-to-end delay over CAODV.

The proposed protocol [25] considers node mobility issue as well as user activity in neighbour coordination of the nodes of the network. MASAR follows neighbour prioritization and assigns different priorities to the forwarding neighbours in order to have reduced overhead. In this author uses short and stable route approaches to estimate back off times using mobility metrics. Author has compared results of MASAR protocol with CAODV protocol and showed that there is a significant increase in the performance and reduction in the control overheads.

Huang et al. [26] have characterized spectrum temperature routing metrics based on PU presence over a connection. The authors have shown the improvement in the protocol by establishing a coolest path.

Table 2, provides the comparison of various routing protocols base on various parameters such as their routing type, metric used.

4   Survey on Solutions to the Security Issues in CRAHNs

Ad hoc routing protocols are susceptible to a variety of security attacks and threats because there is cooperative communication among nodes while routing the data packets. Many security solutions are designed for mitigating security attacks in ad hoc networks. No single protocol is there to handle every possible attack, so there are many solutions, which are designed for handling different types of security attacks like black hole attack, wormhole attack, selfish nodes attack, etc. [27,28, 29].

Security aspect in CRNs requires dealing with both the ad hoc network security threats and threats which are specific to CRNs that includes jamming, primary user emulation, etc. [30, 31, 32, 33, 34, 35]. In [36] authors have given review of attacks of wireless networks and discussed how those can be extended to cognitive scenario. In [37], authors emphasized and demanded a separate set of security measures for CRNs. They have shown a detailed view of the security threats that are currently faced by CRNs classified with respect to different layers of communication stack. Paper [38] addresses the security attacks and their mitigation techniques in CRNs.

There are numerous security attacks [39, 40] in ad hoc networks and CRNs possible which would either disturb the entire working of the network or consume resources making them considerably deficient. However, CRAHNs are more susceptible to security attacks because of certain obvious intrinsic characteristics [41] but very less research has happened in the area [42, 43] specifically on network layer. One of the motivations of this paper is to commence a security attack on a routing protocol for CRAHNs and hence screening its vulnerability.

In the literature, one of the security attacks which are prone to network layer routing protocol is NEPA Attack [44]. The intention of NEPA attack is to increase the influence and interference at heavily loaded channels. NEPA attack can be launched through a malicious node in the communication network. NEPA attack assumes presence of at least one malicious SU that falsely announces its channel choice to its neighbours. In NEPA, an attacker occupies the spectrum portions considered high priority in a network and hides this information from its neighbours. It is possible by broadcasting bogus messages to all the channels which are heavily occupied in a network and hides this information from its neighbours. This would cause interference among the nodes, performance drop of the network and less availability of bandwidth to the nodes. Also, malicious node would not send latest information about

Table 2 Comparison of routing protocols

Protocol Name	Single-channel/Multi-channel/multi-path	Routing Metric used/Routing scheme used	Dedicated control channel
Backup channel and cooperative channel-switching on-demand routing protocol (BCCCS) [14]	Single channel	Back up channels or cooperative channel-switching	Common control channel
Weighted Hop, spectrum-Awareness, and sTability (WHAT) [15]	Multi-channel	Path stability/hop count	Common control channel
Spectrum-aware opportunistic routing (SAOR) [16]	Multi-path	Shortest path/delay, QoS guaranteed throughput	No
Traffic-Aware Routing Protocol for Cognitive Network(TARP) [17]		End to end delay, throughput	No
Interference-aware multipath routing in a cognitive radio ad hoc network [18]	Multi-channel, multi-radio and Multi-path	Shortest path/interference	No
Spectrum-Aware Beaconless geographical routing [19]	Multi-channel	Delivery ratio, throughput	Control channel
Spectrum-aware routing in ad hoc cognitive radio networks [20]	Multi-channel	Delay	No
Novel spectrum-aware routing protocol(SARP) [21]	Multi-channel	Rerouting	Common control channel
Cooperative routing protocol in cognitive radio ad hoc networks [22]	Multi-channel	Relay Availability	Common control channel
Cognitive on-demand distance vector(CAODV) [23]	Multi-channel	Shortest path	No

(continued)

Table 2 (continued)

Protocol Name	Single-channel/Multi-channel/multi-path	Routing Metric used/Routing scheme used	Dedicated control channel
Dual diversity cognitive routing protocol(D2CARP) [24]	Multi-channel and multi-path	Shortest path	No
Mobility assisted spectrum-aware routing protocol(MASAR) [25]	Multi-channel	Shortest path and stable routes	Common control channel
Coolest path spectrum mobility aware routing metrics [26]	Multi-channel	Spectrum availability/stable path	Common control channel

the changes in the network to its neighbouring nodes, i.e., whatever information an attacker node is forwarding is contaminated. If neighbours not getting right and latest information, it will not be receptive about its network. This stressed condition would further lead to significant performance dive.

Algorithm for imposing NEPA attack in multi-channel CRAHNs

1. Implementation of secure Routing Protocol
   a. Input variables as provided in the paper [24]
      i. INPUTS = [CU, PU, CHANNELS]
      ii. OUTPUT = [TRC files]
      iii. Analyze the output
   b. Prioritize the channels on the basis of routing metric design considered in [15]
      Store and sort channels priority wise in decreasing order
      (Channel with least remaining bandwidth is given highest priority and so on)
   For each channel i = 1 to n
         if ith channel free from PU
            Attacker node block ith channel through bogus messages//NEPA attack
   End for
   Where n is the number of channels
2. Analysis can be done using statistical models to compute the distribution and cache threshold [45] of the stressed network.

Hence, there is a need to design, implement and analyze a spectrum-aware routing protocol for CRANHs keeping in view the stressed conditions.

5 Conclusions

CRAHNs are promising solution to the problem of spectrum inefficiency because it shares the spectrum portion between the users (PUs and SUs) in an opportunistic manner. Routing in CRAHNs has gained a lot of attention in recent years and presented new issues and challenges as compared to the long-established ad hoc networks. In this paper, we have exemplified a review of a variety of solutions to the routing and security issues that have been proposed in the literature for CRAHNs. Most of the research has conducted in this area so far majorly focused on spectrum management functions like spectrum sensing, spectrum decision, spectrum sharing, and spectrum mobility. There is a need for the exploration of spectrum-aware routing protocols in CRAHNs in which the routing module is integrated with security aspect. Security aspect in CRAHNs is the imperative subject to be discussed and explored so that the CR nodes can communicate with each other in a secure mode in stressed conditions as well.

References

1. Mitola J (1999) Cognitive radio for flexible mobile multimedia communications. In: Book Cognitive radio for flexible mobile multimedia communications. IEEE, pp 3–10
2. Akyildiz IF, Lee W-Y, Vuran MC, Mohanty S (2006) NeXt generation/dynamic spectrum access/cognitive radio wireless networks: a survey. Comput Netw 50(13):2127–2159
3. Cesana M, Cuomo F, Ekici E (2011) Routing in cognitive radio networks: challenges and solutions. Ad Hoc Netw 9(3):228–248
4. Mitola J (2000) 'Cognitive radio—an integrated agent architecture for software defined radio
5. Haykin S (2005) Cognitive radio: brain-empowered wireless communications. IEEE J Sel Areas Commun 23(2):201–220
6. Akyildiz IF, Lee W-Y, Chowdhury KR (2009) CRAHNs: cognitive radio ad hoc networks. Ad Hoc Netw 7(5):810–836
7. Youssef M, Ibrahim M, Abdelatif M, Chen L, Vasilakos AV (2014) Routing metrics of cognitive radio networks: a survey. IEEE Commun Surveys Tutorials 16(1):92–109
8. Lee WY (2009) Spectrum management in cognitive radio wireless networks
9. Che-Aron Z, Abdalla AH, Abdullah K, Hassan WH, Rahman MA (2015) Racarp: a robustness aware routing protocol for cognitive radio ad hoc networks. J Theor App Info Technol 76(2)
10. Salim S, Moh S (2013) On-demand routing protocols for cognitive radio ad hoc networks. EURASIP J Wireless Commun Networking 2013(1):1
11. Abdelaziz S, ElNainay M (2014) Metric-based taxonomy of routing protocols for cognitive radio ad hoc networks. J Network Comput Appl 40:151–163
12. Chauhan KK, Sanger AKS (2014)Survey of Security threats and attacks in cognitive radio networks. In:: Book Survey of Security threats and attacks in cognitive radio networks. IEEE, pp 1–5
13. Khalifé H, Malouch N, Fdida S (2009) Multihop cognitive radio networks: to route or not to route. IEEE Network 23(4):20–25
14. Zeeshan M, Manzoor MF, Qadir J (2010)Backup channel and cooperative channel switching on-demand routing protocol for multi-hop cognitive radio ad hoc networks (BCCCS). In: Book Backup channel and cooperative channel switching on-demand routing protocol for multi-hop cognitive radio ad hoc networks (BCCCS). IEEE, pp 394–399

15. Chen J, Li H, Wu J (2010) WHAT: a novel routing metric for multi-hop cognitive wireless networks. In: Book WHAT: a novel routing metric for multi-hop cognitive wireless networks. IEEE, pp 1–6
16. Lin S-C, Chen K-C (2010) Spectrum aware opportunistic routing in cognitive radio networks. In: Book Spectrum aware opportunistic routing in cognitive radio networks. IEEE, pp 1–6
17. Xu Y, Sheng M, Zhang Y (2010) Traffic-aware routing protocol for cognitive network. In: Book Traffic-aware routing protocol for cognitive network. IEEE, pp 1–5
18. Khanna B, Ramanathan R (2011) Interference-aware multipath routing in a cognitive radio ad hoc network. In: Book Interference-aware multipath routing in a cognitive radio ad hoc network. IEEE, pp 855–860
19. Kim J, Krunz M (2011) Spectrum-aware beaconless geographical routing protocol for mobile cognitive radio networks. In: Book Spectrum-aware beaconless geographical routing protocol for mobile cognitive radio networks. IEEE, pp 1–5
20. Mastorakis G, Bourdena A, Kormentzas G, Pallis E (2012) Spectrum aware routing in ad-hoc cognitive radio networks. In: Book Spectrum aware routing in ad-hoc cognitive radio networks. IEEE, pp 1–9
21. Sun B, Zhang J, Xie W, Li N, Xu Y (2012) A novel spectrum-aware routing protocol for multi-hop cognitive radio ad hoc networks. In: Book A novel spectrum-aware routing protocol for multi-hop cognitive radio ad hoc networks. IEEE, pp 1–5
22. Sheu J-P, Lao I-L (2012) Cooperative routing protocol in cognitive radio ad-hoc networks. In: Book Cooperative routing protocol in cognitive radio ad-hoc networks. IEEE, pp 2916–2921
23. Cacciapuoti AS, Caleffi M, Paura L (2012) Reactive routing for mobile cognitive radio ad hoc networks. Ad Hoc Netw 10(5):803–815
24. Rahman MA, Caleffi M, Paura L (2012) Joint path and spectrum diversity in cognitive radio ad-hoc networks. EURASIP J Wireless Commun Networking 2012(1):1–9
25. Abedi O, Berangi R (2013) Mobility assisted spectrum aware routing protocol for cognitive radio ad hoc networks. J Zhejiang Uni Sci C 14(11):873–886
26. Huang X, Lu D, Li P, Fang Y (2011) Coolest path: spectrum mobility aware routing metrics in cognitive ad hoc networks. In: 31st International conference on Distributed computing systems (ICDCS) 2011, IEEE, pages 182–191
27. Khurana S, Gupta N (2011) End-to-end protocol to secure ad hoc networks against wormhole attacks. Secur Commun Networks 4(9):994–1002
28. Khurana S (2011) Handling attacks on routing protocols in ad hoc networks. Doctoral dissertation, University of Delhi, India
29. Aneja S, Gupta N (2012) Reliable distance vector routing protocol to handle blackhole and selfish (RDVBS) nodes in ad hoc networks. Int J Next-Generation Comput 3(1)
30. Burbank JL (2008) Security in cognitive radio networks: the required evolution in approaches to wireless network security. In: Book Security in cognitive radio networks: the required evolution in approaches to wireless network security. IEEE, pp 1–7
31. Fragkiadakis AG, Tragos EZ, Askoxylakis IG (2013) A survey on security threats and detection techniques in cognitive radio networks. IEEE Commun Surveys Tutorials 15(1):428–445
32. Cadeau W, Li X (2012) Anti-jamming performance of cognitive radio networks under multiple uncoordinated jammers in fading environment. In: 'Book Anti-jamming performance of cognitive radio networks under multiple uncoordinated jammers in fading environment. IEEE, pp 1–6
33. Mao H, Zhu L (2011) An investigation on security of cognitive radio networks. In: Book An investigation on security of cognitive radio networks. IEEE, pp 1–4
34. Mody AN, Reddy R, Kiernan T, Brown TX (2009) Security in cognitive radio networks: An example using the commercial IEEE 802.22 standard. In: 'Book Security in cognitive radio networks: An example using the commercial IEEE 802.22 standard. IEEE, pp 1–7
35. Zou Y, Wang X, Shen W (2013) Physical-layer security with multiuser scheduling in cognitive radio networks. IEEE Trans Commun 61(12):5103–5113
36. Araujo A, Blesa J, Romero E, Villanueva D (2012) Security in cognitive wireless sensor networks. Challenges and open problems. EURASIP J Wireless Commun Networking 2012(1):1

37. Rizvi S, Mitchell J, Showan N (2014) Analysis of security vulnerabilities and threat assessment in Cognitive Radio (CR) networks. In: Book Analysis of security vulnerabilities and threat assessment in Cognitive Radio (CR) networks. IEEE, pp 1–6
38. Khasawneh M, Agarwal A (2014) A survey on security in cognitive radio networks. In: Book A survey on security in Cognitive Radio networks. IEEE, pp 64–70
39. Feng J, Du X, Zhang G, Shi W (2017) Securing multi-channel selection using distributed trust in cognitive radio ad hoc networks. Ad Hoc Netw 61:85–94
40. Sharma S, Mohan S (2016) Cognitive radio adhoc vehicular network (CRAVENET): architecture, applications, security requirements and challenges. In: 2016 IEEE international conference on advanced networks and telecommunications systems (ANTS), IEEE, pp 1–6
41. Hu X, Cheng J, Zhou M, Hu B, Jiang X, Guo Y, Bai K, Wang F (2018) Emotion-aware cognitive system in multi-channel cognitive radio ad hoc networks. IEEE Commun Mag 56(4):180–187
42. Clancy TC, Goergen N (2008) Security in cognitive radio networks: threats and mitigation. In: Book Security in cognitive radio networks: threats and mitigation. IEEE, pp 1–8
43. Zhang T (2014) Security issues in cognitive radio networks. University of Calgary
44. Mathur CN, Subbalakshmi K (2007) Security issues in cognitive radio networks. In: Cognitive networks: towards self-aware networks. pp 284–293
45. Kr S, Rajanikanth K (2009) Intelligent caching in on-demand routing protocol for mobile adhoc networks. World Acad Sci Eng Technol Int J Electr Comput Energ Electron Commun Eng 3(8):1533–1540
46. Sumathi A, Vidhyapriya R (2012) Security in cognitive radio networks-a survey. In: Book Security in cognitive radio networks-a survey. IEEE, pp 114–118
47. Naveed A, Kanhere SS (2006) NIS07-5: Security vulnerabilities in channel assignment of multi-radio multi-channel wireless mesh networks. In: Book NIS07-5: security vulnerabilities in channel assignment of multi-radio multi-channel wireless mesh networks. IEEE, pp 1-5
48. Arkoulis S, Marias GF, Frangoudis PA, Oberender J, Popescu A, Fiedler M, Meer HD, Polyzos GC (2010) Misbehavior scenarios in cognitive radio networks. Future Internet 2(3):212–237

High Data Rate Audio Steganography

Sangita Roy and Vivek Kapoor

Abstract Steganography is an art to hide information without affecting perceptual transparency of digital media files. Media files include audio, video, image, text, software and so on. Hiding information is required for secure transmission of data. In general, media files are known as cover file which wrap the data to be transmitted. The media file must be imperceptible before and after embedding of data. But when we embed data, we do some modification at different bit plane that introduces error or noisy media. Motivation of this work is based on the property of human auditory system (HAS) which is discernible towards any kind of modification or distortion in audio file. Keeping in mind this HAS property, in this work, we concentrate on hiding a large amount of data in a small size cover file considering reduced noise and minimized bit error rate.

Keywords Steganography · Least Significant Bit (LSB) · Audio

1 Introduction

We are more interested in suspicious data than blank paper. Cryptography provides us unintelligible data to see but steganography makes the data invisible. The word steganography came from two Greek words—"steganos" means secret or covered and "graphein" means writing. There are many stories that exist where steganography technique has been used very skillfully. 9/11 attack is one of the biggest application of steganography where the skill was abused immensely. Steganography is used for secure communication where we take the advantage of human audio-visual perception. We take the help of digital media, e.g. audio, video, text, image to hide the secure data. Our secure data may be anything, e.g. text, image, zip file etc. [1] The media

S. Roy (✉) · V. Kapoor
Thapar Institute of Engineering and Technology, Patiala, Punjab, India
e-mail: sangita.roy@thapar.edu

V. Kapoor
e-mail: vivekkapoor97@gmail.com

© Springer Nature Singapore Pte Ltd. 2020   503
A. Khanna et al. (eds.), *International Conference on Innovative Computing and Communications*, Advances in Intelligent Systems and Computing 1059,
https://doi.org/10.1007/978-981-15-0324-5_43

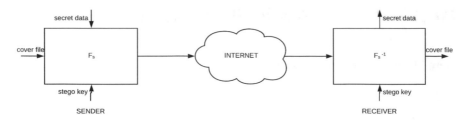

Fig. 1 Components of steganography

through which we want to transfer the secret data is known as "cover file". We can hide/embed the secret data inside the cover file and after processing, the technique produces stego file. Figure 1 shows the overall process of audio steganography. The components of steganography system are

- Cover file: Audio, where we hide/embed data.
- Secret data: Text, what we want to hide/embed.
- Key: How we want to hide/embed secret data in cover file.
- Fs: Audio processing function which takes above-mentioned three inputs (cover file, secret data and key).
- Fs^{-1}: Audio processing function which takes the following three inputs—stego file, secret data and key.

In this work, the audio file has been taken as cover file, and the secret data is text. Steganography mainly suffers from its size. To embed n length data, we need considerable number (greater than n) of samples of audio. The main objective of this work is to hide more number of data in limited sample space. To increase hidden text size, more noise is introduced. But the analysis shows that the noise depends on the quality of the audio. So, the selection of audio file is also a challenging task. The whole work is divided into two parts—first, an LSB technique is performed on multiple audio files; second, the audio file with maximum SNR is chosen to send the data. Data is hidden in one bit as well as two bits format.

After inverse audio processing, we can extract the original audio and secret data. In order to hide data successfully, it must adhere to two key points:

- Cover file should not be suspicious after data hiding.
- Data should not lose its integrity after transmission.

The rest of the paper is organized as follows: Sect. 2 briefs the literature behind the audio steganography and its techniques. In the following, Sect. 2 summarizes related work on LSB-based audio steganography for noise detection and reduction processes. System design and basic components required for audio steganography are discussed in Sect. 3. Section 4 comprises of the proposed method and the work flow. Results of the experiments are discussed in Sect. 5. Section 6 concludes the whole work presented in this paper.

2 Literature

The data to be stealthily transferred is passed through two layers before inducing into cover data at the third layer of tri-layered model proposed by Asad et al. [1] for audio steganography. The approach used takes into account the least significant bit for replacement. Mirrored operations are performed at the receiver's end to recover the secret data after retrieving from the network. Focus is magnified on the confidentiality of the secret data. The shallow side of implementation where capacity, transparency and robustness are parametric issues is also discussed.

The LSB is selected to inculcate minimal noise in the cover media, but this makes the bit positions predictable and more susceptible to attack, hence Rana and Banerjee, in their paper have put forward dual layer randomization technique where sample points are selected randomly for each process unlike selecting same LSB every time [2]. This provides an additional layer of security, and the steganographic technique has been stretched to next level of security.

The message type, message length and initial position are basic elements required for recovering messages from the steganographed audio. Gandhi et al. boost the robustness of audio steganography by proposing an efficient method to audio steganography which used light wavelet transform with modification of LSB and three random keys to protect message type, message length and initial position from being exposed [3].

The data is transmitted over the wire/wireless in form of signals. Under certain circumstances, the transmitted signal might distort which is an imperceptible attack on the signal. With this thought process, Priyanka et al. proposed the usage of genetic algorithm which uses numerous and higher LSB values which leads to robustness as far as attacks for decoding messages or noise addition is concerned [4].

To keep the noise introduced by steganography under limits, only four LSBs of 16-bit audio sample are considered suitable. The novel approach introduced by Divya makes use of RSA algorithm for cryptographic needs and improves the cover audio capacity by making use of seven LSBs for hiding data and hence increasing the amount of data that can be hidden [5].

Kundu and Kaur [6] proposed a modified LSB technique, where hidden text is encrypted first using Vigenere Square Encryption algorithm. Once the encryption is done, then the ASCII conversion takes place of text. After these processes, authors proposed to perform modified LSB technique and then audio transposition encryption technique is applied on audio.

Gadicha introduced an LSB audio steganography method where the fourth bit position is used to hide the data, and according to this paper, the reduced distortion is better than layer 1 LSB [7]. The robustness is reported high in this paper.

Zamani et al. described the substitution technique problem and proposed different solution for these [8]. This paper considered different attack on audio file and the mechanism to handle those attacks. It described how changing position at MSB affects the audio quality instead of changing data at LSB position.

LSB is more susceptible to noise while transferring data; Sridevi et al. suggested a productive system towards audio steganography accompanied by solid encryption for improving the security [9]. Enhanced audio steganography utilizes powerful encryption at the first level followed by bit level encoding for protecting data from hackers.

Djebbar et al. [10] reported the comparison study among different steganographic techniques. These comparison has been done mainly based on robustness, security and hiding capacity of audio. The other contribution of this paper is that authors found the provision of classification based on robustness of different steganographic models.

Cloud computing has gained immense importance in IT sector due to services which are being provided by it, like networking, software, database and Internet. As with great power comes great responsibility, it is essential to provide high level of security. Garg et al. provided a split algorithm for secure transfer of data on cloud servers [11]. This paper also proposes the idea of cryptography followed by steganography and then splitting the stego image.

Gupta and Bhagat [12] proposed the process of camouflaging the secret information into media, be it image, video, audio or file to stealthily transmit between sender and receiver(s) or to prevent information from being exposed to the user of the system on which data is residing. Images are most popular and widely used carriers for concealing sensitive information. The major concern while steganography is that the media used for the purpose should not distort significantly. In this paper, secure steganographic mechanism using pre-shared password is used along with least significant bit substitution method so that there is minimal noise in the image.

3 System Design

3.1 LSB

There are several methods to hide information in audio file. Least significant bit (LSB) is standard and the easiest method to hide data in digital audio file. In this technique, least significant bits of audio sample are substituted by secret data bits. Data transmission rate in LSB coding is 1 kbps per 1 kHz. In standard LSB, first LSB bit position is used in 16-bit sample to hide the secret data bit [14]. Figure 2a shows the standard LSB technique, where we have used the first LSB plane to hide data. The secret data "abc" is converted into binary, and we get 21 bits (considered seven bits ASCII) to hide. For each bit of secret data, we choose one audio sample. We replace the last bit of one audio sample with the corresponding bit of secret data.

Standard LSB coding technique is depicted in the following steps:

1. Read an audio sample from the wave stream.
2. Extract next bit from the current secret data byte.
3. Place the secret data bit in the selected audio sample's first LSB bit position.
4. Concatenate rest of the wave file after completion of embedding secret data.

Audio sample (16 bit) without secret message	"abc" in binary	Audio sample (16 bit) with secret message, 1st LSB is encoded
0000000000100100	1	0000000000100101
1000000001001000	1	1000000001001001
1001100000001111	0	1001100000001110
1111110010000100	0	1111110010000100
1000000001000111	0	1000000001000110
0000000001010011	0	0000000001010010
0000000000001010	1	0000000000001011
1000000000010101	1	1000000000010101
1000000001100010	1	1000000001100011
0000000000101101	0	0000000000101100
0000000001100100	0	0000000001100100
1000000000110011	0	1000000000110010
1110000000000100	1	1110000000000101
1111100010000001	0	1111100010000000
1111110010000100	1	1111110010000101
1011111111111010	1	1011111111111011
1101101001111100	0	1101101001111100
1111001111011101	0	1111001111011100
1111110000101011	0	1111110000101010
0000000000110010	1	0000000000110011
0000000000010101	1	0000000000010101

(a) Standard LSB

Audio sample (16 bit) without secret message	"abc" in binary	Audio sample (16 bit) with secret message, 4th LSB is encoded and rest LSBs are 1's complimented
0000000000100100	1	0000000000101011
1000000001001000	1	1000000001001000
1001100000001111	0	1001100000000000
1111110010000100	0	1111110010000100
1000000001000111	0	1000000001000111
0000000001010011	0	0000000001010011
0000000000001010	1	0000000000001010
1000000000010101	1	1000000000011010
1000000001100010	1	1000000001101101
0000000000101101	0	0000000000100010
0000000001100100	0	0000000001100100
1000000000110011	0	1000000000110011
1110000000000100	1	1110000000001011
1111100010000001	0	1111100010000001
1111110010000100	1	1111110010001011
1011111111111010	1	1011111111111010
1101101001111100	0	1101101001110011
1111001111011101	0	1111001111010010
1111110000101011	0	1111110000100100
0000000000110010	1	0000000000111101
0000000000010101	1	0000000000011010

(b) Modified LSB

Fig. 2 LSB techniques

There are different modified LSB techniques, where different LSB bit positions have been used to store secret data [14]. In modified LSB technique, one bit from sample is embedded in the third or fourth LSB bit of audio sample and rest of the bits are 1's complemented based on some condition. For example, if the original sample is $(0\ldots0101)_2 = (5)_{10}$ and the watermark bit is "1" to be embedded at the fourth LSB layer, then after embedding, the steganographed sample would be $(0\ldots1010)_{10}$. Figure 2b shows the example of modified LSB technique.

Modified LSB coding technique is depicted in the following steps:

1. Read an audio sample from the wave stream.
2. Extract next bit from the current secret data byte.
3. Place the secret data bit in the selected audio sample's fourth LSB bit position.
4. If the fourth LSB of audio sample is not equal to the secret data bit, then do 1's complement of next three LSB bits.
5. Else, Do not perform 1's complement.
6. Concatenate rest of the wave file after completion of embedding secret data.

3.2 Audio Quality

Selection of audio file is a big task in audio steganography. By updating LSB values, we introduce distortion in audio signal. Our main aim in this work is not only to hide the maximum secret data but also to reduce the embedded distortion. Audio file is converted into pulse-code modulation (PCM) when played. PCM is measured using the following properties:

- **Audio channel**: Two types of audio channels are available—Monophonic (Mono), audio with one central channel, e.g. single instrument sound and Stereophonic (Stereo), audio with left and right channels, e.g. most of the mp3 files.
- **Sampling rate**: It is the number of repetition of samples per second. Sampling rate is measured by Hertz (Hz). High sampling rate defines high amount of details to be stored.
- **Bit depth**: It is the number of bits per sample. Common bit depth we have are −8, 16, 24 and 32 bit. In most of the cases, audio file is encoded with 16-bit depth. High bit depth is often used to minimize rounding error.
- **Bit rate**: This is the measurement of space taken by any audio file. It is generally measured by bits per second (bps). For example, Stereophonic sound has almost twice bit rate than Monophonic sound.

4   Proposed Method

This work is divided into two parts. In the first part, we do data hiding in a new approach, and in the second part we choose the final suitable audio file which to be transmitted.

4.1   Data Hiding Algorithm

Our first objective is to hide more data in limited space [13]. In proposed LSB technique, two bits from sample are embedded in the fourth and third LSB bits of audio sample, and rest of the bits are 1's complemented based on some condition. For example, if the original sample is $(0\ldots0101)_2$ and the watermark bit "10" is to be embedded at the fourth and third LSB bit positions, then after embedding, the steganographed sample would be $(0\ldots10\mathit{10})_{10}$. If in this same sample, if the watermark bits are "01" is to be embedded at the fourth LSB layer, then after embedding, the steganographed sample would be $(0\ldots10\mathit{10})_{10}$. Figure 4 shows the example of proposed LSB technique.

The major steps (Fig. 3) are as follows:

1. Read an audio sample from the wave stream.
2. Extract next two bits from the current secret data byte.
3. Place the secret data bits in the selected audio sample's fourth and third LSB bit positions.
4. If the fourth and third LSBs of audio sample are not equal to the secret data bits then do 1's complement of next two LSB bits.
5. Else, Do not perform 1's complement.
6. Concatenate rest of the wave file after completion of embedding secret data.

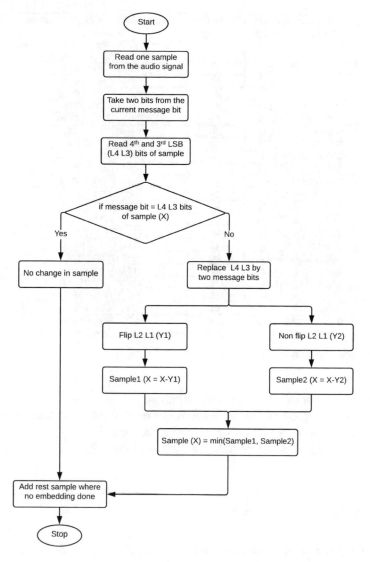

Fig. 3 Flow chart of proposed LSB encoding method

4.2 Audio File Selection

After successful embedding of data, the second task is to select the audio file in such a way that the SNR must be the least. Experimental result shows that for a same size of secret data, SNR is different in different audio file. Audio files are categorized based on sample rate, number of channels and bit depth. The broad level steps are as follows:

Audio sample (16 bit) without secret message	"abcdef" in binary	Audio sample (16 bit) with secret message, 4th and 3rd LSBs are encoded and rest LSBs are 1's complimented
0000000000100100	11	0000000000101111
1000000001001000	00	1000000001000011
1001100000001111	00	1001100000000000
1111110010000100	11	1111110010001111
1000000001000111	10	1000000001001000
0000000001010011	00	0000000001010011
0000000000001010	10	0000000000001010
1000000000010101	11	1000000000011110
1000000001100010	00	1000000001100010
0000000000101101	01	0000000000100110
0000000000100100	11	0000000001101111
1000000000110011	10	1000000000111000
1110000000000100	01	1110000000000100
1111100010000001	00	1111100010000001
1111110010000100	11	1111110010001111
1011111111111010	00	1011111111110001
1101101001111100	10	1101101001111011
1111001111011101	11	1111001111011101
1111110000101011	10	1111110000101011
0000000000110010	01	0000000000110101
0000000000010101	10	0000000000011010

Fig. 4 Proposed LSB encoding technique

1. Select n number of audio files for data hiding.
2. Apply the proposed data hiding algorithm in each of the audio files selected.
3. Calculate SNR of each audio signal after embedding data.
4. Choose the signal to be transmitted having maximum SNR.

5   Results and Discussion

The implementation of proposed model in Sect. 4 follows the given steps:

1. Select different cover audio file of sampling rate 11.05, 22.05 and 44.1 KHz.
2. Keep the audio file size almost same.
3. Keep the text data size as 1 KB and little above.
4. Sampled audio file is treated into audio steganography embedding process using modified two bits LSB.
5. SNR value is calculated between the cover audio and the sampled audio after steganography process.

Results are tabulated in Table 1. The test is performed on Mono and Stereo audio files where the size of the hidden text is 1 Kb. As per our proposed algorithm, we embedded one bit and two bits on different test cases. We also examined the flip bits

Table 1 Comparison study of one and two bits embedding in Mono and Stereo audio

File name	Audio quality	Cover audio size	Number of bits to hide	Flip?	Zero or nonzero sample?	SNR		
						11.025	22.05	44
A	Mono	295 KB	1	Without	Nonzero	52.5	55.47	58.25
A	Mono	295 KB	2	Without	Nonzero	54.77	56.69	58.73
A	Mono	295 KB	1	With	Nonzero	49.87	52.43	54.67
A	Mono	295 KB	2	With	Nonzero	53.81	55.48	57.02
B	Stereo	1 MB	1	Without	Nonzero	55.52	58.46	61.26
B	Stereo	1 MB	2	Without	Nonzero	57.78	59.71	61.74
B	Stereo	1 MB	1	With	Nonzero	52.82	55.33	57.68
B	Stereo	1 MB	2	With	Nonzero	56.77	58.48	60.03

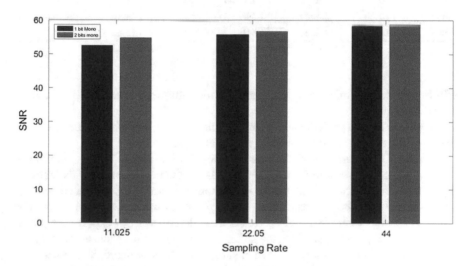

Fig. 5 SNR versus sampling rate of one bit and two bits embedding

or non-flip bits after embedding bits. The meaning of the first row data of Table 1 is—the selected audio file named A is of Mono type and the size of the file is 265 Kb. One bit is embedded inside this audio file, and we have not flipped rest of the bits. For embedding, nonzero sample values of the taken audio have been used. SNR has been calculated on different sampling rate, i.e. 11.025, 22.05 and 44 KHz. Figure 5 depicted the comparison study of one bit and two bits embedding in Mono type audio at different sampling rate.

According to Table 1, it is clear that embedding two bits providing higher SNR than embedding one bit. Another observation is Stereo type audio which is more suitable to hide two bits, and as per the experiment, the SNR is high in this case where no flip of bits is considered. As per this result, audio file B of Stereo type

Table 2 Comparison study with existing work for one bit embedding

File name	Sample rate	Audio quality	Cover audio size	SNR		
				11.025	22.05	44
A	8	Mono	866 KB	25.8	28.6	50
Awo	8	Mono	295 KB	23.7	26.69	48.6
Aw	8	Mono	295 KB	21.83	24.73	45.7
B	16	Mono	1.73 MB	32.8	39.4	68.1
Bwo	16	Mono	295 KB	52.5	55.47	60
Bw	16	Mono	295 KB	49.87	52.43	55.67
C	8	Stereo	1.34 MB	25.0	31.8	44.7
Cwo	8	Stereo	867 KB	26.7	29.73	41.52
Cw	8	Stereo	867 KB	25	36.26	42.28
D	16	Stereo	1.73 MB	37.0	42.9	67.9
Dwo	16	Stereo	1 MB	55.52	58.46	68.4
Dw	16	Stereo	1 MB	52.82	55.33	67.4

will be selected to embed data with consideration that no bits will be flipped after embedding two bits.

In order to test the performance of the designed system, a graph is plotted between the sampling range and corresponding SNR values in Fig. 5.

In Table 2, comparisons of variations among SNR values [15] have been taken under consideration. Audio files A, B, C and D are the files taken from the existing paper. Awo, Bwo, Cwo and Dwo are the files where no bits are flipped after embedding hidden text at the third bit position as per the proposed work. Aw, Bw, Cw and Dw are the audio files where bit flipping is under consideration. In this experiment, single bit is embedded per sample. From the table, we can see that the same amount data is embedded in small size audio, and when sampling rate is changed, SNR values are changed accordingly. The Stereo audio performed better than Mono audio.

Figure 6 depicts the Guitar .wav file and the structure of samples before and after embedding a secret data is depicted in Fig. 7. The experiment is repeated in Piano

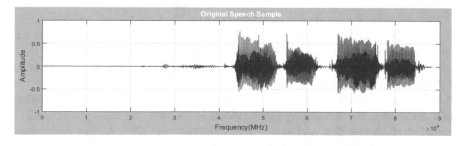

Fig. 6 Guitar .wav file

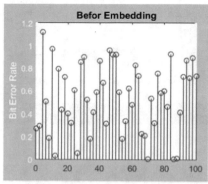
(a) Guitar .wav file before embedding

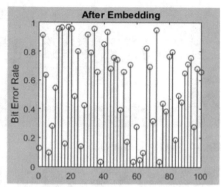
(b) Guitar .wav file aftere mbedding

Fig. 7 Change in samples in Guitar .wav file

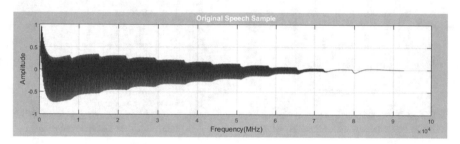

Fig. 8 Piano .wav file

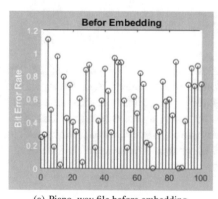
(a) Piano .wav file before embedding

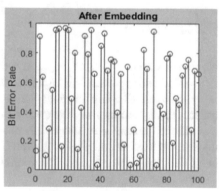
(b) Piano .wav file after embedding

Fig. 9 Change in samples in Piano .wav file

.wav file and a pop music .wav files. Figures 8 and 10 depict the original .wav file of Piano and pop music. Figures 9 and 11 provide sample structure of Piano and pop music before and after embedding secret data.

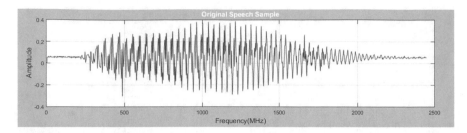

Fig. 10 Pop music .wav file

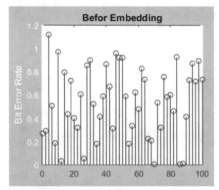

(a) Pop music .wav file before embedding

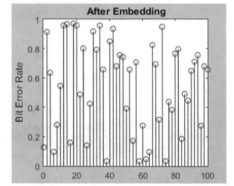

(b) Pop music .wav file after embedding

Fig. 11 Change in samples in Piano .wav file

6 Conclusion

We show a structure for steganography utilizing LSB encoding on audio signal. The idea behind this approach is to minimize distortion where data hiding capacity should be maximized. Keeping in mind the perceptual transparency of human auditory system (HAS), our approach is also to find the audio signal where the SNR is the least. The algorithm takes lower bit error as compared to the existing techniques. The algorithm is flexible to choose required LBS bit positions for a set of audio signals, and it also finds where the error is minimum.

References

1. Asad M, Gilani J, Khalid A (2011) An enhanced least significant bit modification technique for audio steganography. In: 2011 international conference on computer networks and information technology (ICCNIT), pp 143–147
2. Rana L, Banerjee S (2013) Dual layer randomization in audio steganography using random byte position encoding. Int J Eng Innov Technol 2(8)

3. Gandhi K, Garg G (2012) Modified LSB audio steganography approach. Int J Emerg Technol Adv Eng 3(6):158–161
4. Priyanka B, Vrushabh K, Komal P (2012) Audio steganography using LSB. Int J Electron Commun Soft Comput Sci Eng 90–92
5. Divya SS, Reddy MRM (2012) Hiding text in audio using multiple LSB steganography and provide security using cryptography. Int J Sci Technol Res 1:68–70
6. Kundu N, Kaur A (2017) A secure approach to audio steganography. Int J Eng Trends Technol (IJETT) 44:1–7
7. Gadicha AB (2011) Audio wave steganography. Int J Soft Comput Eng (IJSCE) 1:174–177
8. Zamani M et al (2009) A secure audio steganography approach. In: International conference for internet technology and secured transactions
9. Sridevi R, Damodaram A, Narasimham S (2009) Efficient method of audio steganography by modified LSB algorithm and strong encryption key with enhanced security. J Theor Appl Inf Technol 771–778
10. Djebbar F, Ayad B, Meraim KA, Hamam H (2012) Comparative study of digital audio steganography techniques. EURASIP J Audio Speech Music Process
11. Garg P, Sharma M, Agarwal S, Kumar Y (2018) Security on cloud computing using split algorithm along with cryptography and steganography. In: ICICC'18, vol 1, pp 71–79
12. Gupta P, Bhagat J (2018) Image steganography using LSB substitution facilitated by shared password. In: ICICC'18, pp 369–376
13. Roy S, Manasmita M (2011) A novel approach to format based text steganography. In: ICCCS'11, February 12–14
14. Roy S, Parida J, Singh AK, Sairam AS (2012) Audio steganography using LSB encoding technique with increased capacity and bit error rate optimization. In: International conference on computational science engineering and information technology, pp 372–376
15. Ramya Devi R, Pugazhenthi D (2016) Ideal sampling rate to reduce distortion in audio steganography. Procedia Comput Sci 85:418–424

Cognitive Fatigue Detection in Vehicular Drivers Using K-Means Algorithm

Manish Kumar Sharma and Mahesh Bundele

Abstract In vehicular drivers, the cognitive fatigue has been one of the major factors that leads to loss of lives or disabilities due to vehicular accidents (Bundele and Banerjee in Proceedings of the 11th international conference on information integration and web-based applications and services, ACM, pp 739–744, 2009 [1]). The factors governing driver fatigue such as monotonous driving, traffic conditions on road, road conditions, insufficient sleep, anxiety, health conditions, work environment, type of vehicle, and driving comfort do affect largely the driving behavior. Many researchers across the world are working finding best suitable methods to minimize vehicular accidents. In this paper, our proposed approach is an alternative solution to detect cognitive fatigue in vehicular drivers to decrease the number of incidence of vehicular accidents specially those occurred due to cognitive fatigue (Hu and Zheng in Expert Syst Appl 36:7651–7658, 2009 [2]). The objective of this work has been to provide simple classification technique using K-means algorithms. The basic K-means and two modified versions have been proposed and validated to reliably detect cognitive fatigue while driving. It is of paramount importance that the sensory parameters are chosen such that they could be used without causing discomfort to the driver and without creating obstruction while driving. The data for simple physiological signals such as skin conductance (SC), oximetry pulse (OP), and respiration (RSP) for pre- and post-driving state of drivers has been used for sensing change in fatigue level of the drivers (Bundele and Banerjee in 2009 second international conference on emerging trends in engineering & technology. IEEE, pp 934–939, 2009 [3]). All features of statistical and wavelet were extracted and analyzed (Brown in D1 Methodological issues in driver fatigue research. Fatigue and driving: driver impairment, driver fatigue, and driving simulation, p 155, 1995 [4]). Selected features were used as input to the classifiers designed and implemented using basic K-means and two modified versions. Finally, comparative performance analysis of classifiers and the features is discussed. This paper discusses prominent results obtained during experimentation.

M. K. Sharma (✉) · M. Bundele
Computer Engineering, Poornima College of Engineering, Jaipur, Rajasthan, India
e-mail: kumar.manish.sharma786@gmail.com

M. Bundele
e-mail: maheshbundele@poornima.org

© Springer Nature Singapore Pte Ltd. 2020
A. Khanna et al. (eds.), *International Conference on Innovative Computing and Communications*, Advances in Intelligent Systems and Computing 1059,
https://doi.org/10.1007/978-981-15-0324-5_44

Further the maximum classification accuracy could be produced by these features, so as to reduce the computational complexity (Bundele and Banerjee in Proceedings of the 11th international conference on information integration and web-based applications and services, ACM, pp 739–744, 2009 [1]). It could be found that a smaller set of features could provide the correctness of fatigue detection.

Keywords Vehicular driver · Skin conductance · Oximetry pulse signal · Respiration · Cognitive fatigue · Wavelet features · K-means classifier

1 Introduction

Lot of work has been done on cognitive fatigue detection or drowsiness detection of vehicular driver. The researchers have used various physiological signals like EEG, ECG, EMG [5], EOG [2], etc., visual cues like eye blink patterns, eyelid closer, face expressions, vehicular behavioral parameters such as lane departure and acceleration/deceleration pattern. The simple and realistic system could be provided by the use of simple type of physiological signals like skin conductance, oximetry pulse, and respiration [3]. However, there are problems with either in precision of fatigue detection or acceptability by the drivers. The physiological signals and respective extraction of features were obtained as input to many classification techniques using SVM, fuzzy systems, neural network, and statistical methods. Although many classification methodologies have been discussed in the literature, no perfect solution is presented which can be easily accepted by one and all. The proposed work targeted for simplifying the classifier design, its implementation, and performance. K-means is a simple algorithm which makes the classifier simple and fast. We have designed three classifiers using basic K-means and two modifications proposed along with simple physiological parameters SC, OP, and RSP. The performance of these three classifiers is compared.

2 System Architecture

Figure 1 shows general architecture of the proposed cognitive fatigue detection system. The sensory signals like SC, OP, and RSP can be recorded through data acquisition and signal conditioning equipment. In the proposed work, SC sensor NX-GSR1A, OP sensor-NONIN model 8000AA finger clip sensor 1M [1], and RSP sensor-NX-RSP1A were used in conjunction with Nexus-10, a physiological monitoring in 10 channels and signal processing equipment of biofeedback, which utilizes wireless communication with Bluetooth 1.1 class 2 and flash memory technologies [6]. These signals are nonlinear time-varying in nature that makes it difficult to correlate them directly with fatigue. Further, it is necessary to compute certain unique

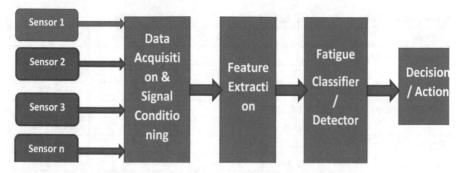

Fig. 1 System architecture

features that can represent the intended relationship. The architecture of fatigue classifier/detector mainly depends upon inputs. There are large number of possibilities for extracting typical features so required. Statistical and wavelet features were after number of iterations could be finalized as input to the classifier. Here, K-means and two modified versions are used for classification of input signals individually and in combination.

3 Data Collection and Processing

Three- to five-minute duration signals of each kind (SC, OP, RSP) under pre- and post-driving states were recorded from 150 vehicular drivers using Nexus 10 and sensors which utilizes wireless communication with the technologies of flash memory and Bluetooth 1.1 class 2 [6] for transferring on to remotely located laptop. Biotrace+™ software tool has been used for interfacing the hardware equipment. The state of drivers as normal (pre-driving) and fatigue (post-driving: 500–600 km) were ensured through questionnaire designed in consultation with medical doctors. The signals were analyzed, and the set (pre and post) of 10 driver's data has been used for experimentation. The recorded signals were segmented into frames of 1 s duration. Two types of features were extracted for each frame, viz statistical and wavelet features. Statistical features incorporated *Frame Energy (FE), Maximum of Signal (MOS), Maximum of Frequency Spectrum (MAXF), Standard Deviation of Signal (STDEVS), Standard Deviation (SD) of Frequency Spectrum (STDFS)*, and *Mean of Frequency Spectrum (MOFS)* [7]. *Further slope and gradient of these features were computed leading to* total 18 statistical features were used for experimental analysis. Wavelet features were extracted for each frame using different mother wavelets such as Daubechies, Bior, and D Meyer with varying order and level of decomposition. Finally, two feature sets of SC, statistical features (SC_Stat) and Daubechies order 3 level 6 (SC_DB3_L6), Two feature sets of OP, with Biorlet order 1.1 level 4 (OP_Bior1.1_L4) and D Meyer level 8 (OP_DMey_L8), and Two feature sets of RSP,

Table 1 Details of feature vectors used

Signals	Feature file	No. of features	Training matrix	Testing matrix
Skin conductance	SC_STAT_1	6	400 × 6	400 × 6
	SC_STAT_2	18	400 × 18	400 × 18
	SC_DB3_L6	77	400 × 77	400 × 77
Oximetry pulse	OP_BIOR1.1_L4	60	400 × 60	400 × 60
	OP_DMey_L8	110	400 × 110	400 × 110
Respiration	RSP_DB6_L3	37	400 × 37	400 × 37
	RSP_DMey_L4	56	400 × 56	400 × 56

Daubechies order 6 Level 3 (RSP_DB6_L3) and D Meyer Level 4 (RSP_DMey_L4) were used. Minimum, maximum, mode, mean, variance, and standard deviation [2] of decomposed coefficients and reconstructed coefficients were considered along with entropy. Table 1 shows details of feature sets used for individual parameters SC, OP, and RSP.

4 Design of Fatigue Classifiers

The feature vectors so derived were used to test three algorithms used as fatigue classifier.

- Basic version of K-means classifier
- Modified version 1 of K-means classifier
- Modified version 2 of K-means classifier.

4.1 Algorithm to Compute the Basic Version of K-Mean Classifier

Design and implementation flow of basic version of K-means classifier can be represented with the following steps:

   (row = 0–399 and 400–799 are replicates; total 1200 rows; first 400 are used as training, and last 400 used as testing dataset)

Step 1: Read an Excel data sheet of two different classes of data of (pre- and post-driving).

Step 2: In each excel sheet, calculate the number of rows and columns.

Step 3: In the starting, the dataset from row 0 to row/3 and column from 0 to total no of columns is stored into 2-D array along with row tag.

Step 4: For row = 0 to rows/3 and for column = 0 to (number of columns-1).

Step 5: The cluster centers of two different classes assign into initial two values of each column.

Step 6: In each column, measure the minimum distance between next data point to the consistent cluster centers.

Step 7: For each column, allocate data point to respective cluster with minimum distance.

Step 8: The values of cluster centers of consistent columns are updated.

Step 9: In each column, repeatedly measure the distance between new cluster centers and assigned data points.

Step 10: Stop the loop if no data point of each column was reassigned; otherwise, go to step 6 for corresponding column.

Step 11: Then, dataset from row $i = $ rows/3 $+ 1$ to 2 $*$ rows/3 and column 0 to total no. of columns are stored into 2-D array along with row tag.

Step 12: Assign the each column data points to resultant cluster on the basis of minimum distance with cluster center.

Step 13: Write an arranged confusion matrix of training data for dataset from row = rows/3 $+ 1$ to 2 $*$ rows/3 and column 0 to total no of columns to output file.

Step 14: Then, the dataset from row $= 2 *$ rows/3 $+ 1$ to 3 $*$ rows/3 and column $= 0$ to total no. of columns are stored into 2-D array along with row tag.

Step 15: Assign the each column data points to resultant cluster on the basis of minimum distance with cluster center.

Step 16: Write an arranged confusion matrix of training data for dataset from row $= 2 *$ rows/3 $+ 1$ to 3 $*$ rows/3 and column 0 to total no of columns to output file.

As shown in Table 1, each feature set has different number of feature vectors depending upon whether it is statistical feature set or level of wavelet decomposition. The two class feature values were randomly placed in the same column and were tagged. Each feature was treated separately in order to know the contribution of the feature in detecting fatigue state of the driver. While implementing basic K-means on these feature vectors, starting two feature values of each feature vector has been considered as initial cluster centers of two classes (normal and fatigue). The distance between the initial centers and the next feature value of corresponding vector was computed, and the feature value has been added to the cluster having minimum of the distances computed. Likewise, the process was repeated for all values of each of the feature vectors. The data was partitioned into training set and testing set.

4.2 Modified Version 1 of K-Means Classifier

The basic version of K-means classifier has been modified through considering percentile value of whole feature vector as initial cluster rather than considering feature value. Firstly the basic statistical parameters like mean, variance, and standard deviation were computed for each feature vector and later the percentile was computed that could fetch initial cluster center values. Next minimum distance data points from the cluster centers of similar category to cluster head of each column is assigned. The process was repeated for all data values. This classifier has been designed to evaluate fatigue state of the driver based on each of the feature vector individually.

The design and implementation flow of the modified version 1 of K-means classifier can be represented with the following steps:

(row = 0–399 and 400–799 are replicates; total 1200 rows; first 400 are used as training, and last 400 used as testing dataset)

Step 1: Read an Excel data sheet of two different classes of data of (pre- and post-driving).

Step 2: In each excel sheet, calculate the number of rows and columns.

Step 3: In the starting, the dataset from row 0 to row/3 and column from 0 to total no of columns is stored into 2-D array along with row tag.

Step 4: For row = 0 to rows/3 and for column = 0 to (number of columns-1).

Step 5: For each column to two cluster centers, calculate the values of mean, variance, standard deviation, and percentile.

Step 6: The cluster centers of two different classes assign the calculated attribute values of each column.

Step 7: For each column, allocate data point to respective cluster with minimum distance.

Step 8: The values of cluster centers of consistent columns are updated.

Step 9: In each column, repeatedly calculate the distance between new cluster centers and assigned data points.

Step 10: Stop the loop if no data point of each column was reassigned; otherwise, go to step 6 for corresponding column.

Step 11: Then, dataset from row $i =$ rows/3 + 1 to 2 * rows/3 and column 0 to total no. of columns is stored into 2-D array along with row tag.

Step 12: Assign the each column data points to resultant cluster on the basis of minimum distance with cluster center.

Step 13: Write an arranged confusion matrix of training data for dataset from row = rows/3 + 1 to 2 * rows/3 and column 0 to total no. of columns to output file.

Step 14: Then, the dataset from row = 2 * rows/3 + 1 to 3 * rows/3 and column = 0 to total no. of columns is stored into 2-D array along with row tag.

Step 15: Assign the each column data points to resultant cluster on the basis of minimum distance with cluster center.

Step 16: Write an arranged confusion matrix of training data for dataset from row = 2 * rows/3 + 1 to 3 * rows/3 and column 0 to total no. of columns to output file.

4.3 Modified Version 2 of K-Means Classifier

In this modification, the decision of fatigue state of the vehicular driver has been assessed on the basis of again individual input feature vector. The basic K-means classifier is implemented through considering the feature values as members of an object array of each class rather than considering each value separately.

The design and implementation flow of the modified version 2 of K-means algorithm as a classifier can be represented with the following steps.

(row = 0–399 and 400–799 are replicates; total 1200 rows; first 400 are used as training, and last 400 used as testing dataset)

Step 1: Read an Excel data sheet of two different classes of data of (pre- and post-driving).

Step 2: Measure the number of the rows.

Step 3: In the starting, make a class of classification that contains member variables of numeric attributes with string.

Step 4: In the classification class, the dataset row = 0 to rows/3 is stored into 1-D object array.

Step 5: For row = 0 to rows/3.

Step 6: In the classification class, assign the initial two values of 1-D object array to corresponding class 1-D cluster centers array of classification class to store the 2-cluster center values.

Step 7: Read and measure the distance between next data element in 1-D object array to the corresponding cluster centers values in 1-D cluster center array.

Step 8: The minimum distance data elements assign to the cluster.

Step 9: Each cluster centers data values updated.

Step 10: Repeatedly calculate the distance between new 1-D array of cluster centers and previously assigned data point of 1-D object array.

Step 11: If no data point of 1-D object array was reassigned, then stop; otherwise, go to step 6.

Step 12: Then store the dataset row = rows/3 + 1 to 2 ∗ rows/3 into 1-D object array along with row tag.

Step 13: Assign each of the data element of 1-D object array to corresponding cluster on the basis of minimum distance with the 1-D current cluster centers object array.

Step 14: Write an arranged confusion matrix of training data for dataset from row = rows/3 + 1 to 2 ∗ rows/3 to output file.

Step 15: Then, store the dataset row = 2 ∗ rows/3 + 1 to 3 ∗ rows/3 into 1-D array along with row tag.

Step 16: Assign each of the data elements of 1-D object array to corresponding cluster on the basis of minimum distance with the element of 1-D current cluster centers object array.

Step 17: Write an arranged confusion matrix of training data for dataset from row = 2 ∗ rows/3 + 1 to 3 ∗ rows/3 to output file.

4.4   Performance Parameters Used

The classifier designed is evaluated by constituting confusion matrix and calculating average percentage classification accuracy (PCLA). It has been used as the main parameter for performance analysis of the classifiers and those of features. It is average of percentage accuracy of normal and fatigue state obtained.

Percentage classification accuracy of individual class is determined by

Classification Accuracy (CA) = Number of Correct Classifications/Total Number of Class Values * 100

PCLA = (CA of Normal Class + CA of Fatigue Class)/2.

5   Results and Discussion

5.1   Comparative Performance Analysis of Classifiers for SC

Three sets of features of SC were used for detecting fatigue state and testing the performance of the classifiers. The feature set-I-SC_STAT_1, II-SC_STAT_2, and III-SC_DB3_L6 were analyzed. For experimental purpose, it is assumed that all 10 drivers exhibit similar parametric values at normal and fatigue states. After mixing, normal and fatigue state features of 10 drivers 400 row (pre and post) were randomly selected as training data while other 400 rows (pre and post) as test data. Table 2 shows PCLA obtained for test data of each of the features in three feature sets when used with three classifiers. It can be seen that for the first set of features where only six features were used, the maximum PCLA of 80% could be achieved with FE, STDFS, and MOS and all three classifiers fetched the same results. In the second set of features where 18 feature vectors were used, FE could give 100% accuracy with modified K-means V2 and MOS could fetch 100% with basic K-means classifier. However, former classifier could give 86.5% PCLA for MOFS, STDFS, and MOS. Wavelet features set III level 6 approximate coefficient of decomposed and reconstructed wave fetched 100% average accuracy for basic and modified version 2 classifiers. Further entropy of the signal could provide better detection accuracy. From the results tabulated, it can be seen that maximum, mean, mode, and minimum values of CA6 and A6 have given 100% classification accuracy for basic and modified version 2 K-means algorithms. With reference to design of cognitive fatigue detection using SC as input physiological parameter, it can be designed with limited best performing features out of three feature sets and decision based on concurrence of results be used as final decision. This would lead to a simple and fast fatigue detection in vehicular drivers.

Table 2 Performance analysis of SC-based classifiers

Features	Basic K-means	Modified K-means version 1	Modified K-means version 2
Skin conductance feature set-I			
FE	80	80	80
MOFS	47	78.75	47
STDFS	80	80	80
MOS	80	80	80
Skin conductance feature set-II			
FE	61.25	17.25	100
MOFS	84.25	53	86.5
STDFS	50	50	86.5
MOS	100	86.5	86.5
STDFS	27.25	74.5	22.75
Skin conductance feature set-III			
MAX CA6	100	0	100
MIN CA6	100	0	100
MEANCA6	100	0	100
VAR CA6	44.5	52.25	44.5
MODE CA6	100	0	100
MAX A6	100	0	100
MIN A6	100	0	100
MEANA6	100	0	100
MODE A6	100	0	100
MIN D1	82.25	69	49.5
MODE D1	85.75	67.5	50.75
MIN D4	76.75	54	67.25
MODE D4	76.75	45	68.75
VAR D6	47.5	79.25	44.75
ENTROPY	100	0	98.5

5.2 Comparative Performance Analysis of Classifiers for OP

For oximetry pulse OP, two feature sets were selected one with bi-orthogonal mother wavelet with 1.1 order decomposed at level 4 and another D Meyer mother wavelet decomposed at level 8. A couple of scaling functions and associative features one for synthesis and one for analysis in bi-orthogonal wavelets feature. For analysis, it offers various number of vanishing moments for the analysis. Meyer wavelet also provides better analysis strategy. Table 3 shows the results obtained for feature set I OP-Bior1.1_L4 when tested with test dataset. Total 60 features were tested for 400

Features	Basic K-means	Modified K-means version 1	Modified K-means version 2
MAX CA4	69.5	45	92
STD CA4	45.25	51	84.5
MAX CD1	78.5	54.75	83
MIN CD1	48.25	50	82.25
STD CD1	65	64.25	77.75
MAX CD2	70.25	71.75	78.25
MIN CD2	48.75	50.25	79
STD CD2	68	35.75	79.75
VAR CD2	60	54.5	72.5
MAX CD3	61.25	48.75	77.75
MIN CD3	86.25	52.5	80
STD CD3	80.75	81.25	79.5
MODE CD3	83.75	50.25	81.5
MAX A4	91.25	50.5	92
STD A4	64.5	49	83.25
MAX D1	76.25	84	76.25
MIN D1	60.5	50	76.25
STD D1	59.5	67.75	79.25
MAX D2	77	81.5	77
MIN D2	67.75	50	77
STD D2	58	74	79.75
MAX D3	78.25	79.75	78.25
MIN D3	61.75	58	78.25
STD D3	73	46.75	78.5
MODE D3	69.5	64.25	76.5
STD D2	58	74	79.75

Table 3 Performance analysis of oximetry pulse set I-based classifiers

frames, and it can be seen that maximum of 92% accuracy could be obtained for Max-CA4 and Max-A4 with third classifier. Basic K-means also performed nearly well for Max-A4 with 91.25% accuracy. If we compare three classifiers, modified version 2 performed better than other two classifiers except Mode CD3 and Min CD3. Here, MAX CD1, MIN CD3, STD CD3, MODE CD3, MAX A4MAX D1, and MAX D3 can be considered as valuable comparative inputs for first classifier leading to more than 75% accuracy, whereas for the third classifier, all the features shown in table give more than 75% accuracy. Keeping 80% as the criteria for selecting the features, numbers can further be reduced making classifiers simpler. The performance of the

Table 4 Performance analysis of oximetry pulse set II-based classifiers

Features	Basic K-means	Modified K-means version 1	Modified K-means version 2
MAX CA8	98.25	98.5	99.5
MIN CA8	72	64.25	85.5
MEAN CA8	93.75	85.5	94.75
MODE CA8	72	64.25	85.5
MAX CD1	81.75	85.75	67
VAR CD1	92	57.25	57.25
MAX CD2	83.75	73	68
MIN CD2	88.25	15.5	68.5
MIN CD5	83.5	50	69.75
MIN CD6	90.25	50	71
MEAN CD7	94	71.75	94.75
MEAN CD8	90	71.75	94.75
MAX A8	97.5	76.25	97.5
MIN A8	91.75	67	93.5
MODE A8	91.75	64.75	93.5
VAR D1	90.75	42	57.25
MAX D2	83.75	42.25	67.75
MEAN D6	49	79	82.25
MIN D7	58.5	50	79
STD D7	42.25	80	67.5
MEAN D7	33	48.75	76.25
MODE D7	58.5	50	79
MIN D8	92.25	50	95.25
MODE D8	92.25	50	95.25
ENTROPY	88.75	50	88.75

second classifier is poor as compared to other two. It could give more than 75% PCLA for only STD CD3, MAX D1, MAX D2, and MAX D3.

Table 4 shows the results obtained for test dataset of set II, i.e., OP_DMey_L8. It has total 110 feature vectors. The table shows only those feature which could contribute well above decided level of accuracy. Here, MAX CA8 has performed the best with all the three classifiers with maximum PCLA of 99.5% for the third classifier, 98.5% with the second classifier, and 98.25% PCLA for the first classifier. Another feature MEAN CA8 also fetched the results above 90% for first and third classifiers, whereas above 85% for second. MAX A8 has given accuracy above 97% for first and third classifiers, whereas below 80% for the second. Here, if we observe the results, the third classifier has given better performance in all the cases but first classifier also fetched comparable results. For designing fatigue detection system based on OP features, it would be best to consider the third classifier for biorthogonal feature set and for Meyers features classifier first of third with appropriate features can be selected. Feature vectors MODE CA8, MEAN D6, MIN D7, STD D7, MEAN D7, and MODE D7 did not perform well with first classifier. Features MAX CD1, VAR CD1, MAX CD2, MIN CD2, MIN CD5, MIN CD6, VAR D1, MAX D2, and STD D7 for modified version 3 classier did not perform well. The performance analysis of three classifiers with two feature sets derived from OP signal of two classes could lead to conclude that either of first and third classifiers with limited number of Biorlet or D Meyers wavelet features can give classification accuracy more than 99%. A multi-parametric approach with selected feature vectors having concurrency of decision may lead to design of simple and robust fatigue detection system for vehicular drivers.

5.3 Comparative Performance Analysis of Classifiers for RSP

Two feature vectors were derived from 1 s RSP signal frame for normal and fatigue drivers. Feature set II includes Daubechies wavelet features with order 6 and level 3 of decomposition, and Feature set I includes D Meyer's wavelet decomposed at level 4. Table 5 shows the results obtained for feature set II. Total 56 features were considered for testing the three classifiers designed. The results for limited number of features those fetched good classification accuracy are shown in the table. It can be seen that classifiers first and third could contribute up to 100% PCLA for MIN CA4, MEAN CA4, MODE CA4, MAX A4, MIN A4, MEAN A4, and MODE A4. All other features and the second classifier performed badly. Table 6 shows performance results of the three classifiers for the second set of features derived from RSP signal. Here, the first and third classifier could produce 100% classification accuracy of maximum, minimum, mean, and mode of CA3 and A3. Second classifier did not perform well. In order to design the fatigue detection system with RSP as input less number of feature vectors can provide better results. It can be concluded that D Meyer's level 4 approximate and reconstructed coefficient carries better distinction of information related to fatigue state of the driver in the first case, whereas in the second, it is level 3 coefficients.

5.4 Comparative Analysis of All Features

Table 7 shows the features those could fetch PCLA above 80%. It can be seen that classifiers first and third can be designed by significant reduction in number feature vectors of SC, OP, and RSP either independently or in combination.

Table 5 Performance analysis of respiration set I-based classifiers

Features	Basic K-means	Modified K-means version 1	Modified K-means version 2
MIN CA4	100	0	100
MEAN CA4	100	0	100
MODE CA4	100	0	100
MAX CD1	74	76.75	72.5
STD CD2	75.25	45.25	75.25
MODE CD2	68	50	75.5
MAX CD3	69.75	49.5	75.25
STD CD3	74	45.25	75.25
MAX A4	100	0	100
MIN A4	100	0	100
MEAN A4	100	0	100
VAR A4	55	71.25	40.75
MODE A4	100	0	100
ENERGY D2	76.5	71.25	58.5
ENERGY D3	73	58.5	58.5
ENTROPY	100	100	100

Table 6 Performance analysis of respiration set II-based classifiers

Features	Basic K-means	Modified K-means version 1	Modified K-means version 2
MAX CA3	100	0	100
MIN CA3	100	0	100
MEAN CA3	100	0	100
MODE CA3	100	0	100
MAX A3	100	0	100
MIN A3	100	0	100
MEAN A3	100	0	100
MODE A3	100	0	100
MIN D3	70	50	62.75
STD D3	72	50	65.25

Table 7 Optimal performing features with classifiers

Feature file	Total features	Basic K-means	Modified K-means version 1	Modified K-means version 2
Feature set I	6	NIL	NIL	NIL
Feature set II	18	NIL	MOS	FE, MOFS, STDFS, MOS
Feature set III SC_DB3_L6	77	MAX, MIN, MEAN and MODE of CA6 and A6, Entropy (9 features)	NIL	MAX, MIN, MEAN and MODE of CA6 and A6 and Entropy (9 features)
Feature set I OP_BIOR1.1_L4	60	MAX A4	NIL	MAX CA4 and MAX A4
Feature set II OP_DMey_L8	110	MAX, MEAN of CA8, VAR, MODE of CD1, MAX, MEAN, MODE of A8, VAR CD2, MODE D8 and entropy (10 features)	MAX CA8	MAX, MEAN of CA8, MEAN CD7, MEAN CD8, MAX, MEAN, MIN, MODE of A8, MIN and MODE of D8 (10)
Feature set I RSP_DMey_L4	56	MIN, MEAN, MODE of CA4 and A4 with MAX A4, and Entropy (8 features)	NIL	MAX, MIN, MEAN, MODE of CA4 and A4 and Entropy (9 features)
Feature set II RSP_DB6_L3	33	MAX, MIN, MEAN and MODE of CA3 and A3 (8 features)	NIL	MAX, MIN, MEAN, MODE of CA3 and A3 (8 features)

6 Conclusion

We were implemented three versions of K-means algorithms in JAVA, and the use of K-means algorithm as cognitive fatigue detection classifier has been exhaustively carried out. We tested the performance of these algorithms with consideration of selected two classes: pre-driving and post-driving physiological parameters. Various statistical and wavelet features including Daubechies wavelet function, Bior's wavelet function, and D Meyer's wavelet function were derived from SC, OP, and RSP signals of these two classes. Three kinds of feature sets were derived from SC files, two kinds in OP and RSP too. We used these feature files as input, and separately the performance of each feature has been tested through these three versions of K-means algorithms in order to find out features set which can perform best and as alternative solution and the best performing version of K-means. Average percentage classification accuracy has been used as performance measures to test the performance of the classifiers. After exhaustive experimentations, it can finally be concluded that the proposed algorithms very well perform in line through the present approaches with some selected features of physiological signals like skin conductance, oximetry pulse, and respiration.

- Many of the features of three physiological parameters like skin conductance, oximetry pulse, and respiration could perform very well with basic K-means algorithm and modified K-means algorithm version 2.
- The performance could be retained with reduced number of features.
- Although this work suggests the reduced set of features to be used for fatigue detection, the performance of these features in combination has not been tested.
- It is recommended that by using multi-input system with multi-parameters with selected features and a committee at last can lead to a better fatigue detection system.

Acknowledgements It is great pleasure to acknowledge the vehicular drivers to allow us for collecting the real-time data in this research work. We also wish to extend our sincere thanks to Poornima University and Poornima College of Engineering, Jaipur, for providing us good environment for work and good management.

References

1. Bundele MM, Banerjee R (2009) Detection of fatigue of vehicular driver using skin conductance and oximetry pulse: a neural network approach. In: Proceedings of the 11th international conference on information integration and web-based applications and services. ACM, pp 739–744
2. Hu S, Zheng G (2009) Driver drowsiness detection with eyelid related parameters by support vector machine. Expert Syst Appl 36(4):7651–7658. www.elsevier.com/locate/eswa
3. Bundele MM, Banerjee R (2009) An SVM classifier for fatigue-detection using skin conductance for use in the BITS-Lifeguard wearable computing system. In: 2009 second international conference on emerging trends in engineering and technology. IEEE, pp 934–939
4. Brown I (1995) D1Methodological issues in driver fatigue research. Fatigue and driving: driver impairment, driver fatigue, and driving simulation, p 155
5. Yeo MVM, Li X, Shen K, Wilder-Smith EPV (2009) Can SVM be used for automatic EEG detection of drowsiness during car driving? Saf Sci 47(1):115–124
6. Yang G, Yingzi L, Bhattacharya P (2005) A driver fatigue recognition model using fusion of multiple features. In: 2005 IEEE international conference on systems, man and cybernetics, vol 2. IEEE, pp 1777–1784
7. Sharma MK, Bundele MM (2005) Design & analysis of performance of K-means algorithm for cognitive fatigue detection in vehicular drivers using skin conductance signal. In: 2015 2nd international conference on computing for sustainable global development (INDIACom), IEEE, pp 707–712

8. Zhang C, Zheng C-X, Yu X-L (2009) Automatic recognition of cognitive fatigue from physiological indices by using wavelet packet transform and kernel learning algorithms. Expert Syst Appl 36(3):4664–4671. www.elsevier.com/locate/eswa

9. Eoh HJ, Chung MK, Kim S-H (2005) Electroencephalographic study of drowsiness in simulated driving with sleep deprivation. Int J Ind Ergon 35(4):307–320. www.elsevier.com/locate/ergon

10. Bundele MM (2008) Identification of body parameters for changes in reflexes of a vehicular driver under drowsiness/fatigue/stress conditions. Published in the proceeding of Frontier, pp 123–131

11. Shen K-Q, Ong C-J, Li X-P, Hui Z, Wilder-Smith EPV (2007) A feature selection method for multilevel mental fatigue EEG classification. IEEE Trans Biomed Eng 54(7):1231–1237

12. Sharma MK, Bundele MM (2015) Design & analysis of k-means algorithm for cognitive fatigue detection in vehicular driver using oximetry pulse signal. In: 2015 international conference on computer, communication and control (IC4). IEEE, pp 1–6

Optimal Route Selection for Error Localization in Real-Time Wireless Sensor Networks (ORSEL)

Anuj Kumar Jain, Sandip Goel, Devendra Prasad and Avinash Sharma

Abstract Wireless sensor networks provide a wide array of applications. They have crucial application areas of intrusion detection, fire detection, or several other critical information reporting applications. In such condition, any information delayed is information of no use. Thus, real-time data delivery is essential. An optimal route for fast data delivery and fault-tolerant, error-free operations are keys to such a real-time wireless sensor networks. In this paper, we propose an optimal route selection for error localization in real-time wireless sensor network (ORSEL) scheme. It can guarantee soft reliability in real-time data delivery and provides improved miss ratio of data packets. It considers the fact that there are a number of reasons that increase the end-to-end delay in data delivery in WSN. A packet en-route can be lost due to channel contention, interference, link break, dead nodes in the path, malicious/compromised node/s in path or simply die in a queue of retransmission in a node on the route. ORSEL uses a communication link trust (CLT) value of a link between two nodes which acts as the factor of deciding the minimum end-to-end delay in delivery of packet while avoiding the interference and malfunctioning nodes in the network, if any. As the results received, it is verified that the number of packets successfully delivered to the sink has increased and the miss ratio and cumulative packets consumption have improved.

Keywords Localization · Fault tolerant · Optimal route selection · Real-time data delivery · Real-time wireless sensor networks

A. K. Jain (✉) · S. Goel · A. Sharma
CSE Department, MM (Deemed to be University), Mullana, Haryana, India
e-mail: jainanuj143@gmail.com

S. Goel
e-mail: skgmmec@gmail.com

A. Sharma
e-mail: asharma@mmumullana.org

D. Prasad
CSE Department, Panipat Institute of Engineering & Technology, Samalkha, Haryana, India
e-mail: devendracad@gmail.com

© Springer Nature Singapore Pte Ltd. 2020   533
A. Khanna et al. (eds.), *International Conference on Innovative Computing and Communications*, Advances in Intelligent Systems and Computing 1059,
https://doi.org/10.1007/978-981-15-0324-5_45

1 Introduction

Numerous applications in wireless sensor networks (WSN) reckon on timely transmitted information. Disaster management system, based on WSN, must broadcast discovered incidents in the stated real-time deadline. Similarly, an intrusion detection system should report detection of intruders to authorities immediately. Considering lossy nature of WSN links, soft probabilistic and real-time guarantee protocols are designed for real-time communication. Point-to-point packet delivery delay between source and destination nodes may be affected by many factors. Retransmission of the packets, induced by channel allocation and unreliable wireless network, is the main issue. Increasing transmission power, to improve the quality of network is a usual way. Due to multiple advantages of modulating transmission power, it helps in reducing point-to-point delay. Number of retransmissions can be reduced by improving signal-to-noise ratio with greater power transmission. Higher power transmission is supported by most of sensor hardware nowadays. Moreover, increase in each node's peripheral area may help in reducing required number of the hops for source-to-destination delivery. Adaptation of transmission power at each node can help in achieving the desired delays, but it can also increase channel conflicts and interference which results in reduction of network capacity. In this case, providing real-time guarantee becomes an issue. In the rest of paper, we discuss the related work in Sect. 2, whereas Sect. 3 gives the complete insight of problem formulation and the entire network operation. Section 4 provides the verification of the work with the help of results received. In Sect. 5, we provide a comprehensive conclusion. References are provided at the end of the paper.

2 Related Work

Shah and Rabaey [1] proposed energy-aware routing (EAR) which has maintenance phase, and it updates sensor nodes' residual energy. EAR protocol finds numerous routes from source to the destination node. Every selected route, based on residual energy, is assigned a probability to transmit packet. In EAR communication, end-to-end delay can increase as it defends every route from getting selected again. In real-time communication, EAR routing cannot be used.

In [2], to update residual energy, dynamic probability scaling (EAR-DPS) provides overhead packet and need not to maintain any maintenance phase. EAR-DPS also finds numerous routes from source to the destination node. Every selected route, based on residual energy, is assigned a probability to transmit packet. EAR-DPS can also not be used in real-time communication as end-to-end delay increases in routing.

In [3], EAR-RT provides a real-time guarantee routing protocol which provides guaranteed delivery while not impacting energy awareness in wireless sensor networks, while network is stable. It is basically real-time routing protocol that provides real-time communication that commits high survivable, energy awareness routing.

Protocol does not impact existing energy, ensures efficient usage of energy by providing sensors networks' longer connectivity, and reduces deadline miss ratio. Data communication phase and setup phase are the two phases of operation in EAR-RT. Data packets are forwarded from source to sink node in data communication phase. To ensure on-time delivery of packets, intermediate nodes send data packet only to the adjacent nodes which can deliver. Routing table is built in setup phase where the sink node explores the time delay, energy cost, and path from source to sink by sending route request. EAR-RT assigns deadline of packets to decrease end-to-end delay in real-time communication.

In robust and real-time data delivery (RRTD) model [4], dynamic ring structure provides the guarantee in hard real time, where low priority stations can be removed and stations with high priority have more chances of admittance in the ring. Destination-sequenced distance vector (DSDV) ensures soft real-time guarantee for discovering the path, on-time delivery, and maintenance by being a proactive routing protocol. RRTD proposed bandwidth reservation and control plane which ensure real-time delivery by providing more bandwidth to traffic with higher priority than traffic with lower priority. In WSN, robust and real-time data delivery (RRTD) model [4] bestows real-time and robust delivery of data. To provide hard real-time guarantee, RRTD applies MAC layers' time token protocol with centralized control plane, and for providing the guarantee in soft real time, reservation of bandwidth method is used. Providing sufficient bandwidth and managing dynamic ring is the main goal of the control plane. RRTD can transmit the data on desired rate when the data is ready to deliver, and real-time task is complete. The protocol also decreased the delay, miss ratio, congestion, and energy consumption.

To achieve real-time communication, multichannel real-time communication protocol (MCRT) [5] uses transmission power modulation and multiple channel allocation, which is flow-based allocation identified on the basis of multichannel realities. Channel allocation problem is a compelled optimization problem in achieving point-to-point data flow communication delay. MCRT design comprises an algorithm to allocate channels established on packet forwarding and well-examined strategy. MCRT reduces the deadline miss ratio by utilizing multiple channels in point-to-point communication.

RPAR [6] dynamically adapts transmission power while taking routing decisions as per the deadlines for packet delivery. RPAR can adjust power of communication from time to time. Unlike MMSPEED, it does not require to define different delivery speeds. It also uses an on-demand mechanism for neighborhood management which reduces the energy dissipation, while SPEED and MMSPEED exchange beacons periodically and thus waste a lot of energy in the process. This scheme is invoked if no choice is valuable in the neighbor table to forward a packet. Simulation results depict that significant energy is saved by this neighborhood management mechanism. It not only saves energy but also provides a considerable amount of reliability of real-time delivery of packet. Only problem is the time taken in neighbor discovery process. Given the fact that there is always a trade-off between the real-time delivery of packets and reliable delivery of packets; this combination of RPAR and neighborhood management scheme is a considerable option for any real-time application with a desire of reliability.

Energy-aware QoS routing protocol [7] by Akkaya and Younis can meet end-to-end delay requirement, and we can discover the path which is energy-efficient. Suggested protocol in [8] discovers delay-constrained and a least cost path while considering nodes' communication parameters, transmission energy, and energy reserve for real-time data. Additionally, while balancing the service rate at sensor nodes for real-time and non-real-time data, throughput is maximized for non-real-time data. A model for class-based queuing is implemented to give best effort and real-time traffic. Classifier checks incoming packet type at each node, and separate priority queues are assigned to reroute real-time traffic and non-real-time traffic. Multiple priorities are not supported for the real-time traffic is the drawback of this approach.

SPEED: SPEED, a soft real-time communication quality of service routing protocol that provides end-to-end guarantees for sensor networks, is presented in [9]. The protocol finds path by using geographic forwarding and requires the every node to maintain information about its neighbors. In addition, SPEED endeavors to maintain a certain delivery speed for every packet in the network. It helps in estimating the end-to-end packet delay by dividing the distance between the sink and the speed of the packet. SPEED also helps in efficiently handling voids and in congestion avoidance with minimal control overhead.

SPEED uses stateless geographic non-deterministic forwarding (SNFG) [9] routing module and at the network layer toils accompanied by other four modules [10]. The information about nodes and location of nodes is collected using beacon exchange mechanism. Elapsed time between transmitted data packet and ACK received from the neighbor as a response is used for estimating delay at each node. SNGF chooses the speed satisfying nodes by observing delay values, and nodes' relay ratio is examined if a right node cannot be found. Providing relay ratio is the responsibility of neighborhood feedback loop (NFL) module, which can be measured by inspecting the miss ratios of the neighbors of a node (nodes which did not cater the desired speed) and is fed to the SNGF module. Packet is likely to be dropped wherever the relay ratio may be less than an arbitrary produced number between 0 and 1. Finally, to prevent voids and to eliminate congestion by sending messages back to the source nodes to pursue new routes, the back pressure-rerouting module is used when a node fails to find a next hop node.

3 System Model

All WSNs can only guarantee soft reliability in real-time data delivery due to the lossy nature of their wireless links. Any information delayed is information of no use. There are a number of reasons that increase the end-to-end delay in data delivery in WSN as shown in Fig. 1. A packet en-route can be lost due to channel contention, interference, link break, dead nodes in the path, malicious/compromised node/s in

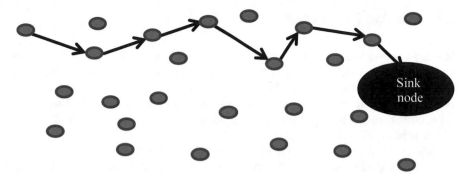

Fig. 1 Data delivery from source node to sink node

path or simply die in a queue of retransmission in a node on the route. Thus, reliability is a hard thing to achieve.

Out of all these reasons, a dead node can be avoided during route discovery (that too only if the proactive approach of route discovery is used). To improve link quality, one can use high transmission power which can again cause interference, channel contention, and energy depletion of a node. To avoid malicious node/ compromised nodes, single-hop model can be used for data delivery to sink. But this again can result in high interference and fast energy depletion. Therefore, multihop model can be preferred over single-hop model.

Assumptions: We assume the following points in our model:

The network is a homogeneous network where all sensor nodes exhibit equal capabilities in communication as well as computation.
The sensor nodes are scattered and positioned arbitrary in the field with high density.
All the sensor nodes and base station or sink cannot move or being stationary.
The base station/sink has no energy constraint and acts as an intermediate between the senor network and the application interface that uses this data for tactical decision making.
The network is based on a continuous reporting model where data is continuously generated by sensor nodes and is communicated to base station.
A multihop data communication model is considered since a single-hop communication model can lead to high interference.
This can lead to an increase in the overall number of pages. We would therefore urge you not to squash your paper.

3.1 Problem Formulation

In this paper, we use a communication link trust (CLT) value of a link between two
nodes. This trust value is inverse of packet reception rate (PRR). Any two nodes A
and B can have a communication link between them for packet exchange. But at the
same time any transmission from A which is not intended for B can interfere with
the transmissions of B despite the case that transmissions from B are intended for A
or not as in Fig. 2.

Any communication link with more than 90% PRR can be too overloaded to
deliver or forward a packet reliably under the required end-to-end delay. This link can
also be a result of a compromised or malfunctioning node simply flooding the network
and will have a very low trust value since its PRR is high. But any communication
link having less than 10% PRR would mean that this link has node/s that may not be
actually communicating but simply interfering. This link will have a very high trust
value. Based on this trust value we can illustrate our network as the weighted graph
$G = (V, E)$ with V being the set of nodes and E being the set of communication
links. Based on the CLT value, we can use the link with a permissible CLT value
so that we neither rely on a CLT value being the result of interference between two
nodes nor this CLT should be a result of a high PRR but unreliable node. We call it
the bounded CLT value. Based on this bounded CLT value, we can create a bounded
CLT graph $G' = (V', E')$ with E' being the communication satisfying bounded CLT
value only and V' being nodes on such links.

To select the optimal route in the given bounded CLT graph G', Dijkstra's algorithm is used to identify the shortest path between any node s to sink t. This gives

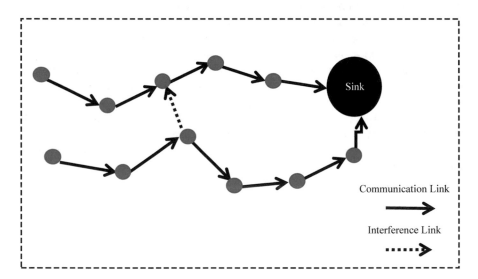

Fig. 2 Communication and interference links

essentially a shortest end-to-end delay route to deliver a packet from any node v belongs to V'.

3.2   Optimal Route Setup and Data Communication

To communicate and set up a route, there are different types of packet exchanged between the nodes.

Route Discovery Request packet: The base station/sink will send a route discovery packet in the network to obtain the CLT values of various links.

Route Discovery Reply packet: On receiving route discovery request packet from the base station, each sensor node will reply with its ID, PRR, and remaining energy.

Route Information packet: After receiving the route discovery reply packet from various sensor nodes, the base station will create a bounded CLT graph $G' = (V', E')$ of the network based on the given bounded CLT value. Then, it uses Dijkstra's algorithm to obtain the shortest path between any node s that belongs to V' to base station t. The base station then sends a route information packet to sensor nodes with all the routing information and having the next hop to deliver a data packet from any node v belongs to V' to u belongs to V' where u in turn hands over the packet to its next hop node.

Data Packets Each sensor node continuously monitors the desired phenomenon and generates the data. Any node v communicates this data to next hop u on its shortest route to base station t.

3.3   Error Localization and Removal in the Network

Since we are using bounded CLT graph $G' = (V', E')$ of the network. Every time the optimal route discovery and setup is done, nodes with CLT value other than the bounded CLT value are isolated. If a node v does not comply with the bounded CLT value, then v does not belong to V'. In such a case, any u has a link with v, i.e., (u, v), then u does not belong to V'. Therefore, if v is a compromised node, then v cannot flood and attack the network. If v is a dead node, then no traffic will be directed through any of the links on which this node is present. Thus, any source of error is completely isolated and the network can have a fault-tolerant operation as shown in Fig. 3. v can also have high PRR due to high traffic through it, or it may be facing high interference from the neighboring nodes. Therefore, isolating it in one round will ease down the interference as well as the high traffic. v can be part of the network in the next round such that v belongs to V' if it complies with the bounded CLT value in the next round. Thus, no node is isolated from the network for always unless it is a dead node or compromised node.

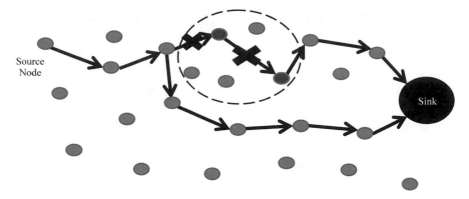

Fig. 3 Interference node isolated for error localization and removal

3.4 Algorithm for the Continuous Operation of the Network

The network, once going through setup phase, can go through continuous operation by following the three phases in the network as shown in Fig. 4. The error localization and optimal route setup phase go hand in hand, whereas the data communication phase is only possible after the optimal route has been set up and the errors in the network have been localized and removed.

ORSEL Network Operation Algorithm (N, R, t):

N   Set of all sensor nodes deployed in the field
R   Number of rounds the network is supposed to operate
t   Sink/base station

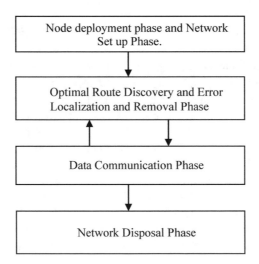

Fig. 4 ORSEL network operation

1. If $(N == 0||R == 0)$
   Exit.
2. Repeat: for $(i = 1 \text{ to } R)$
   i. t sends a RouteDiscoveryRequest to all the nodes of set N.
   ii. When any node v that belongs to set N receives this packet, it sends a RouteDiscoveryReply to t.
   iii. t creates a graph $G = (V, E)$ of the network.
   iv. Based on bounded CLT value t creates $G' = (V', E')$.
   v. t finds the optimal route P for all nodes in the network.
   vi. t sends RouteInformation to all the nodes in V'.
   vii. All the nodes send DataPacket to their next hop node as per P.
3. Exit.

4 Simulation and Results

We compare our scheme ORSEL with three different schemes that provide the real-time data delivery for wireless sensor networks. We compare our scheme with EAR-RT, RPAR, RRTD [3, 4, 6]. We assume a square network field of $100 * 100$ m, size of message as 80 bytes; distance from one node to other nodes is less than the minimum transmission range out of any of them. A WSN of varying network sizes ($|V| + |E|$) with varying no. of sensors is laid in the same field for varying observations. We have developed simulation tool in MATLAB for this work.

The simulation metrics are:

- Cumulative packet consumption by sink versus cumulative packet generation by sensor nodes.
- Proportion of interference nodes versus network lifetime.
- Proportion of interference nodes versus miss ratio.
- Number of packets successfully delivered versus miss ratio.

Figure 5 shows the cumulative number of packets consumed by the sink versus the cumulative number of packets generated by the sensor nodes. Not all the packets generated by the sensor nodes are consumed by sink since most of them are either lost or destroyed by interference. In this case, ORSEL shows that maximum of packets generated by the sensor nodes in the network have reached and consumed by the sink node/base station.

Figure 6 shows that the interfering nodes have an effect on the lifetime of the network. They can either flood the network or cause some nodes to die due to energy depletion. If such nodes in the network increase, then the lifetime of the network

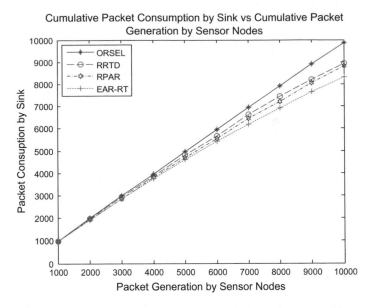

Fig. 5 Cumulative packet consumption by sink versus cumulative packet generated by sensor nodes

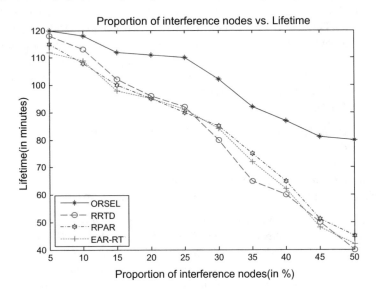

Fig. 6 Proportion of interference nodes versus network lifetime

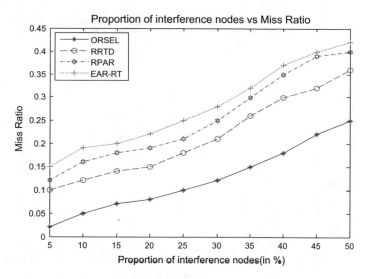

Fig. 7 Proportion of interference nodes versus miss ratio

drastically falls. ORSEL shows that it clearly has a significant resistance to such nodes in comparison with other three schemes. Even with 50% interfering nodes, the network still survives due to ORSEL's error localization and removal policy. It isolates such interfering nodes and provides an optimal route in the network.

Figure 7 shows that ORSEL provides less implicative miss ratio even with the increase in the proportion of interfering nodes. The miss ratio in ORSEL shows a latency in comparison with other three schemes where miss ratio rises sharply with the rise in number of interfering nodes. Miss ratio is the number of packets generated minus the number of packets consumed by sink divided by the number of packets generated.

Figure 8 shows that there is an effect of network density on successful packet delivery of network. Increase in number of total sensor nodes in the network results in more interference and delay in the delivery of packets in the network.

The other three schemes even being energy-efficient routing schemes could not perform better because they do not have error localization policy to deal with dead as well as malicious nodes in the network.

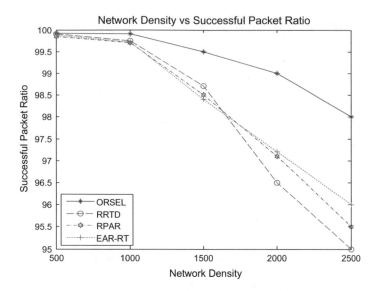

Fig. 8 Number of packets successfully delivered versus miss ratio

5 Conclusion

In this paper, we proposed an optimal route selection for error localization in real-time wireless sensor networks (ORSEL) scheme. ORSEL uses a communication link trust (CLT) value of a link between two nodes which acts as the factor of deciding the minimum end-to-end delay in delivery of packet while avoiding the interference and malfunctioning nodes in the network, if any. This trust value is inverse of packet reception rate (PRR). Any communication link with more than 90% PRR can be too overloaded to deliver or forward a packet reliably under the required end-to-end delay. This link can also be a result of a compromised or malfunctioning node simply flooding the network and will have a very low trust value since its PRR is high. But, any communication link having less than 10% PRR would mean that this link has node/s that may not be actually communicating but simply interfering. This link will have a very high trust value. To select the optimal route in the given bounded CLT graph, it uses Dijkstra's algorithm to find the shortest path from any sensor node to sink. This gives essentially a shortest end-to-end delay route to deliver a packet from any node to sink. If a node does not comply with the bounded CLT value, then no traffic will be directed through any of the links on which this node is present. Thus, any source of error is completely isolated and the network can have a fault-tolerant operation.

References

1. Shah RC, Rabaey M (2007) Energy aware routing for low energy ad hoc sensor networks. In: IEEE WCNW
2. Park G, Yi S, Heo J, Choi WC, Jeon G, Cho Y, Shim C (2005) Energy aware routing with dynamic probability scaling. Lect Notes Comput Sci 3642:662–670
3. Heo J, Yi S, Park G, Cho Y, Hong J (2006) EAR-RT: energy aware routing with real-time guarantee for wireless sensor networks. In: Alexandrov VN, van Albada GD, Sloot PMA, Dongarra J (eds) Computational Science—ICCS 2006. ICCS 2006. Lecture Notes in computer science, vol 3994. Springer, Berlin
4. Virmani D, Jain S (2010) Robust and real time data delivery in wireless sensor networks. In: Das VV, Vijaykumar R (eds) Information and communication technologies, ICT 2010. Communications in computer and information science, vol 101. Springer, Berlin
5. Wang X, Wang X, Fu X, Xing G, Jha N (2009) Flow-based real-time communication in multi-channel wireless sensor networks. In: Roedig U, Sreenan CJ (eds) Wireless sensor networks, EWSN 2009. Lecture Notes in computer science, vol 5432. Springer, Berlin
6. Chipara O, He Z, Xing G et al (2006) Real-time power-aware routing in sensor networks. In: Proceedings of the 14th IEEE international workshop on quality of service (IWQoS 2006), New Haven, CT, June 2006
7. Akkaya K, Younis M (2003) An energy-aware QoS routing protocol for wireless sensor networks. In: Proceedings of the IEEE workshop on mobile and wireless networks (MWN2003), Providence, Rhode Island, May 2003
8. Younis M, Youssef M, Arisha K (2002) Energy-aware routing in cluster-based sensor networks. In: Proceedings of the 10th IEEE/ACM international symposium on modeling, analysis and simulation of computer and telecommunication systems (MASCOTS2002), Fort Worth, TX, Oct 2002
9. He T, Stankovic JA, Lu C et al (2003) SPEED: a stateless protocol for real-time communication in sensor networks. In: Proceedings of international conference on distributed computing systems, Providence, RI, May 2003
10. Al-Karaki J, Kamal A (2004) Routing techniques in wireless sensor networks: a survey. IEEE Wirel Commun 11(6):6–28
11. Liang C-JM, Musaloiu-Elefteri R, Terzis A (2008) Typhoon: a reliable data dissemination protocol for wireless sensor networks. In Proceedings of 5th European conference on wireless sensor networks (EWSN), pp 268–285
12. Bindal AK, Jain A, Prasad D, Patel RB (2015) Two-way energy efficient localization scheme for mobile wireless sensor networks. Int J Appl Eng Res 10(3):7659–7672
13. Venkatesan L, Shanmugavel S, Subramaniam C (2013) A survey on modeling and enhancing reliability of wireless sensor network. Wirel Sens Netw 5:41–51
14. Park S-J, Vedantham R, Sivakumar R, Akyildiz IF (2008) GARUDA: achieving effective reliability for down-stream communication in wireless sensor networks. IEEE Trans Mob Comput 7(2):214–230
15. Lu C, Blum B, Abdelzaher T et al (2002) RAP: a real-time communication architecture for large-scale wireless sensor networks. In: Proceedings of the eighth IEEE real-time and embedded technology and applications symposium (RTAS'02)
16. Liu C, Layland J (1973) Scheduling algorithms for multiprogramming in a hard-real-time environment. J ACM 20(1):46–61
17. Perkins C, Royer E (1999) Ad hoc on-demand distance vector (AODV) routing. In: Proceedings of IEEE WMCSA'99, Feb 1999
18. Johnson D, Maltz D (1996) Dynamic source routing in ad hoc wireless networks (Chap. 5). In: Imielinski T, Korth H (eds) Mobile computing. Kluwer Academic Publisher, pp 153–181. ISBN: 0792396079
19. Zhao L, Kan B, Xu Y et al (2007) A fault-tolerant, real-time routing protocol for wireless sensor networks. In: Wireless communications, networking and mobile computing 2007 (WiCom 2007), 21–25 Sept 2007, pp 2531–2534

A Mobile-Cloud Framework with Active Monitoring on Cluster of Cloud Service Providers

Mani Goyal and Avinash Sharma

Abstract Openstack software can be used to achieve huge groups of computer networking assets and storage. Openstack facilities are structured in such a way every occurrence of a facility (i.e., service worker) is unprotected from end-to-end API available through a remote RPC. Openstack exposes a precise gorgeous API which can be functional to manage every characteristic of your cloud. One of the methods of work together with an Openstack cloud is programmatically. There is a ruby gem named fog that allows such interaction. The planned work targets to realize by letting every mobile node to allocate the scalar data and multimedia data among cloud providers, in order to increase packet transfer rate and to decrease the end-to-end delay by estimating buffer usage of bottleneck cloud providers. Our proposal will greatly alleviate the packet loss problem, thereby achieving significant improvement in end-to-end packet sending performance.

Keywords Remote procedure call · Application program interface · Internet group management protocol · User datagram protocol · Data integrity · Turnaround time

1 Introduction

Notion of trust is predictably hard to accurately measure in a heterogeneous environment, so there is no universally believed description of trust in cloud computing. However, to improve security, one can create incremental enhancements to the credibility of clouds by excluding and allocating trust relations among cloud setup components and clients. The server stores encoded facts, and it is decrypted at client side. Data is secure at server, and the third party cannot access the data. Concern is disloyal behavior of the cloud. The data stored on cloud may be transformed without

M. Goyal (✉) · A. Sharma
Department of Computer Science, Maharishi Markendeshwer Deemed to be University, Ambala, India
e-mail: er.mani.goyal@gmail.com

A. Sharma
e-mail: sh_avinash@yahoo.com

© Springer Nature Singapore Pte Ltd. 2020
A. Khanna et al. (eds.), *International Conference on Innovative Computing and Communications*, Advances in Intelligent Systems and Computing 1059,
https://doi.org/10.1007/978-981-15-0324-5_46

547

the awareness of client. Therefore, a mechanism that verifies the stored should be present at cloud to verify data being retrieved at client is same. At client side, after decrypting the data, certain rules can be set to verify uniqueness of the facts. This verifies that the facts have been transformed or it is same as the genuine data put in storage by the client.

1.1 UDP Attack

A UDP flood does not exploit any vulnerability. The aim of UDP floods is solely making and causation countless contracts of UDP datagrams from spoofed IP are to the target server. Once a server receives this sort of traffic, it is incapable to process every request.

1.2 IGMP Attack

IGMP is a connectionless procedure similar to UDP. IGMP is a protocol used to manage multicast members in TCP/IP. Similar to UDP flood, IGMP flood fixes any vulnerability. By directing, some type of IGMP packets endlessly makes server overwhelmed from annoying to process each request.

2 Related Work

Bonomi et al. [1] Fog is a suitable policy used for a range of serious internet of things (IoT) facilities and requests. Huang et al. [2] talk those restrictions by suggesting more severe mechanisms grounded on evidence, element certification and validation. Recommended a framework to incorporate several confidence mechanisms to expose cuffs of confidence in the cloud. Manuel et al. [3] define how a provision level contract is organized, joining excellence of service necessities of user and abilities of cloud source provider. Stojmenovic et al. [4] study a classic attack, man-in-the-middle attack, for the conversation of safety in fog computing. Explore the silent topographies of this attack by exploratory its CPU and memory intake on fog device. Shaikh et al. [5] present such a dimension by a trust model. A trust model processes the security asset and calculates a trust value. A trust value comprises several constraints that are required measurements alongside which security of cloud facilities can be measured. Aazam et al. [6] discuss the problems of resource estimation, client category based resource approximation, advance reservation, and cost estimating for different and current IoT customers, on the basis of their characteristics. Yi et al. [7] examined the areas and difficulties in fog computing platform and offered platform strategy with numerous example applications. Nandyala et al. [8] suggested design for IoT grounded U-healthcare watching with the inspiration and benefits of Cloud to Fog (C2F) work out which interrelates more by helping nearby to the edge (endpoints) at clever homes and hospitals. Gupta et al. [9] discover the part of early trust based on promised contract named service-level agreement (SLA) and cloud service provider's repute in handling safety of cloud adoption. Gonzalez et al. [10] offer a systematic study of correlated topics, finding the main research parts associated with edge computing. Bruschi et al. [11] suggest an intercommunication level that lets separating the physical resources while handling the relocation of service occurrences rendering to the user's position. Kamath [12] Fog applications dialog in a straight line with movable devices and supports heterogeneity of linked objects. It retains data precise where internet of things wants it. This can be fulfilled through appropriate demonstrating and model toolkits plus programs.

3 Proposed Work

3.1 To Analyze Different Cloud Provider's Services by Considering Three Parameters

Load time—it is taken when the submission of a job and the delivery of the completed job. Data integrity is a comprehensive term, and it comprises security, confidentiality, and accurateness of the data. How to select a fog resource specified its faith and competences. Cloud source reliability is a degree of fruitful accomplishment of acknowledged works by the cloud source. If A_k is total jobs acknowledged by resource R_k, let C_k represent the total of jobs accomplished fruitfully by resource R_k in the time period T. Cloud resource reliability = jobs accomplished fruitfully/jobs acknowledged by resource.

Trust Value of a cloud service provider = $C_1 * I + C_2 * LT + C_3 * R$.

Here C_1, C_2, and C_3 are weights of the reliance limits such that $C_1 + C_2 + C_3 = 1$. These weights of the reliance attributes are determined and centered on their priority. For example, $C_1 = 0.5$, $C_2 = 0.2$, and $C_3 = 0.3$, here, the highest priority is given to data integrity, whereas the lowest priority is given to load time (Table 1).

3.2 Algorithm

Data Integrity ()
Step 1: Checking permissions
Define PERMS = p+u+g+acl+xattrs
Find any variations in files or directories by checking file/directory permissions, user, group, access control permissions, and various file attributes
Step 2: Check content and type of files

Table 1 Different integrated aspects to be considered

```
# These are the default rules.
#
#p:        permissions
#i:        inode:
#n:        number of links
#u:        user
#g:        group
#s:        size
#b:        block count
#m:        mtime
#a:        atime
#c:        ctime
#S:        check for growing size
#acl:           Access Control Lists
#selinux        SELinux security context
#xattrs:        Extended file attributes
#md5:      md5 checksum
#sha1:     sha1 checksum
#sha256:        sha256 checksum
#sha512:        sha512 checksum
#rmd160:   rmd160 checksum
#tiger:    tiger checksum

#haval:    haval checksum (MHASH only)
#gost:     gost checksum (MHASH only)
#crc32:    crc32 checksum (MHASH only)
#whirlpool:     whirlpool checksum (MHASH only)

#R:             p+i+n+u+g+s+m+c+acl+selinux+xattrs+md5
#L:             p+i+n+u+g+acl+selinux+xattrs
#E:             Empty group
#>:             Growing logfile p+u+g+i+n+S+acl+selinux+xattrs
```

Define CONTENT = sha256+ftype
Define CONTENT_EX = sha256+ftype+p+u+g+n+acl+xattrs
This finds comprehensive content, type of file, and access rights
Step 3: Find any type of change in data
Define DATAONLY = p+n+u+g+s+acl+xattrs+sha256
Step 4: Find permissions of all files in directory
/directory/ PERMS
Check all files in the directory for any changes
/directory/ CONTENT_EX
Find any changes in data inside all files/directory
/directory/ DATAONLY

3.3 Setup Credential File

See Fig. 1.

```
# Fog Credentials File
# :aws_access_key_id:           022QF06E7MXBSAMPLE
:default:
    :aws_access_key_id:         AKIAJ3CTYYNH6GIEARLA
    :aws_secret_access_key:     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
```

Fig. 1 Fog credentials file

3.4 Multipath Traffic Distribution

This section presents a multipath traffic distribution method. We will deploy heterogeneous sensors (scalar sensors, audio and video) that communicate directly in a certain schedule with a set of cloud providers and send data to it. We realize the entire network's balance by distributing scalar data and multimedia data in order to increase packet transfer rate to reduce the end-to-end delay by buffer usage of bottleneck cloud providers in the network.

4 Results

See Tables 2, 3, and 4; Figs 2, 3, 4, 5, and 6.

Packet delivery ratio detected in the simulation is shown in figure. The results are compared for existing and proposed. The proposed approach gives PDR from 98.01

Table 2 Load time, data integrity, and reliability computation

Fog default provider	Load time for service (computer)	Data integrity for each provider	Reliability
BareMetalCloud	0.251588154	0.246346387	0.179936452
Clodo	0.182860195	0.182117758	0.131934847
CloudSigma	0.189726229	0.189488845	0.134225562
Cloudstack	0.226091768	0.188608394	0.265447767
DNSMadeEasy	0.186719355	0.183954731	0.133005019
Dreamhost	0.183660132	0.185451817	0.130980951
Fogdocker	0.178029982	0.182016324	0.129314563
GoGrid	0.1824301	0.18367598	0.130599022
Glesys	0.185445678	0.183379977	0.132064085
Linode	0.18542454	0.18395853	0.133704841
OpenNebula	0.178383355	0.179662966	0.129102777
OpenVZ	0.179551109	0.179261301	0.134033084
Ovirt	0.252882568	0.25249464	0.179884509
Vcloud	0.25203738	0.252982008	0.179394711
VcloudDirector	0.265361099	0.26681728	0.242289979

Table 3 Trust computation

Fog default provider	Load time for service (Computer)	Data integrity for each provider	Reliability	Trust factor
BareMetalCloud	0.251588154	0.246346387	0.179936452	0.23358858
Clodo	0.182860195	0.182117758	0.131934847	0.18241473
CloudSigma	0.189726229	0.189488845	0.134225562	0.1895838
Cloudstack	0.226091768	0.188608394	0.265447767	0.20360174
DNSMadeEasy	0.186719355	0.183954731	0.133005019	0.18506058
Dreamhost	0.183660132	0.185451817	0.130980951	0.18473514
Fogdocker	0.178029982	0.182016324	0.129314563	0.18042179
GoGrid	0.1824301	0.18367598	0.130599022	0.18317763
Glesys	0.185445678	0.183379977	0.132064085	0.18420626
Linode	0.18542454	0.18395853	0.133704841	0.18454493
OpenNebula	0.178383355	0.179662966	0.129102777	0.17915112
OpenVZ	0.179551109	0.179261301	0.134033084	0.17937722
Ovirt	0.252882568	0.25249464	0.179884509	0.25264981
Vcloud	0.25203738	0.252982008	0.179394711	0.25260416
VcloudDirector	0.265361099	0.26681728	0.242289979	0.26623481

Table 4 Packet delivery ratio

Number of mobile nodes	PDR% (Existing)	PDR% (Proposed)
25	93.52	99.2
50	92.80	99.1
75	87.45	99.0
100	83.35	98.9
125	77.56	98.21
150	73.78	98.01

to 99.2% which is good. On the other hand, existing PDR has low value from 73.78 to 93.52% (Table 5; Fig. 7).

Throughput measured in the simulation is shown in figure. The results are compared for existing and proposed. Throughput always remains high in proposed approach which is good (Table 6; Fig. 8).

Average end-to-end delay measured in the simulation is shown in figure. The results are compared for existing and proposed. Average end-to-end delays always remain low in proposed approach which is good.

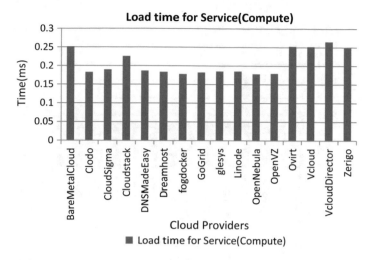

Fig. 2 Load time

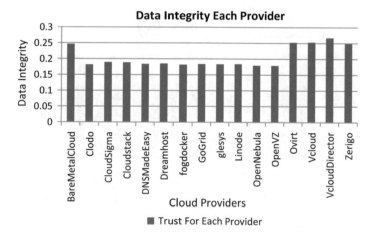

Fig. 3 Data integrity

5 Conclusion and Future Scope

Trust shows a big part in profit-making cloud environments. It is solitary of the main contests of cloud technology. Trust permits shoppers to select the best resources in an exceedingly numerous cloud infrastructure. We tried to implement innovative trust model grounded on previous credentials and current talents of a cloud resource supplier. Trust assessment is taken into account by considering three parameters like load time potency, reliability, and data integrity. Network's transmission rate performance depends upon type of information broadcasted. Packets being ready for

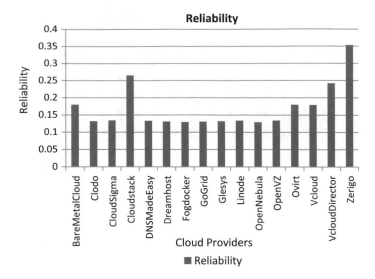

Fig. 4 Reliability

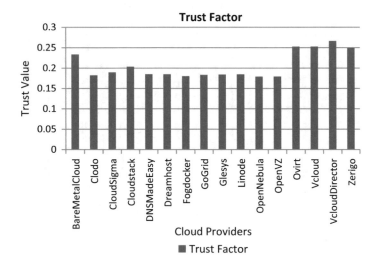

Fig. 5 Trust value of cloud providers

Fig. 6 Packet delivery ratio

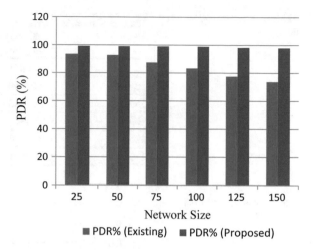

Table 5 Throughput

Number of mobile nodes	Throughput (Kbps) (existing)	Throughput (Kbps) (proposed)
25	5.9070	6.2565
50	4.0915	4.4635
75	3.0250	3.5085
100	2.5240	3.0725
125	2.1460	2.7945
150	1.8320	2.5155

Fig. 7 Throughput (Kbps)

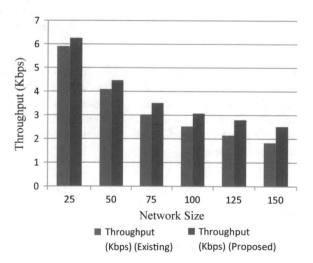

Table 6 Average end-to-end delay

Number of mobile nodes	Average end-to-end delay (ms) (existing)	Average end-to-end delay (ms) (proposed)
25	0.02851	0.001592
50	0.01954	0.001635
75	0.03440	0.001492
100	0.04386	0.001557
125	0.067847	0.001605
150	0.07581	0.001624

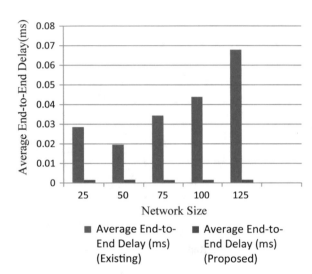

Fig. 8 Average end-to-end delay (ms)

transmission by a mobile node are distributed among cloud providers according to buffer usage of cloud provider. After calculating the buffer usage of each associated cloud provider, a heuristic procedure can be applied to reduce packet drop in performance as against the normal. The suggested scheme shows better outcomes. Optimization using machine learning or nature-inspired algorithm is other challenges for the future work.

References

1. Bonomi F, Milito R, Zhu J, Addepalli,S (2012) Fog computing and its role in the internet of things. In: MCC'12, August 17, 2012, ACM 978-1-4503-1519-7/12/08
2. Huang J, Nicol DM (2013) Trust mechanisms for cloud computing. J Cloud Comput Adv Syst Appl 2:9
3. Manuel P (2013) A trust model of cloud computing based on quality of service. Ann Operations Res. https://www.researchgate.net/publication/241279418

4. Stojmenovic I, Wen S (2014) The fog computing paradigm: scenarios and security issues. In: Proceedings of the 2014 federated conference on computer science and information systems, vol 2, ACSIS, pp 1–8
5. Shaikh R, Sasikumar M (2015) Trust model for measuring security strength of cloud computing service. Procedia Comput Sci 45:380–389. www.sciencedirect.com
6. Aazam M, Huh EN, Fog Computing micro datacenter based dynamic resource estimation and pricing model for IoT. IN: 2015 IEEE 29th international conference on advanced information networking and applications
7. Yi S, Hao Z, Qin Z, Li Q (2015) Fog computing: platform and applications. In: 2015 third IEEE workshop on hot topics in web systems and technologies
8. Nandyala CS, Kim HK (2016) From cloud to fog and IoT-based real-time U-healthcare monitoring for smart homes and hospitals. Int J Smart Home 10(2):187–196
9. Gupta S, Saxena KBC, Saini AK (2016) Towards risk managed cloud adoption a conceptual framework. In: Proceedings of the 2016 international conference on industrial engineering and operations management Kuala Lumpur, Malaysia, March 8–10, 2016
10. Gonzalez NM, Goya WA (2016) Fog computing: data analytics and cloud distributed processing on the network edges. IEEE
11. Bruschi R, Genovese G, Iera A (2017) Openstack extension for fog-powered personal services deployment. In: 29th International Tele traffic Congress
12. Kamath R (2017) Int J Inn Res Comput Commun Eng 5(4)

Estimation of Optimum Number of Clusters in WSN

Nihar Ranjan Roy⊙ and Pravin Chandra⊙

Abstract Grouping of sensor nodes in clusters has several advantages including energy efficiency, network scalability, and efficient data aggregation. Many clustering protocols have been developed till date promising better energy efficiency in comparison with others. In this paper, we have surveyed important clustering techniques with a focus on the estimation of optimum number of clusters. We have also presented a case study on LEACH protocol, suggesting that under certain conditions clustering is not a wise solution, a non-clustered network or a network with mixed approach can give better result. Experimental results show significant improvement in lifetime and throughput.

Keywords Optimum cluster size · Energy efficiency · Lifetime · Wireless sensor network

1 Introduction

A wireless sensor node (WSN) is a resource-constrained node which has limited computational, communication, and power capacity. These nodes are generally deployed to monitor: habitat [12], health care [11], smart cities [8], border surveillance [1], structural monitoring [4], tracking, etc. In certain cases like border surveillance, it may not be possible to place them individually; hence, they are aerially dropped, and thereafter these nodes cooperatively form a network. Due to small size, these nodes have several challenges, and one of them is limited battery capacity. One popular solution for efficient utilization of limited energy is clustering of nodes in the network.

N. R. Roy (✉)
GD Goenka University, Gurgaon 122103, Haryana, India
e-mail: niharranjanroy@gmail.com

N. R. Roy · P. Chandra
USICT, GGSIPU, New Delhi 110078, India
e-mail: chandra.pravin@gmail.com

© Springer Nature Singapore Pte Ltd. 2020
A. Khanna et al. (eds.), *International Conference on Innovative Computing and Communications*, Advances in Intelligent Systems and Computing 1059,
https://doi.org/10.1007/978-981-15-0324-5_47

Fig. 1 Life cycle of a
non-cluster head node

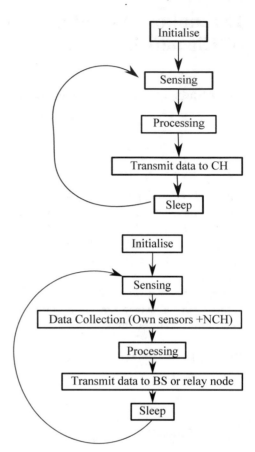

Fig. 2 Life cycle of cluster
head node

In a clustered WSN, sensor nodes (SN) are grouped into disjoint sets called clusters. In each cluster, one of these nodes is selected as cluster head (CH) and rest are called non-cluster head nodes (NCH) or member nodes. All the NCH nodes of a cluster sense their surroundings and send data to its CH. CH collects data from its NCH nodes and does some elementary data processing such as aggregation or compression and forwards it to an intermediate node or base station (BS). Life cycle of a NCH node and CH node is shown in Figs. 1 and 2. It can be seen that cluster heads do more work in comparison with non-cluster head nodes as a result of which they are likely to die early, leading to hole in the network.

This paper is organized as follows: in Sect. 2, we discuss clustering in general, taxonomy, and prominent clustering protocols which have proposed ways for estimating optimum cluster size. In Sect. 3, we present a case study on LEACH protocol along with an extension of our own recent work on LEACH [13] and supported with simulations results. Section 4 concludes this paper.

2 Clustering in Wireless Sensor Network

In WSN, there are different ways to create cluster's such as centralized clustering versus decentralized clustering. In centralized clustering, a central node (preferably base station) helps in cluster formation in comparison with the decentralized approach where nodes cooperatively compete and form cluster [18]. Some of the other characteristics of clustering in WSN includes

- Static versus dynamic clustering: Are clusters formed at the beginning or they are formed dynamically and keep on changing?
- Position of the BS: Normally, BS is either placed at center, outside, corners or on boundary of sensing field.
- Mobility of BS: Is base station static or it is movable? If movable, does it follow a predefined path or a random path?
- Node capabilities: Do all nodes have similar capabilities or some are advanced?
- Number of optimum clusters: What should be the number of optimum clusters for the network in consideration?
- Size of cluster is uniform or non-uniform: Do all the clusters need to have approximately same number of nodes or they can be non-uniform in size?

In this paper, we have critically surveyed research work related to the estimation of number of optimum clusters and presented it in Table 1. In the next subsections, we present a brief taxonomy of WSNs and discuss prominent research work related to optimum cluster size.

2.1 Clustering Taxonomy

Clustering techniques have been classified on the basis of several characteristics. A taxonomy of WSN clustering is shown in Fig. 3. In homogeneous network, all the nodes have identical initial energy levels and hardware capabilities [6]. In heterogeneous networks, nodes may have different level of initial energy and hardware characteristic [5], but popularly heterogeneous network refers to nodes having different initial energy level and having identical hardware capabilities [15]. A survey work on heterogeneous WSN protocols is done in [16]. Probabilistic clustering techniques emphasize on uniform energy utilization in the network and generally select CH on the basis of a randomized CH selection algorithm. In non-probabilistic clustering techniques, cluster heads are selected on characteristics such as degree, connectivity, and information from neighbors. Another way to classify clustering is on the basis of hop counts as shown in Figs. 4 and 5.

Table 1 Comparative analysis of optimum number of clusters estimation techniques

Protocol	Network type	BS location	Optimum K	Advantages	Disadvantages	Dependency
[6]	Homogeneous	Center	$K_{opt} = \sqrt{\dfrac{N}{2\pi}}\sqrt{\dfrac{\epsilon fs}{\epsilon mp}}\dfrac{M}{d_{toBS}^2}$	Energy dissipation is uniform due to rotation of cluster heads, and does data aggregation	It is one of the simplest protocols and over estimates lifetime	N, M, d_{toBS}
[5]	Heterogeneous	Outside	$K_{opt} = \left\lceil \sqrt{\dfrac{N}{6}}\dfrac{M}{d_{toBS}^2}\sqrt{\dfrac{E fs}{D\alpha}}\right\rceil$	Extensive	Complicated	$D\alpha = \epsilon mp + E_{senseCH} + E_{tranCH} + E_{loggCH}$
[14]	Homogeneous	Center	$P_{opt} = \dfrac{\epsilon fs}{\lambda\sqrt{\sqrt{\frac{1}{6}}\frac{M}{r}(2E_{elec}+\epsilon fs\, r^2)-2E_{elec}}}$	Robustness due to biologically inspired algorithm	Considers Poisson distribution	λ and E_{elec}
[17]	Homogeneous	Center	$K_{opt} = \begin{cases}\dfrac{\sqrt{4W-1}}{\lambda} & \text{Horizontal nodes}\\ \sqrt{4Z+2W-3} & \text{Vertical array}\end{cases}$	Grid based nodes	Does not consider other sources of energy dissipation	Nodes in horizontal and vertical grid
[2]	Homogeneous	Center	$K_{opt} = \begin{cases}\sqrt{N}\,\dfrac{\sqrt{6}d_0 M}{M^2+6B^2} & d_{toBS}\geq\sqrt{\frac{\epsilon fs}{\epsilon mf}}\\ \sqrt{N}\,\dfrac{M}{\sqrt{M^2+6B^2}} & d_{toBS}\leq\sqrt{\frac{\epsilon fs}{\epsilon mf}}\end{cases}$ Center of sensing field	Avoids collision and improves performance	Incurs overhead	N and B distance from center of sensing filed to BS
[3]	Homogeneous	Center	$K_{opt} = \left(\dfrac{cN}{2(1-c)}\right)^{\frac{2}{3}}$	Exploits data correlation to maximize lifetime	Computational overhead is not estimated	Correlation factor c
[15]	Heterogeneous	Center	$k_{opt} = \sqrt{\dfrac{N}{2\pi}}\dfrac{2}{0.765}$	Promises better stability due to advance nodes	Node placement could be an issue	N
[10]	Homogeneous	Mobile	$k_{opt} = \sqrt{d D c_0}$	Considers application independent data compression	Does not consider lossy compression	D number of hops to base station and correlation factor c
[13]	Homogeneous	Center	$k_{new} = \sqrt{\dfrac{N}{2\pi}}\sqrt{\dfrac{\epsilon fs}{\epsilon mp}}\sqrt{\dfrac{M}{d_{toBS}^4 - \frac{E_{elecRx}}{\epsilon mp}}}$	Extension of LEACH	Simple algorithm	d_{toBS}, E_{elecRx} and, M
[7]	Homogeneous	Center	$K_{opt} = \left(\sqrt{\dfrac{N}{6}}\,M\sqrt{\dfrac{E fs}{D\alpha}}\right)$	It is more accurate	Bulky and difficult to implement	$D\alpha = (E sensech + Eloggch + Etranssn + Eampd^4)$

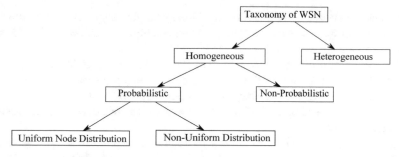

Fig. 3 Clustering taxonomy

Fig. 4 Flat clustering

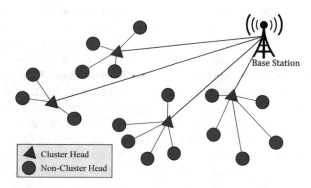

Fig. 5 Multi-hop clustering

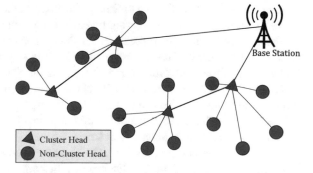

2.2 Optimum Cluster Size

In a clustered WSN, an important question is "What should be the number of clusters in a sensor network?" Answering to this question Heinzelman et al. [6] demonstrated that an optimum cluster size can enhance network lifetime. They found the optimum number of clusters analytically and validated it experimentally, in their work in [6]. They proposed today's most popular and one of the oldest clustering algorithms called LEACH. LEACH is a randomized, self-configuring clustering technique where nodes form clusters periodically. It aims to create energy dissipation balance in the network.

It operates in two steps: setup and steady. During the first step, network is setup and CH selection process takes place where as in the second step all the NCH nodes send their data to the CH as per a TDMA schedule.

Halgamuge et al. [5] proposed an exhaustive energy model for heterogeneous sensor network. The published the relationship between optimum number of clusters and free space fading energy. They also pointed out that for low free space fading energy and large sensing field with less number of sensor nodes, optimum cluster size is less important.

Selvakennedy et al. [14] proposed a biologically inspired clustering technique which converges fast with very limited overhead. Unlike the other traditional applications of such biologically inspired technique, in T-ANT an agent is allowed to influence all nodes within the radio range of the host node and not just this node alone. This is one of the robust techniques for clustering in WSN, keeping in view the unforeseen circumstances in the sensing field and dynamic node failure. A more recent comparison work with similar approach is done in [9].

Vlajic et al. [17] took a special scenario of WSN, where nodes were positioned in a grid fashion. They proposed that five-hop clusters give near-optimal network performance under a range of spatial arrangements. Since some of their assumptions are not always true, it may not work in all scenarios.

Chan et al. [2] proposed fixed optimal cluster (FOC) numbers after first analyzing the network. Further, they proposed two FOC depending on the position of CH. Two scenarios that were considered are BS is a the center of the sensing field and outside the sensing field. This approach includes high overhead.

Chen et al. [3] estimated the optimum number of clusters and CH selection algorithm for aggregation-driven routing by exploiting data correlation.

Smaragdakis et al. [15] proposed stable election protocol for heterogeneous WSN. Here balance energy of nodes is considered while selecting CH. This algorithm is known for its increased stability period, i.e., the time before the first node is dead.

Roy et al. [13] have reworked on Heinzelman et al. [6] work ignoring approximations and assumptions. The new analytical findings proposed are significant. Later in this paper, we have experimentally validated the above findings while presenting a case study on the most popular clustering work.

Kumar et al. [7] have proposed an extensive energy model for homogeneous WSN, covering 11 sources of energy dissipation. In comparison with [5], where seven sources of energy dissipation were considered this model considers homogeneous nodes. It also inherits the disadvantage of being bulky and hard to implement.

From Table 1, it can be seen that no single formula can estimate the number of optimum clusters as there are various factors determining its size. Each technique has certain assumptions pertaining to sensing field, location of base station, nodes, and operational characteristics.

3 LEACH-A Case Study

In this section, we present a case study on LEACH protocol and validate the analytical findings of [13] through simulation. For ease of understanding, we have used the same symbols and values as in [6, 13]. LEACH is one of the most popular and oldest probabilistic clustering techniques. In this approach, it is assumed that N nodes are uniformly and randomly distributed in a sensing region which is square in shape and has dimensions $M \times M$. Here the energy dissipation during transmission of l bits of data is estimated using

$$
\begin{aligned}
E_{Tx}(l, d) &= E_{Tx}(l) + E_{Tx-amp}(l, d) \\
&= \begin{cases} l E_{elec} + l \epsilon_{fs} d^2 & \text{if } d < d_0, \\ l E_{elec} + l \epsilon_{mp} d^4 & \text{if } d \geq d_0, \end{cases}
\end{aligned}
\tag{1}
$$

and energy consumed by receiver is

$$
E_{Rx}(l) = E_{Rx-elec}(l) = l E_{elec}
\tag{2}
$$

Assuming that network consists of k clusters and each cluster has on an average $\frac{N}{k}$ nodes. After CH selection, during initial phase there are $(\frac{N}{k} - 1)$ member nodes in a cluster. As per [13], optimum number of clusters is

$$
K_{new} = \begin{cases} N & : d_{toBS} \leq \text{t-distance} \\ \sqrt{\dfrac{N}{2\pi}} \sqrt{\dfrac{\epsilon_{fs}}{\epsilon_{mp}}} \dfrac{M}{\sqrt{d_{toBS}^4 - \frac{E_{elec}}{\epsilon_{mp}}}} & : \text{otherwise} \end{cases}
\tag{3}
$$

where t-distance is the threshold distance.

$$
\text{t-distance} = \left[\frac{1}{\epsilon_{mp}} \left(\frac{M^2 \epsilon_{fs}}{2\pi N} + E_{elec} \right) \right]^{\frac{1}{4}}
\tag{4}
$$

It is to be noted that if a node is within a t-distance from the base station, then

- Node should directly communicate with the BS.
- Node can participate in CH selection process.

Further average d_{toBS} distance can be estimated as

$$
\begin{aligned}
D[d_{toBS}^4] = {}& \frac{28 M^4}{45} - \frac{5 M^3 a}{3} - \frac{5 M^3 b}{3} + \frac{8 M^2 a^2}{3} \\
& + 2 M^2 a b + \frac{8 M^2 b^2}{3} - 2 M a^3 - 2 M a^2 b \\
& - 2 M a b^2 - 2 M b^3 + a^4 + 2 a^2 b^2 + b^4
\end{aligned}
\tag{5}
$$

where base station is located at (a, b).

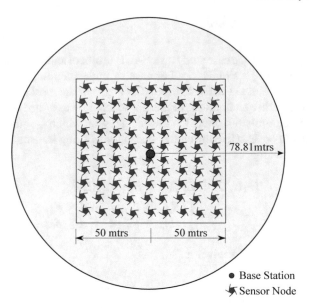

Fig. 6 BS is at the center of the sensing field

Table 2 Simulation results: BS is at the center of the sensing field

	First node dead	Last node dead	Packets to CH	Packets to BS
LEACH	802.85	2302.05	106369.55	12,471
LEACH [13]	1329.7	2854.85	0	189845.55

We simulated analytical work done in [13] using MATLAB. In our first simulation setup, 100 nodes were uniformly and randomly deployed in square sensing area with side length 100 m and BS was kept at the center as shown in Fig. 6. Rest of the initialization parameters are same as in [6]. An average of 30 simulations is presented in Table 2 and plotted in Fig. 8. It is to be noted that in this setup all the nodes are within the range of t-distance, where they are expected to directly communicate to the base station; hence, no clusters are formed. The simulation results presented clearly show that no clustering is required under such scenario. Obtained statistics show that there is an increase in the stability period[1] of the network by 62.62%, and network lifetime by 24% (Fig. 7).

In the second simulation, we moved the base station to (50, 150), i.e., outside the sensing field as shown in Fig. 7. In this setup, it is seen that some of the nodes are with in the range of t-distance (gray circle). These nodes should

– communicate with the base station directly;
– may act as CH for other nodes and perform the task of CH as in [6].

---

[1]Time from network initialization till first node dies.

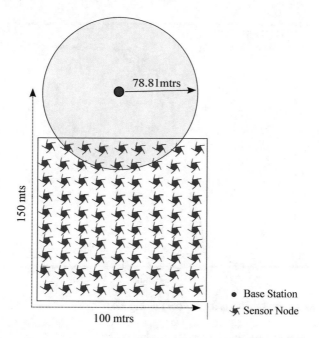

Fig. 7 BS is outside the sensing field

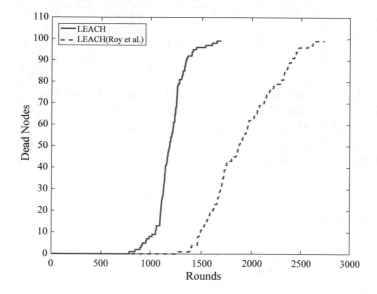

Fig. 8 Lifetime comparison, BS is at the center the sensing field

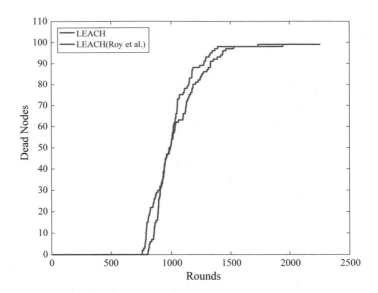

Fig. 9 Lifetime comparison, BS is outside the sensing field

Table 3 Simulation results: BS is outside the sensing field

	First node dead	Last node dead	Packets to CH	Packets to BS
LEACH [6]	768.7	1995.6	90138.1	9220.6
LEACH [13]	770.4	2136.6	93274.5	35325.1

Simulation result is shown in Fig. 9 and an average of 30 simulation is presented in Table 3. It is seen that here number of nodes within the t-distance range is less, which means most of the nodes will form clusters and communicate via CH nodes. The number of packets transmitted is shown in the table. Here stability period is similar to an improvement of 7% in the lifetime.

4 Conclusion

In this paper, we surveyed prominent papers focusing on optimum size of the clusters, the factors defining them and tabulated our findings. It is observed that no single formula can estimate the number of optimum clusters as there are various factors determining its size. Each technique has certain assumptions pertaining to the shape of the sensing field, density of sensors deployed, location of base station, nodes, and operational characteristics (sensing, actuation, etc.).

We also presented a case study on one of the most popular clustering protocols called LEACH and through simulations demonstrated that no clustering is required, if all the nodes are within a range of t-distance from the base station. In certain cases, a mixed approach may be followed. Improved lifetime and throughput is achieved under certain common experimental setup. Clustering still holds its advantages, but clustering is not always a wise solution as demonstrated here.

References

1. Benzerbadj A, Kechar B, Bounceur A, Hammoudeh M (2018) Surveillance of sensitive fenced areas using duty-cycled wireless sensor networks with asymmetrical links. J Netw Comput Appl 112:41–52
2. Chan TJ, Chen CM, Huang YF, Lin JY, Chen TR (2008) Optimal cluster number selection in ad-hoc wireless sensor networks. WSEAS Trans Commun 7(8):837–846
3. Chen H, Megerian S (2006) Cluster sizing and head selection for efficient data aggregation and routing in sensor networks. In: Wireless communications and networking conference, 2006. WCNC 2006, vol 4. IEEE, pp 2318–2323
4. Fang K, Liu C, Teng J (2018) Cluster-based optimal wireless sensor deployment for structural health monitoring. Struct Health Monit 17(2):266–278
5. Halgamuge MN, Zukerman M, Ramamohanarao K, Vu HL (2009) An estimation of sensor energy consumption. Prog Electromagn Res B
6. Heinzelman W, Chandrakasan A, Balakrishnan H (2002) An application-specific protocol architecture for wireless microsensor networks. IEEE Trans Wireless Commun 1(4):660–670. https://doi.org/10.1109/TWC.2002.804190
7. Kumar V, Yadav S, Kumar V, Sengupta J, Tripathi R, Tiwari S (2018) Optimal clustering in Weibull distributed WSNs based on realistic energy dissipation model. In: Progress in computing, analytics and networking. Springer, pp 61–73
8. Lu W, Gong Y, Liu X, Wu J, Peng H (2018) Collaborative energy and information transfer in green wireless sensor networks for smart cities. IEEE Trans Ind Inform 14(4):1585–1593
9. Miranda K, Zapotecas-Martínez S, López-Jaimes A, García-Nájera A (2019) A comparison of bio-inspired approaches for the cluster-head selection problem in WSN. In: Advances in nature-inspired computing and applications. Springer, pp 165–187
10. Pattem S, Krishnamachari B, Govindan R (2008) The impact of spatial correlation on routing with compression in wireless sensor networks. ACM Trans Sensor Netw (TOSN) 4(4):24
11. Rice J, Mechitov K, Sim SH, Spencer B Jr, Agha G (2011) Enabling framework for structural health monitoring using smart sensors. Struct Control Health Monitor 18(5):574–587
12. Ross SRJ, Friedman NR, Dudley KL, Yoshimura M, Yoshida T, Economo EP (2018) Listening to ecosystems: data-rich acoustic monitoring through landscape-scale sensor networks. Ecol Res 33(1):135–147
13. Roy NR, Chandra P (2018) A note on optimum cluster estimation in leach protocol. IEEE Access 6:65690–65696
14. Selvakennedy S, Sinnappan S, Shang Y (2007) A biologically-inspired clustering protocol for wireless sensor networks. Comput Commun 30(14–15):2786–2801
15. Smaragdakis G, Matta I, Bestavros A (2004) SEP: a stable election protocol for clustered heterogeneous wireless sensor networks. Tech. rep., Boston University Computer Science Department
16. Tiwari T, Roy NR (2015) Hierarchical clustering in heterogeneous wireless sensor networks: a survey. In: 2015 international conference on computing, communication & automation (ICCCA). IEEE, pp 1385–1390
17. Vlajic N, Xia D (2006) Wireless sensor networks: to cluster or not to cluster? In: Proceedings of the 2006 international symposium on a world of wireless, mobile and multimedia networks. IEEE Computer Society, pp 258–268

18. Zanjireh MM, Larijani H (2015) A survey on centralised and distributed clustering routing algorithms for WSNs. In: 2015 IEEE 81st vehicular technology conference (VTC Spring). IEEE, pp 1–6

Analysis of Data Aggregation Techniques in WSN

Nihar Ranjan Roy and Pravin Chandra

Abstract Wireless sensor networks (WSNs) produce a huge amount of application-specific data. These data need to be processed and transmitted to base station, which is a costly affair. Since WSN nodes are resource-constrained, efficient data processing and conserving energy are prime challenges. It has been observed that most of the data sensed by the sensors are redundant in nature. If data redundancy can be reduced, then it will lead to an increased lifetime of the network and reduced latency. In this paper, we surveyed different techniques for reducing redundancy in data, and in particular through aggregation. We have discussed data aggregation taxonomy, challenges and critically analysed aggregation techniques proposed in the last 10 years.

Keywords Data redundancy · Data aggregation · Data compression · Lifetime · Wireless sensor network

1 Introduction

Wireless sensor nodes are small-sized, resource-constrained, inexpensive devices deployed in large number to sense data from its field of deployment and send it to base station. WSN finds its application in different areas including healthcare monitoring, border surveillance, habitat monitoring. Due to small size, sensor nodes have small battery source which powers all other components (transmission, reception, sensing and processing). In certain scenarios, it becomes either difficult, costly or almost impossible to recharge or replace their battery. Minimising energy dissipation in WSNs is a primary challenge for researchers [1]. Since last two decades, researchers

N. R. Roy (✉)
GD Goenka University, Gurgaon, Haryana 122103, India
e-mail: niharranjanroy@gmail.com

N. R. Roy · P. Chandra
School of Information, Communication and Technology, Guru Gobind Singh
Indraprastha University, New Delhi 110078, India
e-mail: chandra.pravin@gmail.com

© Springer Nature Singapore Pte Ltd. 2020
A. Khanna et al. (eds.), *International Conference on Innovative Computing
and Communications*, Advances in Intelligent Systems and Computing 1059,
https://doi.org/10.1007/978-981-15-0324-5_48

have adopted different strategies to enhance network lifetime: using energy-efficient clustering, energy-efficient routing, sleep–wakeup cycles, topology optimizations, etc. A good survey work on energy-efficient WSNs is done in [2–5].

Energy dissipation rate can be minimised efficiently if we are aware of all the sources of energy dissipation and optimise them. Several energy models for WSNs have been proposed in the past which include [6–13]. Heinzelman et al. [6] proposed a model, which considers only 3 sources of energy dissipation in comparison with Kumar et al. [13], which considers 11 sources of energy dissipation. These sources include [13], energy dissipation due to communication unit (transmitter, receiver and coding), processing unit (logging, microcontroller and data compression), sensing unit (transducer, signal conditioning, A/D converter), actuation and transient.

In a clustered wireless sensor network, sensor nodes are grouped into disjoint sets, where each set forms a cluster. Inside a cluster, one node is chosen as cluster head (CH) and rest are referred as non-cluster head node (NCH). There are different methods for selection of CH, some of which are; randomly, statically, on the basis of residual energy, on the basis of node degree etc. These sensor nodes continuously and periodically sense their physical environment of deployment and send sensed data to CH node, which forwards either directly to base station (BS) or via some intermediate node(s). General architecture of a single-hop clustered wireless sensor network is shown in Fig. 1. Sensor nodes generate a good amount of data which needs to be transmitted to BS, and it is highly possible that there is redundancy in it. If the size of data to be transmitted and can be minimised, the rate of energy depletion can also be minimised. We have found that less work is done in this direction, with a special focus on energy-efficient data aggregation.

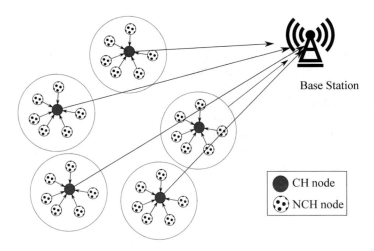

Fig. 1 General architecture of single-hop, clustered wireless sensor network

The rest of the paper is organised as follows: In Sect. 2, we discuss different techniques for data reduction in WSN; in Sect. 3, we discuss data aggregation techniques in WSN; in Sect. 4, we discuss data aggregation issues and challenges; Sect. 5 presents summary and comparison work; and finally Sect. 6 concludes this paper.

2 Data Reduction Techniques in WSN

One of the most popular data reduction techniques is data aggregation. Aggregation techniques are commonly summarisation functions/techniques used in database management systems for query purpose [14]. One of the early usages of data aggregation techniques in clustered WSN was by Heinzelman et al. [6]. In LEACH, it is assumed that l bits of data packet are received from NCH nodes by CH nodes; after processing these data along with their own sensed data, CH node transmits l bits of data to (or towards) base station. Rest of the symbols used in this section are the same as in [6]. The spent by radio unit in sending 'l' bits to a distance 'd' is:

$$
\begin{aligned}
E_{Tx}(l, d) &= E_{Tx}(l) + E_{Tx-amp}(l, d) \\
&= \begin{cases} l E_{elec} + l\epsilon_{fs}d^2 \text{ if } d < d_0, \\ l E_{elec} + l\epsilon_{mp}d^4 \text{ if } d \geq d_0, \end{cases}
\end{aligned}
\tag{1}
$$

Similarly, energy dissipated by receiver is

$$
E_{Rx}(l) = E_{Rx-elec}(l) = l E_{elec}
\tag{2}
$$

On the basis of above two, the energy dissipated inside a cluster by CH node for one round is

$$
E_{CH} = l E_{elec}(\frac{N}{k} - 1) + l E_{DA}\frac{N}{k} + l E_{elec} + l\epsilon_{mp}d^4_{toBS}
\tag{3}
$$

where N is total number of nodes; k is the number of clusters in the network. Here it can be seen that the data reduction ratio is $N_s:1$ where N_s is number of sensor nodes in the cluster which is

$$
N_s = \frac{N}{k}
\tag{4}
$$

Value of k varies with d_{toBS} as shown in Fig. 2 [6, 15].

This type of approach is called summarisation approach, where a single value is extracted from the set. Commonly used functions under this approach are min, max, average and count.

The second technique is data compression. Data compression can be performed on the sensed data by exploiting spatial correlation among the data, but it requires the extra cost of data dissemination [16, 17]. A good survey on data compression techniques for wireless sensor network is done in [18, 19]. Early works by

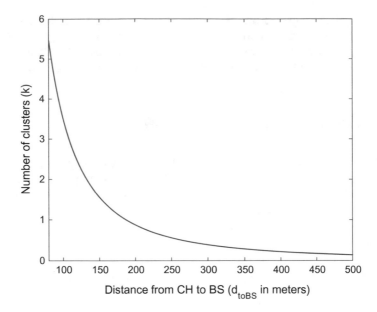

Fig. 2 Optimum cluster size in LEACH

Barr et al. [20] showed that data compression techniques have to trade-off between computational cost- and power-saving from compression ratio. Most of the classical data compression techniques are not WSN friendly because of processing and power constraints.

Compression in WSN can be categorised into two categories: distributed approach-based and local approach-based. The distributed approach-based compression techniques require specific assumptions of the model which are hard to correct over time. As an example, an observation model is required in [21, 22] and a correlation model is required in [23, 24]. In comparison with distributed approach, local approaches are more robust and have less assumptions. Local approaches fail to exploit the benefits of routing and network topology.

3   Data Aggregation Techniques in WSN

Aggregation in WSN can be categorised in four categories: cluster-based, tree-based, in-network and centralised data aggregation.

In a cluster-based approach, whole network is divided into clusters. These clusters work in two phases: first phase is the set-up phase where cluster selection process is done and clusters are chosen, and second phase is steady phase where cluster operates. During steady phase, all the member nodes of the cluster, including the cluster head node, sense their surroundings for application-specific data on a regular and periodic manner. All the member nodes send these sensed data to the cluster head. CH node aggregates it and forwards to the base station. In the literature, it is found that clustered approach is energy efficient, but simultaneously, it creates bottleneck from data collection point of view, leading to an increase in latency [25]. Cluster set-up process further increases the latency as it is done periodically. Some of the popular cluster-based data aggregation techniques are LEACH [26], HEED [27], SEP [28], PAgIoT [29].

In tree-based approach, an aggregation tree is constructed [30, 31], which is a minimum spanning tree. In this approach, base station acts as root of this tree and sensor nodes as leaves. Kyung et al. [30] proposed tree-based clustering (TBC) where CHs of the cluster form tree. Data is collected by the leaves and transmitted towards the root. GSTEB [32] is another tree-based approach where base station broadcasts the details of the root nodes. Clustered GSTEB [33] uses a mixed approach where within a cluster tree is formed, and CH node acts as a sink and forwards the aggregated data to the base station.

In in-network, [34] is a bigger approach, and it gathers details from the network (usually a multi-hop network) such as routing information. Data is aggregated at intermediate points, with the objective of minimising power consumption [35]. Again this approach may either reduce the width[1] of the data, resulting in reduced data to be transmitted further or it may keep the width intact by merging all the packets received. In [36], authors have used a metaheuristic approach with an objective of minimising communication cost and aggregation cost.

Centralised approach [37] is an address-centric approach where base station—a powerful node (gateway) plays a role in collecting and sharing network information with the nodes [25]. Nodes send the sensed data to this node via the shortest possible route.

4 Data Aggregation Issues and Challenges

During the process of data aggregation, there are few challenges that need to be conquered. It is difficult to conquer all of these together, as evident from the literature review. Some of the important among these are:

1. Data redundancy: sensor nodes sense similar kind of data and in certain cases same events, thus the sink node collects data which is redundant. Thus, it is wastage of time, energy and other resources.
2. Delay: In certain cases data from farther nodes arrive late at the sink or root node as a result of which aggregation process starts late. Further, aggregations at intermediate levels further increase the delay time.

---

[1]Packet length.

3. Accuracy: There can be two types of accuracy issues. First, the aggregator[2] function is an approximation function; hence, accuracy is lost while forwarding the data. Secondly, there could be a compromised node which sends false/inappropriate data to the aggregator node. Aggregator node does not ensure its correctness and processes it.
4. Traffic load: In certain scenarios, the aggregator node is overburdened. It happens if the load balancing is not done or the clusters are of unequal sizes.
5. Aggregation freshness [39, 40]: Data from similar frames should be aggregated, old stored data or aggregation of data from multiple frames from different time should be avoided as it loses freshness.

5 Summary and Comparison Work

In this section, we present a summary of the surveyed work and remarks, which is provided in Table 1. We have surveyed on the basis of four key criteria: node type, network type, algorithm and aggregator node mobility.

1. Node type: In WSN, there are two categories of nodes, homogeneous and heterogeneous. Heterogeneous nodes have more capabilities, and here we have considered their capability for more initial energy.
2. Network type: In order to exploit redundancy, network topology is very important. The network could be a clustered network, or a tree-based network or a flat network with no grouping of nodes.
3. Algorithm: Aggregation algorithm may be centralised, where a central node coordinates and aggregates data itself or chooses an aggregator node. Another approach could be that the algorithm operates in a distributed manner, and each node is equally likely to perform the task of aggregator on the basis of a selection criteria. The third category is locally here the algorithm works locally to decide who will aggregate and how to do it.
4. Position of aggregator node: Early research work focused on aggregator being at a fixed position, and it could be any node deployed in the sensing field.[3] Later, researchers adopted an approach where aggregator node is a mobile node, which is not resource-constrained. This node moves and collects data from different points in the network.

We have given remarks against each work surveyed in Table 1. Remarks column briefly focuses on the approach adopted by the respective work.

---

[2]Node which aggregates the incoming data [38].
[3]The sensing area in which sensors are deployed.

Table 1 Overview of data aggregation mechanisms and their main features

Paper	Year	Node Type	Network Type	Algorithm	Aggregator	Remarks
Kumar et al. [41]	2019	Homogeneous	Tree	Distributed	Fixed	Use multi-channel TDMA scheduling algorithms to reduce collision along with meta-heuristic algorithm for energy minimization and reduced latency
Sarangi et al. [42]	2019	Homogeneous	Cluster	Distributed	Mobile	Neural network-based cluster formation and Ant Colony based optimization
Yadav et al. [43]	2019	Homogeneous	Flat	Distributed	Fixed	An aggregator node uses linear classifier based on SVM for identification and elimination of redundant data
SreeRanjani et al. [44]	2018	Homogeneous	Flat	Distributed	Mobile	Adaptive Particle Swarm Optimization is used for optimum path selection of the mobile node
Khriji et al. [45]	2018	Homogeneous	Cluster	Distributed	Fixed	During consecutive iterations transmission of redundant data is reduced by using differential data from sensor
ShivKumar et al. [33]	2016	Homogeneous	Cluster	Distributed	Fixed	It is clustered GSTEB, tree is formed in side a cluster, root of the tree is CH. It has better throughput then GSTEB
Atoui et al. [38]	2016	Homogeneous	Flat	Centralized	Fixed	It has two levels at the first level it filters the data using fitting function,if norm value of data is below a threshold then it will not be forwarded to the aggregator for aggregation
Xiao et al. [37]	2015	Heterogeneous	Cluster, Flat	Centralized, Local	Fixed	Tradeoffs between data quality and energy dissipation to achieve efficient data aggregation. It has two versions Distributed Energy Allocation (DEA) and Centralised Energy Allocation algorithm (CEA). It uses GA based approach and Gibbs Sampling Theory

(continued)

Table 1 (continued)

Paper	Year	Node Type	Network Type	Algorithm	Aggregator	Remarks
Mottaghi et al. [46]	2015	Homogeneous	Cluster	Distributed	Mobile	It has mobile sink and rendezvous node which acts as a store for mobile sink
Han et al. [32]	2014	Homogeneous	Tree	Distributed	Fixed	GSTEB, Selection of root node is broadcasted by the sink, based on itself and neighbours info nodes are binded to parent nodes
Yuan et al. [47]	2014	Homogeneous	Cluster	Distributed	Fixed	It uses data density correlation, which is a spatial correlation between a nodes data and its neighbour
Guo et al. [48]	2013	Homogeneous	Tree	Distributed	Mobile	Uses a framework based on energy replenishment and anchor based data gathering where aggregator is mobile
Jin et al. [49]	2012	Homogeneous	Tree	Distributed	Fixed	Energy-balanced Transmission Protocol (ETP) is proposed which uses slice based energy model. Energy balancing is done using inter- and intra-slice energy balancing
Mathapati et al. [50]	2012	Homogeneous	Cluster	Distributed,Local	Fixed	Randomly a coordinator node (CN) is selected within a cluster. CH sends data to CN, which does loss ratio calculation and modifies the cluster if needed
Zhao et al. [51]	2011	Homogeneous	Tree	Distributed	Mobile	Data is collected by the mobile nodes and uses meta-heuristic approach for selecting polling points
Yang et al. [52]	2011	Homogeneous	Tree	In-Network	Fixed	Does an in-network aggregation of data based on the order compression techniques. Based on history, data may be suppressed
Zhao et al. [53]	2010	Homogeneous	Tree	Distributed	Mobile	Multiple nodes, which are mobile used Multi mobile sink based heuristic approach for division of region and tour-planning for data aggregation

6 Conclusion

In this paper, we surveyed different techniques for data reduction using aggregation. The surveyed techniques are based on some critical factors such as node heterogeneity, network type, algorithm type and mobility of the aggregator node. It has been found that most of these techniques try to improve energy efficiency and/or latency. The tree-based data aggregation technique is better in comparison with clustered approach as it reduces latency and energy consumption, but has a disadvantage of high overhead, in making of tree topology. The cluster-based mechanisms have been able to reduce energy dissipation and increase network scalability. The centralised networks are not preferred because of the overhead and limit the dynamic nature of the network. Tree-based approach is appropriate for carrying aggregation task.

Our survey finding also includes few least explored areas. First observation is that very less work is available on fault tolerance during data aggregation. Second observation is that we could not find any work related to multiple data-type aggregation, which could be used when there are more than one type of sensors mounted on a single node and these sensors generate different types of data. The survey results in this work may be useful in designing new energy-efficient algorithms/protocols for WSN which exploit data reduction.

References

1. Akyildiz I, Su W, Sankarasubramaniam Y, Cayirci E (2002) Wireless sensor networks: a survey. Comput Networks 38(4):393–422
2. Pantazis NA, Nikolidakis SA, Vergados DD (2013) Energy-efficient routing protocols in wireless sensor networks: a survey. IEEE Commun Surveys Tutorials 15(2):551–591
3. Rault T, Bouabdallah A, Challal Y (2014) Energy efficiency in wireless sensor networks: a top-down survey. Comput Networks 67(Suppl C):104–122
4. Zuhra FT, Bakar KA, Ahmed A, Tunio MA (2017) Routing protocols in wireless body sensor networks: a comprehensive survey. J Network Comput Appl 99(Suppl C):73–97
5. Rai R, Rai P (2019) Survey on energy-efficient routing protocols in wireless sensor networks using game theory. In: Advances in communication, cloud, and big data. Springer, Berlin, pp. 1–9
6. Heinzelman W, Chandrakasan A, Balakrishnan H (2002) An application-specific protocol architecture for wireless microsensor networks. IEEE Trans Wireless Commun 1:660–670
7. Wang A, Sodini C (2004) A simple energy model for wireless microsensor transceivers. In: Global telecommunications conference (GLOBECOM '04). IEEE, vol 5, pp 3205–3209, Nov 2004
8. Miller M, Vaidya N (2005) A mac protocol to reduce sensor network energy consumption using a wakeup radio. IEEE Trans Mob Comput 4:228–242
9. Mallinson M, Drane P, Hussain S (2007) Discrete radio power level consumption model in wireless sensor networks. In: IEEE International conference on mobile adhoc and sensor systems (MASS 2007), pp 1–6, Oct 2007
10. Han B, Zhang D, Yang T (2008) Energy consumption analysis and energy management strategy for sensor node. In: International conference on information and automation (ICIA 2008), pp 211–214, June 2008

11. Halgamuge MN, Zukerman M, Ramamohanarao K, Vu HI (2009) An estimation of sensor energy consumption. In: Progress in electromagnetics research B
12. Zhou H-Y, Luo D-Y, Gao Y, Zuo D-C (2011) Modeling of node energy consumption for wireless sensor networks. Wireless Sens Network 3(01):18
13. Kumar V, Yadav S, Kumar V, Sengupta J, Tripathi R, Tiwari S (2018) Optimal clustering in weibull distributed wsns based on realistic energy dissipation model. In: Progress in computing, analytics and networking, pp 61–73. Springer, Berlin
14. Madden S, Franklin MJ, Hellerstein JM, Hong W (2002) Tag: a tiny aggregation service for ad-hoc sensor networks. ACM SIGOPS Oper Syst Rev 36:131–146
15. Roy NR, Chandra P (2018) A note on optimum cluster estimation in leach protocol. IEEE Access 6:65690–65696
16. Hoang AT, Motani M (2005) Exploiting wireless broadcast in spatially correlated sensor networks. In: IEEE international conference on commun (ICC 2005), vol 4, pp 2807–2811. IEEE
17. Hoang AT, Motani M (2007) Collaborative broadcasting and compression in cluster-based wireless sensor networks. ACM Trans Sens Networks (TOSN) 3(3):17
18. Kimura N, Latifi S (2005) A survey on data compression in wireless sensor networks. In: International conference on information technology: coding and computing (ITCC 2005), vol 2, pp 8–13. IEEE
19. Srisooksai T, Keamarungsi K, Lamsrichan P, Araki K (2012) Practical data compression in wireless sensor networks: a survey. J Network Comput Appl 35(1):37–59
20. Barr KC, Asanović K (2006) Energy-aware lossless data compression. ACM Trans Comput Syst (TOCS) 24(3):250–291
21. Oka A, Lampe L (2008) Energy efficient distributed filtering with wireless sensor networks. IEEE Trans Signal Process 56(5):2062–2075
22. Teng J, Snoussi H, Richard C (2010) Decentralized variational filtering for target tracking in binary sensor networks. IEEE Trans Mob Comput 9(10):1465–1477
23. Tang Z, Glover I, Evans A, Monro D, He J (2006) An adaptive distributed source coding scheme for wireless sensor networks. In: 12th European wireless conference, University of Bath
24. Wang W, Peng D, Wang H, Sharif H, Chen H-H (2009) Cross-layer multirate interaction with distributed source coding in wireless sensor networks. IEEE Trans Wireless Commun 8(2):787–795
25. Shao-Liang P, Shan-Shan L, Yu-Xing P, Pei-Dong Z, Nong X (2007) A delay sensitive feedback control data aggregation approach in wireless sensor network. In: International conference on computational science. Springer, Berlin, pp 393–400
26. Heinzelman W, Chandrakasan A, Balakrishnan H (2000) Energy-efficient communication protocol for wireless microsensor networks. In: Proceedings of the 33rd annual Hawaii international conference on system sciences, vol 2, p 10, Jan 2000
27. Younis O, Fahmy S (2004) Heed: a hybrid, energy-efficient, distributed clustering approach for ad hoc sensor networks. IEEE Trans Mob Comput 3(4):366–379
28. Smaragdakis G, Matta I, Bestavros A (2004) Sep: a stable election protocol for clustered heterogeneous wireless sensor networks. Technical report, Boston University, Computer Science Department
29. González-Manzano L, de Fuentes JM, Pastrana S, Peris-Lopez P, Hernández-Encinas L (2016) Pagiot-privacy-preserving aggregation protocol for internet of things. J Network Comput Appl 71:59–71
30. Kim KT, Lyu CH, Moon SS, Youn HY (2010) Tree-based clustering (tbc) for energy efficient wireless sensor networks. In: IEEE 24th international conference on advanced information networking and applications workshops (WAINA). IEEE, pp 680–685
31. Kalpakis K, Dasgupta K, Namjoshi P (2002) Maximum lifetime data gathering and aggregation in wireless sensor networks. In: Networks. World Scientific, pp 685–696
32. Han Z, Wu J, Zhang J, Liu L, Tian K (2014) A general self-organized tree-based energy-balance routing protocol for wireless sensor network. IEEE Trans Nuclear Sci 61(2):732–740
33. Shivkumar S, Kavitha A, Swaminathan J, Navaneethakrishnan R (2016) General self-organizing tree-based energy balance routing protocol with clustering for wireless sensor network. Asian J Inform Technol 15(24):5067–5074

34. Bahi JM, Makhoul A, Medlej M (2012) Frequency filtering approach for data aggregation in periodic sensor networks. In: Network operations and management symposium (NOMS). IEEE, pp 570–573

35. Dasgupta K, Kalpakis K, Namjoshi P (2003) An efficient clustering-based heuristic for data gathering and aggregation in sensor networks. In: Wireless communications and networking (WCNC 2003). IEEE, vol 3, pp 1948–1953

36. Zhang B, Guo W, Chen G, Li J (2013) In-network data aggregation route strategy based on energy balance in wsns. In: WiOpt, pp 540–547

37. Xiao S, Li B, Yuan X (2015) Maximizing precision for energy-efficient data aggregation in wireless sensor networks with lossy links. Ad Hoc Networks 26:103–113

38. Atoui I, Ahmad A, Medlej M, Makhoul A, Tawbe S, Hijazi A (2016) Tree-based data aggregation approach in wireless sensor network using fitting functions. In: 2016 sixth international conference on digital information processing and communications (ICDIPC). IEEE, pp 146–150

39. Yu Y, Prasanna VK, Krishnamachari B (2006) Energy minimization for real-time data gathering in wireless sensor networks. IEEE Trans Wireless Commun 5(11):3087–3096

40. Bagaa M, Younis M, Ouadjaout A, Badache N (2013) Efficient multi-path data aggregation scheduling in wireless sensor networks. In: 2013 IEEE international conference on communications (ICC). IEEE, pp 1560–1564

41. Kumar S, Kim H (2019) Energy efficient scheduling in wireless sensor networks for periodic data gathering. In: IEEE access

42. Sarangi K, Bhattacharya I (2019) A study on data aggregation techniques in wireless sensor network in static and dynamic scenarios. In: Innovations in systems and software engineering, pp 1–14

43. Yadav S, Yadav RS (2019) Redundancy elimination during data aggregation in wireless sensor networks for iot systems. In: Recent trends in communication, computing, and electronics. Springer, Berlin, pp 195–205

44. SreeRanjani N, Ananth A, Reddy LS (2018) An energy efficient data gathering scheme in wireless sensor networks using adaptive optimization algorithm. J Comput Theor Nanosci 15(11–12):3456–3461

45. Khriji S, Raventos GV, Kammoun I, Kanoun O (2018) Redundancy elimination for data aggregation in wireless sensor networks. In: 2018 15th international multi-conference on systems, signals & devices (SSD). IEEE, , pp 28–33

46. Mottaghi S, Zahabi MR (2015) Optimizing leach clustering algorithm with mobile sink and rendezvous nodes. AEU-Int J Electron Commun 69(2):507–514

47. Yuan F, Zhan Y, Wang Y (2014) Data density correlation degree clustering method for data aggregation in wsn. IEEE Sens J 14(4):1089–1098

48. Guo S, Wang C, Yang Y (2013) Mobile data gathering with wireless energy replenishment in rechargeable sensor networks. In: INFOCOM, 2013 Proceedings IEEE. IEEE, , pp 1932–1940

49. Jin N, Chen K, Gu T (2012) Energy balanced data collection in wireless sensor networks. In: 2012 20th IEEE international conference on network protocols (ICNP). IEEE, pp 1–10

50. Mathapati BS, Patil SR, Mytri V (2012) Energy efficient reliable data aggregation technique for wireless sensor networks. In: 2012 international conference on computing sciences (ICCS). IEEE, pp 153–158

51. Zhao M, Ma M, Yang Y (2011) Efficient data gathering with mobile collectors and space-division multiple access technique in wireless sensor networks. IEEE Trans Comput 60(3):400–417

52. Yang C, Yang Z, Ren K, Liu C (2011) Transmission reduction based on order compression of compound aggregate data over wireless sensor networks. In: 2011 6th international conference on pervasive computing and applications (ICPCA). IEEE, pp 335–342

53. Zhao M, Yang Y (2010) Data gathering in wireless sensor networks with multiple mobile collectors and sdma technique sensor networks. In: 2010 IEEE Wireless communications and networking conference (WCNC). IEEE, pp 1–6

Load Balanced Fuzzy-Based Clustering for WSNs

Deepika Agrawal and Sudhakar Pandey

Abstract The wireless sensor networks (WSNs) form an integral part of the Internet of Things (IoT). The prospective use of WSNs in various applications has grown interested in WSNs. Since it is almost not possible to replace or recharge the nodes battery when they are deployed. Hence, energy consumption should be carefully monitored. Minimizing the consumption of the energy of the sensor nodes leads to the prolongation of network lifetime. This paper proposes a clustering protocol based on fuzzy logic which not only prolongs the network life span but also balances the load among nodes. The proposed protocol is evaluated with many protocols. The output obtained proved that the proposed protocol outperforms over existing standard protocols.

Keywords Fuzzy logic · Clustering · WSNs · Network lifetime · Energy efficiency

1 Introduction

WSNs incorporate billions of sensing nodes that collaborate with each other to perform some sensing tasks and convey that information to the sink or base station (BS). The sensor nodes are deployed in either way in an ad hoc manner [1]. WSNs can be used in many applications like battlefield vigilance, healthcare monitoring, forest fire detection, etc. They have a limited constraint of battery, as they are deployed in a harsh environment, so it would be challenging to replace or recharge the node batteries. So, the consumption of the node's energy is to be properly administered so as to lengthen the life span of the WSNs [1].

To recuperate the energy efficacy of WSNs, the typical method includes the routing and clustering algorithms which are energy efficient [2, 3]. Clustering is a technique used for the periodic collection of data. In this technique, the network is parted into

D. Agrawal (✉) · S. Pandey
Department of IT, NIT Raipur, Raipur, India
e-mail: deepika721@gmail.com

S. Pandey
e-mail: spandey.it@nitrr.ac.in

© Springer Nature Singapore Pte Ltd. 2020
A. Khanna et al. (eds.), *International Conference on Innovative Computing and Communications*, Advances in Intelligent Systems and Computing 1059,
https://doi.org/10.1007/978-981-15-0324-5_49

583

clusters with one leader is elected known as cluster head (CH) [4]. All the other sensor nodes that are not elected as CH become cluster members (CM). Nodes sense the physical phenomena and send the information to the CH, and then, the CH gathers the data from all cluster members of its cluster and performs aggregation.

Designing an energy-efficient clustering protocol is an NP-hard problem [5]. To design this, various computational intelligence methods like fuzzy logic, particle swarm optimization, etc. have been employed in WSNs for various issues [6, 7]. To solve the uncertainties in the WSNs, generally fuzzy logic is used. A system can be made more optimized with incomplete information with the usage of fuzzy logic [8].

2 Related Work

Several researchers are carrying their research on clustering issues in WSNs. Some of the protocols developed so far relevant to the proposed protocol are explained in this section. Low-energy adaptive clustering hierarchy (LEACH) [9] is the oldest protocols developed for clustering. It is a TDMA-based protocol. The complete network is partitioned into groups (clusters), and one group leader known as cluster head is chosen. It has several issues. Sometimes, a node having less amount of energy becomes CH and dies quickly.

Hybrid energy-efficient distributed (HEED) protocol [10] is an advancement of LEACH protocol. HEED protocol considers residual energy factor in CHs election. This protocol performs better in comparison with LEACH protocol.

Gupta et al. [11] proposed a protocol which implements fuzzy logic during CH election. This work is not compatible with large-scale WSNs as it suffers from scalability issue. Energy-aware distributed clustering protocol using fuzzy logic [12] elects the tentative CHs by taking the remaining energy of the nodes. It does not use the probabilistic approach. Fuzzy logic-based energy-efficient clustering hierarchy (FLECH) [13] protocol is based on equal clustering. The fuzzy system is employed. This protocol is mainly designed for nonuniform WSNs. A fuzzy energy-aware unequal clustering algorithm (EAUCF) [6] takes into account other factors. Fuzzy logic is used for this purpose.

Multi-objective fuzzy clustering algorithm for wireless sensor networks (MOFCA) [14] is an advancement of EAUCF. MOFCA considers node density as an additional parameter for computation of cluster size. Final CHs selections are based on the single parameter. This leads to a nonuniform distribution of CHs.

Fuzzy-based unequal clustering algorithm (FUCA) [15] is a protocol which considers other factors also at the time of calculation of the size of clusters. The fuzzy logic methodology is utilized. The outputs produced are competition radius and rank. However, it suffers from intra-cluster communication overhead. A K-means technique is proposed for clustering, and purity method is used to validate this clustering in WSNs [16]. This protocol meets the demand of WSNs that are of large-scale and able to reduce power consumption. Another protocol is proposed for large-scale

WSNs where the problem of routing is taken in low-density area in WSNs. This protocol takes azimuthal routing protocol into consideration [17].

To solve the intra-cluster communication overhead, some other parameters are also considered in the proposed protocol. This reduces the intra-cluster communication cost of a node, saving dissipation of energy.

3 Proposed Protocol

The proposed protocol is designed for WSNs with homogeneous nodes. This section is divided into two parts.

3.1 Fuzzy System

The fuzzy logic controller is employed to select the CHs and calculates the chance. Four inputs are known to the system, and one output is produced. Inputs are residual energy, node degree, distance to the BS, and centrality, while the output is a chance.

Input variables are:

(i) Residual energy: It is the most essential criterion for the choice of CH. The considerable higher amount of energy is spent by the cluster head than other nodes since it collects information from the CMs, aggregates, and transmits the accumulated data to the sink.

(ii) Distance to the BS: For sending the data, the energy utilization by the nodes rises with the rise in distance amidst transmitter and receiver nodes. Hence, the distance amid CH and BS is less.

(iii) Centrality: Centrality is a measure of how well the sensor node is positioned at the middle of its neighbors in the whole network. This is an important measure to reduce the intra-cluster communication cost. Less quantity of energy is needed by the cluster members to transfer the information to the CHs if the value of CH's centrality is lower.

(iv) Node's degree: Node's degree is the number of neighbor nodes within a range of transmission. A higher value of the node's degree gives more chance to a node for being selected as a CH.

The output variable is:

Chance: It defines the eligibility of a node to be elected as a CH. The higher the significance of a chance output, the higher the probability of a sensor node to be elected as CH.

Nodes that have high residual energy, less distance to the BS, close centrality, a high number of neighbors' nodes will have a high probability to become a CH (Table 1).

The fuzzy set for each variables, i.e., input and output are shown in Figs. 1, 2, 3, 4, and 5, respectively.

Boundary variables follow membership function, i.e., trapezoidal in nature, while the middle variable follows membership function, i.e., triangular in nature as shown in figures.

In FIS, the crisp values of input are transformed into fuzzy linguistic variables. The fuzzy if-then rules are applied depending on the Mamdani method to translate the input variables to the corresponding fuzzy variables. The method, i.e., center of area (CoA) is applied for defuzzification, i.e., to convert linguistic variables into a crisp output value.

Table 1 Variables and their membership functions

	Variables	Membership function
Inputs	Residual energy	Low, medium, and high
	Distance to BS	Far, adequate, and close
	Node degree	Low, medium, and high
	Centrality	Far, adequate, and close
Output	Chance	Very very weak, very weak, weak, little lower medium, lower medium, higher medium, medium, little higher medium, little strong, strong, very strong, very very strong

Fig. 1 'Residual energy' fuzzy set

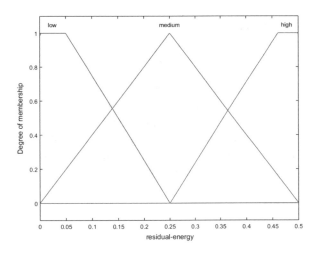

Fig. 2 'Degree' fuzzy set

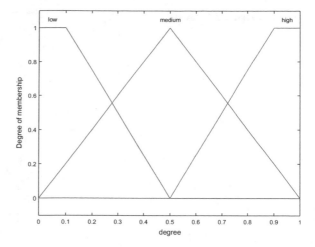

Fig. 3 'Distance to BS' fuzzy set

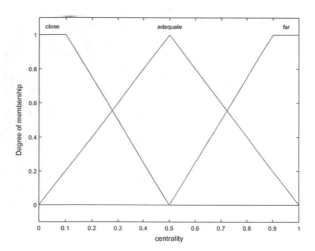

3.2 Clustering Process

This phase is separated into rounds like LEACH protocol. The proposed protocol operates in two phases:

Formation of clusters. At the beginning of each round, each sensor node picks an arbitrary number amid 0 and 1. If the created number is below the predetermined threshold, then the node will be eligible for the election of CH. Then, the eligible node will compute chance. The node will transmit a message to all the neighbors inside its range of transmission. This message will comprise of the details such as the node id and the value of chance. The nodes whose chance value is higher within the neighbor will be declared as final CH. The nodes who will not become CH and the nodes that are not eligible will now broadcast a message of join to the nearest

Fig. 4 'Centrality' fuzzy set

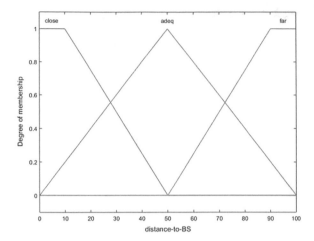

Fig. 5 'Chance' fuzzy set

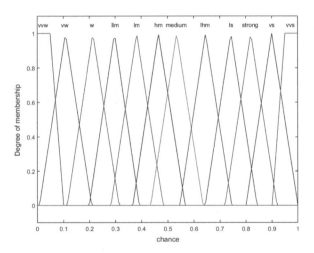

CH. After cluster formation, CH nodes will create the schedule for TDMA and send it to all the members.

Data collection. After the cluster formation and generation of TDMA schedules, CH nodes wait for the data. A non-CH node senses the data and transmits to the respective CH within the allotted time frame. Nodes will go to the sleep state in other time slots. Aggregation of all the collected data into a single message will be done by CH nodes. The accumulated data will then ultimately be transmitted to BS through CHs. The CHs need to be in wakeup mode all the time, and this leads to the consumption of higher energy by the CHs. The role of CHs is switched among the sensor nodes so as to balance the depletion of the energy.

4 Evaluation of the Performance of the Proposed Protocol

4.1 Simulation Setting

The proposed protocol is simulated by MATLAB. To prove the efficacy of the proposed algorithm, different scenarios are created for the simulations. The random generation of nodes brings some coincidence factors which may influence the results of the experiment. Hence, the results of the average of 50 experiments are taken.

100 and 200 nodes are taken for 100 × 100 region of interest. The sink position is sited first at the middle of the area then at the corner of the area. So, in total, 4 scenarios are created that are represented in Table 2.

4.2 Simulation Parameters

Table 3 listed the parameters of the simulation. The radio energy model of first order is used for calculating the depletion of energy. The proposed protocol's performance is evaluated with some standard equal and unequal clustering protocols like MOFCA [14], FLECH [13] and FUCA [15].

Table 2 Scenarios for evaluation of the proposed protocol

Representation	Number of nodes	Sink position
SN#1	100	(50,50)
SN#2	200	(50,50)
SN#3	100	(100,100)
SN#4	200	(100,100)

Table 3 Simulation Parameters

Parameters	Values
Located area (m^2)	100 * 100
Deployment of sensor nodes	Random
Placement of sink	Center and corner
Total deployed nodes	100 and 200
Energy	0.5 J
E_{elec}	50 nJ/bit
ε_{amp}	0.0010 pJ/bit/m^4
E_{DA}	5 nJ/bit

5 Results and Discussion

5.1 Network Life span Evaluation

The death of first node (FND) value for each of the protocol is represented in Fig. 6. The network life span is stated as the round in which the first coverage hole or the death of the first node occurs. From Fig. 6, it is well-defined that the proposed protocol enhances the time of FND and hence prolongs the life span of the network.

5.2 Evaluation of the Average Remaining Energy

Figure 7 depicts the average of the remaining energy by all the protocols in each and every scenario. This energy calculation includes all the cost, i.e., the cluster formation

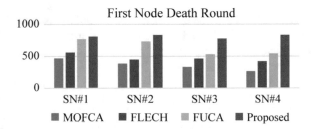

Fig. 6 First node death round for scenario 1, the proposed protocol extended the network lifetime by 41.97% with the comparison to MOFCA, with FLECH it is 30.86%, and with FUCA it is 4.94%. For scenario 2, the proposed protocol extended the network lifetime by 53.71% with the comparison to MOFCA, with FLECH it is 46.04%, and with FUCA it is 12.11%. For scenario 3, the proposed protocol extended the network lifetime by 57.14% with the comparison to MOFCA, with FLECH it is 40.28%, and with FUCA it is 31.27%. For scenario 4, the proposed protocol extended the network lifetime by 68.14% with the comparison to MOFCA, with FLECH it is 49.27%, and with FUCA it is 34.73%

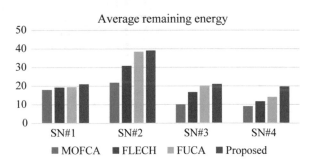

Fig. 7 Average remaining energy of protocols

cost, the intra-cluster communication cost as well as the inter-cluster communication cost during a round.

It can be depicted from the figure that the average remaining energy of the proposed protocol is higher when competed to other protocols like MOFCA, FLECH, and FUCA. The proposed protocol balances the load among the nodes as well as prolongs the network lifetime.

6 Conclusion

In this paper, load balancing and minimization of energy dissipation problem are considered for WSNs. Fuzzy logic is engaged to elect the cluster heads. Cluster heads should be selected properly to reduce the utilization of the energy and also to balance the load among nodes. The proposed protocol is competed with some standard equal and unequal clustering protocols over diverse network scenarios. In all the scenarios, a significant improvement in terms of network life span is observed. In the future, the protocol, i.e., proposed can be evaluated by incorporating the mobility of nodes and obstacle in the region of the interest.

References

1. Akyildiz IF, Weilian S, Sankarasubramaniam Y, Cayirci E (2002) Wireless sensor networks: a survey. Comput Netw 38(4):393–422
2. Akkaya K, Younis M (2005) A survey on routing protocols for wireless sensor networks. Ad Hoc Netw 3(3):325–349
3. Abbasi AA, Younis M (2007) A survey on clustering algorithms for wireless sensor networks. Comput Commun 30(14):2826–2841
4. Afsar MM, Tayarani NM H (2014) Clustering in sensor networks: a literature survey. J Netw Comput Appl 46:198–226
5. Agarwal PK, Procopiuc CM (2002) Exact and approximation algorithms for clustering. Algorithmica 33(2):201–226
6. Bagci H, Yazici A (2013) An energy aware fuzzy approach to unequal clustering in wireless sensor networks. Appl Soft Comput 13(4):1741–1749
7. Zungeru AM, Ang LM, Seng KP (2012) Classical and swarm intelligence based routing protocols for wireless sensor networks: a survey and comparison. J Netw Comput Appl 35(5):1508–1536
8. Zadeh LA (1965) Information and control. Fuzzy Sets 8(3):338–353
9. Heinzelman WR, ChandrakasanA, Balakrishnan H (2002) nergy-efficient communication protocol for wireless microsensor networks. In: Proceedings of the 33rd annual Hawaii international conference on system sciences, 2000. IEEE, p 10
10. Younis O, Fahmy S (2004) HEED: a hybrid, energy-efficient, distributed clustering approach for ad hoc sensor networks. IEEE Trans Mob Comput 3(4):366–379
11. Gupta I, Riordan D, Sampalli S (2005) Cluster-head election using fuzzy logic for wireless sensor networks. In: Proceedings of the 3rd Annual Communication Networks and Services Research Conference, 2005. IEEE pp 255–260

12. Taheri H, Neamatollahi P, Younis OM, Naghibzadeh S, Yaghmaee MH (2012) An energy-aware distributed clustering protocol in wireless sensor networks using fuzzy logic. Ad Hoc Netw 10(7):1469–1481
13. Balakrishnan B, Balachandran S (2017) FLECH: fuzzy logic based energy efficient clustering hierarchy for nonuniform wireless sensor networks. Wireless Commun Mobile Comput
14. Sert SA, Bagci H, Yazici A (2015) MOFCA: Multi-objective fuzzy clustering algorithm for wireless sensor networks. Appl Soft Comput 30:151–165
15. Agrawal D, Pandey S (2018) FUCA: Fuzzy-based unequal clustering algorithm to prolong the lifetime of wireless sensor networks. Int J Commun Syst 31(2)
16. Almajidi AM, Pawar VP, Alammari A (2019) K-means-based method for clustering and validating wireless sensor network. In: International conference on innovative computing and communications. Springer, Singapore, pp 251–258
17. Agrawal P, Anand V, Tripathi S, Pandey S, Kumar S (2019) A solution for successful routing in low–mid-density network using updated Azimuthal protocol. In: International conference on innovative computing and communications. Springer, Singapore, pp 339–347

Proposing a Framework for Citizen's Adoption of Public-Sector Open IoT Data (OIoTD) Platform in Disaster Management

Muhammad Mahboob Khurshid, Nor Hidayati Zakaria, Ammar Rashid, Muhammad Noman Shafique, Ashish Khanna, Deepak Gupta and Yunis Ali Ahmed

Abstract Disaster management revolves around preparing, mitigating, responding, and recovering from a sudden disruption that has catastrophic effects on human lives, economy, and infrastructure. Data of Internet-connected things (termed as Internet of things) openly available in big quantities digitally by the government leverages data-driven innovations and efficient decision-making in disaster management. However, research on exploring the predictors affecting the usability and benefits of such platforms (i.e., Open IoT Data (OIoTD) platform) from the perspective of individual unit of adoption is currently lacking. Therefore, lack of research on predictors affecting the adoption of OIoTD platform from the citizen's perspective stimulated us to

M. M. Khurshid (✉)
School of Computing, Faculty of Engineering, Universiti Teknologi Malaysia, Johor Bahru,
Malaysia
e-mail: mehboob.khursheed@vu.edu.pk

N. H. Zakaria
Azman Hashim International Business School, Universiti Teknologi Malaysia, Kuala Lumpur,
Malaysia
e-mail: hidayati@utm.my

A. Rashid
College of Engineering and IT, Ajman University, Ajman, UAE
e-mail: a.rashid@ajman.ac.ae

M. N. Shafique
Dongbei University of Finance and Economics, Dalian, China
e-mail: shafique.nouman@gmail.com

A. Khanna · D. Gupta
Computer Science and Engineering, Maharaja Agrasen Institute of Technology, Delhi, India
e-mail: ashishkhanna@mait.ac.in

D. Gupta
e-mail: deepakgupta@mait.ac.in

Y. A. Ahmed
Faculty of Computing, SIMAD University, Mogadishu, Somalia
e-mail: yunisali@simad.edu.so

© Springer Nature Singapore Pte Ltd. 2020
A. Khanna et al. (eds.), *International Conference on Innovative Computing
and Communications*, Advances in Intelligent Systems and Computing 1059,
https://doi.org/10.1007/978-981-15-0324-5_50

conduct this research. A better understanding of these predictors can help policy-makers to determine the policy instrument to increase adoption of OIoTD platform for a wider community. In order to evaluate its adoption from citizen's perspective in disaster management scenario, a framework has been proposed in this study.

Keywords Internet of things · Open data · Big data · Public sector · Open Internet of things data · Disaster management · Technology adoption

1 Introduction

Disaster is referred to as a complex issue and an emergency which demands for urgent actions that need to be taken to avoid its devastating effects on human, animals, buildings, infrastructure, and economy. Disaster management, which revolves around preparing, preventing and mitigating, responding, rehabilitating, reconstructing, and recovering, is the contemporary and complex phenomenon in Pakistan [1]. It is complex in terms of physical, social, and environmental intersections locally and globally. Government is considered as the primary stakeholder in handling all disaster management activities where data, information, and technology are the key pillars for right decision-making to tackle for the disaster. Therefore, Government of Pakistan (GoP) has given much attention on proactive approaches for effective disaster management [2, 3], with the rise of disaster frequency and its impact especially after occurrence of major earthquake in 2005 and flood in 2010.

With the advent of new information technologies on each coming day, new solutions are being proposed for tackling complex issues such as disaster management. Technology can play a significant role in pre-disaster, disaster, and post-disaster situations. Besides limitations, a technology itself also bears a lot of potential to aid disaster management effectively and efficiently. Internet of things (IoT) technology has been considered a mature and useful technology that has potentials in all disaster phases. Since data is the basic resource of organizations for effective decision-making, the data of things connected through Internet (owned by government agencies) openly available in big quantities online offers opportunities for citizens, private organizations, developers, and all types of actors involved in emergency management. Therefore, the platform where the IoT data is available is referred to as OIoTD platform. OIoTD platform would provide with the opportunities for data-driven innovation and efficient decision-making in disaster management. The adoption of OIoTD platform has been less explored in the literature from the citizen's perspective in disaster management and become the aim of the authors. Therefore, this study explores the predictors of citizen's adoption of OIoTD platform so that implementation of Open IoT Data initiative can be successful. This study helps the policy-makers in making right policy instrument for OIoTD adoption for wider community.

2 An Overview of OIoTD Platform

Internet of things (IoT) is interpreted as a composition of three concepts: an interaction between people to people, the interaction between people to things, and an interaction between things to things. However, these interactions must be established through the Internet. IoT may also be described as an existence of effective network of things or objects physically connected through the Internet and communicating using standard protocols without interventions of human. The IoT paradigm can also be envisioned as both effective strategies for data collection as well as data sharing. The IoT technology can also be considered as a complex DSS which can perform several activities like generating, collecting, visualizing, and sharing data of connected objects more precisely, systematically, and intelligently.

Open Internet of Things Data (OIoTD) is conceptualized as a combination of different ingredients including Internet of connected objects, big data, open data, intelligence tools and analysis, government, and different stakeholders. Figure 1 elaborates the complete flowchart of OIoTD where objects are connected through Internet and generating large amount of heterogenous data rapidly. Since government is among the largest producers of data, it openly provides big data using intelligence tools onto online platform referred to as OIoTD platform. Stakeholders will then use, reuse, and distribute data of connected objects without any kind of restrictions to support the government in better decision-making on complex issues, e.g., disaster management. Therefore, OIoTD platform has been defined, in this study, as a platform, delivered via information and communication technology (ICT), onto which large amount of data generated by connected objects is openly provided by

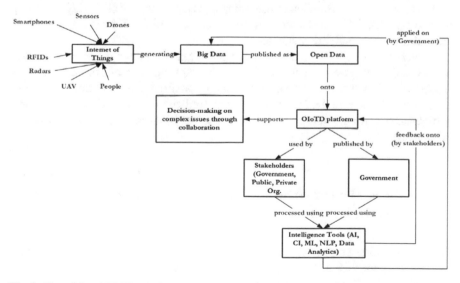

Fig. 1 Flowchart of OIoTD platform in disaster management

the government which can then be used, reused, and distributed by the stakeholders without any kind of restrictions or legal obligations.

The OIoTD platform bears some goals such as enabling collaboration for data-driven innovations by the stakeholders locally and globally, better and efficient decision-making in complex issues, etc. However, it is pertinent to note that the data would be collecting, generating, processing, publishing, and owning by the governmental agencies which will then be available to the public for open use. The OIoTD initiative is quite similar to different concepts such as Open Source, Open Government Data, Open Access, and Big Open Data [4]. Undergoing this initiative, public-sector organizations are publishing data online so that the public can access data of devices connected through Internet on real-time basis. The key motivator to encourage public-sector organizations in increasing the availability of OIoTD platforms is to address complex problems and to bring respective better solutions through collaborative innovations.

Several OIoTD platforms are being proposed such as SensorCentral, COAST, Open Geospatial Data infrastructure. SensorCentral is a platform which provides a framework to enable interoperability with a large range of agnostic sensor devices along with the supporting features to do research. COAST (abbreviated as Connected Open plAtform for Smart objecTs) introduces a platform to provide the capability to IoTs in terms of autonomously negotiating and deploying respective additional resources to all connected objects for rescue operations in disaster management. The Open Geospatial Data infrastructure introduces a design to improve interoperability issues by employing sensors' capabilities and connecting objects for data-driven precision agriculture purpose in the agriculture sector. Moreover, there are also several Open IoT platforms exist such as Nimbits (for developing applications using open source Java library, SensorCloud (for acquiring, visualizing, and analyzing data using proprietary MATLAB tool), and GroveStreams IoT platform for providing some of the real-time data analytics techniques.

Open IoT Data (OIoTD) platform makes it easy for the public to access to disaster data owned by the government agencies obtained through sensors, unmanned aerial vehicle (UAV), actuators, drones, robots, radars, smartphones, near field communication (NFC), and RFIDs. This data is further utilized for data-driven innovations, participatory and democratic governance, and decision-making in better disaster management.

3   Measures Development for Evaluating the Acceptance of OIoTD Platform

This study aims at empirically investigating the OIoTD platform adoptability from the citizen's perspective. An integrated approach is adopted to evaluate the citizen's adoption of OIoTD platform and proposed a framework (Fig. 2). The performance of OIoTD platform is evaluated empirically in few studies. The use of IoT technology in

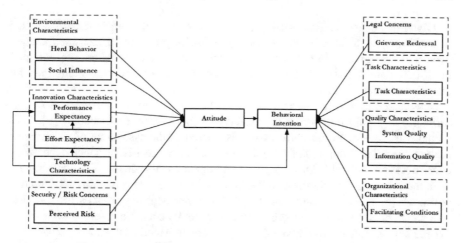

Fig. 2 Proposed model for Open IoT Data platform

health care has been evaluated by employing UTAUT alongside health belief model (HBM) as an added theory. Carcary et al. [4] developed IoT adoption matrix based on three constructs of UTAUT including performance expectancy, effort expectancy, and social influence.

From the perspective of OIoTD platform as innovation because, in Pakistan, it is the technology which is new from the perspective of its unit of adoption, as per definition of innovation presented by Rogers [5]. In IS context, the literature has largely employed different theoretical models to measure the adoption of service, technology, Web sites, or products. The models used mostly in the literature include TPB theory [6], TAM model [7], UTAUT theory [8], and unified model of electronic government [9]. The UMEGA model is employed to be the main theory for adoption of OIoTD platform. The framework is specifically developed on the system offered by government, which will then be empirically tested and validated employing an extensive pool of questions. However, in order to broaden the acceptance and use predictors of OIoTD platform, the task-technology fit (TTF) theory [10] and IS success model [11] are regarded suitable.

The TTF theory has widely been used in different contexts such as measuring the strategic value of using Internet of things technology in better management of disaster [12]. The UMEGA technology acceptance model comprises six attributes and perceived to have significant impact on technology acceptance offered by governments. These attributes are similar to UTAUT theory including one addition construct: perceived risk and with two dependent variables—attitude and behavioral intention [9]. Similarities are detected such that the constructs of UMEGA are an improved, modified as well as extended version of UTAUT technology acceptance model with an inclusion of perceived risk and attitude attributes and excluding use behavior (UB) attribute. Moreover, factors from the D&M model presented by William and Ephraim [11] have also been included in the proposed framework.

Given due consideration on adoption characteristics, the attributes are selected depending upon their relevance to the government platform in disaster management and presented in Fig. 2. These are basic four constructs which are common in UMEGA and UTAUT theory including perceived risks, and task-technology fit: information quality and system quality.

The degree to which a system is expected to accomplish improvement in performance of job for individuals is referred to as performance expectancy [9]. Individual's expectancy in job performance is treated as the assistance of individual's attitude toward the adoption of an IT/IS system. Therefore, performance expectancy construct is considered to have a significant impact on attitude of individuals toward the adoption of OIoTD platform. The notion of this construct is related to the relative advantage [5], perceived usefulness [7], and outcome expectation of earlier theories.

The simplicity which is attached with a system's usage is referred to as effort expectancy [9]. Evidences are found in large number of studies that effort expectancy has a significant positive impact on attitude of adopting IT/IS systems. Therefore, this construct is considered to have an impact on individuals' attitude toward adopting OIoTD platform.

In measuring the attitude of individuals toward the use of an IT/IS system, the individual's influence like friends, family member, and colleagues in his/her circle matters [9]. Social influence is found to be the significant positive predictor of attitude in prior research. Therefore, social influence has been included in measuring individual's attitude toward adopting OIoTD platform.

Individual's attitude toward adoption of a system may be influenced by the infrastructure (whether it is organizational or technical) facilities available to them for supporting the system's usage [9]. The available infrastructure will also reduce the efforts an individual put in using the system. In this respect, the level to which an individual believes on available infrastructure facilities is referred to as facilitating conditions. An individual is intending to use a system if sufficient infrastructure is available. Therefore, facilitating conditions construct is considered to have positive effects on both individual's use intention and effort expectancy to use such system, i.e., OIoTD platform in the context of this study.

Attitude of an individual is affected when there are both perceptions of insecurities: behavioral (unfriendly nature of Internet) and environmental (unpredictable nature of Internet-based technology), referred to as perceived risk [9]. Since OIoTD platform is a technological application based on the Internet, individual's belief on insecurities would have significant negative impact on attitude. Therefore, perceived risk is considered to have a negative impact on person's attitude toward the adoption of OIoTD platform.

Herd behavior is referred to as a social phenomenon in which an individual follows other individuals in adopting an IT/IS system even when his or her own personal knowledge suggests adopting some other systems [13]. In this respect, an individual prioritizes and follows to adopt a system by disregarding his/her own private knowledge and imitating others because of lack of information and experience. Thus, herd behavior is considered to be the significant predictor of individual's behavioral intention to use OIoTD platform.

There may some legal disputes arise between government agencies and users in using government ICT-enabled systems. Grievance redressal is a mechanism to address and resolve these kinds of disputes. This mechanism is very effective in creating positive impact on users. Therefore, grievance redressal has been taken as a significant positive predictor of individual's use intention of OIoTD platform.

Individual tends to use a system, technology, or service if it is fit according to the tasks that an individual can perform using it. Users will intend to use a system if it is fit to use with respect to the characteristics or requirements of a task, and technology as well [10]. Therefore, to what extent a system or technology becomes the cause of different activities an individual performs for the required tasks is referred to as task-technology fit. The task-technology fit is extended by including information and system quality [14]. Aligning to these concepts, task-technology fit is perceived to be a significant factor that influences individual's intention to use OIoTD platform.

Individual bears positive or negative appraisals about system's adoption [9]. These positive or negative appraisals form the individual's attitude toward adoption of a system. Since attitude is considered as the influencing factor of behavioral intention to use a system in earlier studies [9], it is also regarded as the influencing factor of behavioral intention to use OIoTD platform.

Behavioral intention is referred to as the probability of an individual being involved in certain behavior [15]. The behavioral intention, alternatively called as use intention, is found to be the best immediate predictor of use behavior/actual behavior [8]. Based on evidences of prior studies, the behavioral intention is considered as the significant influencing factor of OIoTD platform use behavior.

Information quality is referred to the quality and the desired characteristics of report contents and form that an information system generates [11]. A system may be perceived as relevance to practical affairs/tasks if it will be providing data and information of high quality to the individuals [14]. Therefore, information quality is perceived as a high fit between OIoTD platform and disaster data to be used for data-driven innovation and decision-making which will then lead to individual's behavioral intention to use OIoTD platform.

System quality refers to the quality of the functionality and the desired characteristics of an information system itself [11]. In this regard, an individual may find a system reliable and perceive its quality as good as it corresponds to tasks' suitability [14]. Therefore, system quality is perceived as a high fit between OIoTD platform and disaster operations to be performed which will then lead to individual's behavioral intention to use OIoTD platform.

4 Discussion and Conclusion

In this study, we are focusing on proposing a framework on Open IoT Data platform to build an understanding on its adoption in disaster management scenario from the citizen's perspective. Open IoT Data is a part of an action plan for Pakistan vision-2025 [16, 17], so that data-driven innovation and efficient decision-making may be leveraged enabling collaboration with stakeholders especially in the context of contemporary complex issue like disaster management.

This study has some theoretical implications broadly capturing the open and big data, Internet of things, information and communication technologies, digital government, and disaster management notions from the literature. The framework is developed based on the well-established and newly used information theories for technology adoption [9, 10]. The proposed framework is likely to implement in the context of open data and digital government domains to evaluate the adoption of technology for data-driven innovations and efficient decision-making. Other researchers can modify and extend the presented research framework to get more deeper insights according to the context.

As theoretical implications, this study bears some practical implications. Since the innovation, i.e., free sharing of IoT data digitally in the public sector, is at nascent stage, policy- and decision-makers have poor understanding of the usability factors of OIoTD platform. Therefore, the proposed framework will provide an understanding of the benefits of OIoTD platform to the policy- and decision-makers, developers, and IoT data and service owners. Moreover, the policy- and decisions-makers will be able to evaluate the user's behavior toward adoption of OIoTD platform by finding the technological, organizational, task, environment, quality, security, and requirement concerns. Thus, knowledge about these concerns will assist policy- and decisions-makers to focus on a policy instrument which helps in successful implementation of the technology.

The study has the limitation that no empirical evidence is collected regarding the influencing factors and their impact on attitude, intention, and use behavior of OIoTD platform. This study is in progress, and, in future, the influence of independent variables on dependent variables will be evaluated in the context of Pakistan in the disaster management scenario. Future researchers are also invited to extend, modify, or derive new relationships and to apply other information system's theories relating to adoption of technology.

References

1. http://www.un-spider.org/risks-and-disasters/the-un-and-disaster-management
2. Maqbool MY, Hussain S, Khan MB (2017) National framework of disaster risk management in Pakistan: issues, challenges & policy recommendations. Abasyn J Soc Sci 10:182–192
3. NDMA (2017) National Disaster Management Authority Annual Report 2017
4. Carcary M, Maccani G, Doherty E, Conway G (Year) Exploring the determinants of IoT adoption: findings from a systematic literature review. Springer International Publishing, pp 113–125
5. Rogers EM (2003) Diffusion of innovations. The Free Press, New York
6. Fishbein M, Ajzen I (1977) Belief, attitude, intention, and behavior: an introduction to theory and research. Addison-Wesley, Reading
7. Davis FD (1989) Perceived usefulness, perceived ease of use, and user acceptance of information technology. MIS Q 13:319–340

8. Venkatesh V, Morris MG, Davis GB, Davis FD (2003) User acceptance of information technology: toward a unified view. MIS Q 27:425–478
9. Dwivedi YK, Rana NP, Janssen M, Lal B, Williams MD, Clement M (2017) An empirical validation of a unified model of electronic government adoption (UMEGA). Gov Inf Q 34:211–230
10. Goodhue DL, Thompson RL (1995) Task-technology fit and individual performance. MIS Q 19:213–236
11. William HD, Ephraim RM (2003) The DeLone and McLean model of information systems success: a ten-year update. J Manag Inf Syst 19:9–30
12. Sinha A, Kumar P, Rana NP, Islam R, Dwivedi YK (2017) Impact of internet of things (IoT) in disaster management: a task-technology fit perspective. Ann Oper Res
13. Sun H (2013) A longitudinal study of herd behavior in the adoption and continued use of technology. MIS Q 37
14. Cheng Y-M (2018) A hybrid model for exploring the antecedents of cloud ERP continuance. Int J Web Inf Syst
15. Ajzen I, Fishbein M (1980) Understanding attitudes and predicting social behaviour. Prentice-Hall, Engle-wood-Cliffs
16. DPP (2018) Digital Pakistan Policy. In: Telecom, M.o.I.T. (ed)
17. NA (2017) Pakistan Right of Access to Information Act, 2017. Federal Government, National Assembly of Pakistan

Author Index

Printed in the United States
By Bookmasters